S0-ASW-695

# The Single-Particle Density in Physics and Chemistry

# The Single-Particle Density in Physics and Chemistry

*Edited by*

N.H. MARCH

*Coulson Professor of Theoretical Chemistry, University of Oxford, Oxford, UK*

B.M. DEB

*Professor of Theoretical Chemistry, Panjab University, Chandigarh, India*

1987

ACADEMIC PRESS

*Harcourt Brace Jovanovich, Publishers*

London San Diego New York

Boston Sydney Tokyo Toronto

Academic Press Limited
24–28 Oval Road
London NW1

*US edition published by*
Academic Press Inc.
San Diego, CA 92101



**British Library Cataloguing in Publication Data**

March, N.H.
The single-particle density in physics
and chemistry. — (Techniques of physics).
1. Particles (Nuclear physics) 2. Density
functionals
I. Title II. Deb, B.M. III. Series
539.7'21 QC793.3.D4/

ISBN 0-12-470518-9

Printed in Northern Ireland by The Universities Press (Belfast) Ltd

# Contributors

**L.J. Bartolotti,** Department of Chemistry, University of Miami, Coral Gables Florida 33124, USA

**B.M. Deb**, Department of Chemistry, Panjab University, Chandigarh, India

**D.J.W. Geldart**, Department of Physics, Dalhousie University, Halifax, Nova Scotia, Canada B3H 3J5

**S.K. Ghosh**, Heavy Water Division, Bhabha Atomic Research Centre, Bombay 400 085, India

**R.O. Jones,** Institut für Festkörperforschung, Kernforschungsanlage Jülich, D-5170 Jülich, Federal Republic of Germany

**M. Levy**, Department of Chemistry, Tulane University, New Orleans, Louisiana 70118, USA

**N.H. March**, Department of Theoretical Chemistry, University of Oxford, 1 South Parks Road, Oxford OX1 3TG, UK

**P. Politzer**, Department of Chemistry, University of New Orleans, New Orleans, Louisiana 70118, USA

**M. Rasolt**, Solid State Division, Oak Ridge National Laboratory, Oak Ridge, Tennessee 37831, USA

**A.K. Theophilou,** "Demokritos", Scientific Research Centre, Aghia Parakevi, Attikis, Greece

# Techniques of Physics

*Editor*

N.H. MARCH

*Department of Theoretical Chemistry, University of Oxford, Oxford, England*

Techniques of physics find wide application in biology, medicine, engineering and technology generally. This series is devoted to techniques which have found and are finding application. The aim is to clarify the principles of each technique, to emphasize and illustrate the applications, and to draw attention to new fields of possible employment.

1. D.C. Champeney: *Fourier Transforms and their Physical Applications*
2. J.B. Pendry: *Low Energy Electron Diffraction*
3. K.G. Beauchamp: *Walsh Functions and their Applications*
4. V. Cappellini, A.G. Constantinides and P. Emiliani: *Digital Filters and their Applications*
5. G. Rickayzen: *Green's Functions and Condensed Matter*
6. M.C. Hutley: *Diffraction Gratings*
7. J.F. Cornwell: *Group Theory in Physics*
8. N.H. March and B.M. Deb: *The Single-Particle Density in Physics and Chemistry*
9. D.B. Pearson: *Quantum Scattering and Spectral Theory*

# Preface

The object of this book is to focus attention on a general scientific viewpoint which has acquired considerable importance in the last two decades. This involves the microscopic examination and understanding of physico-chemical phenomena in terms of the single-particle density as the basic variable, rather than the wave function. There are two interlinked approaches in this viewpoint, namely, density functional theory (DFT), or the theory of the inhomogeneous electron gas, and quantum fluid dynamics (QFD). DFT considers all properties of a system in the ground state, or the lowest state of a given symmetry, to be unique functionals of the charge density. QFD considers the evolution of the system to be determined by two "classical" fluid dynamical equations, namely, the continuity equation and an equation of motion. The basic quantities involved in these two equations, obtained by transforming the Schrödinger equation, are the quantum mechanical charge density and the current density; the latter vanishes in the ground state and some excited states. In contrast to the many-particle wave function, or the corresponding one- and two-particle density matrices, the basic quantities in DFT and QFD are functions of only three spatial variables, apart from time. This not only leads to a considerable reduction in the mathematical and computational complexities in dealing with many-particle systems, but also to great conceptual simplicity and visual clarity. Further, in applications to atoms, molecules and solids, there have been many instances where the density viewpoint has led to new, detailed insights as well as numerical results which were not accessible through the standard formalisms based on wave functions or density matrices. Such cases are well documented in the scientific literature. On an epistemological level, of equal importance is the fact that the density viewpoint provides rigorous foundations for a classical interpretation of quantum mechanics.

Historically, it is interesting to note that the above developments were initiated in the very early days of quantum mechanics, but it took many years for their full significance to dawn upon physicists and chemists. Thus, the earliest precursor to DFT was suggested by Thomas and Fermi, while the fluid dynamical significance was pointed out by Madelung, Schrödinger and de Broglie. In particular, de Broglie's concept of non-linear wave propagation was much ahead of its time.

This book presents a number of important recent developments in DFT and QFD as well as their applications to structures, properties and dynamics of atoms, molecules and solids. Chapters 1 to 6 review various developments in DFT for such systems. It would be clear from these chapters that although the exact total energy functional for the ground-state energy in terms of the charge density is still unknown, and is indeed equivalent to exactly solving the full many-body problem, an internally consistent ground-state DFT exists already. Further, an interlinking between DFT and QFD is expected eventually to remove two main drawbacks of present-day DFT, namely, the difficulties in dealing with excited states and time-dependent processes, e.g. molecular collisions and reactions. Chapters 7 to 9 pay particular attention to these problems. In addition to the developments in the quantum theory of atoms, molecules and solids, there have also been parallel developments in semi-classical and quantum statistical mechanics, especially in the treatments of inhomogeneous liquids and their phase transitions. Such developments are discussed in Chapter 10.

This book is addressed to graduate students and other research workers in physics and chemistry. The background required for understanding this book is primarily a basic course in non-relativistic quantum mechanics, although relativistic theory is also embraced here. A course in basic condensed matter theory will also be helpful, equivalent to the book on solids by Ashcroft and Mermin (*Solid State Physics*; Holt, Rinehart and Winston, New York, 1976) as well as either the liquid-state treatment of Hansen and McDonald (*Theory of Simple Liquids, 2nd Ed.*; Academic Press, London and San Diego, 1986), or of March and Tosi (*Atomic Dynamics in Liquids*; Macmillan, London, 1976).

In a work of this type it is perhaps inevitable that an occasional error of fact, and of judgement, might creep in. Should this happen, the Editors would be very grateful to have their attention drawn by the readers to such errors.

Finally, the Editors owe a debt of gratitude to the other contributors to the chapters in this volume, especially for their meticulous adherence to the deadlines which the Editors had inevitably to set.

N.H. March

B.M. Deb

# Contents

# Contents

# 1 Electron Density Theory of Atoms and Molecules

N.H. MARCH

*Theoretical Chemistry Department, University of Oxford, 1 South Parks Road, Oxford OX1 3TG, England*



Single-Particle Density in Physics and Chemistry ISBN 0-12-470518-9

## 1.1 Background

The aim of this Chapter is to provide an introduction to the density functional theory of atoms and molecules. This theory has its origins in the statistical treatment given independently by Thomas[1] and Fermi.[2] Since this Thomas–Fermi theory has already contained in it all the important ideas, it is the natural starting point for development of density functional theory.

## 1.2 Phase-space description of many-electron atoms

The idea behind the Thomas–Fermi statistical theory is to treat the electron cloud in an atom (molecule) as an inhomogeneous electron gas.

Consider a small region of the electron cloud around position vector $\mathbf{r}$, the number of electrons per unit volume at $\mathbf{r}$ being denoted by $\rho(\mathbf{r})$. This quantity, the electron density, is the basic tool we shall work with in the treatment below. Its connection with one-electron wave functions $\psi_i(\mathbf{r})$ is that the electron density is given by the sum of the squares of the wave functions over the occupied states:

$$\rho(\mathbf{r}) = \sum_{\substack{\text{occupied}\\ \text{states}}} \psi_i^*(\mathbf{r})\psi_i(\mathbf{r}). \tag{1.1}$$

Suppose these one-electron wave functions were eigenfunctions of the single-particle Hamiltonian,

$$H = -\tfrac{1}{2}\nabla^2 + V(\mathbf{r}), \tag{1.2}$$

where $V(\mathbf{r})$ is the assumed common potential energy in which all the electrons move. Then it is clear that the $\psi_i$ are determined by

$$H\psi_i = \varepsilon_i \psi_i, \tag{1.3}$$

where the $\varepsilon_i$ are the one-electron energies. Evidently, the $\psi_i$ are determined once $V(\mathbf{r})$ is specified, and hence $\rho$ is found from eqn (1.1). Since $V(\mathbf{r})$ is the "input" information, it would obviously be helpful if $\rho$ could be expressed directly in terms of $V(\mathbf{r})$, rather than through the intermediate step of calculating all the one-electron wave functions. This is just the achievement of Thomas–Fermi statistical theory; it provides an approximate, but quite explicit, route to $\rho(\mathbf{r})$, from a given $V(\mathbf{r})$. Naturally, in applications to atoms and molecules, it will be essential to require that $V(\mathbf{r})$ is the "self-consistent" potential energy but this is not a difficult matter, at least in principle, once the $\rho$–$V$ relation is known.

### 1.2.1 Thomas–Fermi density–potential relation

Returning to the electron density $\rho(\mathbf{r})$, let us recall that if we have $N$ electrons in volume $\mathcal{V}$, then these occupy in the ground state the lowest states in momentum space, i.e. a sphere of radius $p_{\max} \equiv p_{\text{Fermi}} = p_{\text{F}}$. The phase-space result that a cell of volume $h^3$ is equivalent to one energy level* gives that

$$N = 2\frac{\frac{4}{3}\pi p_{\text{F}}^3 \mathcal{V}}{h^3}, \tag{1.4}$$

the factor 2 representing double occupancy with opposed spins, and if we now apply this Fermi gas relation locally at $\mathbf{r}$ we have

$$\rho(\mathbf{r}) = \frac{8\pi}{3h^3} p_{\text{F}}^3(\mathbf{r}), \tag{1.5}$$

where evidently, since the electron gas is now inhomogeneous, the Fermi momentum $p_{\text{F}}$ becomes a function of position, i.e. $p_{\text{F}} \to p_{\text{F}}(\mathbf{r})$.

The next step is to write down the classical energy equation for the fastest electron; namely

$$\mu = \frac{p_{\text{F}}^2(\mathbf{r})}{2m} + V(\mathbf{r}), \tag{1.6}$$

where the energy of the fastest electron has been written as $\mu$, anticipating its later identification with the chemical potential of the electronic assembly.

Equation (1.6) is of considerable importance conceptually. For we see that, on the right-hand side, we have a sum of kinetic and potential energy terms, which evidently depend separately on position $\mathbf{r}$ in the charge cloud. But the left-hand side must be a constant, independent of $\mathbf{r}$, since otherwise electrons could redistribute themselves to lower the energy. This is equivalent to saying that the condition that $\mu$ is constant throughout the charge cloud is the statement that all charge redistribution is complete. Especially in molecules, this latter statement is clearly a very significant one for chemical bonding theory.

If we now replace $p_{\text{F}}(\mathbf{r})$ in eqn (1.6) by $\rho(\mathbf{r})$ using eqn (1.5), we find the basic equation of Thomas–Fermi[1,2] statistical theory:

$$\mu = \tfrac{5}{3} c_{\text{k}} \rho^{2/3}(\mathbf{r}) + V(\mathbf{r}), \tag{1.7}$$

where we have found it convenient to introduce the constant $c_{\text{k}}$ which, as we

*Roughly, the Uncertainty Principle will not permit one to divide phase space more finely than $\Delta x \Delta p_x \Delta y \Delta p_y \Delta z \Delta p_z \sim h^3$.

shall show below, fixes the magnitude of the kinetic energy in terms of the density $\rho(\mathbf{r})$. Evidently, from eqns (1.5), (1.6) and (1.7)

$$c_k = \frac{3h^2}{10m}\left(\frac{3}{8\pi}\right)^{2/3}. \tag{1.8}$$

In a neutral atom, we expect that $V(\mathbf{r})$ tends to zero as $r$ tends to infinity faster than $r^{-1}$, while $\rho(\mathbf{r})$ also tends to zero at infinity. This is disappointing, in the sense that the chemically interesting quantity $\mu$ must be zero, since we can evaluate it at any point $\mathbf{r}$, and therefore in particular at the point of infinity. This result, which we shall use below, is only an approximation to the truth because in eqn (1.6) we gave the classical energy equation of the fastest electron. Thus, eqn (1.7) is semi-classical in nature, as is evident from its solution:

$$\rho(\mathbf{r}) = \text{const}\,[\mu - V(\mathbf{r})]^{3/2}. \tag{1.9}$$

Clearly unless $\mu - V(\mathbf{r}) > 0$, we must take $\rho(\mathbf{r}) = 0$, corresponding to classically forbidden regions. We shall see below that this situation, with classically forbidden regions, obtains for heavy positive ions, in the Thomas–Fermi statistical theory. For these, the chemical potential $\mu$ is non-zero, though naturally it tends to zero as we pass to the limit of the neutral atom. Negative ions are not stable in this theory; this is again connected with $\mu = 0$ for the neutral atom.

One other point is immediately evident from eqn (1.9), which is indeed the explicit density–potential relation we were seeking. The density $\rho$ at $\mathbf{r}$ is determined solely by the potential energy $V$ at this point (together with knowledge of the constant $\mu$). That this is not correct in wave mechanics is clear from the Coulomb-potential example. Here we know the electron density at the nucleus is finite, whereas $V(r)$ is infinite. Therefore, in a complete theory, the relation between $\rho(\mathbf{r})$ and $V$ must be non-local. We know this non-local relation exactly only in a perturbation expansion of $\rho(r)$ to all orders in $V(\mathbf{r})$ based on plane waves, as given by March and Murray[3] (see also Appendix A 5.1 of Chapter 5).

### 1.2.2 Variational principle

We can regard eqn (1.7) as the Euler equation of a variational principle, in which the total ground-state energy $E$ is made stationary with respect to variations in the electron density $\rho(\mathbf{r})$. To see how to construct this, let us set up the expression for the kinetic energy of the inhomogeneous electron gas in terms of the electron density $\rho(\mathbf{r})$. To do so, we can use the standard result that for a free (uniform) Fermi gas of $N$ electrons the mean energy per electron $E/N$ is $\frac{3}{5}E_F$, $E_F$ being the Fermi energy. Thus we can write for the kinetic energy per unit volume, $t$, in such a Fermi gas (for which $E$ is entirely kinetic):

$$t=\frac{T}{\mathscr{V}}=\frac{3}{5}\frac{p_F^2}{2m}\frac{N}{\mathscr{V}}, \tag{1.10}$$

and using eqn (1.4) we find

$$t=c_k\left(\frac{N}{\mathscr{V}}\right)^{5/3}, \tag{1.11}$$

where $c_k$ is given by eqn (1.8). Taking over the result (1.11) into the inhomogeneous electron gas we have for the local kinetic energy density

$$t_r(\rho)=c_k\{\rho(\mathbf{r})\}^{5/3} \tag{1.12}$$

and hence in Thomas–Fermi statistical theory the total kinetic energy $T$ is given by

$$T=\int t_r(\rho)\,d\mathbf{r}=c_k\int \rho^{5/3}\,d\mathbf{r}. \tag{1.13}$$

If we now add on to eqn (1.13) the usual classical potential energy terms, we can write

$$E=c_k\int \rho^{5/3}\,d\mathbf{r}+\int \rho V_N\,d\mathbf{r}+\tfrac{1}{2}\int \rho V_e\,d\mathbf{r}, \tag{1.14}$$

where $V_N+V_e=V$ is the self-consistent potential energy contributed by the nuclear framework $V_N$ and the electron cloud $V_e$. Evidently

$$V_e(\mathbf{r})=e^2\int\frac{\rho(\mathbf{r}')\,d\mathbf{r}'}{|\mathbf{r}-\mathbf{r}'|}, \tag{1.15}$$

and we now vary $E$ with respect to $\rho$ using eqns (1.14) and (1.15), subject to

$$\int \rho(\mathbf{r})\,d\mathbf{r}=N, \tag{1.16}$$

where $N$ is the given number of electrons in the atom or ion. Then we regain eqn (1.7), with $\mu$ playing the role of the Lagrange multiplier taking care of the auxiliary condition (normalization) (1.16). We can regard the variational principle

$$\delta(E-\mu N)=0 \tag{1.17}$$

as telling us that $\mu=\partial E/\partial N$, confirming the interpretation of $\mu$ as the chemical potential.

Evidently, the solution of the Euler equation (1.7) presents a self-consistent field problem, since $V(\mathbf{r})$ depends on the density $\rho(\mathbf{r})$ through eqn (1.15).

For atoms, a universal solution of eqn (1.7) has been obtained (see, for example, March;[4] also Appendix A 1.1); but for molecules, though some distance scaling is again possible, eqn (1.7) still presents a complicated multicentre problem which has to be solved by largely numerical methods because of the essential non-linearity implied by the kinetic term proportional to $\rho^{2/3}$.

However, it is important at this point to recognize that any solution based on the Euler equation (1.7) has some severe approximations underlying it. These are

(i) the use of a (symmetrized) Hartree field; and
(ii) the use of the statistical approximation (1.13) for the kinetic energy density.

In spite of these approximations, we shall see below that the Euler equation (1.7) already contains relations of considerable interest in the theory of atoms and molecules. Nevertheless, this is the point at which we must turn to the formal completion of the above Thomas–Fermi theory by Hohenberg and Kohn.[5]

## 1.3 Proof that ground-state energy is a unique functional of electron density*

Suppose that we have an arbitrary number $N$ of electrons moving under the influence of an external potential $V_{\text{ext}}(\mathbf{r})$ and their mutual Coulomb repulsion. Let the many-electron ground-state wave function be denoted by $\Psi(1,2,\ldots,N)$, where space and spin coordinates of the electrons are denoted by $1,2,\ldots,N$. Evidently by suitable integration of $\Psi^*\Psi$ over space, and summation over spin we can construct $\rho(\mathbf{r})$, the electron density.

The first step in the argument of Hohenberg and Kohn[5] is to show that $V_{\text{ext}}(\mathbf{r})$ is a unique functional of $\rho(\mathbf{r})$, apart from an (unimportant for present purposes) additive constant. The proof proceeds by *reductio ad absurdum.*

Assume that another external potential, $V'_{\text{ext}}(\mathbf{r})$ say, with ground-state wave function $\Psi'(1,\ldots,N)$, gives rise to the same density $\rho(\mathbf{r})$ as $V_{\text{ext}}(\mathbf{r})$. Now clearly (unless $V'_{\text{ext}}-V_{\text{ext}}$ is constant) $\Psi'\neq\Psi$, because they satisfy different Schrödinger equations. Hence, if we denote the Hamiltonian and ground-state energies associated with $\Psi$ and $\Psi'$ by $H$, $H'$ and $E$, $E'$ we have by the

*The treatment of a degenerate ground state is referred to by Kohn[76]; see also ref. 35.

minimal property of the ground-state energy that

$$E' = \int \Psi'^* H' \Psi' \, d\tau < \int \Psi^* H' \Psi \, d\tau$$

$$= \int \Psi^* (H + V' - V) \Psi \, d\tau. \tag{1.18}$$

Thus it follows that*

$$E' < E + \int (V' - V) \rho(\mathbf{r}) \, d\tau. \tag{1.19}$$

Using the same argument on $E$ is equivalent to interchanging primed and unprimed quantities, and we then have

$$E < E' + \int (V - V') \rho(\mathbf{r}) \, d\tau, \tag{1.20}$$

which involves the assumption that the same $\rho(\mathbf{r})$ can be generated by two different external potentials. Adding (1.19) and (1.20) we obtain

$$E' + E < E + E', \tag{1.21}$$

which is inconsistent. Therefore $V_{\text{ext}}(\mathbf{r})$ is (to within a constant) a unique functional of $\rho(\mathbf{r})$. Since, in turn, $V_{\text{ext}}(\mathbf{r})$ fixes $H$ we see that the many-electron ground-state energy is a unique functional of $\rho(\mathbf{r})$.

## 1.4 Generalized Euler equation of Hartree theory

Utilizing the above result within, however, the Hartree framework, the first step in the generalization of the Euler equation (1.7) is to write for the total kinetic energy $T$ the result

$$T = \int t[\rho] \, d\mathbf{r}. \tag{1.22}$$

Here, we have now indicated by the square brackets that, in general, the kinetic energy density $t[\rho]$ is a functional of $\rho(\mathbf{r})$, unlike eqn (1.13) where $t$ is the function $c_k \rho^{5/3}$.

We use again the variational principle (1.17) to find the new Euler equation:

$$\mu = \frac{\delta T}{\delta \rho(\mathbf{r})} + V_N(\mathbf{r}) + V_e(\mathbf{r}). \tag{1.23}$$

*In eqns (1.18)–(1.20), $V$ is written for $V_{\text{ext}}$ etc. for notational convenience.

If we knew the exact form of $t[\rho]$ in the above one-electron theory, then eqn (1.23) could be solved for the density. Obviously, if we now insert the approximation (1.13) for $t$ into eqn (1.23) we regain the Euler equation (1.7) of Thomas–Fermi statistical theory.

## 1.5 Energy relations for molecules in equilibrium

Though, as already mentioned, for a molecule, the multicentre non-linear problem posed by eqn (1.7) is formidable, it is well worth studying what information we can gain from eqns (1.7) and (1.23) without explicitly solving them. Since eqn (1.23) contains (1.7) as a special case, let us work at first with eqn (1.23) before passing to the limiting case of Thomas–Fermi statistical theory.

We first multiply both sides of eqn (1.23) by the density $\rho(\mathbf{r})$ and integrate over the whole of space. Since the Lagrange multiplier $\mu$ is simply a number, the left-hand side becomes $N\mu$ when we use the normalization condition (1.16), and we find

$$N\mu = \int \rho \frac{\delta T}{\delta \rho} \mathrm{d}\mathbf{r} + \int \rho V_{\mathrm{N}}(\mathbf{r}) \, \mathrm{d}\mathbf{r} + \int \rho V_{\mathrm{e}}(\mathbf{r}) \, \mathrm{d}\mathbf{r}. \tag{1.24}$$

But it is clear from eqn (1.14) that we can write eqn (1.24) in terms of the electron–nuclear and electron–electron potential energies $U_{\mathrm{en}}$ and $U_{\mathrm{ee}}$, respectively, as

$$N\mu = \int \rho \frac{\delta T}{\delta \rho} \mathrm{d}\mathbf{r} + U_{\mathrm{en}} + 2U_{\mathrm{ee}}. \tag{1.25}$$

This is as far as we can get without making some assumptions about $\mu$ and about $t$. If we take $t$ from eqn (1.12), then

$$\int \rho \frac{\delta T}{\delta \rho} \mathrm{d}\mathbf{r} = \frac{5}{3} c_{\mathrm{k}} \int \rho^{5/3} \, \mathrm{d}\tau \cong \frac{5}{3} T,$$

where $T$ is the total kinetic energy, which in terms of $t$ is evidently

$$T = \int t \, \mathrm{d}\mathbf{r}. \tag{1.26}$$

The above argument suggests that we add and subtract $\frac{5}{3}T$ from the right-hand side of eqn (1.25), and using eqn (1.26) we then obtain

$$N\mu = \int \rho \left[ \frac{\delta T}{\delta \rho} - \frac{5}{3} \frac{t}{\rho} \right] \mathrm{d}\mathbf{r} + \frac{5}{3} T + U_{\mathrm{en}} + 2U_{\mathrm{ee}}. \tag{1.27}$$

In the statistical theory, as argued above, the first term on the right-hand side is identically zero. In the same theory, for a neutral atom, we have already seen that the chemical potential $\mu$ must be zero. Thus, we obtain the first (approximate) energy relation of the density functional theory, namely

$$\frac{U_{en}+2U_{ee}}{T}=-\frac{5}{3}. \tag{1.28}$$

Though the arguments on which eqn (1.28) rests are statistical, Mucci and March[6] have taken a variety of self-consistent field calculations for light molecules and plotted $U_{en}+2U_{ee}$ against $T$ for some 20 of these. An excellent straight line is thereby obtained, with a slope in agreement with the value $-\frac{5}{3}$ given in eqn (1.28). This is an immediate demonstration that regularities do indeed exist in these light molecules, exactly as the simplest density description would indicate.

We can make further progress if we add to the result (1.28) the virial theorem for molecules at equilibrium. If $E$ is the total energy, this tells us that

$$E=-T; \tag{1.29}$$

and using the fact that, in Hartree theory, with $U_{nn}$ the nuclear–nuclear potential energy

$$E=T+U_{en}+U_{ee}+U_{nn}, \tag{1.30}$$

we obtain from eqn (1.29) the result that

$$-2T=U_{en}+U_{ee}+U_{nn}. \tag{1.31}$$

Dividing this equation by $T$, and eliminating $U_{en}$ by means of eqn (1.28) yields

$$\frac{U_{nn}-U_{ee}}{T}=-\frac{1}{3}. \tag{1.32}$$

Again, for the same twenty or so light molecules, this result (1.32) has been confirmed by Mucci and March[6] for the equilibrium molecular geometry.

If one chooses to eliminate $U_{ee}$, then the third relation

$$\frac{2U_{nn}+U_{en}}{T}=-\frac{7}{3} \tag{1.33}$$

is obtained, which is equivalent to a relation proposed by Politzer[7] (see also Chapter 3). Again, eqn (1.33) works well for light molecules at equilibrium.

We must stress that these three equations (1.28), (1.32) and (1.33) are such that any two imply the third. They rest on the neglect of

(i) exchange and correlation (to be discussed below);
(ii) $N\mu$ in eqn (1.27); and

(iii) the $\int\left[\rho\frac{\delta T}{\delta\rho}-\frac{5}{3}t\right]d\mathbf{r}$ term in eqn (1.27).

Some examples in which approximation (iii) can be checked for model potentials and, in particular, for a bare Coulomb field and a harmonic-oscillator potential are recorded in ref. 8.

More interesting for molecular theory is the reason why (ii) is a useful starting point. We shall discuss the chemical interest of $\mu$ in some detail later in this Chapter. First, however, let us turn to the problem of the total energy of neutral atoms.

## 1.6 Total energy of neutral atoms and relation to 1/Z expansion

Evidently eqns (1.28) and (1.29) are valid as they stand in the statistical theory of neutral atoms. In eqns (1.32) and (1.33) we merely set $U_{nn}=0$ to obtain

$$T=-E=-\tfrac{3}{7}U_{en}=3U_{ee}. \tag{1.34}$$

But $U_{en}$ is the interaction energy of the nucleus of charge $Ze$ with the electron cloud. Instead of using $\int\rho V_N\,d\mathbf{r}$ we could calculate the electrostatic potential (see Chapter 3) due to the electrons at the position of the nucleus, say $\phi_e(0)$, and then multiply this by the charge $Ze$ to obtain

$$U_{en}=Ze\phi_e(0). \tag{1.35}$$

As is discussed fully by March,[4] we can calculate $\phi_e(0)$ self-consistently from eqns (1.7) and (1.15), putting $V_N(r)=-Ze^2/r$ and $\mu=0$, when we find by numerical calculation (see Appendix A 1.1)

$$U_{en}=-1.79\,Z^{7/3}\frac{e^2}{a_0}. \tag{1.36}$$

Inserting this into eqn (1.34) we obtain the total energy of the neutral atom with atomic number $Z$ as

$$E=-0.769\,Z^{7/3}\frac{e^2}{a_0}. \tag{1.37}$$

It should be stressed that this result that the energy is proportional to $Z^{7/3}$ follows, for sufficiently large $Z$, for electrons moving in a bare Coulomb field. In that case one finds that the total energy (now simply the sum of the hydrogenic energy levels) is given by[4]

$$E_{\mathrm{Coulomb}}=-\left(\frac{3}{2}\right)^{1/3}Z^{7/3}\frac{e^2}{a_0}=-1.1\,Z^{7/3}\frac{e^2}{a_0}. \tag{1.38}$$

Thus for heavy neutral atoms, treated non-relativistically, we find that the effect of the self-consistent field is to reduce the binding of the outer electrons, which changes the coefficient 1.1 in eqn (1.38) to the value 0.77 in eqn (1.37).

More interesting for atomic theory is to effect the generalization of the above argument to positive atomic ions. The result can be shown to take the form

$$E(Z,N)=Z^{7/3}f\left(\frac{N}{Z}\right)\frac{e^2}{a_0}, \tag{1.39}$$

for $N \leqslant Z$. Negative ions, however, are found not to be stable in Thomas–Fermi statistical theory as already mentioned above. The function $f(N/Z)$ in eqn (1.39) can be obtained numerically from the Euler equation (1.7), combined with (1.15), although $\mu$ is now non-zero when $N \neq Z$.

Whereas the above statistical theory of positive ions is evidently true only for large numbers of electrons $N$, an independent development of $E(Z,N)$ by Layzer,[9] following the earlier work of Hylleraas[10], has proved of considerable utility even for small $N$. This is the so-called $1/Z$ expansion, in which one writes

$$E(Z,N)=Z^2\left[\varepsilon_0+\frac{1}{Z}\varepsilon_1+\frac{1}{Z^2}\varepsilon_2+\frac{1}{Z^3}\varepsilon_3+\cdots\right]. \tag{1.40}$$

Here, evidently all the $N$ dependence is contained in the coefficients $\varepsilon_n$ and convergence is guaranteed for sufficiently large $Z$, as shown by Kato.[11]

These coefficients are available, in low order, through the work of a number of groups, for the $N=2$ to 10 ground states and their results are summarized in Table 1.1.

**Table 1.1** Energy coefficients in eqn (1.40) for the $Z^{-1}$ expansion.

| $N$ | *Configuration* | | $-\varepsilon_0$ | $\varepsilon_1$ | $-\varepsilon_2$ | $\varepsilon_3$ |
|---|---|---|---|---|---|---|
| 2 | $^1S$ | $1s^2$ | 1 | 0.625 | 0.158 | 0.0087 |
| 3 | $^2S$ | 2s | 1.125 | 1.0228 | 0.408 | −0.0230 |
| 4 | $^1S$ | $2s^2$ | 1.25 | 1.5593 | 0.882 | |
| 5 | $^2P$ | 2p | 1.375 | 2.3275 | 1.86 | |
| 6 | $^3P$ | $2p^2$ | 1.50 | 3.2589 | 3.29 | |
| 7 | $^4S$ | $2p^3$ | 1.625 | 4.3535 | 5.26 | |
| 8 | $^3P$ | $2p^4$ | 1.75 | 5.6619 | 8.13 | |
| 9 | $^2P$ | $2p^5$ | 1.875 | 7.1343 | 11.76 | |
| 10 | $^1S$ | $2p^6$ | 2.00 | 8.7708 | 16.27 | |

March and White[12] compared the statistical form (1.39), valid for large $N$ and $Z$, with the form (1.40) taken in the limit of large $N$. It is straightforward, as they pointed out, to show from hydrogen-like atom theory that for large $N$,

$$\varepsilon_0(N) \sim N^{1/3}. \tag{1.41}$$

This is trivially rewritten as $Z^{1/3}(N/Z)^{1/3}$ and similarly it is clear that in the same limit of large $N$,

$$\varepsilon_1(N) \propto N^{4/3} \tag{1.42}$$

or

$$\frac{1}{Z}\varepsilon_1(N) \propto \left(\frac{N}{Z}\right)^{4/3} Z^{1/3}. \tag{1.43}$$

In general, we expect therefore

$$\frac{1}{Z^n}\varepsilon_n(N) \propto Z^{1/3}\left(\frac{N}{Z}\right)^{n+1/3}. \tag{1.44}$$

March and White[12] made a least-squares fit of the results in Table 1.1 to obtain $\varepsilon_1$ and $\varepsilon_2$ in the form (cf. eqns (1.42) and (1.44))

$$\varepsilon_1(N) = a_1 N^{4/3} + b_1 N^{1/3} + \cdots \tag{1.45}$$

and

$$\varepsilon_2(N) = a_2 N^{7/3} + b_2 N^2 + \cdots. \tag{1.46}$$

They thereby estimated numerically the coefficients $a_1 - b_2$ in eqns (1.45) and (1.46). More accurate calculations have been made by Plindov and Dmitrieva[13,12] to whose work the reader is referred for an extension of the values in Table 1.1 to large $N$.

## 1.7 Introduction of exchange

So far, the theory we have discussed has remained within the Hartree framework. At this point, we shall turn to treat exchange within the density functional theory. The first treatment was given by Dirac.[14] The aim of the approach must clearly be to obtain a formula for the exchange energy analogous to eqn (1.13) for the kinetic energy.

We note first that whereas Hartree theory is based on a product wave function, Hartree–Fock theory incorporates exchange by working with a single Slater determinant. By analogy with the kinetic energy derivation, consider the problem of the exchange energy in a uniform electron gas. For high densities, the one-electron wave functions are plane waves and we form the appropriate Slater determinant from those plane waves with allowed momentum vectors falling within the Fermi sphere of radius $p_F$. The exchange energy is, essentially, the average value of the electron–electron interaction terms of the form $e^2/r_{ij}$ representing the Coulomb repulsion of electrons $i$ and $j$ at separation $r_{ij}$ with respect to the above Slater determinant built out of plane waves. This calculation can be carried out precisely and will be referred to again below. But the important point to emphasize first of all is that whereas the kinetic energy operator involves $h^2/m$, which must therefore appear in this combination in eqn (1.13) for the kinetic energy, the exchange energy, obtained from the average of $e^2/r_{ij}$, must evidently depend on $e^2$.

### 1.7.1 Form of free-electron exchange energy from dimensional analysis

It is easy to check that the form of eqn (1.12) follows on dimensional grounds once it is accepted that (i) the kinetic energy density is proportional to a power of the electron density $\rho$, and (ii) as discussed above, $h^2/m$ is the combination of fundamental constants that must enter the formula.

Similarly therefore, let us now examine how the exchange energy varies with the density, since we have argued it depends on $e^2$, provided again we assume a power-law dependence on the density $\rho$. Thus we write

$$\text{exchange energy density} = (e^2)^l (\rho)^n. \tag{1.47}$$

The dimensions of energy density in terms of mass M, length L and time T are evidently given by $\mathrm{ML^2T^{-2}L^{-3}}$, and hence we have, since $e^2/\mathrm{L}$ is also an energy

$$\mathrm{ML^{-1}T^{-2}} = (\mathrm{ML^3T^{-2}})^l\,\mathrm{L}^{-3n}. \tag{1.48}$$

We see immediately by equating powers of M in eqn (1.48) that $l=1$, as we could have anticipated from the first-order perturbation calculation of $e^2/r_{ij}$ averaged with respect to a determinant of plane waves. From powers of L, we find

$$\mathrm{L^{-1}} = \mathrm{L^3L^{-3n}} \tag{1.49}$$

or $n=4/3$. This is readily shown then to lead to a consistent result with equating powers of T in eqn (1.48). The conclusion is that the exchange energy density, $\varepsilon_x$ say, has the form

$$\varepsilon_x = -\text{const}\,\rho^{4/3} = -c_e\,\rho^{4/3} \tag{1.50}$$

and we see that whereas the kinetic energy density in eqn (1.12) is proportional to $\rho^{5/3}$, for the exchange energy the power is $\frac{4}{3}$; a result first obtained by Dirac.[14] Of course, dimensional analysis cannot give the value of the (exchange) constant $c_e$, written such that it is positive in eqn (1.50). The determination of $c_e$ will now be discussed briefly.

## 1.7.2 Fermi (exchange) hole and constant in exchange energy density

Let us briefly consider the relative motion of electrons in the approximation of a single Slater determinant of plane waves. If we choose to sit on an electron at the origin of coordinates then this electron creates a hole around itself because the Pauli exclusion principle insists that another electron with spin parallel to the spin of the electron at the origin cannot also sit at the origin. Thus the density of other electrons, which we write as $\rho_0 g(r)$ as a function of distance $r$ from the electron at the origin, must rise from a value less than the mean density $\rho_0$ at $r=0$ to $\rho_0$ as $r$ tends to infinity. The quantity $g(r)$ therefore describes the Fermi or exchange hole. In the Hartree–Fock approximation and for a non-magnetic electron gas with equal numbers of upward and downward spins, there is no correlation between antiparallel spin electrons (the corrections are wholly from the Fermi statistics, the Coulombic repulsions not being included since the wave function is a single determinant of plane waves inside the Fermi sphere), and hence $g(r)$ goes to $\frac{1}{2}$ at $r=0$.

In fact the pair function $g(r)$ can be calculated from the above Slater determinant of plane waves, the result being[4]

$$g(r)=1-\frac{9}{2}\left[\frac{j_1(k_{\mathrm{F}}r)}{k_{\mathrm{F}}r}\right]^2, \tag{1.51}$$

where $j_1(x)$ is the first-order spherical Bessel function $(\sin x - x\cos x)/x^2$ and $k_{\mathrm{F}}$ is the Fermi wavenumber $(p_{\mathrm{F}}=\hbar k_{\mathrm{F}})$.

Using eqn (1.51) it is easy to calculate the mean potential energy per electron (simply exchange energy, because in the free-electron model a neutralizing background of positive charge is implicit, which cancels off the classical Coulomb potential energy terms). We note that the energy is essentially the interaction of the "Fermi hole" electron density described by eqn (1.51) with the electron, of charge $-e$, at the origin. Thus, as in eqn (1.15), we calculate the electrostatic potential, but now at the origin (see also chapter 3), due to the Fermi hole density $\rho_0[g(r)-1]$ to obtain

$$\text{exchange energy per electron} = \frac{e^2}{2}\int \rho_0 \frac{[g(r)-1]}{r}\,\mathrm{d}\mathbf{r}, \tag{1.52}$$

where the factor $\frac{1}{2}$ avoids double counting of the electron–electron interactions. It is readily shown, by using this free-electron result locally, that the constant $c_{\mathrm{e}}$ in eqn (1.50) is given by

$$c_{\mathrm{e}}=\frac{3}{4}e^2\left(\frac{3}{\pi}\right)^{1/3}. \tag{1.53}$$

In summary, to the same approximation as implied by eqn (1.13) for the kinetic energy density, we have obtained the exchange energy density in terms of the electron density $\rho$. Equation (1.50), with $c_e$ given by eqn (1.53), is an important result in the theory of many-electron systems and is the basis of the so-called Dirac–Slater exchange potential which will be discussed below. The relation between the Dirac–Slater exchange energy* and the mean momentum $\langle p \rangle$ for molecules is set out in Appendix A 1.4.

## 1.8 Correction for correlation

The exchange energy density (1.50) was obtained by neglecting the correlation between antiparallel-spin electrons. But of course, because electrons suffer Coulombic repulsion, this means that the pair function $g(r)$ in an electron gas has a deeper hole near $r=0$ than the Fermi hole (1.51). The difference between the Hartree–Fock energy and the exact energy is generally referred to as "correlation energy".

This correlation energy can be calculated to a good approximation for the above model of a uniform interacting electron gas in a uniform positive neutralizing background. For high densities ($\rho_0 = N/V = 3/4\pi r_s^3$, where $r_s$ is evidently the mean inter-electronic spacing), $\rho_0 \to \infty$ or $r_s \to 0$, many-body perturbation theory can be used. The correlation energy per electron is found to have the form[15]

$$\frac{E_c}{N} = -A \ln r_s + B + \cdots, \tag{1.54}$$

where $A$ and $B$ are known constants.

In the low-density limit, $r_s \to \infty$, an argument due to Wigner[16] shows how to calculate the correlation energy to high accuracy. One argues that the kinetic energy per electron is proportional to $r_s^{-2}$ (essentially the Laplacian operator involves a length squared, whereas the potential (exchange) energy per electron is proportional to $r_s^{-1}$). Thus, in the high-density limit ($r_s \to 0$), the kinetic energy dominates and the filled Fermi sphere of plane-wave states is the appropriate description. But in the low-density regime ($r_s \to \infty$), the kinetic energy becomes unimportant and one must find the configuration of electrons that minimizes the potential energy. As Wigner[16] pointed out, electrons then avoid one another as efficiently as possible by going onto a lattice, and the Madelung energy of point electrons crystallized out in a uniform positive-charge background must be

*Chen and Spruch[77] have reopend the matter of correcting eqn (1.37) for exchange energy and density gradients (see Appendix A1.3). These contribute $O(Z^{5/3})$ and $O(Z^2)$ corrections respectively. For comparison with Hartree–Fock energies, see also ref. 78.

minimized according to classical electrostatic considerations. The body-centred cubic lattice has the lowest Madelung energy and leads to the total potential energy per electron of the form, in Ry,

$$\text{potential energy per electron} = -\frac{1.8}{r_s}; \quad r_s \to \infty. \tag{1.55}$$

Subtracting the exchange energy $(-0.92/r_s)$ yields

$$\text{correlation energy per electron} = -\frac{0.88}{r_s}; \quad r_s \to \infty. \tag{1.56}$$

Approximate calculations have been carried out for intermediate densities and interpolation formulae are now available which join the limiting forms (1.54) and (1.56). (See for example, Gordon and Kim,[17] and also later references[18–20] based on the computer simulation studies of Ceperley and Alder.[21])

## 1.9 Euler equation for ground-state electron density

This is the point at which we can now construct a formally exact Euler equation for the ground-state density $\rho(\mathbf{r})$. We must modify the argument of Section 1.4 to include the exchange and correlation energy density $\varepsilon_{xc}[\rho]$ say. Then we can write the total ground-state energy $E$ as

$$\begin{aligned} E[\rho] &= T_s \qquad + U_{en} \qquad + U_{ee} \qquad + E_{xc} \qquad + U_{nn} \\ &= \int t[\rho]\,\mathrm{d}r + \int \rho V_N\,\mathrm{d}\tau + \tfrac{1}{2}\int \rho V_e\,\mathrm{d}\tau + \int \varepsilon_{xc}[\rho]\,\mathrm{d}\tau + U_{nn}. \end{aligned} \tag{1.57}$$

Here, we have chosen to write the total energy $E[\rho]$ in terms of the exact single-particle kinetic energy $T_s = \int t[\rho]\,\mathrm{d}\mathbf{r}$ for a system with electron density $\rho$. This means that we must incorporate the correlation kinetic energy into the quantity $\varepsilon_{xc}[\rho]$. Again we use the variational principle (1.17) to find the new (formally exact) Euler equation:

$$\mu = \frac{\delta T_s}{\delta \rho} + V_N(\mathbf{r}) + V_e(\mathbf{r}) + \frac{\delta E_{xc}}{\delta \rho(\mathbf{r})}. \tag{1.58}$$

We emphasize that (i) eqn (1.58) differs from eqn (1.23) in that the exchange and correlation are now built in through $E_{xc}[\rho]$; and (ii) $V_N + V_e \equiv V_{\mathrm{Hartree}}$ is to be calculated (in principle) with the exact electron density $\rho(\mathbf{r})$ of the ground state for the interacting electronic charge cloud in the atom or molecule under discussion.

### 1.9.1 Form of one-body potential to generate ground-state density

We now note that eqn (1.58) is like the single-particle Euler equation (1.23), since $t$ still represents a single-particle kinetic energy. Thus, following Slater,[22] and later Kohn and Sham,[23] we can regard eqn (1.58) as leading to a one-body potential $V(\mathbf{r})$ which will, again in principle, generate the exact many-body ground-state density $\rho(\mathbf{r})$. This one-body potential energy $V(\mathbf{r})$ is evidently

$$V(\mathbf{r}) = V_{\mathrm{N}} + V_{\mathrm{e}} + \frac{\delta E_{\mathrm{xc}}}{\delta\rho(\mathbf{r})}$$

$$\equiv V_{\text{Hartree}} + \frac{\delta E_{\mathrm{xc}}}{\delta\rho(\mathbf{r})}. \tag{1.59}$$

### 1.9.2 Dirac–Slater exchange potential

For a given nuclear framework of the molecule, and a given $\rho(\mathbf{r})$, it is clearly a straightforward matter to calculate the Hartree potential contribution in eqn (1.59). However, all the many-electron effects are now subsumed into the additional term in the potential and it will be clear that exact knowledge of the exchange and correlation energy density as a functional of $\rho$ is equivalent to exact solution of the many-body problem, which is presently not feasible. Thus, the whole theory for calculating the ground-state density and energy is in fact a matter of judicious approximation to $\varepsilon_{\mathrm{xc}}$ (see, especially, Chapter 6). We need not, if we are willing to solve one-electron Schrödinger equations with the potential $V(\mathbf{r})$, appeal to approximations to the single-particle kinetic energy density.

The simplest approximation by means of which we construct $V(\mathbf{r})$ in eqn (1.59) is to insert for $E_{\mathrm{xc}}[\rho]$ the exchange energy result (1.50), when we find

$$V_{\text{exchange}} = \frac{\delta E_{\mathrm{x}}}{\delta\rho} = -\frac{4}{3}c_{\mathrm{e}}\{\rho(r)\}^{1/3}. \tag{1.60}$$

Sometimes it has been argued that some account of correlation can be incorporated into eqn (1.60) by multiplying it by a constant factor to be determined semi-empirically. But the point of view taken in this Chapter is that it is more fundamental to transcend eqn (1.60) by adding the correlation energy density for a locally uniform electron gas, through the interpolation formulae referred to above.[16–21]

Thus, the important conclusion is that the exact electron density can be obtained from a single-particle potential $V(\mathbf{r})$ provided the exchange and correlation correction to the Hartree term is included, as displayed in eqn (1.59). Once the electron density is known, the theorem of Hohenberg and Kohn,[5] discussed in Section 1.3, assures us that the ground-state energy can be calculated, since it is a unique functional of the electron density.

## 1.10 Chemical meaning of the Lagrange multiplier $\mu$

Having given a formally exact Euler equation for the density (see also Appendix A 1.2) through eqn (1.58), we wish to consider further the interpretation of the Lagrange multiplier $\mu$. In order to gain some insight into this, let us return to the simplest density description for a moment; that is to the Thomas–Fermi statistical theory. Then we find, with $t$ approximated by eqn (1.12), the Euler equation (1.17). But if we now replace $\rho(\mathbf{r})$ in the kinetic term by the maximum momentum $p_F(\mathbf{r})$ using eqn (1.5) we find immediately eqn (1.6). As already emphasized this is the classical energy equation of the fastest electron, since $p_F^2(r)/2m$ is evidently the kinetic energy of this electron, while $V_{\text{Hartree}}(\mathbf{r})$ is the potential energy in which it moves. The right-hand side of eqn (1.6) comprises two terms which individually vary with position $\mathbf{r}$ in the inhomogeneous electron cloud of the atom or molecules: yet the left-hand side is a constant independent of $\mathbf{r}$.

Here we have an important merit of the density functional theory. Evidently if the energy of the fastest electron (to be identified with $\mu$ in the Thomas–Fermi limit) varied with position $\mathbf{r}$, electrons could redistribute themselves in the molecule to reduce the total energy. So the constancy of $\mu$ is the condition of equilibrium; that charge flow is complete on molecular formation.

It would be of considerable chemical interest if the chemical potential $\mu$ in the Euler eqn generated by eqn (1.59) could be calculated precisely. However, this is a matter that will require knowledge of the functionals $t[\rho]$ and $\varepsilon_{xc}[\rho]$. For example, since $V_N(\mathbf{r})+V_e(\mathbf{r})\to 0$ as $r\to\infty$, i.e. when we are far from all nuclei in the molecule, we see that

$$\mu = \left.\frac{\delta T_s}{\delta\rho}\right|_{\mathbf{r}\to\infty} + \left.\frac{\delta E_{xc}}{\delta\rho}\right|_{\mathbf{r}\to\infty}. \tag{1.61}$$

Elementary local approximations to $t$ such as eqn (1.12) lead to $\delta T_s/\delta\rho \propto \rho^{2/3}$ which tends to zero as $\mathbf{r}$ tends to infinity in say a neutral atom or molecule. In spite of the difficulty in calculating $\mu$, as evidenced by eqn (1.61), some progress can be made, as discussed, for example, in ref. 38.

## 1.11 Foundations of Walsh's rules for molecular shape

As an example of the use of density functional theory, we turn to discuss the basic foundation of Walsh's rules[24] for molecular shape. These rules remain of considerable interest and are supported by variety of data. Their foundation in first principles theory, however, has been lacking. In particular, the assumption was, at the very least, implicit in Walsh's arguments that the total energy was simply the sum of one-electron orbital energies. Then, in say HAH, the whole argument depended on following the eigenvalue or orbital energy changes as one went across a correlation diagram from say the 90° angle to the linear molecule. As Walsh emphasized, in this example everything then hinged on the number of valence electrons. If the lowest two levels only were fully occupied, which required four valence electrons, then the sum of the one-electron energies would be lower for the linear configuration than for the smaller angles. But if there were more than four electrons, the bent molecule would have a lower orbital energy sum than the linear molecule.

A number of workers[25,26] pointed out in the early development of Walsh's rules that the total energy could not in elementary single-particle theories be the sum of orbital energies. This is clear because in self-consistent field theories the electrostatic self-energy of the electronic charge cloud of the molecule is counted twice in the sum of the orbital energies.

We start out from the Euler equation based on eqn (1.59) and as in section 1.5 we multiply both sides by $\rho$ and integrate over all space to obtain, for a molecule with $N$ electrons,

$$N\mu = \int \rho \frac{\delta T_s}{\delta \rho}\, d\mathbf{r} + U_{en} + 2U_{ee} + \int \rho \frac{\delta E_{xc}}{\delta \rho}\, d\mathbf{r}. \tag{1.62}$$

Next we turn to the sum of the orbital energies $\varepsilon_i$ generated by the one-body potential (1.59). The orbital energy sum

$$E_s = \sum_{\substack{\text{occupied} \\ \text{levels}}} \varepsilon_i \tag{1.63}$$

is clearly, using eqn (1.59),

$$\begin{aligned} E_s &= \int t\, d\mathbf{r} + \int \rho V(\mathbf{r})\, d\mathbf{r} \\ &= \int t\, d\mathbf{r} + U_{en} + 2U_{ee} + \int \rho \frac{\delta E_{xc}}{\delta \rho}\, d\mathbf{r}. \end{aligned} \tag{1.64}$$

Thus, by subtracting eqn (1.64) from (1.62) we obtain

$$N\mu - E_s = \int\left[\rho\frac{\delta T_s}{\delta\rho} - t\right]d\mathbf{r}. \tag{1.65}$$

But as we stressed in Section 1.5,

$$\int\left[\rho\frac{\delta T_s}{\delta\rho} - \frac{5}{3}t\right]d\mathbf{r}$$

is zero in the Thomas–Fermi limit and hence it is useful to rewrite eqn (1.65) as

$$N\mu - E_s = \int\left[\rho\frac{\delta T_s}{\delta\rho} - \frac{5}{3}t\right]d\mathbf{r} + \frac{2}{3}\int t\, d\mathbf{r}. \tag{1.66}$$

Now we would like again to employ the virial theorem $T = -E$ for molecules at equilibrium. Unfortunately $t$ is by definition the single-particle kinetic energy density and its integral $T_s = \int t\, d\mathbf{r}$ is therefore not equal to $T$, the exact total kinetic energy, which appears in the virial theorem. It is therefore helpful to define the correlation kinetic energy, $\Delta T$ say, by

$$\Delta T = T - T_s. \tag{1.67}$$

Hence the sum of the orbital energies $E_s$ and the total energy $E$ in the full many-electron theory are related at equilibrium by[27]

$$E = \frac{3}{2}E_s - \frac{3}{2}N\mu + \frac{3}{2}\int\left[\rho\frac{\delta T_s}{\delta\rho} - \frac{5}{3}t\right]d\mathbf{r} - \Delta T. \tag{1.68}$$

In the simplest density description, as discussed earlier in this Chapter, the chemical potential $\mu$ is zero, the integral vanishes identically and the correlation kinetic energy $\Delta T$ is obviously zero in one-electron theory. Thus $E = \frac{3}{2}E_s$ and, though Walsh's identification of $E$ with $E_s$ is not confirmed, the arguments leading to his rules are still all right since $E$ is simply proportional to $E_s$. This relation $E = \frac{3}{2}E_s$ was first given for neutral atoms by March and Plaskett[28] and for neutral molecules at equilibrium by Ruedenberg;[29] see also March.[30] We should mention here the work by Mehrotra and Hoffmann[31] on Walsh's rules. This is also concerned with the relation between total energy and the orbital energy sum but it does not have electron correlation included explicitly.

What has been demonstrated above is that at equlibrium there is a fundamental relation (1.68) between the one-electron orbital energy sum $E_s$ and the total energy $E$. The considerable successes of one-electron theories of molecules show that many-electron effects are not dominant but they do appear explicitly through the correlation kinetic energy $\Delta T$ in eqn (1.68). It is worthy of note that the deviations from the relation $E=\frac{3}{2}E_s$ displayed in eqn (1.68) are qualitatively of the form that the chemical potential correction term $-\frac{3}{2}N\mu$ is positive, since $\mu$ is negative for bound molecules, while both the kinetic energy contributions are expected to be negative. This latter point is supported by using a low-order density gradient expansion of $t$, the lowest-order result being

$$\int\left[\rho\frac{\delta T_s}{\delta\rho}-\frac{5}{3}t\right]\mathrm{d}\tau=-\frac{1}{108}\int\frac{(\nabla\rho)^2}{\rho}\mathrm{d}\tau, \tag{1.69}$$

which is evidently negative to this lowest order.[32]

As for the correlation kinetic energy $\Delta T$, this is expected to be positive, because, in forming a wave function from, say, the single-particle basis generated by the one-body potential $V(\mathbf{r})$ defined in eqn (1.59), one will expect the correlated wave function to involve the creation of particles and holes, and hence to increase the total kinetic energy above its single-particle value $T_s$.

Thus, the first part of our argument in displaying the first principles foundations of Walsh's rules is achieved in eqn (1.68) at equilibrium, for as we have emphasized above, the corrections to $E=\frac{3}{2}E_s$ are identically zero in the simplest density description and are, on the general grounds discussed above, expected to be small in the full many-body treatment. Furthermore the chemical potential and kinetic energy corrections at least partly cancel, as argued above.

But there remains a lacuna in the argument needed for Walsh's rules. Evidently for the HAH example referred to above, there is one equilibrium geometry. Away from that, if we take an arbitrary geometry, we cannot use the virial theorem in the form $T=-E$. To be explicit, consider the simpler case of a diatomic molecule, with internuclear separation $R$. Then if $U$ is the total potential energy, such that $T+U=E$, the virial theorem for arbitrary $R$ reads

$$2T+U=-R\frac{\mathrm{d}E}{\mathrm{d}R}, \tag{1.70}$$

where the term on the right-hand side is the virial of the forces needed to hold fast the nuclei at separation $R$. But this is clearly only zero, for a diatomic molecule, at one configuration, the equilibrium separation, $R_e$ say, at which $\mathrm{d}E/\mathrm{d}R$ is zero. Thus only for this one case in a diatomic molecule can we use $T=-E$ which was, of course, employed above to reach the basic equation (1.68).

But now the message for the discussion of molecular shapes following Walsh is clear if we note, from the generalization of eqn (1.68) for polyatomics, the importance, in HAH say, of minimizing with respect to bond lengths at each angular configuration chosen. Provided this is done, in the example HAH, for each angle considered then as Nelander[33] has proved, for every set of angles for which the energy is minimized with respect to bond distances the virial theorem holds in its simple form $T=-E$.

In short, eqns (1.62) and (1.64) follow quite generally from density functional theory. But to reach eqn (1.68) which is required as a basis for Walsh's rules, one must add to these the virial theorem in the simple form $T=-E$. We have emphasized that while this certainly does not hold for aribirary geometries, provided we regard the correlation diagram for say HAH as plotted at each HAH angle at the bond length which minimizes the energy at that angle, then eqn (1.68) holds. Thus Walsh's arguments go through as before. We stress that, if and when the need arises to discuss possible deviations from Walsh's rules, the present argument points out the directions in which such a discussion would go; they reside in chemical-potential and kinetic-energy corrections spelt out below (eqn (1.68)).

In this respect, Mucci and March[34] have noted that $E-\frac{3}{2}E_s$ for a variety of light molecules correlates with electronegativity. Equation (1.68) shows it should correlate with the chemical potential and this, at least qualitatively, supports the connection proposed by Parr *et al.*[35] that the chemical potential $\mu$ of the density functional theory is the negative of (Sanderson's) electronegativity.[36]

Although clearly, correlation cannot be treated exactly, Alonso and March[37] have argued that in the Hartree–Fock approximation the chemical potential $\mu$ is identically equal to the Koopmans eigenvalue $\varepsilon_k$. This result rests on two points:

(i) the kinetic energy density at large $r$ has the von Weizsäcker form $(\hbar^2/8m)\,(\nabla\rho)^2/\rho$, while the asymptotic density is given by $\rho(r)=Ar^n \exp[-2(2|\varepsilon_k|)^{1/2}r]$, for atoms in the Hartree–Fock theory;
(ii) the density matrix of Hartree–Fock theory is idempotent.

Of course, correlation will naturally have to be included approximately in any completely satisfactory theory of the chemical potential.[38]

## 1.12 Molecular dissociation and electron density gradients

Mucci and March[39] drew qualitative conclusions about the importance of electron density gradients for molecular dissociation energy prediction, on the basis of Teller's theorem.[40] This states that molecular binding cannot occur in any local electron density theory; "local" in this present context meaning that all electron density gradient corrections, not only those in exchange and correlation energy, but also in kinetic energy, are neglected. Thus, Mucci and March suggested that in particular the dissociation energies of diatomic molecules should correlate with the inhomogeneity kinetic energy $(\hbar^2/72m)\int(\nabla\rho)^2/\rho\, d\mathbf{r}$ (see Appendix A 1.3).

Allan *et al.*[41] have therefore evaluated the above energy for a set of light diatomic molecules, using molecular electron densities, at the equilibrium spacing, of Hartree–Fock–Roothaan quality. Pursuing the above idea, they have plotted the dissociation energy $D$, divided by $N^2$ (where $N$ is the total number of electrons), against $T_2$ which is defined by

$$T_2 = \frac{\hbar}{72m}\int\frac{(\nabla\rho)^2}{\rho}\, d\mathbf{r}. \tag{1.71}$$

There is seen (Fig. 1.1, taken from ref. 42) to be a considerable degree of correlation between these quantities. These workers[41] cautioned against any claim that $T_2$—essentially a property of the equilibrium molecular density—can lead to any absolute determination of $D$.

Independent support for such a correlation comes from the subsequent calculations of Lee and Ghosh;[42] using a larger variety of data than Allan *et al*, they demonstrate a rather linear relation between $D/N^2$ and $T_2$ in eqn (1.71) for a variety of molecules (cf. Fig. 1.1).

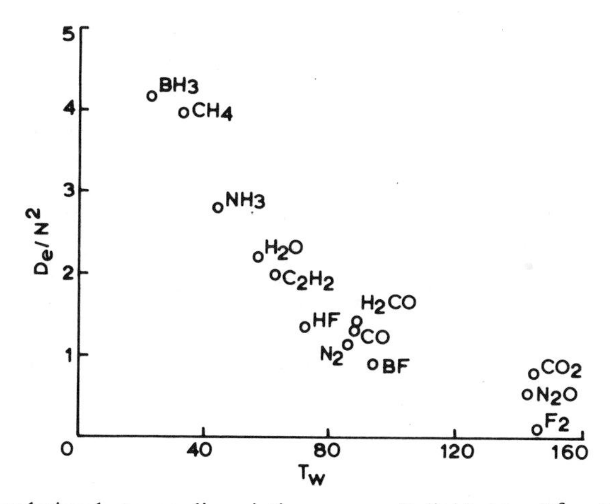


**Fig. 1.1** Correlation between dissociation energy $D$ divided by $N^2$, with $N$ the total number of electrons and the inhomogeneity kinetic energy $T_2$ defined in eqn (1.71) (N.B. $T_2 = \frac{1}{9} T_W$, where $T_W$ denotes the von Weizsäcker kinetic energy correction). (After Lee and Ghosh[42].)

In view of the grave difficulties that always face conventional attempts to determine dissociation energies of heavy molecules from the difference between the molecular energy and the sum of the energies of the constituent atoms, it would seem that this is an area in which further work is likely to be well worth while.

## 1.13 Asymptotic predictions for tetrahedral and octahedral molecules

There is one molecular model, of almost spherical molecules, for which the simplest density description; namely the Thomas–Fermi theory, can be fully worked out.[43] It has the additional merit that it exemplifies the scaling relations (1.28), (1.32) and (1.33) exactly, as well as the relation $E=\frac{3}{2}E_s$. The model is most appropriate to represent the tetrahedral series of molecules $CH_4$, $SiH_4, \ldots, PbH_4$. What will be demonstrated below, following the work of Pucci and March,[44] is that in this model, in which the four outer protons in $XH_4$ are smeared uniformly over the surface of a sphere of radius $R$ equal to the X–H bond length, as the nuclear charge $Z_2 e$ of the X atom is allowed to tend to infinity, the equilibrium bond length $R_e$ tends to a finite limit. Pucci and March[44] have confronted this prediction of a finite asymptotic bond length with the experimental facts not only for the series $CH_4, \ldots, PbH_4$, but also for tetrahedral fluorides, chlorides and bromides and for octahedral molecules.

By way of background to this work, Mucci and March[45] have presented evidence that for homonuclear diatomic molecules it is unlikely that there is any homonuclear chemistry for molecules heavier than around $Pb_2$. However, when they studied heteronuclear diatoms, their conclusions were quite different; namely that even very heavy molecules of this kind would be stabilized by charge transfer.

It is therefore a matter of some interest for first principles theory to study some aspects of the theoretical chemistry of very heavy heteronuclear molecules. In ref. 44, such a theoretical study was made on what is, in essence, the basis of the non-relativistic Schrödinger equation. Later in this Chapter we consider briefly the effects of relativity, but no account of them is attempted at the present stage.

### 1.13.1 Central-field model for the series $CH_4, \ldots, PbH_4$

We re-emphasize that, because of the high symmetry of such a series $XH_4$, one can contemplate expanding the nuclear potential field of the four outer protons about the central nucleus carrying charge $Z_2 e$. If one retains only the first (s) term of such an expansion, one is led immediately to the model in which the outer nuclear charges, say with total nuclear charge $Z_1 e$, are spread uniformly over a sphere of radius $R$ equal to, say, the Pb–H bond length.

Within this model, a number of simple results follow.[43] First of all, the total self-consistent potential energy $V(r)$, scaled in accord with the nuclear charge $Z_2 e$ at the origin, can be written in the form

$$V(r) = -\frac{Z_2 e^2}{r}\phi(x), \tag{1.72}$$

where $x$ is a dimensionless measure of the distance $r$ from the point charge $Z_2 e$, and is given by (see Appendix A 1.1)

$$r = bx = \left(\frac{3}{32\pi^2}\right)^{2/3} \frac{h^2}{2me^2 Z_2^{1/3}} = \frac{0.88534}{Z_2^{1/3}} a_0, \tag{1.73}$$

where $a_0$ is the Bohr radius $\hbar^2/me^2$. The dimensionless dependent variable $\phi$ satisfies the usual non-linear Thomas–Fermi equation (cf. Appendix A 1.1)

$$\frac{d^2\phi}{dx^2} = \frac{\phi^{3/2}}{x^{1/2}}. \tag{1.74}$$

Owing to the presence of the surface charge at distance $R = bX$ from the origin, the electric field suffers a discontinuity determined by the magnitude of the surface charge density, and this, in turn, is reflected by a discontinuity at $X$ in the derivative of $\phi(x)$, as depicted schematically in Fig. 1.2 the discontinuity being given precisely by

$$\left.\frac{\partial \phi_1}{\partial x}\right|_X - \left.\frac{\partial \phi_2}{\partial x}\right|_X = \frac{Z_1}{XZ_2}. \tag{1.75}$$

The equilibrium bond length $R_e = bX_e$ is determined by the usual condition

$$\left.\frac{dE}{dR}\right|_{R_e} = 0, \tag{1.76}$$

$E$ being the total molecular energy, including of course the nuclear–nuclear potential energy $U_{nn}$ given by

$$U_{nn} = \frac{Z_1 e^2}{R}[Z_2 + dZ_1], \tag{1.77}$$

where $d$ is determined by the molecular geometry and for tetrahedral molecules has the value

$$d_{\text{tetrahedral}}=\frac{3\sqrt{6}}{32}. \tag{1.78}$$

As shown in ref. 43, $dE/dR$ is zero when, in the notation of Fig. 1.2,

$$X_e \frac{\partial \phi_1}{\partial x}\bigg|_{X_e} - \phi_1(X_e)=\frac{dZ_1}{Z_2}. \tag{1.79}$$

Equations (1.75)–(1.79) constitute the essential basis for the derivation, in the following section, of an analytical formula for $X_e$ and hence the equilibrium bond length $R_e=bX_e$ in the limiting case when $Z_2\to\infty$ for fixed $Z_1$. To lead into this derivation, the numerical calculations of ref. 43, which can be presented by plotting $X_e$ against $Z_1/Z_2$, leave no doubt that as $Z_1/Z_2\to 0$ in accord with the desired limiting process discussed above, $X_e\to\infty$. It is the detailed way in which $X_e$ goes to infinity that is our main concern in what follows.[44]

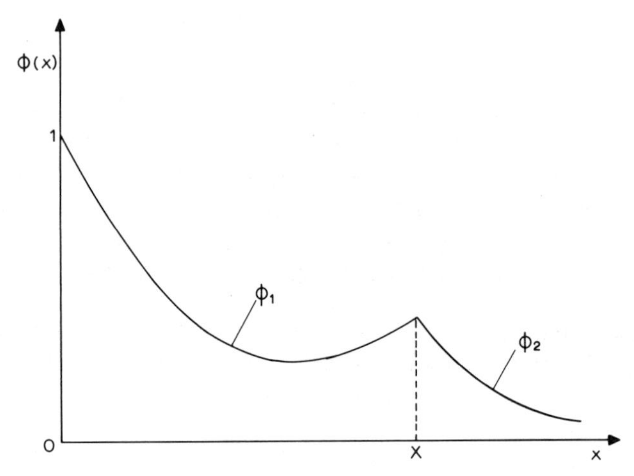


**Fig. 1.2** Screening function $\phi(x)$ defined in eqn (1.72), the function being defined in region 1 ($0<x\leqslant X$) as $\phi_1(x)$ and in region 2 ($X<x<\infty$) as $\phi_2(x)$. The discontinuity in the slope at $X$ reflects the discontinuity in the electric field across a surface charge distribution such as occurs in the one-centre expansion for a molecule like $PbH_4$.

### 1.13.2 Finite equilibrium bond length in limit $Z_2/Z_1$ tends to infinity

Appeal to Fig. 1.2 shows that, as $X$ tends to infinity, we require for $x>X$ a solution $\phi_2$ of the Thomas–Fermi equation (1.74) which tends to zero at infinity. In fact, the general form of the solutions of eqn (1.74) which tend to zero at infinity was established by Coulson and March,[46] and their work allows one to write, for sufficiently large $x$,

$$\phi_2(x)=\frac{144}{x^3}\left[1-\frac{F_1}{x^c}+\cdots\right], \tag{1.80}$$

where in the present problem $F_1=F_1(Z_1/Z_2)$ and the exponent $c$ is given by

$$c=\frac{\sqrt{73}-7}{2}=0.772. \tag{1.81}$$

Here we shall suppose that $X$, and hence evidently $x$, is so large that we can neglect the term in $F_1$ in eqn. (1.80): its effect is considered in detail in ref. 44. Inserting therefore the approximation $\phi_2=144/x^3$ valid as $Z_1/Z_2\to 0$ into eqn (1.75) we can evidently calculate both $\partial\phi_1/\partial x|_{X_e}$ and $\phi_1(X_e)=\phi_2(X_e)$ entering the equilibrium condition (1.79) and we readily find, in the limit $Z_1/Z_2\to 0$:

$$X_e^\circ=\left(\frac{Z_2}{Z_1}\right)^{1/3}\frac{12}{[3(1-d)]^{1/3}}, \tag{1.82}$$

confirming the assertion made above that $X_e\to\infty$ as $Z_2\to\infty$ for finite fixed $Z_1$. Forming the asymptotic equilibrium bond length $R_e^\circ=bX_e^\circ$ in this same limit, we find from eqn (1.73) that $Z_2$ cancels out and we are left with the finite equilibrium bond length as $Z_1/Z_2\to 0$:

$$R_e^\circ=\frac{1}{Z_1^{1/3}}\frac{12}{[3(1-d)]^{1/3}}\left(\frac{3}{32\pi^2}\right)^{2/3}\frac{h^2}{2me^2}. \tag{1.83}$$

In this molecular model, therefore, the limiting bond length is determined solely by the total nuclear charge $Z_1e$ of the outer atoms and the geometrical factor $d$ introduced in eqn (1.77).

As a numerical example, we take the tetrahedral series $CH_4,\ldots,PbH_4$, when $Z_1=4$ and $d$ is given by eqn (1.78). The numerical value of $R_e^\circ$ from eqn (1.83) is 2.68 Å.

### 1.13.3 Relation of prediction of finite asymptotic bond length $R_e^\circ$ to experiment

Though, as we have stressed, the entire theory presented here is based on non-relativistic quantum mechanics, it is of obvious interest to enquire whether experiment indicates agreement, or otherwise, with the model prediction that the equilibrium bond length $R_e$ tends to a finite asymptotic value as $Z_1/Z_2 \to 0$. As Pucci and March[44] discuss in some detail, it seems possible for the series $CH_4, \ldots, PbH_4$, which seems the most appropriate testing ground for the model's prediction, that the prediction may be quantitatively correct in spite of neglect of (*a*) non-spherical terms *p*, *d*, etc. in the one-centre expansion and (*b*) effects of relativity. Even for molecules with atoms with well defined cores in the outer positions, e.g. $UF_6$, it seems still possible that the model prediction of a finite asymptotic bond length is qualitatively correct.

In summary, the work of Mucci and March[45] makes it unlikely that there will be any homonuclear chemistry beyond say $Pb_2$. In contrast, charge transfer stabilizes heteronuclear molecules. The work of Pucci and March[44] discussed above makes a start on some aspects of the theoretical chemistry of very heavy heteronuclear molecules. However, relativistic effects should plainly be considered in such systems; this matter will be referred to in the following section.

## 1.14 Spin and relativistic generalizations

The Hohenberg–Kohn theorem proved in section 1.3 is, of course, important in formally completing the original Thomas–Fermi theory. However, as presented, it is the basis of a non-relativistic scheme and the purpose of this section is twofold. First, we shall consider how the theorem has been extended to remove the limitation that *c* tends to infinity, *c* being the velocity of light. Secondly, we shall follow this (rather formal) discussion with an account of the relativistic Thomas–Fermi theory and its application to atomic and, briefly, to a class of molecular systems.

The presentation of relativistic density functional theory below follows that of Callaway and March.[47]* By starting from a relativistic point of view, one derives the result that the ground-state energy is a functional of the four-vector current density $j_\mu(\mathbf{r})$, provided that this vector is constrained by an equation of continuity (see Chapter 8) which attains its minimum value when $j_\mu$ is correct. The result involving spin densities only is then obtained in the non-relativistic limit (see also Appendix A 5.4, p. 146).

The starting point is the time-dependent Schrödinger equation:

$$[i\hbar(\partial/\partial t) - H]|\Psi\rangle = 0. \tag{1.84}$$

The Hamiltonian $H$ contains four parts:

$$H = H_0 + H_c + H_1 + H_{ext}. \tag{1.85}$$

*This section is the most advanced in this Chapter. The reader interested in practical results can pass to Section 1.14.1 on a first reading.

The operator $H_0$ describes non-interacting Dirac and electromagnetic fields:

$$H_0 = H_{em} + \int \psi^{\dagger}(x) h(x) \psi(x)\, d\mathbf{r}, \tag{1.86}$$

where $x$ denotes coordinates and time. In eqn (1.86), $H_{em}$ describes the free electromagnetic field, $\psi$ and $\psi^{\dagger}$ are annihilation and creation operators, respectively, for the free Dirac field, while $h(x)$ is the Hamiltonian of this field:

$$h(x) = c\boldsymbol{\alpha}\cdot\mathbf{p} + \beta mc^2. \tag{1.87}$$

The electromagnetic field is described in the radiation gauge so that the Coulomb interaction between electrons appears explicitly:

$$H_c = \frac{e^2}{2}\int \psi^{\dagger}(x)\psi(x)\frac{1}{|\mathbf{r}-\mathbf{r}'|}\psi^{\dagger}(x')\psi(x')\, d\mathbf{r}\, d\mathbf{r}'. \tag{1.88}$$

The interaction between matter and the transverse radiation field is contained in $H_1$:

$$H_1 = \frac{e}{c}\int \mathbf{j}(x)\cdot\mathbf{A}(x)\, d\mathbf{r}, \tag{1.89}$$

where $\mathbf{j}$ refers to the space components of the current operator:[48]

$$j^{\mu}(x) = c\psi^{\dagger}(x)\gamma^{0}\gamma^{\mu}\psi(x) \tag{1.90}$$

which satisfies the continuity equation

$$\partial_{\mu} j^{\mu} = 0. \tag{1.91}$$

Finally one has the interaction between the current and a (classical) external electromagnetic field $A^{\mu}_{ext}(x)$:

$$H_{ext} = -\frac{e}{c}\int j_{\mu}(x) A^{\mu}_{ext}(x)\, d\mathbf{r}. \tag{1.92}$$

In principle, operator products in these equations must be normally ordered, as discussed, for example, in the book by Schweber,[48] but this will not be indicated explicitly in what follows.

Now one wishes to consider the ground-state energy of the Hamiltonian $H$. There is a technical problem because relativistic Hamiltonians are not bounded from below and therefore do not have a ground state. Since physical systems, of course, do have ground states, in the conventional way it will be supposed that all negative-energy states are filled.

Let $|G\rangle$ therefore denote the ground state in this sense. One describes the current density in the ground state as

$$J_{\mu}(x) = \langle G | j_{\mu}(x) | G \rangle. \tag{1.93}$$

The expectation value in eqn (1.93) is taken with respect to the Fock-space operator so that $J_\mu(x)$ is an ordinary function of position. The four components of $J_\mu$ are not all independent since the equation of continuity,

$$\partial_\mu J^\mu(x)=0, \tag{1.94}$$

must be satisfied. It is now possible to repeat arguments like those of Hohenberg and Kohn (cf. Section 1.3) to show that the ground-state energy of the system is a unique functional of $J_\mu(x)$.

If we finally seek to write the functional in compact form, it is useful (cf. ref. 47) to introduce the following:*

(1) the particle density $\rho$,

$$\rho(x)=\langle G|\psi^\dagger(x)\psi(x)|G\rangle; \tag{1.95}$$

(2) the spin density vector

$$s^l(x)=\langle G|\tilde{\psi}\sigma^l\psi(x)|G\rangle; \tag{1.96}$$

(3) the ordinary (plus polarization) current vector $j_l(x)$,

$$\begin{aligned} J_l(x)=(2m)^{-1}\langle G|\tilde{\psi}(x)(p_l-eA_l/c)\psi(x) \\ -[(p_l+eA_l/c)\tilde{\psi}(x)]\psi(x)|G\rangle; \end{aligned} \tag{1.97}$$

(4) a vector without classical analogue, namely

$$g^l(x)=(-i/2m)\langle G|\tilde{\psi}(x)\alpha^l\psi(x)|G\rangle. \tag{1.98}$$

The contribution to $\langle G|H_1|G\rangle$ from the spin density can be transformed into

$$\frac{e}{2m}\int \varepsilon^{k}{}_{j}{}^{l}\,\frac{\partial s(x)}{\partial x^j}A_{(\mathrm{ext})k}(x)\,\mathrm{d}\mathbf{r}=-\frac{e}{2m}\int \mathbf{s}(x)\cdot\mathbf{B}_{\mathrm{ext}}(x)\,\mathrm{d}\mathbf{r} \tag{1.99}$$

in which $\mathbf{s}(x)$ is the ordinary spin vector and $\mathbf{B}_{\mathrm{ext}}(x)=\nabla\times\mathbf{A}_{\mathrm{ext}}$. Finally, the ground-state energy functional can be re-expressed as

$$\begin{aligned} E=F[J]+\int \mathrm{d}\mathbf{r}\Big[\rho(x)V_{\mathrm{ext}}(x)+\frac{e}{2m}\mathbf{s}(x)\cdot\mathbf{B}_{\mathrm{ext}}(x)-\frac{e}{c}\mathbf{J}(x)\cdot\mathbf{A}_{\mathrm{ext}}(x) \\ +\frac{e}{2mc}\frac{\mathrm{d}\mathbf{g}}{\mathrm{d}t}\cdot\mathbf{A}_{\mathrm{ext}}(x)\Big] \end{aligned} \tag{1.100}$$

and $V_{\mathrm{ext}}=-e\phi_{\mathrm{ext}}(x)$, where $\phi$ is the scalar potential, i.e. the zeroth component of $\mathbf{A}$. This result generalizes the results of the earlier part of this Chapter by describing in a more comprehensive way the interaction between a system and an external field (see also Chapter 7).

The non-relativistic limit of this treatment, namely spin density theory, will be referred to again in the treatment of magnetic solids in Chapter 5.

*Below, the notation $\tilde{\psi}=\psi^\dagger\gamma^0$ is used, and $\varepsilon^{kjl}$ is the Levi–Civita notation $\varepsilon^{123}=1, \varepsilon^{213}=-1$ etc.

### 1.14.1 Relativistic Thomas–Fermi theory

The above treatment remains, of course, at a rather formal level. Therefore, we shall conclude this Chapter by discussing in some detail the generalization of the simplest density functional theory, namely that of Thomas and Fermi, to incorporate the requirements of special relativity.

One writes a generalization of the non-relativistic Thomas–Fermi equation (1.6) by using the relativistic expression for the kinetic energy of a particle of rest mass $m$ having momentum $p$, to yield

$$\mu = [c^2 p_F^2(\mathbf{r}) + m^2 c^4]^{1/2} - mc^2 + V(\mathbf{r}), \tag{1.101}$$

and if, as in the non-relativistic theory, we use a dimensionless screening function (cf. Appendix A 1.1 for the precise definition), then we reach the equation, due to Vallarta and Rosen[49]

$$\frac{d^2\phi}{dx^2} = \frac{\phi^{3/2}}{x^{1/2}}\left(1 + \lambda\frac{\phi}{x}\right)^{3/2}, \tag{1.102}$$

where

$$\lambda = (4/3\pi)^{2/3}\alpha^2 Z^{4/3}, \tag{1.103}$$

with $\alpha$ the fine structure constant $e^2/\hbar c$. This reduces to the non-relativistic counterpart (1.74) as we let $\alpha$ tend to zero.

As Senatore and March[50] have shown, whereas eqn (1.74) has the asymptotic solution (1.80), the exact solution $144/\alpha^3$ of Sommerfeld in eqn (1.74) must be recalculated for $\lambda \neq 0$.

Senatore and March[50] obtain an exact solution of eqn (1.102) of the form

$$\phi(\lambda, x) = \frac{144}{x^3} f(\lambda/x^4), \tag{1.104}$$

where $f(s)$ has a simple pole at a critical value $s_c$, which is in order of magnitude about $10^{-2}$. This evidently corresponds to a critical length $r_c = bx_c$ where these workers[50] show that

$$x_c = \text{const}\,\lambda^{1/4}; \quad \text{const} \doteqdot 2.9, \tag{1.105}$$

yielding $r_c \cong 2.2\,\alpha^{1/2} a_0 \doteqdot 0.19 a_0$.

Senatore and March[50] demonstrate how this affects the analyticity of the total energy of an atomic ion with $N$ electrons, namely $E(Z, N, \alpha)$. In particular, these authors propose a Laurent expansion

$$E(Z, N, \alpha, R) = Z^2 \sum_{n=-\infty}^{\infty} \varepsilon_n(N, \alpha, R) Z^{-n}, \tag{1.106}$$

where $R$ is the nuclear radius.

They demonstrate that this result (1.106) includes the generalization of the $1/Z$ expansion in eqn (1.40) by Layzer and Bahcall[51] to incorporate the fine-structure constant $\alpha$, namely

$$E_{\mathrm{LB}}(Z, N) = Z^2 \sum_{n=0}^{\infty} \sum_{m=0}^{\infty} E_{nm}(N)\varepsilon^m Z^{-n} \tag{1.107}$$

with $\varepsilon = \alpha^2 Z^2$. As written in the above equation (1.107), the coefficients $E_{nm}(N)$ also have a weak dependence on $\varepsilon Z$ in the Layzer–Bahcall formulation.

Returning at this point to the relativistic Thomas–Fermi theory, Marconi and March[52] demonstrate that this corresponds, for large $N$, to the coefficients

$$E_{nm}(N) \sim c^0_{nm} N^{n-2m/3-1/3}, \tag{1.108}$$

which contains the non-relativistic result (1.44) for $m=0$.

In addition to such analytical results, Hill *et al.*[53] have solved numerically the relativistic Thomas–Fermi equation (1.102) for neutral atoms and positive ions. They have thereby brought the predictions of this theory into direct contact with relativistic Dirac–Fock calculations.[53]

The use of relativistic Thomas–Fermi theory for almost spherical molecules to display corrections to eqn (1.83) arising from $\alpha \neq 0$ is discussed in ref. 54.

### 1.14.2 Relativistic Thomas–Fermi equation in an extremely strong magnetic field

The discovery of very strong magnetic fields in compact galactic objects, namely white dwarf stars, as reviewed for example by Garstang,[55] and neutron stars,[56,57] has led to substantial effort in understanding the electronic structure of atoms subjected to such intense fields.

Therefore it is of importance to consider the relativistic generalization of the Thomas–Fermi equation in an extremely strong magnetic field, following the work of Hill *et al.*[58]

The starting point in setting up such an equation is the relativistic expression for the chemical potential $\mu$ of the electron cloud. This equation, written in terms of the Fermi momentum $p_{\mathrm{F}}$, will then be combined with a relation valid in the extreme high-field limit between $p_{\mathrm{F}}$, the electron density $\rho$ and the magnetic field $H$.

Using the relativistic expression for the Fermi energy, equal to the constant chemical potential $\mu$, one has

$$\mu = (c^2 \Pi_{\mathrm{F}}^2 + m_0^2 c^4)^{1/2} - m_0 c^2 + V(\mathbf{r}) \tag{1.109}$$

where $\Pi$ is the magnitude of the canonical momentum $\mathbf{p}(\mathbf{r})+e\mathbf{A}/c$, $m_0$ is the rest mass of the electron and $V(\mathbf{r})$ is the self-consistent Hartree potential energy. As usual, the magnetic field is introduced via the vector potential $\mathbf{A}$ (cf. eqn (1.99)).

It is useful to work in the case of atomic ions with the dimensionless quantities $\phi$ and $x$ (cf. Appendix A 1.1 for the case $H=0$) defined by

$$\left.\begin{aligned} &\mu - V(r) = (Ze^2/r)\phi(x); \\ &r = b'x, \quad b' = \tfrac{1}{2}(\gamma/2\pi)^{-2/5} Z^{1/5} a_0, \end{aligned}\right\} \tag{1.110}$$

where $a_0$ is the Bohr radius $\hbar^2/me^2$, while $\gamma = (\mu_B H/\mathrm{Ry})$.

Equation (1.109) can now be combined with the density–momentum relation in the high-field limit, given by, for instance, Banerjee *et al.*:[59]

$$\rho(\mathbf{r}) = \frac{2eH}{h^2} \Pi_F(\mathbf{r}), \tag{1.111}$$

which, using eqn (1.110), gives

$$\frac{d^2\phi}{dx^2} = (x\phi)^{1/2}[1 + \lambda(\phi/x)]^{1/2}, \tag{1.112}$$

where

$$\lambda = (\gamma/2\pi)^{2/5} \alpha^2 Z^{4/5}, \tag{1.113}$$

$\alpha$ being the fine-structure constant, $e^2/\hbar c$, as before. In the limit as $c$ tends to infinity, it is clear that eqn (1.112) yields $\phi'' = (x\phi)^{1/2}$ which is the corresponding non-relativistic dimensionless Thomas–Fermi equation.[60]

Hill *et al.*[58] have obtained a neutral-atom solution of eqn (1.112) for $H = 10^{14}$ G. By the methods they employ, it is straightforward to obtain specific results for, say, the Crab pulsar's field[55] estimated at $6 \times 10^{12}$ G, should precise data for specific fields eventually be needed.

To conclude this discussion of atoms in high magnetic fields, notable work is that of Jones,[61] who used the density functional theory by employing correct generalizations of exchange and correlation energy in a uniform electron liquid in the presence of a strong magnetic field. His conclusions are important for the nature of matter under extreme conditions, and these conclusions follow by comparing atomic configurations with the solid state. This latter point will be taken up again in Chapter 5.

Finally, preceding the references for this Chapter, we list below useful books and reviews on the single-particle density, for the convenience of the reader.

## Appendix A1.1

Scaling of the electron density and self-consistent field in heavy atomic ions and in almost spherical molecules

The purpose of this appendix is to expose the scaling properties of the Thomas–Fermi electron density for heavy atoms and positive ions, and, more briefly, for almost spherical molecules and positive molecular ions.

If one combines Poisson's equation of electrostatics with the density–potential relation (1.9), then one can put the problem of atoms into universal form by introducing a screening function $\phi(x)$ which converts the self-consistent potential energy $V(r)$, measured relative to the chemical potential $\mu$, from the bare nuclear potential energy $-Ze^2/r$ through

$$\mu - V(r) = \frac{Ze^2}{r}\phi(x). \tag{A 1.1.1}$$

The Poisson equation is then readily scaled into the form (1.74) to be solved for the screening function $\phi(x)$, where $x$ represents the scaled, dimensionless distance from the nucleus through

$$r = bx; \quad b = 0.88534\, a_0/Z^{1/3}, \tag{A 1.1.2}$$

displaying the scaling with the cube root of the atomic number $Z$.

For neutral atoms, the dimensionless Thomas–Fermi equation (1.74) has to be solved subject to the boundary conditions

$$\phi(0) = 1 \tag{A 1.1.3}$$

and

$$\phi(\infty) = 0. \tag{A 1.1.4}$$

The first of these is evident from eqn (A 1.1.1) while the second requires that the self-consistent potential energy falls off at large $r$ faster than $1/r$ in the neutral atom. In fact, eqn (1.74) is readily shown to have an exact solution $144/x^3$, due to Sommerfeld. This satisfies the boundary condition (A 1.1.4); to construct the general solution of eqn (1.74) which tends to zero at infinity, one can use the form (1.80). The neutral atom solution satisfying both boundary conditions (A 1.1.3) and (A 1.1.4) corresponds, it turns out, to the choice $F_1 = 13.27$.

This deals with the neutral-atom case, when supplemented with the small-$x$ solution of eqn (1.74) due to Baker,[62] namely

$$\phi(x) = 1 + a_2 x + a_3 x^{3/2} + a_4 x^2 + \cdots, \tag{A 1.1.5}$$

where $a_3$, $a_4$, etc., are known. To join on to the neutral-atom solution at large $x$ discussed above, one must choose the value $a_2 = -1.58805$. This, in turn, determines the electrostatic potential due to the electron cloud at the nucleus, given by eqns (1.35) and (1.36).

For positive ions with $N(<Z)$ electrons, one can employ Gauss' theorem to find that the ions have a positive radius, $r_0 = bx_0$, and this is determined by the boundary condition

$$-x_0\left(\frac{\partial \phi}{\partial x}\right)_{x_0} = 1 - \frac{N}{Z} \qquad \text{(A 1.1.6)}$$

which reduces to eqn (A 1.1.4) in the limit of the neutral atom, in which $N$ tends to $Z$, and $x_0$ tends to infinity. For positive ions, the chemical potential $\mu$ is readily shown from electrostatics to be

$$\mu = -\frac{(Z-N)e^2}{r_0}, \qquad \text{(A 1.1.7)}$$

which again reduces to $\mu = 0$ for the neutral Thomas–Fermi atom.

For the model of almost spherical molecules discussed in Section 1.13, the scaling property (A 1.1.2) becomes eqn (1.73), while application of electrostatics to give the correct discontinuity in the electric field across the surface charge distribution yields eqn (1.75). The corresponding scaling properties for positive molecular ions in the same model have been fully treated by Lawes and March.[63]

## Appendix A1.2

### Exact density–potential relation for the ground state of the Be atom

The exact ground-state density $\rho(r)$ of the Be atom can, at least in principle, be generated from a Hartree-like one-body potential energy $V(r)$, provided this includes an appropriate contribution from exchange and correlation interactions. In this appendix an exact relation between $\rho(r)$ and this given $V(r)$ is exhibited by making use of the density matrix variational method set up by Dawson and March[64] for two occupied levels in a one-dimensional system. In the course of the present treatment, their Lagrange multipliers, taking care of the constraints imposed by orthonormality, are given a quite explicit interpretation following March and Nalewajski.[65]

#### *(a) Relation of Be density to an equivalent two-level problem in one dimension*

Given the central potential energy $V(r)$, we can obviously calculate the radial wave functions $R_1(r)$ and $R_2(r)$ of the 1s and 2s electrons respectively in the ground state of Be from the Schrödinger equations

$$\frac{1}{r}\frac{\mathrm{d}^2}{\mathrm{d}r^2}(rR_{\mathrm{n}})+2(\varepsilon_n-V(r))R_n=0, \quad n=1,2. \tag{A 1.2.1}$$

Provided we adopt the normalization

$$\int_0^\infty R_n^2\, 4\pi r^2\, \mathrm{d}r=1 \tag{A 1.2.2}$$

below, then the ground-state density is evidently

$$\rho(r)=2(R_1^2(r)+R_2^2(r)). \tag{A 1.2.3}$$

Instead of setting up the off-diagonal density matrix generalization of eqn (A 1.2.3) for Be, below we transform instead the calculations of Dawson and March[64] in which a variational density matrix was set up for two occupied levels in a one-dimensional system.

To make contact with their work, all that is needed is to make the simple transformation

$$\left.\begin{aligned} rR_{\mathrm{n}}(r)\rightarrow\phi_{\mathrm{n}}(r), \quad n=1,\ 2; \\ r\rightarrow x. \end{aligned}\right\} \tag{A 1.2.4}$$

It must then be noted that the "one-dimensional density $\rho_1(x)$" they calculate is in fact such that

$$\rho_1(x)=4\pi(\phi_1^2(x)+\phi_2^2(x)). \tag{A 1.2.5}$$

Once one has calculated $\rho_1(x)$ from the Dawson–March variational procedure, the Be atom density $\rho(r)$ given in eqn (A 1.2 3) is simply related to this one-dimensional density $\rho_1(r)$ by

$$4\pi r^2\rho(r)=2\rho_1(r). \tag{A 1.2.6}$$

Clearly, if $\int_0^\infty \rho_1(r)\mathrm{d}r=2$, as we shall assume, then from eqn (A 1.2.6) it follows that

$$\int_0^\infty \rho(r)4\pi r^2\,\mathrm{d}r=2\int_0^\infty \rho_1(r)\,\mathrm{d}r=4 \tag{A 1.2.7}$$

as it must be for the ground state of Be.

Thus, the burden of part (*b*) of this appendix is to employ the Dawson–March equations to yield an explicit relation between $\rho(r)$ and the given $V(r)$ which, as we have stressed in eqn (1.59), involves many-electron contributions which are not yet known exactly. However the procedure given in (*b*) below relates $\rho(r)$ exactly for Be to the given $V(r)$.

### (b) *Exact density–potential relation for a two-level one-dimensional case*

Consider a two-level one-dimensional independent-particle problem for an external potential $V(x)$, e.g. the harmonic oscillator with the first two states occupied, as discussed in an appendix to ref. 65. Following ref. 64, we adopt polar coordinates in which to express the one-body orbitals $\phi_1$ and $\phi_2$ as well as the single-particle density matrix $\gamma(x', x)$, in terms of the density $\rho(x)$, given simply by $\rho(x)=\gamma(x,x)= 2\phi_1^2+2\phi_2^2$:

$$2^{1/2}\,\phi_1(x)=\rho^{1/2}\cos\theta(x), \tag{A 1.2.8}$$

$$2^{1/2}\,\phi_2(x)=\rho^{1/2}\sin\theta(x) \tag{A 1.2.9}$$

and

$$\gamma(x',x)=[\rho(x')]^{1/2}[\rho(x)]^{1/2}\cos[\theta(x')-\theta(x)]. \tag{A 1.2.10}$$

Hence one obtains the expression for the kinetic energy density

$$\begin{aligned} t(x) &= -\frac{1}{2}\frac{\partial^2}{\partial x^2}\gamma(x',x)\bigg|_{x'=x} \\ &= -\frac{1}{4}\rho''(x)+\frac{1}{8}\frac{\rho'^2}{\rho}+\frac{1}{2}\rho\theta'^2 \end{aligned} \tag{A 1.2.11}$$

satisfying the one-dimensional virial theorem[66,67]

$$t(x)=-\tfrac{1}{8}\rho''-\tfrac{1}{2}\int^x \rho V'\,\mathrm{d}t. \tag{A 1.2.12}$$

Combining eqns (A 1.2.11) and (A 1.2.12) immediately yields

$$u=(\theta')^2=\frac{1}{4}\left[\frac{\rho''}{\rho}-\left(\frac{\rho'}{\rho}\right)^2\right]-\frac{1}{\rho}\int^x \rho V'\,\mathrm{d}t, \tag{A 1.2.13}$$

which gives $\theta'^2$ in terms only of $\rho$ and $V$.

The equations relating $\rho=\chi^2$ and $V$ for the two-level independent-particle problem in one dimension have been derived from the Euler–Lagrange equations corresponding to a minimum of the appropriate energy functional for the two-level case:

$$E_2[\rho,\theta]=\int(t+V\rho)\,\mathrm{d}x,$$

subject to the density matrix idempotency constraints

$$\int\rho\cos 2\theta\,\mathrm{d}x=\int\rho\sin^2\theta\,\mathrm{d}x=2;\qquad \int\rho\sin\theta\cos\theta\,\mathrm{d}x=0. \tag{A 1.2.14}$$

In this way, one obtains the equations

$$\frac{1}{2}\theta'' + \left(\frac{\chi'}{\chi}\right)\theta' = \xi \sin 2\theta - \eta \cos 2\theta \tag{A 1.2.15}$$

and

$$-\tfrac{1}{2}\chi'' + [\tfrac{1}{2}(\theta')^2 + V]\chi = (\lambda + \xi \cos 2\theta + \eta \sin 2\theta)\chi, \tag{A 1.2.16}$$

where $\chi^2 = \rho$, and $\lambda, \xi$ and $\eta$ are undetermined Lagrange multipliers.

Clearly, for given exact orbitals $\phi_1$ and $\phi_2$ satisfying the one-body Schrödinger equations,

$$\phi_i'' + 2(\varepsilon_i - V(x))\phi_i = 0; \quad i = 1, 2, \tag{A 1.2.17}$$

the idempotency conditions (A 1.2.14) are automatically fulfilled by the orthonormality of orbitals. Therefore we can alternatively derive eqns (A 1.2.15) and (A 1.2.16) using the Schrödinger equations. This allows the following identification of the Lagrange multipliers:[65]

$$\eta = 0; \quad \xi = \tfrac{1}{2}(\varepsilon_1 - \varepsilon_2); \quad \lambda = \tfrac{1}{2}(\varepsilon_1 + \varepsilon_2). \tag{A 1.2.18}$$

As demonstrated by Dawson and March,[64] differentiation of eqn (A 1.2.16) and substitution of eqn (A 1.2.15) yields

$$\theta''\theta' + \frac{\chi'}{\chi}\theta'^2 + \frac{1}{2}\left[V' - \frac{1}{2}\left(\frac{\chi''}{\chi}\right)'\right] = 0, \tag{A 1.2.19}$$

which can be transformed into a solvable Bernoulli differential equation for $u = \theta'^2$, with general solution

$$\begin{aligned} u &= \frac{C_1}{\chi^2} - \frac{1}{\chi^2}\int^x \left(\frac{\chi'\chi''}{2\chi^2} - \frac{\chi'''}{2\chi} + V'\right)\chi^2 \, dt \\ &= \frac{C_1}{\chi^2} + \frac{1}{2\chi^2}\int^x \left(\frac{\chi''}{\chi}\right)'\chi^2 \, dt - \frac{1}{\chi^2}\int^x \chi^2 V' \, dt. \end{aligned} \tag{A 1.2.20}$$

A simple integration by parts gives eqn (A 1.2.13), since

$$\begin{aligned} u &= \frac{C_1}{\chi^2} + \frac{1}{2}\frac{\chi''}{\chi} - \frac{1}{2}\left(\frac{\chi'}{\chi}\right)^2 - \frac{1}{\chi^2}\int^x \chi^2 V' \, dt \\ &= \frac{C_1}{\rho} + \frac{1}{4}\left[\frac{\rho''}{\rho} - \left(\frac{\rho'}{\rho}\right)^2\right] - \frac{1}{\rho}\int^x \rho V' \, dt, \end{aligned} \tag{A 1.2.21}$$

and hence eqn (A 1.2.13) is regained provided $C_1 = 0$.

Equations (A 1.2.15), (A 1.2.16) and (A 1.2.18) allow, after a somewhat lengthy algebraic manipulation, $\theta$ to be eliminated completely, and in this way one arrives at an exact equation explicitly relating $\rho$ and $V$. This achieves then a basic aim of density functional theory in the two-level case.

The explicit form of this equation is

$$u^3+\left[4F+\frac{1}{4}\left(\frac{\rho'}{\rho}\right)^2\right]u^2+\left[4(F^2-\xi^2)-\frac{1}{2}\left(\frac{\rho'}{\rho}\right)F'\right]u+\frac{1}{4}(F')^2=0, \qquad \text{(A 1.2.22)}$$

where $u$ is given by eqn (A 1.2.13), $\xi$ and $\lambda$ by eqn (A 1.2.18), while $F$ is defined by

$$F=V-\frac{1}{2}\left(\frac{\chi''}{\chi}\right)-\lambda=V+\frac{1}{8}\left[\left(\frac{\rho'}{\rho}\right)^2-2\left(\frac{\rho''}{\rho}\right)\right]-\lambda. \qquad \text{(A 1.2.23)}$$

An alternative way of writing the two-level density–potential relation is to treat eqn (A 1.2.22) as a quadratic equation in $F'$, which can be solved to yield, with an ambiguity of sign to be resolved through physical boundary constraints,

$$F'=\left[\frac{\rho'}{\rho}u\pm 2u^{1/2}(4F^2-4Fu-u^2-4\xi^2)^{1/2}\right]. \qquad \text{(A 1.2.24)}$$

If we now insert $u$ from eqn (A 1.2.13) and $F$ from eqn (A 1.2.23) into the right-hand side of eqn (A 1.2.24), we can write an explicit expression for $F$ as $\int^x F'\,dx$ in terms of only $\rho, V$ and their derivatives. One can then view eqn (A 1.2.23) as a Schrödinger equation for $\chi=\rho^{1/2}$, to be solved self-consistently for a given $V$. (Related, but now approximate, differential equations for $\rho$ were considered earlier in refs 68 and 69).

## Appendix A 1.3

### Low-order gradient corrections in an inhomogeneous electron gas

The Thomas–Fermi expression (1.13) for the kinetic energy of an electron gas is strictly valid only in the limit of a very slowly varying electron density (see also Chapter 6). In atoms and molecules, the electron density varies quite rapidly in the vicinity of nuclei, and at the periphery of the charge cloud, and in these regions corrections to the Thomas–Fermi approximation are therefore called for.

The first attempt to correct eqn (1.13) was made by von Weizsäcker.[70] Essentially, his argument was correct for one level being occupied; in which case, one has that the density $\rho$ is the square of the wave function corresponding to the occupied level. Assuming the wave function $\psi$ to be real, one can therefore write

$$\psi=\rho^{1/2}, \qquad \text{(A 1.3.1)}$$

and hence the quantum mechanical kinetic energy

$$T=\frac{\hbar^2}{2m}\int(\nabla\psi)^2\,d\mathbf{r} \tag{A 1.3.2}$$

becomes, using eqn (A 1.3.1),

$$T=\frac{\hbar^2}{8m}\int\frac{(\nabla\rho)^2}{\rho}\,d\mathbf{r}. \tag{A 1.3.3}$$

Berg and Wilets[71] recognized that a better choice would be

$$T=\frac{\lambda\hbar^2}{8m}\int\frac{(\nabla\rho)^2}{\rho}\,d\mathbf{r},\quad \lambda<1, \tag{A 1.3.4}$$

and later work of Kirzhnitz[72] showed that in a slowly varying electron gas the correct choice is $\lambda=\frac{1}{9}$.

However, it should be noted that at the periphery of an atom, the von Weizsäcker functional comes into its own; without however the Thomas–Fermi term (1.13). Higher-order gradient corrections have been calculated by Hodges[73] and others.

## Appendix A 1.4

### Statistical theory relating average momentum $\langle p\rangle$ to Dirac–Slater exchange energy for molecules

Let us briefly summarize the argument relating the mean momentum $\langle p\rangle$ to the Dirac–Slater exchange energy for atoms, before turning to the proof for molecules.[74] The statistical relation (1.5), where $p_F(\mathbf{r})$ is the maximum momentum at $\mathbf{r}$, is spherical, and allows a quantity $r(p)$ to be extracted for a given $\rho(r)$. Then the relation analogous to eqn (1.5) in $p$-space is ($\hbar=1$)

$$\rho(p)=\frac{1}{3\pi^2}r^3(p). \tag{A 1.4.1}$$

The generalization of eqn (A 1.4.1) for molecules, where $\rho(\mathbf{r})$ is no longer spherical, is readily obtained by noting from eqn (1.5) that each surface of constant electron density defines a value of $p$. In the statistical theory, therefore, there is a unique relation between $p$ and the volume enclosed by the corresponding constant-electron-density surface, say $\Omega(p)$. The spherically averaged momentum density $\rho(p)$ in the molecule is then given by the generalization of eqn (A 1.4.1) to read

$$\rho(p)=\frac{1}{4\pi^3}\Omega(p), \tag{A 1.4.2}$$

reducing to eqn (A 1.4.1), as it must, in the spherically symmetric case when $\Omega(p)=\frac{4}{3}\pi r^3(p)$.

Obviously the mean momentum is given by

$$\langle p\rangle=\frac{1}{N}\int_0^\infty \rho(p)4\pi p^3\,\mathrm{d}p, \tag{A 1.4.3}$$

where $N$ is the total number of electrons in the molecule.

The exchange energy in the Dirac–Slater approximation is, from eqn (1.50),

$$A=-c_e\int\rho^{4/3}\,\mathrm{d}\mathbf{r}. \tag{A 1.4.4}$$

But we now use eqn (1.5) to write

$$A=-c_e\int\frac{1}{(3\pi^2)^{4/3}}p^4(\mathbf{r})\,\mathrm{d}\mathbf{r}, \tag{A 1.4.5}$$

and since we have seen above that surfaces of constant $\rho$ are characterized by their enclosed volume in $\mathbf{r}$-space, $\Omega(p)$, we can rewrite eqn (A 1.4.5) as

$$A=\frac{-c_e}{(3\pi^2)^{4/3}}\int p^4(\Omega)\,\mathrm{d}\Omega. \tag{A 1.4.6}$$

Integrating by parts, and noting that there are no contributions from the limits, one finds

$$A=\frac{c_e}{(3\pi^2)^{4/3}}\int_\infty^0 4\Omega p^3(\Omega)\,\mathrm{d}p. \tag{A 1.4.7}$$

One now utilizes eqn (A 1.4.2) to find

$$A=-\frac{4c_e}{(3\pi^2)^{4/3}}\int_0^\infty 4\pi^3\rho(p)p^3\,\mathrm{d}p$$

$$=-\frac{c_e 4\pi^2}{(3\pi^2)^{4/3}}N\langle p\rangle, \tag{A 1.4.8}$$

which in atomic units yields the same result as for atoms. This relation (A 1.4.8) has been employed in ref. 74 to calculate $A$ for $H_2O$ and $N_2$, using Compton profile data for $\langle p\rangle$.

The relation of other moments of momentum to the electron density in molecules has been studied by Allan *et al.*[75] These workers[41] also consider density gradient corrections to such formulae (cf. Appendix A 1.3).

## Further reading

Bamzai, A.S. and Deb, B.M., (1981). *Rev. Mod. Phys.*, **53**, 95 and 593.

Dahl, J.P. and Avery, J. (eds) (1984). *Local Density Approximations in Quantum Chemistry and Solid State Physics*, Plenum, New York.

Dreizler, R.M. and da Providencia, J. (eds) (1985). *Density Functional Methods in Physics*, Nato ASI Series B, Vol. 123, Plenum, New York.

Erdahl, R. and Smith, V.H. (eds) (1987). *Density Matrices and Density Functionals*, Reidel, Dordrecht.

Gombás, P. (1949). *Die Statistische Theorie des Atoms und ihre Anwendungen*, Springer-Verlag, Berlin.

Keller, J. and Gázquez, J.L. (eds) (1987). *Density Functional Theory*, Lecture Notes in Physics, Vol. 187, Springer-Verlag, Berlin.

Lieb, E.H., (1981). *Rev. Mod. Phys.*, **53**, 603; (1982). *Ibid.*, **54**, 311.

Lundqvist, S. and March, N.H. (eds) (1983). *Theory of the Inhomogeneous Electron Gas*, Plenum, New York.

March, N.H. (1957). *Adv. Phys.*, **6**, 1.

Parr, R.G. (1983). *Ann. Rev. Phys. Chem.*, **34**, 631.

## References

1. Thomas, L.H., (1926). *Proc. Camb. Phil. Soc.*, **23**, 542.
2. Fermi, E. (1928). *Z. Phys.*, **48**, 73.
3. March, N.H. and Murray, A.M. (1961), *Proc. Roy. Soc.*, **A261**, 119.
4. March, N.H. (1975). *Self-Consistent Fields in Atoms*, Pergamon, Oxford.
5. Hohenberg, P.C. and Kohn, W. (1964). *Phys. Rev.*, **136**, B864.
6. Mucci, J.F. and March, N.H. (1979). *J. Chem. Phys.*, **71**, 5270.
7. Politzer, P. (1976). *J. Chem. Phys.*, **64**, 4239.
8. March. N.H. (1981). *J. Chem. Phys.*, **74**, 2376.
9. Layzer, D. (1959). *Ann. Phys.*, **8**, 271.
10. Hylleraas, E. (1930). *Z. Phys.*, **65**, 209.
11. Kato, T. (1957). *Commun. Pure Appl. Math.*, **10**, 151.
12. March, N.H. and White, R.J. (1972). *J. Phys.*, **B5**, 466; Tal, Y. and Levy, M. (1981). *Phys. Rev.*, **A23**, 408; Senatore, G. and March, N.H. (1985). *J. Chem. Phys.*, **83**, 1232.
13. Plindov, G.I. and Dmitrieva, J.K. (1977). *J. Phys. (Paris)*, **38**, 1061; see also (1975). *Phys. Lett.*, **55A**, 3; (1977). *Opt. Spectrosc.* **42**, 3.
14. Dirac, P.A.M. (1930). *Proc. Camb. Phil. Soc.*, **26**, 376.
15. Gell-Mann, M. and Brueckner, K.A. (1957), *Phys. Rev.*, **106**, 364.
16. Wigner, E.P. (1934), *Phys. Rev.*, **46**, 1002; (1938), *Trans. Faraday Soc.*, **34**, 678.
17. Gordon, R.G. and Kim, Y.S. (1972). *J. Chem. Phys.*, **56**, 3122.
18. Vosko, S.H., Wilk, L. and Nusair, M. (1980). *Can. J. Phys.*, **58**, 1200; see also Painter, G.S. (1981). *Phys. Rev.*, **B24**, 4264.
19. Perdew, J.P. and Zunger, A. (1981). *Phys. Rev.*, **B23**, 5048.
20. Herman, F. and March, N.H. (1984). *Solid St. Commun.*, **50**, 725.
21. Ceperley, D.M. and Alder, B.J. (1980). *Phys. Rev. Lett.*, **45**, 566.
22. Slater, J.C. (1951). *Phys. Rev.*, **81**, 385.
23. Kohn, W. and Sham, L.J. (1965). *Phys. Rev.*, **140**, A1133.

24. Walsh, A.D. (1953), *J. Chem. Soc.*, 2260.
25. See, for example, Coulson, C.A. and Beb, B.M. (1971). *Int. J. Quant. Chem.*, **5**, 411.
26. Stenkamp, L.Z. and Davidson, E.R. (1973). *Theoret. Chim. Acta*, **30**, 283.
27. March. N.H. (1981). *J. Chem. Phys.*, **74**, 2973.
28. March, N.H. and Plaskett, J.S. (1956). *Proc. Roy. Soc.*, **A235**, 419.
29. Ruedenberg, K. (1977). *J. Chem. Phys.*, **66**, 375.
30. March, N.H. (1977). *J. Chem. Phys.*, **67**, 4618.
31. Mehrotra, P.K. and Hoffmann, R. (1978). *Theoret. Chim. Acta*, **48**, 301.
32. March, N.H. (1981). *Specialist Periodical Reports: Theoretical Chemistry*, Vol. 4, p. 92, Royal Society of Chemistry, London.
33. Nelander, B. (1969). *J. Chem. Phys.*, **51**, 469.
34. Mucci, J.F. and March, N.H. (1979). *J. Chem. Phys.*, **71**, 1495.
35. Parr, R.G., Donnelly, R.A., Levy, M. and Palke, W.E. (1978). *J. Chem. Phys.*, **68**, 3801.
36. Sanderson, R.T. (1955). *Science*, **121**, 207; (1978). *Chemical Bonds and Bond Energy*, Academic Press, New York.
37. Alonso, J.A. and March, N.H. (1983). *J. Chem. Phys.*, **78**, 1382.
38. See March, N.H. and Bader, R.F.W. (1980). *Phys. Lett.*, **78A**, 242.
39. Mucci, J.F. and March. N.H. (1983). *J. Chem. Phys.*, **78**, 6187.
40. Teller, E. (1962). *Rev. Mod. Phys.*, **34**, 627.
41. Allan, N.L., West, C.G., Cooper, D.L., Grout, P.J. and March, N.H. (1985). *J. Chem. Phys.*, **83**, 4562.
42. Lee, C. and Ghosh, S.K. (1986). *Phys. Rev.*, **A33**, 3506.
43. March, N.H. (1952). *Proc. Camb. Phil. Soc.*, **48**, 665.
44. Pucci, R. and March, N.H. (1986). *Phys. Rev.*, **A33**, 3511.
45. Mucci, J.F. and March. N.H. (1985). *J. Chem. Phys.*, **82**, 5099.
46. Coulson, C.A. and March, N.H. (1950). *Proc. Phys. Soc.*, **A63**, 367.
47. Callaway, J. and March, N.H. (1984). In *Solid State Physics*, Vol. 38, eds Ehrenreich, H. and Turnbull, D., p. 135, Academic Press, New York.
48. Schweber, S.S. (1961). *An Introduction to Relativistic Quantum Theory*, Harper & Row, New York.
49. Vallarta, M.S. and Rosen, N. (1932). *Phys. Rev.*, **41**, 708.
50. Senatore, G. and March, N.H. (1985). *Phys. Rev.*, **A32**, 3277.
51. Layzer, D. and Bahcall, J. (1962). *Ann. Phys.*, **17**, 177.
52. Marconi, U.B.M. and March, N.H. (1981). *Int. J. Quant. Chem.*, **20**, 693.
53. Hill, S.H., Grout, P.J. and March, N.H. (1984). *J. Phys.*, **B17**, 4819; (1987). *J. Phys.*, **B20**, 11.
54. March, N.H. (1986). *Phys. Rev.*, **A34**, 5104.
55. Garstang, R.H. (1982). *J. Phys.* (*Paris*), Coll., **C2**, Suppl. 11, **43**, 19.
56. Trumper, J., Pietsch, W., Reppin, C., Voges, W., Staubert, R. and Kendziorra, E. (1978). *Astrophys. J. Lett.*, **219**, L105.
57. Maurer, G.S., Johnson, W.N., Kurfess, J.D. and Strickman, S.S. (1982) *Astrophys. J.*, **254**, 271.
58. Hill, S.H., Grout, P.J. and March, N.H. (1985). *J. Phys.*, **B18**, 4665.
59. Banerjee, B., Constantinescu, D.H. and Rehák, P. (1974). *Phys. Rev.*, **D10**, 2384.
60. See, for example, Hill, S.H., Grout, P.J. and March, N.H. (1983). *J. Phys.* **B16**, 2301.
61. Jones, P.B. (1986). *Mon. Not. Roy. Astron. Soc.*, **218**, 477.
62. Baker, E.B. (1930). *Phys. Rev.*, **36**, 630.
63. Lawes, G.P. and March, N.H. (1982). *J. Chem. Phys.*, **76**, 458.

64. Dawson, K.A. and March, N.H. (1984). *J. Chem. Phys.*, **81**, 5850.
65. March, N.H. and Nalewajski, R.F. (1987). *Phys. Rev.*, **A35**, 525.
66. March, N.H. and Young, W.H. (1959). *Nucl. Phys.*, **12**, 237.
67. March, N.H. (1979). *J. Chem. Phys.*, **70**, 587.
68. Lawes, G.P. and March, N.H. (1980). *Physica Scripta*, **21**, 402.
69. Deb, B.M. and Ghosh, S.K. (1983). *Int. J. Quant. Chem.*, **23**, 1.
70. von Weizsäcker, C.F. (1935). *Z. Phys.*, **96**, 431.
71. Berg, R. and Wilets, L. (1955). *Proc. Phys. Soc.*, **A68**, 229.
72. Kirzhnitz, D.A. (1957). *Sov. Phys. JETP*, **5**, 64.
73. Hodges, C.H. (1973). *Can. J. Phys.*, **51**, 1428.
74. Allan, N.L. and March, N.H. (1983). *Int. J. Quant. Chem.*, Symp. **17**, 227.
75. Allan, N.L., Cooper, D.L., West, C.G., Grout, P.J. and March, N.H. (1985). *J. Chem. Phys.*, **83**, 239.
76. Kohn, W. (1966). In *Proc. Midwest Conf. on Theoretical Physics*, Indiana University, Bloomington, Indiana.
77. Chen, Z. and Spruch, L. (1987). *Phys. Rev.*, **A35**, 4035.
78. March, N.H. and Parr, R.G. (1980). *Proc. Natl. Acad. Sci. U.S.A.*, **77**, 6285.

# 2 The Coordinate Scaling Requirements in Density Functional Theory

M. LEVY

*Department of Chemistry and Quantum Theory Group, Tulane University, New Orleans, Louisiana 70118, U.S.A.*



## 2.1 Introduction

Consider $N$ interacting electrons in a local-multiplicative spin-independent external potential $v$. The Hamiltonian is

$$\hat{H}(\mathbf{r}_1,\ldots,\mathbf{r}_N)=\hat{T}(\mathbf{r}_1,\ldots,\mathbf{r}_N)+\hat{V}_{ee}(\mathbf{r}_1,\ldots,\mathbf{r}_N)+\sum_{i=1}^{N} v(\mathbf{r}_i), \tag{2.1}$$

where $\hat{T}$ and $\hat{V}_{ee}$ are, respectively, the kinetic and electron–electron repulsion operators. In atoms, molecules or solids, $v(\mathbf{r})$ is the electron–nuclear attraction operator.

Assume that one is interested in the variational determination of the ground-state energy and density of $\hat{H}$. For this purpose, there exists a universal density functional,[1] the Hohenberg–Kohn functional $F[\rho]$ (cf. Chapter 1, Section 1.3), for $\langle \hat{T}+\hat{V}_{ee}\rangle$ of each trial density $\rho(\mathbf{r})$. Since only the three-dimensional electron density need be employed, the use of $F[\rho]$ represents, in principle, a drastic simplification of the many-electron problem. In actual calculations, however, $F[\rho]$ has to be approximated.

Single-Particle Density in Physics and Chemistry ISBN 0-12-470518-9

Accordingly, the more properties of the parts of $F[\rho]$ that are known to us, the better are we able to approximate $F[\rho]$. It is consequently the purpose of this Chapter to review the necessary coordinate-scaling requirements of the parts of $F[\rho]$.

The scaling requirements to be reviewed here were derived in the paper by Levy and Perdew[2] and essentially all but Sections 2.2, 2.6 and 2.11 follow closely the presentation of these workers, although the proofs, at points, have been modified somewhat to make them more "streamlined." The discussion of the lowering of the trial energy by scaling and the discussion of the new functional of $\rho$ and $\lambda$, $F[\rho, \lambda]$, is the subject of Section 2.4 and follows the work of Levy, Yang and Parr.[3] Finally, the correlation-energy scaling equality in Section 2.11 constitutes unpublished work of Levy and Perdew. The constrained-search formulations of $F[\rho]$ will be used throughout the Chapter.

## 2.2 Formal display of the Hohenberg–Kohn functional $F[\rho]$

Let $\Psi_\rho(\mathbf{r}_1, \ldots, \mathbf{r}_N)$ be any antisymmetric function that yields $\rho$. Then by the constrained-search formulation, $F[\rho]$ is identified in a formal sense by[4]

$$F[\rho] = \langle \Psi_\rho^{\min} | \hat{T} + \hat{V}_{ee} | \Psi_\rho^{\min} \rangle, \tag{2.2}$$

where $\Psi_\rho^{\min}(\mathbf{r}_1, \ldots, \mathbf{r}_N)$ is that antisymmetric function which is constrained to yield $\rho(\mathbf{r})$ and simultaneously minimizes $\langle T + V_{ee} \rangle$. Harriman[5] has shown that $N$-representability is no problem and Lieb[6] has shown that $\Psi_\rho^{\min}$ always exists. In addition, Valone[7] has provided an ensemble-search generalization to eqn (2).

$F[\rho]$ is variationally valid in that

$$\begin{aligned}
&\operatorname*{Inf}_{\rho} \left\{ \int v(\mathbf{r})\rho(\mathbf{r})\, \mathrm{d}^3 r + F[\rho] \right\} \\
&\quad = \operatorname*{Inf}_{\rho} \langle \Psi_\rho^{\min} | \sum_i v(\mathbf{r}_i) + \hat{T} + \hat{V}_{ee} | \Psi_\rho^{\min} \rangle \\
&\quad = \operatorname*{Inf}_{\Psi} \langle \Psi | \sum_i v(\mathbf{r}_i) + \hat{T} + \hat{V}_{ee} | \Psi \rangle \\
&\quad = E_{\mathrm{G.S.}}
\end{aligned} \tag{2.3}$$

The definition of $F[\rho]$ in eqn (2.2) and its ensemble-search generalization follow in the spirit of Percus' definition[8] of a universal kinetic-energy functional for independent fermion systems.

## 2.3 Scaled wave functions and densities

I begin by defining what is meant by a scaled density and a scaled wave function. Consider an arbitrary antisymmetric wave function $\Psi_\rho(\mathbf{r}_1,\ldots,\mathbf{r}_N)$ and corresponding density $\rho(\mathbf{r})$. The scaled density $\rho_\lambda(\mathbf{r})$ is defined by

$$\rho_\lambda(\mathbf{r})=\lambda^3\rho(\lambda\mathbf{r}), \tag{2.4}$$

where $\lambda$ is our coordinate scale factor. The factor $\lambda^3$ ensures that $\rho_\lambda(\mathbf{r})$ is properly normalized to $N$ electrons:

$$\int\rho_\lambda(\mathbf{r})\,\mathrm{d}^3r=\lambda^3\int\rho(\lambda\mathbf{r})\,\mathrm{d}^3r=\int\rho(\lambda\mathbf{r})\,\mathrm{d}^3(\lambda r)=\int\rho(\mathbf{r})\,\mathrm{d}^3r=N. \tag{2.5}$$

The normalized scaled wave function corresponding to $\Psi_\rho(\mathbf{r}_1,\ldots,\mathbf{r}_N)$ is $\lambda^{3N/2}\Psi_\rho(\lambda\mathbf{r}_1,\ldots,\lambda\mathbf{r}_N)$. Note that the scaled wave function gives the scaled density:

$$\begin{aligned}
&\int\cdots\int\lambda^{3N}\Psi_\rho(\lambda\mathbf{r}_1,\ldots,\lambda\mathbf{r}_N)^*\Psi_\rho(\lambda\mathbf{r}_1,\ldots,\lambda\mathbf{r}_N)\,\mathrm{d}s_1\cdots\mathrm{d}s_N\,\mathrm{d}^3r_2\cdots\mathrm{d}^3r_N\\
&\quad=\lambda^3\int\cdots\int\Psi_\rho(\lambda\mathbf{r}_1,\ldots,\lambda\mathbf{r}_N)^*\Psi_\rho(\lambda\mathbf{r}_N,\ldots,\lambda\mathbf{r}_N)\,\mathrm{d}s_1\cdots\mathrm{d}s_N\,\mathrm{d}^3(\lambda r_2)\cdots\mathrm{d}^3(\lambda r_N)\\
&\quad=\rho_\lambda(\mathbf{r}_1).
\end{aligned} \tag{2.6}$$

For the remainder of this Chapter, unless otherwise specified, it will be assumed that $\lambda\neq1$ when $\lambda$ is shown.

## 2.4 Scaling requisites of $T[\rho]$ and $V_{ee}[\rho]$.

We now define $\Psi_{\rho\lambda}^{\min}(\mathbf{r}_1,\ldots,\mathbf{r}_N)$ as that antisymmetric wave function which yields the scaled density $\rho_\lambda(\mathbf{r})$ and simultaneously minimizes $\langle\hat{T}(\mathbf{r}_1,\ldots,\mathbf{r}_N)+\hat{V}_{ee}(\mathbf{r}_1,\ldots,\mathbf{r}_N)\rangle$. Then, since for any arbitrary $\Psi_\rho(\mathbf{r}_1,\ldots,\mathbf{r}_N)$,

$$\begin{aligned}
&\lambda^{-2}\int\cdots\int\lambda^{3N/2}\Psi_\rho(\lambda\mathbf{r}_1,\ldots,\lambda\mathbf{r}_N)^*[T(\mathbf{r}_1,\ldots\mathbf{r}_N)+\lambda V_{ee}(\mathbf{r}_1,\ldots,\mathbf{r}_N)]\\
&\qquad\times\lambda^{3N/2}\Psi_\rho(\lambda\mathbf{r}_1,\ldots,\lambda\mathbf{r}_N)\,\mathrm{d}x_1\cdots\mathrm{d}x_N\\
&\quad=\int\cdots\int\Psi_\rho(\lambda\mathbf{r}_1,\ldots,\lambda\mathbf{r}_N)^*[\hat{T}(\lambda\mathbf{r}_1,\ldots,\lambda\mathbf{r}_N)+\hat{V}_{ee}(\lambda\mathbf{r},\ldots,\lambda\mathbf{r}_N)]\\
&\qquad\times\Psi_\rho(\lambda\mathbf{r}_1,\ldots,\lambda\mathbf{r}_N)\,\mathrm{d}^3(\lambda\mathbf{r}_1)\cdots\mathrm{d}^3(\lambda\mathbf{r}_N)\,\mathrm{d}s_1\cdots\mathrm{d}s_N\\
&\quad=\int\cdots\int\Psi_\rho(\mathbf{r}_1,\ldots,\mathbf{r}_N)^*[\hat{T}(\mathbf{r}_1,\ldots,\mathbf{r}_N)+\hat{V}_{ee}(\mathbf{r}_1,\ldots,\mathbf{r}_N)]\\
&\qquad\Psi_\rho(\mathbf{r}_1,\ldots,\mathbf{r}_N)\,\mathrm{d}x_1\cdots\mathrm{d}x_N,
\end{aligned} \tag{2.7}$$

the definition of $\Psi_\rho^{\min}(\mathbf{r}_1,\ldots,\mathbf{r}_N)$ implies that $\lambda^{3N/2}\Psi_\rho^{\min}(\lambda\mathbf{r}_1,\ldots,\lambda\mathbf{r}_N)$ is that antisymmetric function which yields $\rho_\lambda(\mathbf{r})$ and simultaneously minimizes $\lambda^{-2}\langle\hat{T}(\mathbf{r}_1,\ldots,\mathbf{r}_N)+\lambda\hat{V}_{ee}(\mathbf{r}_1,\ldots,\mathbf{r}_N)\rangle$, or simultaneously minimizes $\langle\hat{T}(\mathbf{r}_1,\ldots,\mathbf{r}_N)+\lambda\hat{V}_{ee}(\mathbf{r}_1,\ldots,\mathbf{r}_N)\rangle$.[9] (It is important to observe the factor $\lambda$ multiplying $\hat{V}_{ee}$.) Clearly then, $\lambda^{3N/2}\Psi_\rho^{\min}(\lambda\mathbf{r}_1,\ldots,\lambda\mathbf{r}_N)$ does *not* minimize $\langle\hat{T}(\mathbf{r}_1,\ldots,\mathbf{r}_N)+\hat{V}_{ee}(\mathbf{r}_1,\ldots,\mathbf{r}_N)\rangle$. Hence, although $\lambda^{3N/2}$ $\Psi_\rho^{\min}(\lambda\mathbf{r}_1,\ldots,\lambda\mathbf{r}_N)$ and $\Psi_{\rho\lambda}^{\min}(\mathbf{r}_1,\ldots,\mathbf{r}_N)$ both belong to the same density $\rho_\lambda(\mathbf{r})$, it is obvious that

$$\lambda^{3N/2}\Psi_\rho^{\min}(\lambda\mathbf{r}_1,\ldots,\lambda\mathbf{r}_N)\neq\Psi_{\rho\lambda}^{\min}(\mathbf{r}_1,\ldots,\mathbf{r}_N). \tag{2.8}$$

Consequently, from the minimization statements in the constrained-search definitions

$$\langle\lambda^{3N/2}\Psi_\rho^{\min}(\lambda\mathbf{r}_1,\ldots,\lambda\mathbf{r}_N)|\,\hat{T}(\mathbf{r}_1,\ldots,r_N)+\hat{V}_{ee}(\mathbf{r}_1,\ldots,\mathbf{r}_N)|\lambda^{3N/2}\Psi_\rho^{\min}(\lambda\mathbf{r}_1,\ldots,\lambda\mathbf{r}_N)\rangle$$
$$>\langle\Psi_{\rho\lambda}^{\min}(\mathbf{r}_1,\ldots,\mathbf{r}_N)|\,\hat{T}(\mathbf{r}_1,\ldots,\mathbf{r}_N)+\hat{V}_{ee}(\mathbf{r}_1,\ldots,\mathbf{r}_N)|\,\Psi_{\rho\lambda}^{\min}(\mathbf{r}_1,\ldots,\mathbf{r}_N)\rangle \tag{2.9}$$

and

$$\langle\lambda^{3N/2}\Psi_\rho^{\min}(\lambda\mathbf{r}_1,\ldots,\lambda r_N)|\,\hat{T}(\mathbf{r}_1,\ldots,\mathbf{r}_N)+\lambda\hat{V}_{ee}(\mathbf{r}_1,\ldots,\mathbf{r}_N)|\lambda^{3N/2}\,\Psi_\rho^{\min}(\lambda\mathbf{r}_1,\ldots,\lambda\mathbf{r}_N)\rangle$$
$$<\langle\Psi_{\rho\lambda}^{\min}(\mathbf{r}_1,\ldots,\mathbf{r}_N)|\,\hat{T}(\mathbf{r}_1,\ldots,\mathbf{r}_N)+\lambda\hat{V}_{ee}(\mathbf{r}_1,\ldots,\mathbf{r}_N)|\,\Psi_{\rho\lambda}^{\min}(\mathbf{r}_1,\ldots,\mathbf{r}_N)\rangle. \tag{2.10}$$

Equations (2.9) and (2.10) give, respectively,

$$\lambda^2 T[\rho]+\lambda V_{ee}[\rho]>T[\rho_\lambda]+V_{ee}[\rho_\lambda] \tag{2.11}$$

and

$$\lambda^2 T[\rho]+\lambda^2 V_{ee}[\rho]<T[\rho_\lambda]+\lambda V_{ee}[\rho_\lambda]. \tag{2.12}$$

Notice that eqns (2.11) and (2.12) actually connect the kinetic and repulsion functionals. The following inequalities involve $T$ and $V_{ee}$ separately and result directly from combinations of eqns (2.11) and (2.12):

$$V_{ee}[\rho_\lambda]<\lambda V_{ee}[\rho] \quad (\lambda<1), \tag{2.13}$$

$$V_{ee}[\rho_\lambda]>\lambda V_{ee}[\rho] \quad (\lambda>1), \tag{2.14}$$

$$T[\rho_\lambda]>\lambda^2 T[\rho] \quad (\lambda<1), \tag{2.15}$$

$$T[\rho_\lambda]<\lambda^2 T[\rho] \quad (\lambda>1). \tag{2.16}$$

In contrast with the *inequalities* in eqns (2.11)–(2.16), there are equalities in wave functional theory, so that the above density functional results are, perhaps, somewhat counterintuitive.

## 2.5 Virial theorem

From the scaling *inequalities* in eqns (2.11) and (2.12), it might seem that the virial theorem does not hold, in apparent contradiction to the fact that the virial theorem has to hold. But there is in fact no contradiction because the inequality in eqn (2.11) is always in one direction.

Let us see how the virial theorem comes out of eqn (2.14). Consider for example an atom. By the variational principle and using eqn (2.14) we have

$$E_{\mathrm{G.S.}} \leqslant -\int Zr^{-1}\rho_\lambda(\mathbf{r})\,\mathrm{d}^3r + T[\rho_\lambda] + V_{\mathrm{ee}}[\rho_\lambda]$$
$$\leqslant -\lambda\int Zr^{-1}\rho(\mathbf{r})\,\mathrm{d}^3r + \lambda^2 T[\rho] + \lambda V_{\mathrm{ee}}[\rho], \tag{2.17}$$

so that

$$E_{\mathrm{G.S.}} \leqslant -\lambda\int Zr^{-1}\rho(\mathbf{r})\,\mathrm{d}^3r + \lambda^2 T[\rho] + \lambda V_{\mathrm{ee}}[\rho]. \tag{2.18}$$

When $\lambda = 1$ and $\rho = \rho_{\mathrm{G.S.}}$, the right-hand side of eqn (2.18) equals $E_{\mathrm{G.S.}}$. Therefore

$$\frac{\partial}{\partial\lambda}\left\{-\lambda\int Zr^{-1}\rho_{\mathrm{G.S.}}(\mathbf{r})\,\mathrm{d}^3r + \lambda^2 T[\rho_{\mathrm{G.S.}}] + \lambda V_{\mathrm{ee}}[\rho_{\mathrm{G.S.}}]\right\}_{\lambda=1} = 0, \tag{2.19}$$

which gives, correctly, the following familiar virial expression:

$$V_{\mathrm{ee}}[\rho_{\mathrm{G.S.}}] - \int Zr^{-1}\rho_{\mathrm{G.S.}}(\mathbf{r})\,\mathrm{d}^3r = -2T[\rho_{\mathrm{G.S.}}]. \tag{2.20}$$

## 2.6 Lowering the trial energy by optimization of $\lambda$ as a multiplicative parameter: a new universal functional of $\lambda$ and $\rho$; $F[\rho, \lambda]$

In an atom, for any given trial density $\rho$, an upper bound to $E_{\mathrm{G.S.}}$ is $E[\rho]$, where

$$E[\rho] = -\int Zr^{-1}\rho(\mathbf{r})\,\mathrm{d}^3r + T[\rho] + V_{\mathrm{ee}}[\rho]. \tag{2.21}$$

But, the bound in eqn (2.18) allows us to improve upon $E[\rho]$ through a minimization of the right-hand side of eqn (2.18) by simply optimizing $\lambda$ as a multiplicative parameter, without affecting $\rho$. In other words,

$$E_{\mathrm{G.S.}} \leqslant \min_{\lambda}\left\{-\lambda\int Zr^{-1}\rho(\mathbf{r})\,\mathrm{d}^3r + \lambda^2 T[\rho] + \lambda V_{\mathrm{ee}}[\rho]\right\} \leqslant E[\rho], \tag{2.22}$$

which gives

$$E_{\mathrm{G.S.}} \leqslant -\frac{\left\{\int -Zr^{-1}\rho(\mathbf{r})\,\mathrm{d}^3r + V_{\mathrm{ee}}[\rho]\right\}^2}{4T[\rho]} \leqslant E[\rho]. \tag{2.23}$$

Comparable results hold for molecules. For all systems the minimizing $\lambda$ is given by

$$2\lambda_{\mathrm{min}} T[\rho] + V_{\mathrm{ee}}[\rho] - \lambda_{\mathrm{min}}^{-1}\int \mathrm{d}^3r\,\rho(\mathbf{r})\mathbf{r}\cdot\nabla v(\mathbf{r}/\lambda_{\mathrm{min}}) = 0. \tag{2.24}$$

In other words, the trial energy may be lowered by varying $\lambda$ just as if $T[\rho]$ and $V_{\mathrm{ee}}[\rho]$ possessed homogeneous scaling. The approach of Ghosh and Parr[11] is thereby extended to functionals that include correlation.

The essence of this section is the existence of a universal functional of $\rho$ *and* $\lambda$, $F[\rho, \lambda]$, given by

$$F[\rho, \lambda] = \lambda^2 T[\rho] + \lambda V_{ee}[\rho] \tag{2.25}$$

such that

$$E_{G.S.} = \operatorname*{Inf}_{\lambda, \rho} \{ \int v(\mathbf{r}) \rho_\lambda(\mathbf{r})\, d^3r + F[\rho, \lambda] \}. \tag{2.26}$$

For actual calculations, it is perhaps best to use at present the following:

$$T[\rho] = T_s[\rho] + T_c[\rho], \tag{2.27a}$$

$$T[\rho] = T_s[\rho] - E_c[\rho] - \int d^3r\, \rho(\mathbf{r}) \mathbf{r} \cdot \nabla v_c[\rho(\mathbf{r}), \mathbf{r}] \tag{2.27b}$$

and

$$V_{ee}[\rho] = U[\rho] + E_x[\rho] + E_c[\rho] - T_c[\rho], \tag{2.28a}$$

$$V_{ee}[\rho] = U[\rho] + E_x[\rho] + 2E_c[\rho] + \int d^3r\, \rho(\mathbf{r}) \mathbf{r} \cdot \nabla v_c[\rho(\mathbf{r}), \mathbf{r}] \tag{2.28b}$$

for $T[\rho]$ and $V_{ee}[\rho]$ in eqn (2.25).

There are other relations besides eqn (2.26). Since $T_c \geqslant 0$, we have

$$E_{G.S.} = \operatorname*{Inf}_{\lambda \leqslant 1, \rho} \{ \lambda^2 T_s[\rho] + \lambda(U[\rho] + E_x[\rho] + E_c[\rho] + \int v(\mathbf{r}) \rho_\lambda(\mathbf{r})\, d^3r) \}. \tag{2.29}$$

Also, from exchange-only (XO) theory, we have

$$E_{G.S.} \leqslant E_{XO} = \operatorname*{Inf}_{\lambda, \rho} \{ \lambda^2 T_s[\rho] + \lambda U[\rho] + \lambda E_x[\rho] + \int v(\mathbf{r}) \rho_\lambda(\mathbf{r}) d^3r \}. \tag{2.30}$$

## 2.7 Definitions of exchange and correlation

In Kohn–Sham theory,[12,13]

$$E_{G.S.} = \operatorname*{Inf}_{\rho} \{ \int v(\mathbf{r}) \rho(\mathbf{r})\, d^3r + T_s[\rho] + U[\rho] + E_x[\rho] + E_c[\rho] \}, \tag{2.31}$$

where

$$U[\rho] = \tfrac{1}{2} \iint \rho(\mathbf{r}) \rho(\mathbf{r}_2) r_{12}^{-1}\, d^3r_1\, d^3r_2.$$

There are several constrained-search versions[14,15] of the Kohn–Sham procedure, but virtually all will give the same scaling properties for $T_s[\rho]$, $E_x[\rho]$ and $E_c[\rho]$. For the non-interacting kinetic-energy functional $T_s[\rho]$, we shall employ here[4b,14,15]

$$T_s[\rho]=\langle\Phi_\rho^{\min}|\hat{T}|\Phi_\rho^{\min}\rangle, \tag{2.32}$$

where $\Phi_\rho^{\min}$ is that antisymmetric wave function which yields $\rho$ and minimizes $\langle\hat{T}\rangle$. This means that the exchange-correlation functional is[4b,14,15]

$$E_{xc}[\rho]=\langle\Psi_\rho^{\min}|\hat{T}+\hat{V}_{ee}|\Psi_\rho^{\min}\rangle-U[\rho]-\langle\Phi_\rho^{\min}|\hat{T}|\Phi_\rho^{\min}\rangle. \tag{2.32$'$}$$

Accordingly, the universal exchange-energy functional is defined by[16,17]

$$E_x[\rho]=\langle\Phi_\rho^{\min}|\hat{V}_{ee}|\Phi_\rho^{\min}\rangle-U[\rho], \tag{2.33}$$

so that the universal correlation-energy functional is defined by

$$E_c[\rho]=V_{ee}[\rho]-\langle\Phi_\rho^{\min}|\hat{V}_{ee}|\Phi_\rho^{\min}\rangle+T_c[\rho], \tag{2.34}$$

where $T_c[\rho]$ is the kinetic-energy contribution to $E_c[\rho]$:

$$T[\rho]=T_s[\rho]+T_c[\rho] \tag{2.35}$$

or

$$T_c[\rho]=\langle\Psi_\rho^{\min}|\hat{T}|\Psi_\rho^{\min}\rangle-\langle\Phi_\rho^{\min}|\hat{T}|\Phi_\rho^{\min}\rangle. \tag{2.36}$$

Further, the definition of $\Phi_\rho^{\min}$ as the constrained minimizing function for $\langle T\rangle$ dictates that

$$T_c[\rho]\geqslant 0. \tag{2.37}$$

Remember that $\Psi_\rho^{\min}$ is that constrained minimizing function for $\langle\hat{T}+\hat{V}_{ee}\rangle$ and not for $\langle\hat{T}\rangle$ alone. Equation (2.37) was proved by Theophilou[18] for non-interacting $v$-representable $\rho$.

## 2.8 The homogeneous scaling properties of $T_s[\rho]$ and $E_x[\rho]$

How do $T_s$ and $E_x$ scale? Define $\Phi_{\rho\lambda}^{\min}(\mathbf{r}_1,\ldots,\mathbf{r}_N)$ as that antisymmetric wave function which yields the scaled density $\rho_\lambda(\mathbf{r})=\lambda^3\rho(\lambda\mathbf{r})$ and simultaneously minimizes $\langle\hat{T}(\mathbf{r}_1,\ldots,\mathbf{r}_N)\rangle$. From the definitions in Section 2.7,

$$T_s[\rho_\lambda(\mathbf{r})]=\langle\Phi_{\rho\lambda}^{\min}(\mathbf{r}_1,\ldots,\mathbf{r}_N)|\hat{\mathrm{T}}|(\mathbf{r}_1,\ldots,\mathbf{r}_N)|\Phi_{\rho\lambda}^{\min}(\mathbf{r}_1,\ldots,\mathbf{r}_N)\rangle \tag{2.38}$$

and

$$E_x[\rho_\lambda(\mathbf{r})]=\langle\Phi_{\rho\lambda}^{\min}(\mathbf{r}_1,\ldots,\mathbf{r}_N)|\hat{V}_{ee}(\mathbf{r}_1,\ldots,\mathbf{r}_N)|\Phi_{\rho\lambda}^{\min}(\mathbf{r}_1,\ldots,\mathbf{r})\rangle-\lambda U[\rho]. \tag{2.39}$$

Now, since

$$\lambda^{-2}\int\cdots\int\lambda^{3N}\Phi_\rho^{\min}(\lambda\mathbf{r}_1,\ldots,\lambda\mathbf{r}_N)^*\hat{T}(\mathbf{r}_1,\ldots,\mathbf{r}_N)\Phi_\rho^{\min}(\lambda\mathbf{r}_1,\ldots,\lambda\mathbf{r}_N)\,\mathrm{d}x_1\cdots\mathrm{d}x_N$$
$$=\int\cdots\int\Phi_\rho^{\min}(\lambda\mathbf{r}_1,\ldots,\lambda\mathbf{r}_N)^*\hat{T}(\lambda\mathbf{r}_1,\ldots,\lambda\mathbf{r}_N)\Phi_\rho^{\min}(\lambda\mathbf{r}_1,\ldots,\lambda\mathbf{r}_N) \tag{2.40}$$
$$\rightarrow \mathrm{d}^3(\lambda\mathbf{r}_1)\cdots\mathrm{d}^3(\lambda\mathbf{r}_N)\,\mathrm{d}s_1\cdots\mathrm{d}s_N$$
$$=\int\cdots\int\Phi_\rho^{\min}(\mathbf{r}_1,\ldots,\mathbf{r}_N)^*\,\hat{T}(\mathbf{r}_1,\ldots,\mathbf{r}_N)\Phi_\rho^{\min}(\mathbf{r}_1,\ldots,\mathbf{r}_N)\,\mathrm{d}x_1\cdots\mathrm{d}x_N,$$

we have that

$$\Phi_{\rho\lambda}^{\min}(\mathbf{r}_1,\ldots,\mathbf{r}_N)=\lambda^{3N/2}\Phi_\rho^{\min}(\lambda\mathbf{r}_1,\ldots,\lambda\mathbf{r}_N). \tag{2.41}$$

Consequently $\lambda^{3N/2}\Phi_\rho^{\min}(\lambda\mathbf{r}_1,\ldots,\lambda\mathbf{r}_N)$ may be used for $\Phi_{\rho\lambda}^{\min}(\mathbf{r}_1,\ldots,\mathbf{r}_N)$ in eqns (2.34) and (2.35) to yield

$$T_s[\rho_\lambda]=\lambda^2 T_s[\rho] \tag{2.42}$$

and

$$E_x[\rho_\lambda]=\lambda E_x, \tag{2.43}$$

so $T_s$ and $E_x$ show homogeneous scaling; (2.42) and (2.43) are equalities Equations (2.42) and (2.43) were suggested without proof in ref. 19, corraborated in ref. 20, and proved in ref. 2.

All popular approximations to $E_x$ obey eqn (2.43). For instance, the local-density approximation to $E_x$, $a\int\rho^{4/3}(\mathbf{r})\,\mathrm{d}^3r$, scales according to eqn (2.43):

$$a\int\rho_\lambda^{4/3}(\mathbf{r})\,\mathrm{d}^3r=a\int\lambda^4\rho^{4/3}(\lambda\mathbf{r})\,\mathrm{d}^3r$$
$$=\lambda a\int\rho^{4/3}(\lambda\mathbf{r})\,\mathrm{d}^3(\lambda r)=\lambda a\int\rho^{4/3}(\mathbf{r})\,\mathrm{d}^3r. \tag{2.44}$$

The Thomas–Fermi kinetic energy

$$T_F[\rho]=b\int\rho^{5/3}(\mathbf{r})\,\mathrm{d}^3r$$

scales as eqn (2.42):

$$b\int[\lambda^3\rho(\lambda\mathbf{r})]^{5/3}\,\mathrm{d}^3r=\lambda^2 b\int\rho^{5/3}(\mathbf{r})\,\mathrm{d}^3r. \tag{2.45}$$

$T_F[\rho]$ does not scale as the $T[\rho]$ of eqns (2.15) and (2.16). Consequently, it is perhaps best to look upon $T_F[\rho]$ as an approximation to the non-interacting $T_s[\rho]$ and not as an approximation to the interacting $T[\rho]$.

## 2.9 Scaling inequalities of $E_c$

From eqns (2.33) and (2.34),

$$V_{ee}[\rho]=U[\rho]+E_x[\rho]+E_c[\rho]-T_c[\rho]. \tag{2.46}$$

Substituting eqns (2.35) and (2.46) for $T[\rho]$ and $V_{ee}[\rho]$ into eqns (2.11) and

(2.12), and using the fact that $T_s[\rho]$, $U[\rho]$ and $E_x[\rho]$ scale homogeneously, one obtains

$$\lambda E_c[\rho]-E_c[\rho_\lambda]\geqslant\lambda(1-\lambda)T_c[\rho] \quad (2.47)$$

and

$$\lambda^2 E_c[\rho_\lambda]-\lambda^3 E_c[\rho]\geqslant\lambda(\lambda-1)T_c[\rho_\lambda]. \quad (2.48)$$

Next, using $T_c[\rho]\geqslant 0$ and $T_c[\rho_\lambda]\geqslant 0$ to eliminate $T_c$, one finds

$$E_c[\rho_\lambda]<\lambda E_c[\rho] \quad (\lambda<1) \quad (2.49)$$

and

$$E_c[\rho_\lambda]>\lambda E_c[\rho] \quad (\lambda>1). \quad (2.50)$$

In the local-density approximation, eqns (2.47)–(2.50) will be obeyed if

$$\lambda\varepsilon_c(\rho)-\varepsilon_c(\lambda^3\rho)-\lambda(1-\lambda)t_c(\rho)>0, \quad (2.51)$$

$$\lambda\varepsilon_c(\lambda^3\rho)-\lambda^2\varepsilon_c(\rho)+(1-\lambda)t_c(\lambda^3\rho)>0 \quad (2.52)$$

and

$$(1-\lambda)[\lambda\varepsilon_c(\rho)-\varepsilon_c(\lambda^3\rho)]>0, \quad (2.53)$$

where

$$E_c^{\text{LDA}}[\rho]=\int d^3r\,\rho(\mathbf{r})\varepsilon_c[\rho(\mathbf{r})] \quad (2.54)$$

and where[21]

$$t_c(\rho)=3\mu_c(\rho)-4\varepsilon_c(\rho), \quad (2.55)$$

with

$$\mu_c(\rho)=\frac{d}{d\rho}[\rho\varepsilon_c(\rho)]. \quad (2.56)$$

The inequalities in eqns (2.51)–(2.53) were derived in ref. 2 and then tested there with $r_s$ and $r_s/\lambda$ between 0.001 and 3200 for four common parametrizations of $\varepsilon_c(\rho)$: Wigner,[22] Gunnarsson–Lundquist,[23] random-phase approximation,[24] and Ceperley–Alder.[25] All inequalities were found to be obeyed.

## 2.10 Scaling insensitivity requirement for $E_c[\rho]$

Take the derivative of both sides of eqn (2.47) about $\lambda=1$ to obtain

$$(\partial E_c[\rho_\lambda]/\partial\lambda)_{\lambda=1}=E_c[\rho]+T_c[\rho]. \quad (2.57)$$

In the tight-binding limit

$$E_c[\rho]+T_c[\rho]=0, \tag{2.58}$$

so that in this limit

$$(\partial E_c[\rho_\lambda]/\partial\lambda)_{\lambda=1}=0. \tag{2.59}$$

This means that the exact $E_c[\rho_\lambda]$ is relatively insensitive to scaling for many densities suitable for atoms and molecules.

Equation (2.58) should not be surprising because we know that $E_c=-T_c$ in atoms when $E_c$ and $T_c$ are defined according to the traditional quantum chemistry definition for the correlation energy.

The local-density approximation for $E_c[\rho]$ does *not* obey the scaling invariance for compact densities predicted by eqn (2.59). One should take this into account in the quest for improved correlation-energy functionals.

## 2.11 Scaling equality for the correlation energy: scaling of electronic charge

The exact correlation-energy functional does obey a coordinate-scaling equality when a variation of charge is employed. It was shown in Section 2.4 that $\lambda^{3N/2}\,\Psi_\rho^{\min}(\lambda\mathbf{r}_1,\ldots,\lambda\mathbf{r}_N)$ is that antisymmetric function which yields $\rho_\lambda(\mathbf{r})$ and simultaneously minimizes $\langle\hat{T}(\mathbf{r}_1,\ldots,\mathbf{r}_N)+\lambda\hat{V}_{ee}(\mathbf{r}_1,\ldots,\mathbf{r}_N)\rangle$. It then follows from Sections 2.7 and 2.8 that

$$\begin{aligned}T_{s,\lambda}[\rho_\lambda]+U_\lambda[\rho_\lambda]+E_{x,\lambda}[\rho_\lambda]+E_{c,\lambda}[\rho_\lambda]\\ =\langle\lambda^{3N/2}\Psi_\rho^{\min}(\lambda\mathbf{r}_1,\ldots,\lambda\mathbf{r}_N)|\hat{T}(\mathbf{r}_1,\ldots,\mathbf{r}_N)\\ +\lambda\hat{V}_{ee}(\mathbf{r}_1,\ldots,\mathbf{r}_N)|\lambda^{3N/2}\Psi_\rho^{\min}(\lambda\mathbf{r}_1,\ldots,\lambda\mathbf{r}_N)\rangle,\end{aligned} \tag{2.60}$$

where the subscript $\lambda$ in the functionals on the left-hand side indicates that the functionals pertain to an electronic charge which is $\lambda^{1/2}$ times the actual charge.

Equation (2.60) gives

$$\begin{aligned}\lambda^2T_s[\rho]+\lambda^2U[\rho]+\lambda^2E_x[\rho]+E_{c,\lambda}[\rho_\lambda]\\ =\lambda^2T_s[\rho]+\lambda^2U[\rho]+\lambda^2E_x[\rho]+\lambda^2E_c[\rho],\end{aligned} \tag{2.61}$$

which, in turn, yields the *equality*

$$E_{c,\lambda}[\rho_\lambda]=\lambda^2E_c[\rho]. \tag{2.62}$$

Further, this means that $T_\lambda$, $V_{ee\,\lambda}$, $T_{s,\lambda}$, $U_\lambda$, $E_{x,\lambda}$ and $E_{c,\lambda}$ all possess the same coordinate-scaling properties when charge scaling is invoked. The universal relation is

$$G_{\lambda}[\rho_{\lambda}]=\lambda^{2}G[\rho], \tag{2.63}$$

where $G$ signifies any one of the above six functionals.

In the local-density approximation (LDA),

$$E_{c,\lambda}^{LDA}[\rho]=\int d^{3}r\, \rho_{\lambda}(\mathbf{r})\varepsilon_{c,\lambda}[\rho_{\lambda}(\mathbf{r})], \tag{2.64}$$

where $\varepsilon_{c,\lambda}[\rho_{\lambda}(\mathbf{r})]$ is the correlation energy per electron in a uniform gas of density $\rho_{\lambda}(\mathbf{r})$. The exact LDA will satisfy eqn (2.64) if

$$\varepsilon_{c,\lambda}(\lambda^{3}\rho)=\lambda^{2}\varepsilon_{c}(\rho). \tag{2.65}$$

With $k_{F}=(3\pi^{2}\rho)^{1/3}$ and $(e')^{2}=\lambda e^{2}$, the random-phase approximation to the exact LDA, $\varepsilon_{c,\lambda}$, is

$$\varepsilon_{c,\lambda}^{RPA}(\rho)=(e')^{4}f[k_{F}/(e')^{2}], \tag{2.66}$$

which satisifies eqn (2.65), and hence eqn (2.62).

It sould be clear that eqns (2.49) and (2.50) can be generalized to arbitrary electronic charge. Namely

$$(1-\lambda')E_{c,\lambda}[\rho_{\lambda'}]=\lambda'(1-\lambda')E_{c,\lambda}[\rho]. \tag{2.67}$$

Finally, set $\lambda'=\lambda$ and utilize eqn (2.62) with eqn (2.67) to obtain

$$(1-\lambda)E_{c,\lambda}[\rho]\geqslant(1-\lambda)\lambda E_{c}[\rho]. \tag{2.68}$$

It is of note that eqn (68) involves a fixed $\rho$ while the electronic charge is varied.

## 2.12 Summary

For the purpose of approximating the universal variational density functionals for electronic structure, the coordinate-scaling requirements were reviewed for the universal kinetic ($T[\rho]$ and $T_{s}[\rho]$), electron–electron repulsion ($V_{ee}[\rho]$), exchange ($E_{x}[\rho]$) and correlation energy ($E_{c}[\rho]$) density functionals. $T_{s}[\rho]$ and $E_{x}[\rho]$ exhibit homogeneous scaling, but $T[\rho]$, $V_{ee}[\rho]$ and $E_{c}[\rho]$ do *not* possess this property. Instead, they satisfy scaling *inequalities*. Approximations to these functionals must embody this fact, and explicit inequalities were reviewed. For instance, $(1-\lambda)E_{c}[\rho_{\lambda}]<\lambda(1-\lambda)E_{c}[\rho]$, for scale factor $\lambda\neq1$, where $\rho_{\lambda}(\mathbf{r})=\lambda^{3}\rho(\lambda\mathbf{r})$. A trial energy may be lowered by scaling, however, in a manner as if $T[\rho]$ and $V_{ee}[\rho]$ do exhibit homogeneous scaling. In other words, the trial energy may be lowered by optimization of the coordinate scale factor $\lambda$, when $\lambda$ acts as a multiplicative parameter in $F[\rho,\lambda\}=\lambda^{2}T[\rho]+\lambda V_{ee}[\rho]$. In fact,

$F[\rho, \lambda]$ is a new universal functional of $\rho$ *and* $\lambda$. In the tight-binding limit, the correlation-energy functional $E_c$ must be insensitive to a coordinate scaling of the density. The local-density approximation for $E_c$ does not exhibit this required insensitivity to a scaling of $\rho$, and one should take this into account in the quest for improved exchange-correlation functionals. Common parametrizations of the local-density approximation for $E_c$, on the other hand, do satisfy the required scaling inequalities. Finally, the exact correlation-energy functional does obey a coordinate-scaling equality when the electronic charge is simultaneously scaled.

## Acknowledgment

Acknowledgment is made to the Donors of the Petroleum Research Fund, administered by the American Chemical Society, for support of this research.

## References

1. Hohenberg, P. and Kohn, W. (1964). *Phys. Rev.*, **136**, B864.
2. Levy, M. and Perdew, J.P. (1985). *Phys. Rev.*, **A32**, 2010.
3. Levy, M., Yang, W. and Parr, R.G. (1985). *J. Chem. Phys.*, **83**, 2334.
4. (a) Levy, M. (1979). *Proc. Natl Acad. Sci. U.S.A.*, **76**, 6062; (b) (1979). *Bull. Am. Phys. Soc.*, **24**, 626 (Abstracts EI 15 and EI 16); (1982). *Phys. Rev.*, **A26**, 1200.
5. Harriman, J.E. (1981). *Phys. Rev.*, **A24**, 680.
6. Lieb, E.H. (1982). In *Physics as Natural Philosophy: Essays in Honor of Laszlo Tisza on his 75th Birthday*, eds Feshbach, H. and Shimony, A., MIT Press, Cambridge, Massachusetts; (1983). *Int. J. Quant. Chem.*, **24**, 224.
7. Valone, S.M. (1980). *J. Chem. Phys.*, **73**, 1344; (1980). *Ibid.*, **73**, 4653.
8. Percus, J.K. (1978). *Int. J. Quant. Chem.*, **13**, 89.
9. In addition to ref. 2, see also ref. 3 and the ensemble-search density and one matrix equivalents to eqn (2.7) in ref. 10.
10. (a) Yang, W. (1987). In *Density Matrices and Density Functionals*, eds Erdahl, R. and Smith, V.H., Reidel, Dordrecht; (b) Levy, M. (1987). *Ibid.*
11. Ghosh, S.K. and Parr, R.G. (1985). *J. Chem. Phys.*, **82**, 3307.
12. Kohn, W. and Sham, L.J. (1965). *Phys. Rev.*, **140**, A1133.
13. Lundquist, S. and March, N.H. (eds) (1983). *Theory of the Inhomogeneous Electron Gas*, Plenum, New York.
14. Levy, M. and Perdew, J.P. (1985). In *Density Functional Methods in Physics*, eds Dreizler, R.M. and da Providencia, J., Plenum, New York, and references therein.
15. Perdew, J.P. and Levy, M. (1985). *Phys. Rev.*, **B31**, 6264, and references therein.
16. Sahni, V., Gruenbaum, J. and Perdew, J.P. (1982). *Phys. Rev.*, **B26**, 4371.
17. See also the discussion in Langreth, D.C. and Mehl, M.J. (1983). *Phys. Rev.*, **B28**, 1809.

18. Theophilou, A.K. (1978). *J. Phys.*, **C12**, 5419; see also Hadjisavvas, N. and Theophilou, A.K. (1984). *Phys. Rev.*, **A30**, 2183.
19. Sham, L.J. (1970). *Phys. Rev.*, **A1**, 969.
20. Talman, J.D. (1983). Lecture presented at the First Summer Institute in Theoretical Physics, Queen's University, Kingston, Ontario.
21. Williams, A.R. and von Barth, U. (1983). In *Theory of the Inhomogeneous Electron Gas*, eds Lundqvist S. and March, N.H., Plenum, New York; Averill F.W. and Painter, G.S. (1981). *Phys. Rev.*, **B24**, 6795; see also Section 4 of ref. 2.
22. Wigner, E.P. (1934). *Phys. Rev.*, **46**, 1002; eqn (3.58) of Pines, D. (1983). *Elementary Excitations in Solids*, Benjamin, New York.
23. Gunnarsson, O. and Lundqvist, B.I. (1976). *Phys. Rev.*, **B13**, 4274.
24. Cole, L.A. and Perdew, J.P. (1982). *Phys. Rev.*, **A25**, 1265.
25. Ceperley, D.M. and Alder, B.J. (1980). *Phys. Rev. Lett.*, **45**, 566.

# 3 Atomic and Molecular Energy and Energy Difference Formulae Based Upon Electrostatic Potentials at Nuclei

P. POLITZER

*Department of Chemistry, University of New Orleans, New Orleans, Louisiana 70148, U.S.A.*



## 3.1 Introduction

A primary objective in our work has been to develop new approaches to the calculation of atomic and molecular energies. We have found that these energies can be related, through both exact and approximate formulae, to the electrostatic potentials at the nuclei in these systems.[1,2] These potentials can be determined rigorously from a knowledge of the electronic density function of the system, $\rho(\mathbf{r})$, and its geometry (in the case of a molecule). Thus the electrostatic potential at any nucleus A, having charge $Z_A$ and located at $\mathbf{R}_A$, is given exactly by the following:

$$V_{0,A} = \sum_{B \neq A} \frac{Z_B}{|\mathbf{R}_B - \mathbf{R}_A|} - \int \frac{\rho(\mathbf{r})\,d\mathbf{r}}{|\mathbf{r} - \mathbf{R}_A|}. \tag{3.1}$$

Among the reasons for seeking to develop useful relationships between the energy and the electrostatic potentials at the nuclei,

$$E = f[\{V_{0,A}\}], \tag{3.2}$$

may be cited the following:

Single-Particle Density in Physics and Chemistry ISBN 0-12-470518-9

(a) As is well known, the Hartree–Fock and approximate Hartree–Fock procedures predict total energies less accurately than they do one-electron properties, such as electronic densities and electrostatic potentials.[3] A particularly relevant analysis has been carried out by Levy, Tal and Clement.[4–6] They showed that perturbation expansions of the exact and Hartree–Fock energies of atoms begin to differ in the second-order term. However when derivatives with respect to $Z$ are taken in order to obtain the electrostatic potential at the nucleus (Hellmann–Feynman theorem[7]),

$$V_0 = (\partial E/\partial Z)_N \quad (N = \text{number of electrons}), \tag{3.3}$$

the second-order term vanishes; hence the exact and Hartree–Fock values of $V_0$ begin to differ only in the third-order term. As an example, for the beryllium atom, the Hartree–Fock error in $V_0$ is only 0.016 a.u. (0.19%), whereas the energy is in error by 0.094 a.u. (0.64%).[8] This means that a sufficiently accurate formula relating the total energy directly to the electrostatic potentials at the nuclei should make it possible to obtain energies from Hartree–Fock $V_{0,A}$s that are more accurate than the Hartree–Fock energies. This argument has been advanced earlier, and has indeed been shown to be valid on both the atomic[4–6] and the molecular[9] levels, as we shall discuss.

(b) Electrostatic potentials at nuclei can be measured experimentally, for example from electronic densities obtained by X-ray diffraction or, more directly, from electron diffraction studies.[10] There are also close relationships between $V_{0,A}$ and both core-electron binding energies[11,12] and diamagnetic shielding;[13,14] accordingly, $V_{0,A}$ is linked to the chemical shifts observed in electron spectroscopy and in nuclear magnetic resonance, respectively.

The availability of accurate and practical formulae of the type of eqn (3.2) would therefore permit useful energies to be derived from such experimental measurements, even for very large systems. In an early demonstration of this, Bentley used an approximate energy expression that we have developed earlier[15] (eq. (3.10) below), together with X-ray diffraction data for beryllium and for diamond, to estimate the energy changes, due to polarization and bond formation, that are associated with the formation of these solids from the free atoms.[16] Very recently, Blomquist has discussed extending this type of approach to electron spectroscopy data.[17]

Equation (3.7) is not nearly at the level of accuracy that is sought, since it involves an average error in the total molecular energy of about 2%. This equation does not, therefore, permit the calculation of reliable interaction energies, which is one of our goals, because these are very small differences in total energies, of the order of 1% of the latter. Nevertheless, Bentley's work demonstrates a very promising and encouraging synthesis of experimental and theoretical methods.

## 3.2 Exact total energy relationships

The non-relativistic total energy of an $N$-electron atom having nuclear charge $Z$ can be expressed exactly as a functional of the electrostatic potential at the nucleus, $V_0$:[18]

$$E^{\mathrm{at}}=\frac{1}{2}ZV_0-\frac{1}{2}\int_0^Z\left[Z'\left(\frac{\partial V_0}{\partial Z'}\right)-V_0\right]_N \mathrm{d}Z'. \tag{3.4}$$

The molecular counterpart of eqn (3.4) can be obtained by simply summing over all of the constituent atoms:[18]

$$E^{\mathrm{mol}}=\sum_{\mathrm{A}}\frac{1}{2}Z_{\mathrm{A}}V_{0,\mathrm{A}}-\sum_{\mathrm{A}}\frac{1}{2}\int_0^{Z_{\mathrm{A}}}\left[Z'_{\mathrm{A}}\left(\frac{\partial V_{0,\mathrm{A}}}{\partial Z'_{\mathrm{A}}}\right)-V_{0,\mathrm{A}}\right]_N \mathrm{d}Z'_{\mathrm{A}}. \tag{3.5}$$

Equation (3.5) is also exact.

Equations (3.4) and (3.5) show that the total energy of an atomic or molecular system can be related rigorously to the electrostatic potentials at the positions of the nuclei. As is shown by eqn (3.1), this potential, at any nucleus A, includes contributions from all of the other nuclei (in the case of a molecule) as well as the electrons. A particularly interesting and notable feature of the molecular formula, eqn (3.5), is the complete absence of any explicit interaction or cross-terms; there are only atomic-like terms.

## 3.3 Approximate relationships based on eqns (3.4) and (3.5)

### 3.3.1 Assumption of zero chemical potential:

The accurate evaluation of the integrals in eqns (3.4) and (3.5) is in general not feasible. A particular complication is the requirement that the integrations be carried out over isoelectronic series in which one nuclear charge varies from zero to its final value. It has been shown, however, that this restriction can be removed by introducing the assumption that the chemical potential of the system,

$$\mu^{\mathrm{at}}=(\partial E/\partial N)_Z \quad \text{or} \quad \mu^{\mathrm{mol}}=(\partial E/\partial N)_{R_i,Z_j}, \tag{3.6}$$

is zero.[19] ($R_i$ and $Z_j$ represent the various internuclear distances and nuclear charges, in the case of a molecule.) Equations (3.4) and (3.5) then become

$$E^{\mathrm{at}}=\frac{1}{2}ZV_0-\frac{1}{2}\int_0^Z\left[Z'\left(\frac{\partial V_0}{\partial Z'}\right)-V_0\right]\mathrm{d}Z' \tag{3.7}$$

and

$$E^{\mathrm{mol}}=\sum_{\mathrm{A}}\frac{1}{2}Z_{\mathrm{A}}V_{0,\mathrm{A}}-\sum_{\mathrm{A}}\frac{1}{2}\int_{0}^{Z_{\mathrm{A}}}\left[Z'_{\mathrm{A}}\left(\frac{\partial V_{0,\mathrm{A}}}{\partial Z'_{\mathrm{A}}}\right)-V_{0,\mathrm{A}}\right]\mathrm{d}Z'_{\mathrm{A}}. \tag{3.8}$$

While eqns (3.7) and (3.8) still require a knowledge of the $Z$-dependence of $V_0$ or $V_{0,\mathrm{A}}$, now this need not be over a path for which $N$ is fixed. Thus, for example, $V_0-Z$ or $V_{0,\mathrm{A}}-Z_{\mathrm{A}}$ relationships for free atoms or neutral molecules could be used; good approximations to these can be obtained from available Hartree–Fock results. For atoms, this yields energies that differ from the experimental values by about 1%, and from the Hartree–Fock value by much less.[18] For molecules, we have tested eqn (8) by first establishing a $V_{0,\mathrm{A}}-Z_{\mathrm{A}}$ dependence from near-Hartree–Fock calculations for a group of six homonuclear diatomic molecules:[19] $Li_2$, $B_2$, $C_2$, $N_2$, $O_2$ and $F_2$. When this was used in eqn (3.8), the results for $Li_2$ and $B_2$ were poor, but for $C_2$ to $F_2$ the average error, relative to the experimental total energy, was only 0.5%. The same equation produced even better energies when it was applied to four heteronuclear diatomics: LiF, CO, BF and NO.

Equations (3.7) and (3.8) are rigorously valid for any system in which the chemical potential is zero. This includes, for instance, neutral atoms and molecules within the context of the Thomas–Fermi theory.[20-22] Since Thomas–Fermi theory becomes exact (in a relative sense) as the sum of the nuclear charges in a system approaches infinity,[23] it may be that eqns (3.7) and (3.8) will become more accurate as the atoms or molecules increase in size.

### 3.3.2 Simplified approximate energy formulae

While the assumption of zero chemical potential does facilitate the integration required by eqns (4) and (5), it is also possible to put these formulae into an approximate form which avoids altogether the need for evaluating the difficult integrals. For atoms, Thomas–Fermi theory predicts that[21,24]

$$E^{\mathrm{at}}=\tfrac{3}{7}ZV_0. \tag{3.9}$$

Equation (3.9) is equivalent to taking the integral term in eqn (4) to equal $\frac{1}{7}$ of the first term.

When $V_0$ is evaluated in accordance with the Thomas–Fermi theory, very poor atomic energies are obtained with eqn (3.9); they are in error by as much as 30%.[5,18] However there is a dramatic improvement in these results when Hartree–Fock atomic wave functions are used to compute $V_0$; the errors are reduced to about 2%.[18,25]

The success of eqn (3.9) raised the possibility of extending it to molecules.

Since the exact molecular formula, eqn (3.5), can be viewed as simply a summation of the atomic counterpart over the constituent atoms, the same approach was tried on the approximate level, yielding[15]

$$E^{\mathrm{mol}} = \sum_{\mathrm{A}} \tfrac{3}{7} Z_{\mathrm{A}} V_{0,\mathrm{A}}. \tag{3.10}$$

Equation (3.10) is surprisingly accurate, considering its very simple nature. It has now been applied to nearly one hundred molecular systems, neutral and ionic and in both ground and excited states, using *ab initio* self-consistent-field wave functions to determine $V_{0,\mathrm{A}}$.[15,26,27] The error is usually less than 2%. Again, eqn (3.10) corresponds to setting the integral term in eqn (3.5) equal to $\frac{1}{7}$ of the first term.

If one seeks to improve the accuracy of eqns (3.9) and (3.10), there are at least two possible approaches that can be investigated. One of these is to derive additional "correction" terms, which will account for some portion of the discrepancies between the results of these equations and the correct energies; the second is to modify eqns (3.9) and (3.10) themselves. Thus far, we have used the first approach in the case of eqn (3.9) and the second for eqn (3.10), as will now be briefly summarized.

It was already mentioned that eqn (3.9) comes out of the Thomas–Fermi theory. It is based upon a particular derived relationship between the electrostatic potential $V(\mathbf{r})$ and the electronic density $\rho(\mathbf{r})$ within an atom. We used Hartree–Fock atomic wave functions to obtain a more realistic representation of this relationship, one that treats separately the core and valence regions of the atom.[28] (We have shown that a physically meaningful boundary between the core and valence regions of a ground-state neutral atom is the outermost minimum in the radial electronic density function, $4\pi r^2\rho(\mathbf{r})$.[28–30]) Using the more realistic $V-\rho$ dependence, we repeated the derivation of eqn (3.9) and obtained[30]

$$E = \frac{3}{7} Z V_0 + \frac{3N_{\mathrm{V}}^2}{112 r_{\mathrm{m}}} - 0.030\,315\, r_{\mathrm{m}}^{1/2} N_{\mathrm{V}}^{5/2} + \frac{Z^{2/3} r_{\mathrm{m}} N_{\mathrm{V}}^2}{32}. \tag{3.11}$$

In eqn (3.11), $r_{\mathrm{m}}$ is the radial distance to the core–valence boundary and $N_V$ is the integrated quantity of electronic charge beyond $r_{\mathrm{m}}$ (the valence electronic charge), which is almost invariably non-integral.[29,31] For $Z > 7$, the energies calculated with eqn (3.11) are markedly better than those resulting from eqn (3.9), the accuracy increasing as $Z$ becomes larger.[2,30]

Turning now to eqn (3.10) and ways in which it can be modified and improved, one obvious starting point is the constant factor $\frac{3}{7}$. This universal parameter could be replaced by one that depends upon the particular molecule:

$$E^{\mathrm{mol}} = k_{\mathrm{mol}} \sum_{\mathrm{A}} Z_{\mathrm{A}} V_{0,\mathrm{A}}. \tag{3.12}$$

Another possibility would be

$$E^{\mathrm{mol}} = \sum_{\mathrm{A}} k_{\mathrm{A}} Z_{\mathrm{A}} V_{0,\mathrm{A}}, \tag{3.13}$$

in which $k_A$ is different for each atom A. We have investigated both types of formulas.

To explore the potential of eqn (3.13), the atomic parameters $k_A$ were initially assigned such magnitudes as would cause the equation

$$E^{\mathrm{at}} = kZV_0 \tag{3.14}$$

to be satisfied exactly for each atom, using Hartree–Fock values for $E^{\mathrm{at}}$ and $V_0$.[27] The resulting $k_A$s are naturally very close to $\frac{3}{7}$ (=0.42857); they begin at 0.4335 for lithium, and after first decreasing rapidly with increasing $Z$ and reaching a minimum at neon,[32] they level off in the neighbourhood of phosphorus and subsequently oscillate gently about 0.420. (For hydrogen, $k_H = E/V_0 = 0.5000$.) These oscillations have been discussed by Levy and Tal.[32] When these $k_A$ values were used to calculate molecular energies by means of eqn (3.13) and Hartree–Fock $V_{0,A}$s, a distinct improvement over eqn (3.10) was observed;[27] for the most part, the errors were now less than 0.4%.

It would be desirable, of course, that each $k_A$ reflect at least to some extent the molecular environment of atom A. In this spirit, Anno assigned the $k_A$ values by requiring that eqn (3.13) be satisfied for the homonuclear diatomic $A_2$ molecules, using Hartree–Fock data for $E^{\mathrm{mol}}$ and $V_{0,A}$.[33] This does produce better molecular energies, on the whole, than did the use of the free-atom $k_A$s, although the former are not yet at a level of accuracy sufficient to permit the calculation of interaction energies. A related analysis has been carried out by Fliszar *et al.*, who argued that the $k_A$s determined from molecular data should be essentially independent of the nature of the molecule.[34,35]

With regard to eqn (3.12), one approach to its implementation is to seek ways of predicting the $k_{\mathrm{mol}}$s from the $k_A$s of the free atoms. We have accordingly used Hartree–Fock energies and $V_{0,A}$s to calculate, for about 60 molecules, the $k_{\mathrm{mol}}$ values that satisfy eqn (3.12); these were then compared with our free-atom $k_A$s. (Fliszar and Beraldin have obtained $k_{\mathrm{mol}}$s for some hydrocarbons in this manner.[35]) Using this approach, as can be seen by combining eqns (3.12) and (3.13),

$$k_{\mathrm{mol}} = \frac{\sum_{\mathrm{A}} k_{\mathrm{A}} Z_{\mathrm{A}} V_{0,\mathrm{A}}}{\sum_{\mathrm{A}} Z_{\mathrm{A}} V_{0,\mathrm{A}}}.$$

$k_{\mathrm{mol}}$ should be a weighted average of the respective $k_A$s.[35] It may be anticipated that hydrogens, having very small $Z_A$ and $V_{0,A}$, will contribute little to this average. These expectations were confirmed by our results. In those cases in which one atom greatly predominates in the molecule, or when the atoms are in different rows of the periodic table, then $k_{mol}$ tends to reflect primarily the predominant or the heavier atom(s). For example, for $C_3H_6O$, $k_{mol} = 0.4267$, while $k_C = 0.4276$ and $k_O = 0.4201$; for $C_3H_6S$, on the other hand, $k_{mol} = 0.4221$, with $k_S = 0.4198$.

Taking the $k_A$s as a basis is certainly not the only conceivable means for predicting $k_{mol}$ values. Thus far, however, we have not found any mechanism whereby eqn (3.12) would consistently yield better molecular energies than eqn (3.13).

## 3.4 "Better than Hartree–Fock" approximate energies

As was pointed out in Section 3.1, it should be possible to use Hartree–Fock electrostatic potentials at nuclei to obtain energies more accurate than the Hartree–Fock one, provided that sufficiently accurate relationships between these properties are available. A theoretical basis for this statement has been presented by Levy, Tal and Clement,[4–6] as was discussed earlier. In this section, we review some confirming calculated results.

Levy, Tal and Clement have derived a group of remarkable atomic energy formulae, which yield nearly exact non-relativistic total energies.[4–6] They began with an expansion of the energy in powers of $Z$ that comes from perturbation theory[36–38] or from the Hellmann–Feymann and virial theorems:[39]

$$E(Z, N) = \sum_{j=0}^{\infty} \varepsilon_j^N Z^{(2-j)}. \tag{3.16}$$

This was combined with the Hellmann–Feynman theorem in the form of eqn (3.3) to produce, eventually, relationships such as eqns (3.17)–(3.19) for the energy of an $N$-electron neutral atom:

$$E(N) = -\sum_{M=1}^{N} [V_0^M + \varepsilon_0^M]. \tag{3.17}$$

$$E(N) = -2\sum_{M=1}^{N/2} [V_0^{2M-1} + 0.5(I_{2M} + A_{2M-2})] \quad \text{(for } N \text{ even)}, \tag{3.18}$$

$$E(N) = -2\sum_{M=1}^{(N+1)/2} [V_0^{2M-2} + 0.5(I_{2M-1} + A_{2M-3})] \quad \text{(for } N \text{ odd)} \tag{3.19}$$

In these equations, $V_0^M$, $I_M$ and $A_M$ are, respectively, the electrostatic potential at the nucleus, the ionization potential and the electron affinity of a ground-state neutral atom having $M$ electrons.

Equations (3.17)–(3.19) express the energy of any neutral atom as summations over the electrostatic potentials at the nuclei and other properties of the series of atoms smaller than itself. The results are extremely accurate, well beyond the Hartree–Fock level.[4-6] Thus, by means of these and other similar formulae (which have been termed "trapezoid recursion relations"[4-6]), it is possible to use a property, $V_0$, that is obtained from the Hartree–Fock electronic density to determine most of the correlation energy. In this approach, the atom of interest is reached by progressing through the series of smaller neutral atoms, rather than by starting with the final number of electrons and building up the nuclear charge from zero, as is required by eqn (3.4).

Very accurate formulae for energy differences within certain classes of systems have also been developed. These involve *only* electrostatic potentials at nuclei; no other terms are included. Levy derived an approximate expression for the difference between the energies of two isoelectronic atomic or molecular systems, A and B, which in some manner differ in their external potentials;[40] this difference may be in some nuclear charges, for example, or in an internuclear distance. Then, to a good approximation,

$$E^{\mathrm{B}} - E^{\mathrm{A}} = \int 0.5[\rho_{\mathrm{A}}(\mathbf{r}) + \rho_{\mathrm{B}}(\mathbf{r})]\Delta v(\mathbf{r})\,\mathrm{d}\mathbf{r} + \Delta V_{\mathrm{NN}}, \tag{3.20}$$

where $\Delta v(\mathbf{r})$ and $\Delta V_{\mathrm{NN}}$ are the differences in external potentials and nuclear repulsion energies, respectively.

If A and B are isoelectronic monatomic ions (or an atom and an ion), then $\Delta v(r) = -Z_{\mathrm{B}}/r + Z_{\mathrm{A}}/r$ and eqn (3.20) takes the form,[41]

$$E^{\mathrm{B}} - E^{\mathrm{A}} = 0.5(Z_{\mathrm{B}} - Z_{\mathrm{A}})(V_0^{\mathrm{A}} + V_0^{\mathrm{B}}). \tag{3.21}$$

It has been shown, by numerical examples, that eqn (3.21) is extremely accurate; indeed nearly exact in many instances.[40-42]

For isoelectronic molecular systems A and B which differ only in the nature of the $i$th nucleus, we have independently derived, by another approach, the energy difference formula expressed by eqn (3.22),[9] which is a special case of eqn (3.20):

$$E^{\mathrm{B}} - E^{\mathrm{A}} = 0.5(Z_i^{\mathrm{B}} - Z_i^{\mathrm{A}})(V_{0,i}^{\mathrm{A}} + V_{0,i}^{\mathrm{B}}). \tag{3.22}$$

Equation (3.22) is based on the assumption that corresponding bond lengths in A and B are the same.

This equation was tested for nine pairs of isoelectronic diatomic molecules and molecular ions.[9] Hartree–Fock wave functions were used to compute the $V_{0,i}$. The calculated energy differences were, on the average, within 0.28% of the experimental values, and were in every instance more accurate than the Hartree–Fock ones.

By properly choosing the pairs of molecules and molecular ions, the energy differences obtained with eqn (3.22) can be made to yield molecular electron affinities and ionization potentials. We used this approach to calculate the electron affinities of $C_2$, CN and OF, and obtained overall good results, again better than the Hartree–Fock ones in each case.[9] This is a particularly useful application of eqn (3.22), since electron affinities are often difficult to determine by experimental or other theoretical techniques.

A rather unusual but highly important and desirable feature of eqns (3.21) and (3.22) is that they express energy differences as *sums* of terms. This avoids the problem of having to obtain a small quantity by taking the difference of two large numbers, a procedure that often results in a very sizeable error magnification.

While it is very pleasing that eqns (3.21) and (3.22) involve summations rather than differences, it is also somewhat surprising. One of the conclusions that comes out of the earlier derivations and analyses that have been summarized here, both exact and approximate, is that atomic and molecular energies can be related directly to electrostatic potentials at nuclei. Thus it would be expected that differences between the energies of atomic and molecular systems will be functions of the differences between these potentials. However the opposite is found to be the case.

Some insight into this situation can be obtained by combining the original derivations of eqns (3.21) and (3.22),[9,41] and invoking the Hellmann–Feynman theorem to write

$$E^{\mathrm{B}} - E^{\mathrm{A}} = \int_{Z_i^{\mathrm{A}}}^{Z_i^{\mathrm{B}}} (V_{0,i})_{\mathrm{c}}\, \mathrm{d}Z_i. \tag{3.23}$$

Here A and B are isoelectronic atomic or molecular systems; in the latter case, $Z_i$ refers to the nucleus that is not common to A and B. The subscript "c" indicates that the number of electrons and (for molecules) all internuclear distances and nuclear charges except $Z_i$ are held constant. Equation (3.23) is exact.

Following Levy,[41] the mean-value theorem can now be used to give, rigorously,

$$E^{\mathrm{B}} - E^{\mathrm{A}} = (Z_i^{\mathrm{B}} - Z_i^{\mathrm{A}}) V_{0,i}^{*}, \tag{3.24}$$

where $V_{0,i}^{*}$ is the value of $V_{0,i}$ for some $Z_i$ between $Z_i^{\mathrm{A}}$ and $Z_i^{\mathrm{B}}$. Thus eqn (3.24) is an exact formula giving an energy difference in terms of the electrostatic potential at a nucleus in an intermediate system; this already contradicts the expectation that $E^{\mathrm{B}} - E^{\mathrm{A}}$ will be given by a difference of two quantities.

Levy's next step in deriving the atomic formula, eqn (3.21), was to introduce the approximation[41]

$$V^*_{0,i} = 0.5(V^A_{0,i} + V^B_{0,i}). \tag{3.25}$$

In the present treatment, this would lead to eqns (3.21) and (3.22).

$V^*_{0,i}$ does indeed equal $0.5(V^A_{0,i} + V^B_{0,i})$ if $V_{0,i}$ is a linear function of $Z_i$. This can readily be verified by means of a geometrical analysis of the integral in eqn (23), or alternatively, by inserting $V_{0,i} = \alpha Z_i + \beta$ into the integrand; this leads, ultimately, to eqns (3.21) and (3.22), without any need for assigning numerical values to the parameters $\alpha$ and $\beta$. By virtue of the Hellmann–Feynman theorem, a linear dependence of $V_{0,i}$ on $Z_i$ is equivalent to the total energy being a quadratic function of $Z_i$. Accordingly, eqns (3.21) and (3.22) are exact formulae for isoelectronic energy differences if the atoms and monatomic ions satisfy the three-term truncated form of eqn (3.16),

$$E(Z, N) = \varepsilon^N_0 Z^2 + \varepsilon^N_1 Z + \varepsilon^N_2, \tag{3.26}$$

and if the energies of the molecular systems are related in an analogous manner to the charge on the non-common nucleus. Levy has empirically arrived at the same conclusion in the case of atoms and monatomic ions.[40] In general, these requirements seem to be satisfied fairly well.[37-40]

## 3.5 Some future directions

### 3.5.1 Non-isoelectronic systems

It can easily be demonstrated empirically that the condition of a quadratic $E$–$Z$ dependence is also fairly well satisfied by certain classes of non-isoelectronic systems.[43] These include the neutral ground-state free atoms, and groups of neutral diatomic molecules in which one atom is common to all of the members (e.g. the diatomic hydrides AH; the series CO, NO, $O_2$, OF, etc.). Furthermore, it has been found to be possible to derive formulae that are formally identical with eqns (21) and (22) but are not limited to isoelectronic series.[43] They do, however, require quadratic $E-Z$ relationships.

The energy differences obtained by applying these latter formulae to various pairs of atoms and molecules are presented in Table 3.1.[43] Hartree–Fock wave functions were used to compute the electrostatic potentials at the nuclei. For the atoms and the first-row molecules, the results are reasonably good, with errors (relative to the experimental values) in the general neighbourhood of 1%. There is a marked improvement for the

**Table 3.1** Energy differences between some non-isoelectronic pairs of atoms and molecules.[a–b]

| *System: A, B* | $E^B - E^A$ (eqns (3.21), (3.22))[b] | $E^B - E^A$ (*Hartree–Fock*[c]) | $E^B - E^A$ (*experimental*[d]) |
|---|---|---|---|
| *Atoms* | | | |
| B, N | −29.68 (0.91%) | −29.87 (0.27%) | −29.95 |
| C, Mg | −163.79 (0.81%) | −161.93 (0.34%) | −162.48 |
| O, F | −24.38 (1.26%) | −24.60 (0.38%) | −24.70 |
| Na, Mg | −37.68 (0.57%) | −37.76 (0.36%) | −37.89 |
| Al, P | −98.63 (0.65%) | −98.84 (0.43%) | −99.27 |
| Si, Cl | −170.42 (0.68%) | −170.63 (0.56%) | −171.59 |
| Cl, Ar | −67.05 (1.09%) | −67.34 (0.67%) | −67.79 |
| *Molecules* | | | |
| CN, CO | −20.30 (1.58%) | −20.57 (0.23%) | −20.62 |
| BF, $F_2$ | −75.66 (0.99%) | −74.60 (0.42%) | −74.92 |
| OF, $F_2$ | −24.33 (1.79%) | −24.57 (0.81%) | −24.77 |
| BH, CH | −13.48 (2.14%) | −13.15 (0.39%) | −13.20 |
| CH, NH | −17.00 (1.45%) | −16.70 (0.38%) | −16.76 |
| OH, FH | −25.00 (1.03%) | −24.65 (0.41%) | −24.75 |
| NaH, SiH | −127.97 (0.35%) | −127.04 (0.38%) | −127.52 |
| MgH, SiH | −89.80 (0.19%) | −89.28 (0.39%) | −89.63 |
| AlH, SH | −156.63 (0.13%) | −155.64 (0.50%) | −156.42 |
| SH, ClH | −62.21 (0.32%) | −62.01 (0.63%) | −62.40 |
| PH, ClH | −119.34 (0.16%) | −118.82 (0.60%) | −119.54 |

[a] All energies are in hartrees. The percentage error, relative to the experimental values, are given in parentheses.

[b] Ref. 43.

[c] Atoms: Clementi, E. (1965). *Tables of Atomic Functions*, IBM Corp., San Jose. Molecules: (a) hydrides: Cade, P.E. and Huo, W.M. (1967). *J. Chem. Phys.*, **47**, 614, 649; (b) others: Cade, P.E. and Wahl, A.C. (1974). *At. Data Nucl. Data Tables*, **13**, 339; Cade, P.E. and Huo, W.M. (1975). *Ibid.*, **15**, 1; O'Hare, P.A.G. and Wahl, A.C. (1970). *J. Chem. Phys.*, **53**, 2469.

[d] Atoms: Veillard, A. and Clementi, E. (1968). *J. Chem. Phys.*, **49**, 2415. Molecules; (a) hydrides: see footnote *c*; (*b*) others: experimental values based on data taken from Herzberg, G. (1950). *Spectra of Diatomic Molecules*, 2nd edn, Van Nostrand, Princeton; Gaydon, A.G. (1968). *Dissociation Energies and Spectra of Diatomic Molecules*, 3rd edn, Chapman & Hall, London; and experimental atomic energies from Veillard and Clementi (above).

second-row hydrides; the errors are now less than 0.4%, and our calculated energy differences are in every instance better than the Hartree–Fock ones. Overall, these results can be viewed as encouraging. There is clearly a need for further studies to determine what other classes of systems can be treated by this approach, and what are its potentials and its limitations.

### 3.5.2 Dissociation energies:

Suppose that one bond is ruptured in a molecular system A, producing atom B and leaving A′. The dissociation energy of this bond is

$$D = -[E^{\mathrm{A}} - (E^{\mathrm{B}} + E^{\mathrm{A'}})]. \tag{3.27}$$

Taking the derivative with respect to $Z_{\mathrm{B}}$ and applying the Hellmann–Feynman theorem[7] leads to

$$\left(\frac{\partial D}{\partial Z_{\mathrm{B}}}\right)_{\mathrm{c}} = -\left[\left(\frac{\partial E^{\mathrm{A}}}{\partial Z_{\mathrm{B}}}\right)_{\mathrm{c}} - \left(\frac{\partial E^{\mathrm{B}}}{\partial Z_{\mathrm{B}}}\right)_{\mathrm{c}}\right] \tag{3.28}$$

$$= -[V^{\mathrm{A}}_{0,\mathrm{B}} - V^{\mathrm{B}}_{0,\mathrm{B}}]_{\mathrm{c}}. \tag{3.29}$$

The subscript "c" has the same meaning as in eqn (3.23). Equation (3.28) is similar to one that has been presented by Fliszar.[44] It can be integrated to yield

$$D(Z_{\mathrm{B}}) - D(Z_{\mathrm{B}}=0) = -\int_{Z_{\mathrm{B}}=0}^{Z_{\mathrm{B}}} [V^{\mathrm{A}}_{\mathrm{O,B}} - V^{\mathrm{B}}_{\mathrm{O,B}}]_{\mathrm{c}}\, \mathrm{d}Z_{\mathrm{B}}. \tag{3.30}$$

In this integration, the upper limit is the nuclear charge of the neutral atom B. There is some question concerning the value to be assigned to $D(Z_{\mathrm{B}}=0)$. However by extending the reasoning of Wilson[45] and of Levy, Tal and Clement,[6] it can be argued that $E^{\mathrm{B}}(Z_{\mathrm{B}}=0)=0$ and that $E^{\mathrm{A}}(Z_{\mathrm{B}}=0)=E^{\mathrm{A'}}$, so that $D(Z_{\mathrm{B}}=0)=0$. The evaluation of the integral in eqn (3.30) would then give the dissociation energy of the A′–B bond, within the limitation that all internuclear distances remain constant.

A possible practical difficulty in applying eqn (3.30) is that $V^{\mathrm{A}}_{\mathrm{O,B}}$ and $V^{\mathrm{B}}_{\mathrm{O,B}}$ are likely to be very similar in magnitude.[1] Since the integrand now involves the difference between these quantities, any errors in either of them may be greatly magnified in the result. Despite these various problems, however, the potential usefulness of eqn (3.30) certainly makes it worthy of further study.

## 3.6 Summary

A variety of atomic and molecular energy formulae, both exact and approximate, have been presented and discussed. Their unifying feature is the emphasis upon the primary importance of electrostatic potentials at nuclei. Some of these formulae have led to significant qualitative and conceptual insights; these have been summarized elsewhere.[1, 2] The present focus has been upon examining their quantitative accuracy and applicability. It is now clearly established that electrostatic potentials calculated from Hartree–Fock electronic densities can be used to obtain energies that are more accurate than the Hartree–Fock energy. (It should be noted that this cannot be achieved by the use of the Hellmann–Feynman theorem *alone*, even though this theorem plays an essential role in the derivation of many of the formulae that have been presented, because of the fact that the theorem applies to Hartree–Fock as well as to exact wave functions.[46,47]) With continuing developments in formalism along the lines that have been described, it can be anticipated that increasingly accurate treatment of more classes of systems will become feasible, using Hartree–Fock or experimentally measured electrostatic potentials.

## Acknowledgments

I should like to express my appreciation to Dr Mel Levy and Dr Jane S. Murray for very helpful discussions, and to the U.S. Army Research Office for its support of this work.

## References

1. Politzer, P. (1980). *Israel J. Chem.*, **19**, 224.
2. Politzer, P. (1981). In *Chemical Applications of Atomic and Molecular Electrostatic Potentials*, eds Politzer, P. and Truhlar, D.G., Chap. 2, Plenum, New York.
3. It has been demonstrated that one-electron properties computed with Hartree–Fock wave functions are correct through first order: Møller, C. and Plesset, M.S. (1934). *Phys. Rev.*, **46**, 618; Cohen, M. and Dalgarno, A. (1961). *Proc. Phys. Soc.*, **77**, 748; Pople, J.A. and Seeger, R. (1975). *J. Chem. Phys.*, **62**, 4566. Furthermore, experience has shown that relatively accurate results can be obtained for many one-electron properties even with self-consistent-field wave functions that are not near the Hartree–Fock limit. See, for example: Politzer, P. (1966). *J. Phys. Chem.*, **70**, 1174; Neumann, D. and Moskowitz, J.W. (1968). *J. Chem. Phys.*, **49**, 2056; (1969). *Ibid.*, **50**, 2216; Dunning, Jr., T.H., Pitzer, R.M. and Aung, S. (1972). *J. Chem. Phys.*, **57**, 5044; Truhlar, D.G. and Van-Catledge, F.A. (1973). *J. Chem. Phys.*, **59**, 3207; Rosenberg, B.J. and Shavitt, I. (1975)., *J. Chem. Phys.*, **63**, 2162; Perahia, D., Pullman, A. and Berthod, H. (1975). *Theoret. Chim. Acta*, **40**, 47; Politzer, P. (1979). *J. Chem. Phys.*, **70**, 1067.
4. Levy, M. and Tal, Y. (1980). *J. Chem. Phys.*, **72**, 3416.
5. Levy, M., Clement, S.C. and Tal, Y. (1981). In *Chemical Applications of Atomic and Molecular Electrostatic Potentials*, eds Politzer, P. and Truhlar, D.G., Chap. 3, Plenum, New York.
6. Levy, M., Tal, Y. and Clement, S.C. (1982). *J. Chem. Phys.*, **77**, 3140.
7. Hellmann, H. (1937). *Einfuhrung in die Quantenchemie*, Deuticke, Leipzig; Feyman, R.P. (1939). *Phys. Rev.*, **56**, 340; Epstein, S.T. (1981). In *The Force Concept in Chemistry*, ed. Deb, B.M. Chap 1, Van Nostrand Reinhold, New York.

8. The Hartree–Fock data are from Froese, C. (1972). *Atomic Data*, **4**, 301; (1973). *At. Data Nucl. Data Tables*, **12**, 87; and Mann, J.B. (1973). *At. Data Nucl. Data Tables*, **12**, 1. The exact data are from Politzer, P. and Daiker, K.C. (1978). *J. Quant. Chem.*, **14**, 245.
9. Politzer, P. and Sjoberg, P. (1983). *J. Chem. Phys.*, **78**, 7008.
10. Politzer, P. and Truhlar, D.G. (eds) (1981). *Chemical Applications of Atomic and Molecular Electrostatic Potentials*, Plenum, New York. See chapters by D.G. Truhlar (Chap. 6); M. Fink and R.A. Bonham (Chap. 7); M.A. Spackman and R.F. Stewart (Chap. 17); and G. Moss and P. Coppens (Chap. 18).
11. Basch, H. (1970). *Chem. Phys. Lett.*, **5**, 337.
12. Schwartz, M.E. (1970). *Chem. Phys. Lett.*, **6**, 631.
13. Ramsey, N.F. (1950). *Phys. Rev.*, **78**, 699.
14. Flygare, W.H. (1972). *J. Am. Chem. Soc.*, **94**, 7277.
15. Politzer, P. (1976). *J. Chem. Phys.*, **64**, 4239.
16. Bentley, J. (1979). *J. Chem. Phys.*, **70**, 159.
17. Blomquist, J. (1985). *J. Elect. Spectrosc. Rel. Phenom.*, **36**, 69.
18. Politzer, P. and Parr, R.G. (1974). *J. Chem. Phys.*, **61**, 4258.
19. Politzer, P. (1984). *J. Chem. Phys.*, **80**, 380.
20. Thomas, L.H. (1927). *Proc. Camb. Phil. Soc.*, **23**, 542; Fermi, E. (1928). *Z. Phys.*, **48**, 73; March, N.H. (1957). *Adv. Phys.*, **6**, 1.
21. March, N.H. (1982). *J. Phys. Chem.*, **86**, 2262.
22. Lieb, E.H. (1981). *Rev. Mod. Phys.*, **53**, 603.
23. Lieb, E.H. and Simon, B. (1973). *Phys. Rev. Lett.*, **31**, 681; (1977). *Adv. Math.*, **23**, 22.
24. Milne, E.A. (1927). *Proc. Camb. Phil. Soc.*, **23**, 794.
25. Fraga, S. (1964). *Theoret. Chim. Acta*, **2**, 406.
26. Politzer, P. (1978). *J. Chem. Phys.*, **69**, 491.
27. Politzer, P. (1979). *J. Chem. Phys.*, **70**, 1067.
28. Politzer, P. (1980). *J. Chem. Phys.*, **72**, 3027.
29. Politzer, P. and Parr, R.G. (1976). *J. Chem. Phys.*, **64**, 4634.
30. Politzer, P. (1980). *J. Chem. Phys.*, **73**, 3264.
31. Boyd, R.J. (1977). *J. Chem. Phys.*, **66**, 356.
32. Levy, M. and Tal, Y. (1980), *J. Chem. Phys.*, **73**, 5168.
33. Anno, T. (1980). *J. Chem. Phys.*, **72**, 782. See also Teruya, H. and Anno, T. (1980). *J. Chem. Phys.*, **72**, 6044; (1984). *Mem. Fac. Sci. Kyushu Univ.*, Ser. C., **14**, 221.
34. Fliszar, S., Foucrault, M., Beraldin, M.-T. and Bridet, J. (1981). *Can. J. Chem.*, **59**, 1074.
35. Fliszar, S. and Beraldin, M.-T. (1980). *J. Chem. Phys.*, **72**, 1013.
36. Hylleraas, E.A. (1930). *Z. Phys.*, **65**, 209.
37. Lowdin, P.-O. (1959). *J. Molec. Spectrosc.*, **3**, 46.
38. Linderberg, J. and Shull, H. (1960). *J. Molec. Spectrosc.*, **5**, 1.
39. Politzer, P. and Daiker, K.C. (1978). *Int. J. Quant. Chem.*, **14**, 245.
40. Levy, M. (1979). *J. Chem. Phys.*, **70**, 1573.
41. Levy, M. (1978). *J. Chem. Phys.*, **68**, 5298.
42. Sen, K.D. (1979). *J. Chem. Phys.*, **71**, 3551.
43. Politzer, P. and Levy, M. (1987). *J. Chem. Phys.* (in press).
44. Fliszar, S. (1980). *J. Am. Chem. Soc.*, **102**, 6946.
45. Wilson, Jr. E.B. (1962). *J. Chem. Phys.*, **36**, 2232.
46. Hurley, A.C. (1954). *Proc. Roy. Soc.*, **A226**, 179.
47. Hall, G.G. (1961). *Phil. Mag.*, **6**, 249.

# 4 Density Functional Calculations for Atoms and Molecules: With Implications for Solid-State Physics

R.O. JONES

*Institut für Festkörperforschung der Kernforschunganlage Jülich, D-5170 Jülich, F.R. Germany*



## 4.1 Introduction

Since the pioneering work of Slater,[1] the majority of electronic structure calculations in real solid-state materials have used a description of exchange and correlation effects in terms of the local electron density $\rho(\mathbf{r})$. In Slater's original work, for example, the exchange potential was approximated by the expression (atomic units)

$$v_{\mathrm{x}}^{\mathrm{Sl}} = -C[\rho(\mathbf{r})]^{1/3}, \tag{4.1}$$

where $C = 3(3/4\pi)^{1/3}$. A potential of this form follows from simple dimensional arguments related to the exchange hole[1] and does *not* depend on assumptions concerning the exchange energy in a homogeneous electron gas. The use of a local potential of this form has distinct advantages numerically over the non-local Hartree–Fock (HF) exchange potential, and it has found widespread use in calculations of the electronic structure of extended systems.

The formal justification for using an effective potential with a local

Single-Particle Density in Physics and Chemistry ISBN 0-12-470518-9

dependence on the electron density was provided by the original papers on the density functional formalism. Hohenberg and Kohn[2] showed that the ground-state properties of a system of interacting electrons in an external field $v_{ext}(\mathbf{r})$ are determined by the density alone, i.e. they are "functionals" of the density. The ground-state energy $E$ is such a quantity, and Hohenberg and Kohn showed that there is a functional $E[\rho(\mathbf{r})]$ which is minimized by the exact ground-state density. In the following year, Kohn and Sham[3] suggested a very convenient separation of this functional:

$$E[\rho]=T_0[\rho]+\int d\mathbf{r}\,\rho(\mathbf{r})[v_{ext}(\mathbf{r})+\tfrac{1}{2}\Phi(\mathbf{r})+E_{xc}[\rho]. \tag{4.2}$$

Here $T_0[\rho]$ is the kinetic energy that a system having density $\rho$ would have if there were no electron–electron interactions, $\Phi(\mathbf{r})$ is the Coulomb potential, and $E_{xc}$ defines the exchange-correlation energy. The variational principle of Hohenberg and Kohn then yields

$$\frac{\delta E[\rho]}{\delta\rho(\mathbf{r})}=\frac{\delta T_0[\rho]}{\delta\rho(\mathbf{r})}+v_{ext}(\mathbf{r})+\Phi(\mathbf{r})+\frac{\delta E_{xc}[\rho]}{\delta\rho(\mathbf{r})}=\mu, \tag{4.3}$$

where $\mu$ is the Lagrange multiplier associated with the requirement of constant particle number.

The corresponding equation for a system with the same density distribution and an effective potential $V(\mathbf{r})$ but *without* electron–electron interactions is

$$\frac{\delta E[\rho]}{\delta\rho(\mathbf{r})}=\frac{\delta T_0[n]}{\delta\rho(\mathbf{r})}+V(\mathbf{r})=\mu. \tag{4.4}$$

As a result, if

$$V(\mathbf{r})=V_{ext}+\Phi(\mathbf{r})+\frac{\delta E_{xc}[\rho]}{\delta\rho(\mathbf{r})}, \tag{4.5}$$

the energy minimum for the interacting system can be found by solving the Schrödinger equation for non-interacting particles:

$$[-\tfrac{1}{2}\nabla^2+V(\mathbf{r})]\psi_i(\mathbf{r}_i)=\varepsilon_i\psi_i(\mathbf{r}_i), \tag{4.6}$$

yielding

$$n(\mathbf{r})=\sum_{i=1}^{N}|\psi_i|^2. \tag{4.7}$$

It is necessary to satisfy the condition (4.5), and this can be achieved in a self-consistent manner. The eigenvalues $\varepsilon_i$ and the eigenfunctions $\psi_i$ of eqn (4.6) apply to the fictitious system of non-interacting particles, and do not have immediate relevance for the interacting system.

The density functional formalism was a very fortunate development for band structure theorists, since it justified rigorously one of their main activities, the self-consistent solution of single-particle Hartree-like equations with a local exchange-correlation potential, in this case $v_{xc} \equiv \delta E_{xc}[\rho]/\delta\rho(\mathbf{r})$. Apart from the problems in interpreting the $\varepsilon_i$ as excitation energies, the "only" fly in the ointment is that $E_{xc}$ and $v_{xc}$ are known exactly only if one knows the exact wave function of the interacting system, and approximations for these quantities are currently unavoidable. For this purpose, it has been common to use an expression of the form

$$\begin{aligned} E_x^{X\alpha} &= \tfrac{3}{2}\alpha \int d\mathbf{r}\, \rho(\mathbf{r})\varepsilon_x[\rho(\mathbf{r})] \\ &= -\tfrac{3}{2}\alpha C \int d\mathbf{r}\, \{[\rho(\mathbf{r})]^{4/3}\}, \end{aligned} \tag{4.8}$$

where $\varepsilon_x(\rho)$ is the exchange energy per particle of a homogeneous electron gas of density $\rho$. The parameter $\alpha$ has historical origins,[4] but it is common to use values close to that corresponding to the exchange energy ($\alpha = \frac{2}{3}$), and we use "X$\alpha$" here to mean exchange-only local density calculations. Another common approximation uses results derived from the exchange-correlation energy for a homogeneous electron system;

$$E_{xc}^{LD} = \int d\mathbf{r}\, \rho(\mathbf{r})\varepsilon_{xc}[n(\mathbf{r})], \tag{4.9}$$

where $\varepsilon_{xc}(\rho)$ is the exchange and correlation energy per electron of a uniform gas of density $\rho$. For systems with a net spin, the local density (LD) approximations (4.8) and (4.9) may be generalized[5] to the case of a spin-polarized electron gas, with spin-up and spin-down densities $\rho_\uparrow$ and $\rho_\downarrow$, respectively. In the case of the electron gas parametrization (4.9), this refinement is known as the local spin density (LSD) approximation, and it forms the basis of many of the results discussed below.

Although these approximations are very simple, their application to extended systems has resulted in remarkable successes.[6] Internuclear geometries are given reliably in a wide range of contexts and, although absolute values of cohesive energies are generally overestimated, the observed trends in the periodic table are reproduced very well. In the context of surfaces, the minimization of the total energy as a function of interlayer spacing leads to results for the relaxation of the W(001) surface[7] which represent a challenge to the experimentalist. There is little doubt that local density approximations will find continuing and increasing application to a wide range of problems in solid-state physics.

In spite of these successes (not infrequently overstated in the solid-state literature) local spin density calculations are *approximate*, irrespective of the method used to solve eqn (4.6) or the numerical accuracy attainable. As with any approximate theory, it is essential to study its reliability and the possible errors that arise from its use. For this purpose, atoms and small molecules are ideal test cases. Experimental data are usually more extensive and more precise than in solids, and in a number of cases it is possible to compare the results of density functional calculations with essentially exact solutions of the Schrödinger equation of the interacting many-particle system.

We discuss here the insight into the local density descriptions of exchange and correlation which can be found by comparing the results of density functional calculations for atoms and molecules with both exact energies and the results of other calculations. The emphasis throughout is on total energy differences, such as ionization, excitation and bonding energies. In Section 4.2, we discuss the application of local density calculations to atoms and ions and, in particular, to exchange energy differences. We show that many of the known defects in such calculations can be attributed to errors in describing the exchange energy. In Section 4.3, we discuss LD results for selected small molecules, where some of the larger observed errors can also be traced to the same origin. Consequences for solid-state applications and concluding remarks are given in Section 4.4.

## 4.2 Density functional description of atoms

### 4.2.1 Total energies and energy differences

The focus of the density functional formalism as outlined above is the total energy, and it is natural to ask how accurately local density approximations reproduce this quantity. In atoms, a knowledge of all the ionization energies is sufficient for an estimate of the total (electronic) energy, and this can be carried out for light atoms. A comparison with the LSD and Hartree–Fock results[8] shows that the latter are not only substantially better, but that the deviations from experiment in *both* are much greater than acceptable errors in, for example, binding or cohesive energies. The usefulness of either method for calculations of these quantities then depends crucially on the cancellation of errors.

With the help of some examples we shall now show that the cancellation of errors can be very significant. Ionization energies can be determined from the energy differences between atomic and ionic ground states, and the LSD approximation gives significantly better agreement with experiment than Hartree–Fock values.[8] This is particularly true in cases involving half-filled shells, an example being the break between N and O (or between P and Si) in an otherwise smooth trend. In N, the p-electron removed has its spin parallel to the others ($p\uparrow\uparrow\uparrow \rightarrow p\uparrow\uparrow$), whereas in O it is antiparallel (p↑↑↑↓→p↑↑↑) and more weakly bound. The HF approximation, which neglects correlations between antiparallel spins, does not describe such energy differences adequately.

A related situation is found in the energy differences between the ground state and the first-excited configuration in the carbon atom [$^3$P: 2s(↑↓)2p(↑↑)→$^5$S:2s(↑)2p(↑↑↑)]. These are the lowest-lying states with these symmetries, and a straightforward extension of the above arguments allows one to perform calculations for such states.[9] In this case, an electron is transferred from 2s(↓), where there is no exchange interaction with the other valence electrons, to 2p(↑), where there is an exchange interaction with all of them. The experimental energy difference (4.18 eV)[10] is much greater than the HF value (2.43 eV).[11] The sp-transfer energy using the electron gas parametrization of Vosko *et al.*[12] is 4.10 eV. This is a further example where the local density approximation gives *better* values for energy differences than HF, although the separate values of the total energy are less accurate. This is an important consideration, since it is energy *differences* that are experimentally accessible.

Although the sp-transfer energy is described satisfactorily by LSD calculations for C, this is not true in all cases. In Fig. 4.1, we compare experimental and LSD values for first-row atoms and ions, and also show a comparison of HF (exact exchange) and X$\alpha$ (local density exchange) values.[13] For each configuration, we consider the lowest-lying term and perform density functional calculations for states with total angular and spin quantum numbers corresponding to single determinants. Two features of Fig. 4.1 are striking: (i) the local density approximations reproduce energy differences up to N(O$^+$) very well, i.e. the LD descriptions of changes in both exchange and correlation energies are satisfactory; (ii) both comparisons are extremely poor in O–F (F$^+$–Ne$^+$), i.e. LD calculations describe *neither* of these changes adequately. Similar calculations for the sd-transfer energy in the iron-series transition atoms show a more regular behaviour. In all these atoms, the LSD calculations overestimate the stability of states with greater d-occupancy by approximately 1 eV, with a similar but smaller error in the ions.[13]

These examples show that the use of local density approximations can lead to errors in energy differences that are strongly state-dependent and occasionally unacceptably large. Since sp- and sd-transfer often accompany bonding in molecules and solids, it is important to understand the origin of these errors. In the years since the errors in such transfer energies were first identified, there have been several suggestions to account for them. It is only recently, however, that Gunnarsson and Jones[13] demonstrated that errors in the local density description of exchange energies could explain many of the known deficiencies.

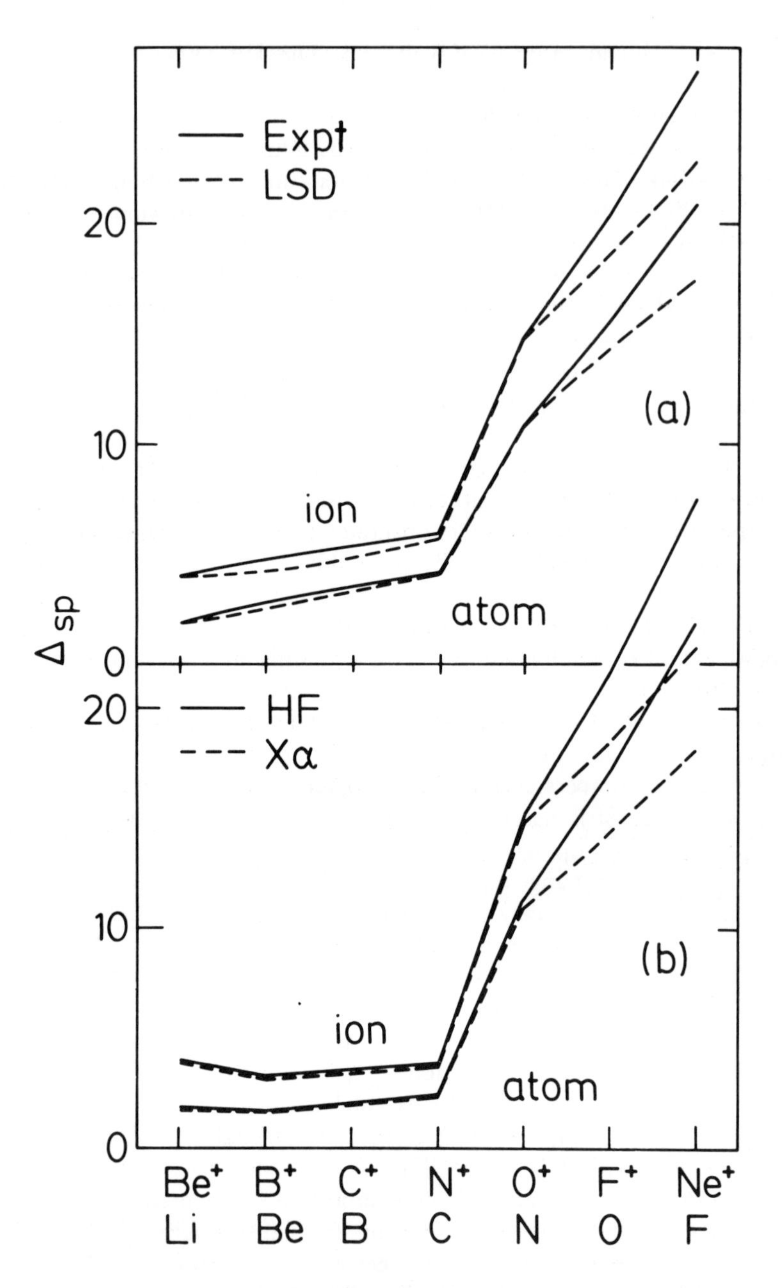


**Fig. 4.1** Transfer energies $\Delta_{sp}$ in first-row atoms and ions: (a) experimental and LSD values; (b) Hartree–Fock and X$\alpha$ values. Energies in eV. (Ref. 13.)

### 4.2.2 Local density description of exchange energies

The exchange energy is numerically larger than the correlation energy in atoms, and it is natural to examine the local density description of exchange in some detail. Since there is an explicit expression for the exact (Hartree–Fock) exchange energy, it should be possible to determine the reason for discrepancies between HF and Xα energy differences, and to gain an understanding of the departures of LSD results from experiment.

If $\Phi_i(\mathbf{r})$ and $\Phi_j(\mathbf{r})$ are Hartree–Fock orbitals, the exchange energy can be expressed in terms of exchange integrals:[14]

$$T_{ij}=e^2\int \mathrm{d}\mathbf{r}\int \mathrm{d}\mathbf{r}'\,\frac{\Phi_i^*(\mathbf{r})\Phi_j(\mathbf{r})\Phi_i(\mathbf{r}')\Phi_j^*(\mathbf{r}')}{|\mathbf{r}-\mathbf{r}'|}. \tag{4.10}$$

These integrals depend strongly on the nodal structure of $\Phi_i$ and $\Phi_j$. If, for example, the orbitals have different $l$ and $m$ quantum numbers, the integrand oscillates, and $I_{ij}$ is reduced. Such details are outside the scope of approximations (4.8) and (4.9), so they cannot describe the effects of different nodal structures reliably. The transfer of electrons between subshells, as in the case of the sp-transfer in C discussed above, involves changes in the angular nature of the orbitals, and provides a sensitive test of local density approximations.

We have noted that sp-transfer from the ground state of the fluorine atom,

$$1s^2 2s{\uparrow} 2p{\uparrow}^3 2s{\downarrow} 2p{\downarrow}^2 \rightarrow 1s^2 2s{\uparrow} 2p{\uparrow}^3 2p{\downarrow}^3, \tag{4.11}$$

is poorly described by LD approximations. In this case, the change in the exchange interaction between the valence electrons when a s↓-electron is transferred to a p↓-orbital is[14]

$$\Delta I=-\tfrac{9}{25}G^2(2p,2p)+\tfrac{2}{3}G^1(2p,2s), \tag{4.12}$$

where the Slater integrals $G^k$ are defined by

$$G^k(i,j)=2\int_0^\infty \mathrm{d}r\, r^2\int_0^\infty \mathrm{d}r'\, r'^2\,\frac{r_<^k}{r_>^{k+1}}\,\phi_i(r)\phi_j(r)\phi_i(r')\phi_j(r'). \tag{4.13}$$

Here $r_>$ ($r_<$) is the larger (smaller) of $r$ and $r'$, and $\phi_i(r)$ is the radial part of $\Phi_i(r)$. Inserting appropriate values for the Slater integrals leads to $\Delta I \sim 6\,\mathrm{eV}$. On the other hand, the radial structures of the 2s- and 2p-orbitals are very similar.[13] If they are assumed to be *identical* and we neglect the small non-spherical corrections to the total energy, the Xα estimate of the exchange energy is unchanged by sp-transfer. It is therefore not surprising that the Xα prediction of the transition (4.11) differs from the HF result by about 6 eV.[11] The LSD and Xα approximations give very similar results for the total sp-transfer energy. The deviation between the LSD result and experiment is only 2.6 eV, since the large error in the exchange energy difference is compensated by a change in the correlation energy of opposite sign.

A similar situation occurs in the sd-transfer energy in the manganese atom:

$$(core\ 3d\uparrow^5 4s\uparrow 4s\downarrow) \rightarrow (core\ 3d\uparrow^5 3d\downarrow 4s\uparrow). \tag{4.14}$$

Although the electrons involved in the transfer ($4s\downarrow \rightarrow 3d\downarrow$) are both spin-down and do not have an exchange interaction with the other valence electrons, the HF and X$\alpha$ results differ by 1.9 eV, a discrepancy which cannot be attributed to the self-interaction corrections.[13] There is, however, an important additional effect, the exchange interaction between the valence electrons and the $1s^2 2s^2 2p^6 3s^2 3p^6$ core. For a single 3d electron interacting with the core, the HF approximation gives a value (12.4 eV) which is substantially less than the X$\alpha$ value (14.1 eV).[13] Since the 4s-orbital is more extended than the 3d, the exchange interaction between the 4s-electron and the core is substantially weaker, being 1.8 and 1.3 eV in the X$\alpha$ and HF approximations, respectively. Most of the discrepancy between the X$\alpha$ and HF estimates of $\Delta_{sd}$ in Mn can then be attributed to 3d-core exchange.

These examples are particularly simple, but they illustrate the general feature that errors in $\Delta_{sp}$ and $\Delta_{sd}$ arise largely from an incorrect description of exchange energy differences. These can be particularly large if there is a change in the *angular* nodal structure of the atomic orbitals. To investigate this point, Gunnarsson and Jones[13] performed a model calculation in which the angular dependence of the atomic orbitals was isolated by assuming that the s, p and d atomic radial functions were identical. They found that the accuracy of the interelectronic exchange energy depends sensitively on the way in which the orbitals were occupied. For occupancies in the order $s \rightarrow p \rightarrow d$, the local density result agreed well with the HF value, but the results were considerably less reliable if the s- or sp-shells were empty. This is perhaps to be expected, since the local density exchange approximation is obtained from the *ground* state of the homogeneous electron gas, and good results cannot necessarily be expected for excited states.

The detailed comparison between local density and exact results for atoms and ions identified a class of systems for which the LD results should not be expected to reproduce experimental energy differences, namely when there is a change in the angular nodal structure of the one-electron orbitals. With hindsight it may well appear obvious that a local density approximation should not be able to reproduce the details of differences in exchange energies, which are intrinsically non-local.

### 4.2.3 Exchange-correlation energies and potentials—comparison of LSD and exact results

In systems where the *exact* wave function is known, it is possible to construct not only the density, but also the correct forms of $E_{xc}$ and $v_{xc}$. Such calculations then provide an additional perspective on the local density and Hartree–Fock approaches to electron structure calculations. Wave functions of sufficient precision are known only for a few light atoms, but Almbladh, von Barth and Pedroza[15–17] have performed an interesting series of calculations on these systems. A comparison of the exact HF and LD densities, for example, showed that LD results are slightly better than HF for 2s functions, and only slightly inferior for the 1s-shell.[15] The resulting error in the Hartree potential is then rather small.

The exchange-correlation potential $v_{xc}$ can be found from the single-particle equation (4.6). In the case of the He atom, with a single doubly occupied orbital, a straightforward inversion gives

$$v_{xc}=\tfrac{1}{2}[\nabla^2\phi(\mathbf{r})]/\phi(\mathbf{r})-v_{ext}-\Phi(\mathbf{r})-\varepsilon_i, \qquad (4.15)$$

where $\phi(\mathbf{r})=[\tfrac{1}{2}\rho(\mathbf{r})]^{1/2}$. In atoms with more than two electrons, $v_{xc}(\mathbf{r})$ may be parametrized and varied until the resulting density agrees with the exact distribution. A similar comparison can be made between the local density exchange-only potential and the Hartree–Fock value. Almbladh and Pedroza[16] found that, although $E_{xc}[\rho]$ is described rather accurately by LD approximations (the errors are $\sim 10\%$), $v_{xc}$ is substantially in error. This is illustrated in Fig. 4.2, where the LD and exact $v_{xc}$ are shown for He, $Li^+$ and $Be^{++}$. Although the error in $v_{xc}$ results in a large error in the eigenvalue, a well-known feature in atoms, we see that the deviation of $v_{xc}$ from the exact value is largely independent of the distance from the nucleus, and results in small errors in the LD density profiles. As numerically accurate densities become available for more systems, similar comparisons would be very valuable in constructing improved non-local energy functionals.

## 4.3 Density functional description of small molecules

There have been numerous density functional calculations for small molecules and, in the spirit of the title of this chapter, I shall survey only representative cases. As in the case of solid-state applications, the geometries corresponding to the energy minima of most molecular states are reproduced remarkably well, as are vibration frequencies. Energy *differences*, on the other hand, are reproduced with variable quality. In this

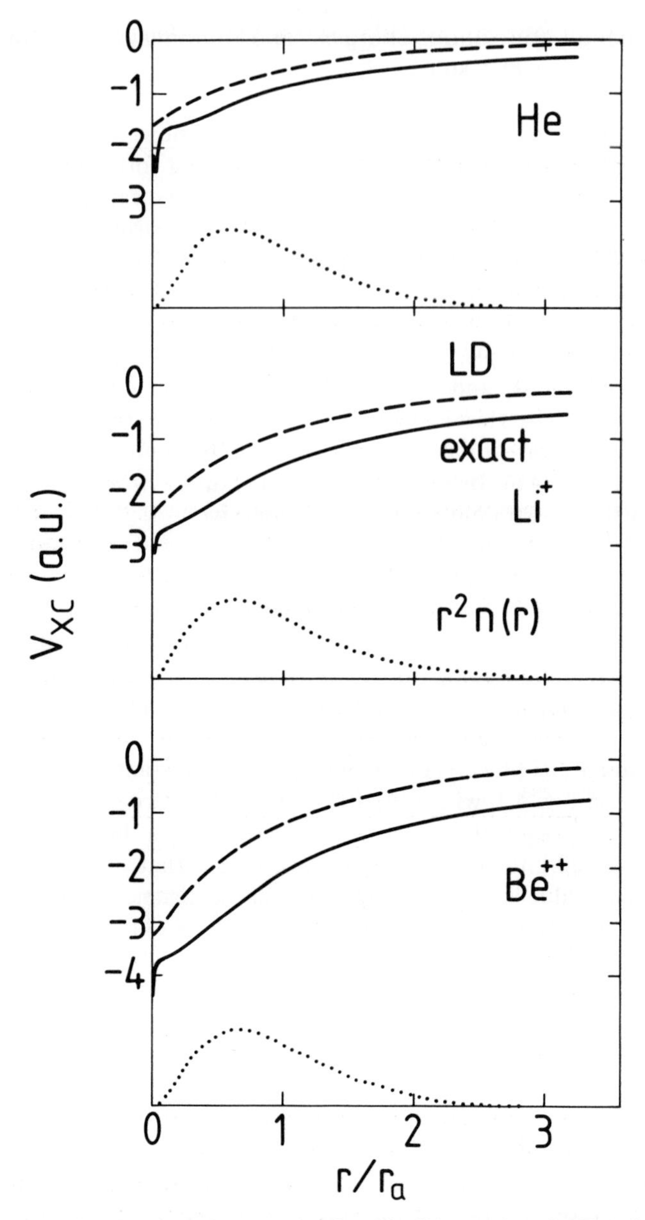


**Fig. 4.2** Exchange-correlation potential $v_{xc}$ (solid curve) and LD value (broken curve) for He, $Li^+$ and $Be^{++}$. The lengths are in units of atomic radii, $r_a$ (0.929, 0.573 and 0.414 a.u., respectively). The radial density is also shown (dotted). (After ref. 16.)

section, we shall attempt to understand the systematics behind these variations.

### 4.3.1 Exchange energies in molecules containing first-row atoms

Diatomic molecules containing atoms with atomic numbers $Z < 10$ have provided popular tests of methods of electronic structure calculations in molecules. This is also the case with local density calculations, and it is encouraging that the most recent calculations using different basis sets and different numerical methods lead to very similar results.[18-21] The equilibrium separations calculated using the LSD and $X\alpha$ approximations are generally in good agreement with experiment, with an overestimate of 1–2% being common. There is a corresponding underestimate in the ground-state vibration frequencies. In cases where it leads to an energy minimum, the HF approximation usually leads to a small underestimate of $r_e$. We note here that the dipole moment of CO and its variation with internuclear separation are given significantly better by the LSD approximation than by Hartree–Fock calculation.[22]

**Table 4.1** Experimental and calculated well-depths for the experimental ground states of first-row dimers. HF calculations for $Be_2$ give a purely repulsive energy curve.

| | *Experiment* (*ref.* 23) | *LSD* (*ref.* 21) | $X\alpha$ (*ref.* 21) | *HF* (*ref.* 24) |
|---|---|---|---|---|
| $H_2$ | 4.75 | 4.91 | 3.59 | 3.64 |
| $Li_2$ | 1.07 | 1.01 | 0.21 | 0.17 |
| $Be_2$ | 0.10 | 0.50 | 0.43 | — |
| $B_2$ | 3.09 | 3.93 | 3.79 | 0.89 |
| $C_2$ | 6.32 | 7.19 | 6.00 | 0.79 |
| $N_2$ | 9.91 | 11.34 | 9.09 | 5.20 |
| $O_2$ | 5.22 | 7.54 | 7.01 | 1.28 |
| $F_2$ | 1.66 | 3.32 | 3.04 | −1.37 |

In Table 4.1, we compare measured well-depths[23] for first-row dimers with values calculated using Hartree–Fock,[24] LSD and $X\alpha$[21] approximations. The HF approximation results in substantial underestimates of the binding energies, particularly for singlet ground states. The LSD values generally *overestimate* the stability of these molecules, although the deviations from experiment are small for $H_2$ and $Li_2$. The LSD[21] and $X\alpha$ approximations give similar energy differences in cases where the change in spin density on bonding is small.

Although the LSD approximation results in binding energy overestimates, closer examination shows some important differences. As noted above, local density exchange-only calculations reproduce HF binding energies well in the ground ($^1\Sigma_g^+$) states of $H_2$ and $Li_2$, and there is also good agreement between LSD and experimental results. This indicates that the X$\alpha$ approximation gives an adequate description of the change in self-interaction energy on the formation of a $\sigma_g(\uparrow\downarrow)$ bond, and the LSD approximation accounts for the large change in correlation energy on bonding. The situation is different in the excited states of these molecules, where there is an exchange interaction between the valence electrons. For an internuclear separation corresponding to the ground-state equilibrium, a comparison of X$\alpha$ and HF calculations shows much greater discrepancies in states with single nodal planes ($^3\Sigma_u^+$, $^3\Pi_u$ and $^3\Delta_u$) than in those with two ($^3\Pi_g$ and $^3\Sigma_g^-$).[13]

As in the case of atomic energy differences, it is instructive to examine the angular nature of the molecular orbitals and the effect on exchange energy differences. The change in the angular nodal structure on dimer formation can be discussed by correlating the atomic orbitals of the constituent atoms with those for zero separation, e.g. in the case of $O_2$ (atomic number $Z=8$) with those of the sulphur atom ($Z=16$). The $\sigma_g$-orbitals, for example, go over smoothly into s-orbitals, the $\sigma_u$ and $\pi_u$ into p, and $\pi_g$-orbitals into orbitals with d-symmetry. In the cases where these functions have nodal planes not present in the constituent atoms, we may anticipate overestimates in the calculated exchange energies. It is important to note, however, that the *radial* parts of the molecular orbitals for separations near equilibrium may be quite different from those at large separations, so a quantitative correlation between the errors in binding energies and the errors in transfer energies in the constituent atoms is difficult.

It is also striking that LSD calculations overestimate the binding energies in the second half of the first-row dimers by larger amounts than in the first half. C and O atoms, for example, have the same ground state ($^3$P) and differ from each other only in replacing electrons in the p-shell by holes. The situation in the $^3\Sigma_g^-$ states of $C_2$ ($\pi_u(\uparrow\uparrow)\sigma_g(\uparrow\downarrow)$) and $O_2$ ($\sigma_g(\uparrow\downarrow)\pi_u(\uparrow\uparrow\downarrow\downarrow)\pi_g(\uparrow\uparrow)$) is analogous, and yet the binding energy overestimate in the former ($\sim 1$ eV)[21] is much less than in the latter (over 2 eV). A similar situation occurs in the corresponding trimers, where the overestimate in the dissociation energy in $O_3$ ($\sim 2$ eV)[25] is much greater than in $C_3$.[26]

In Fig. 4.3, we show the orbitals of the ground states of $O_2$ and $O_3$,[27] together with the occupation numbers. The $C_3$ bond is linear, but the orbitals are qualitatively the same and the occupation numbers for $C_2$ and $C_3$ are also shown. The formations of the dimer and trimer bonds show clear

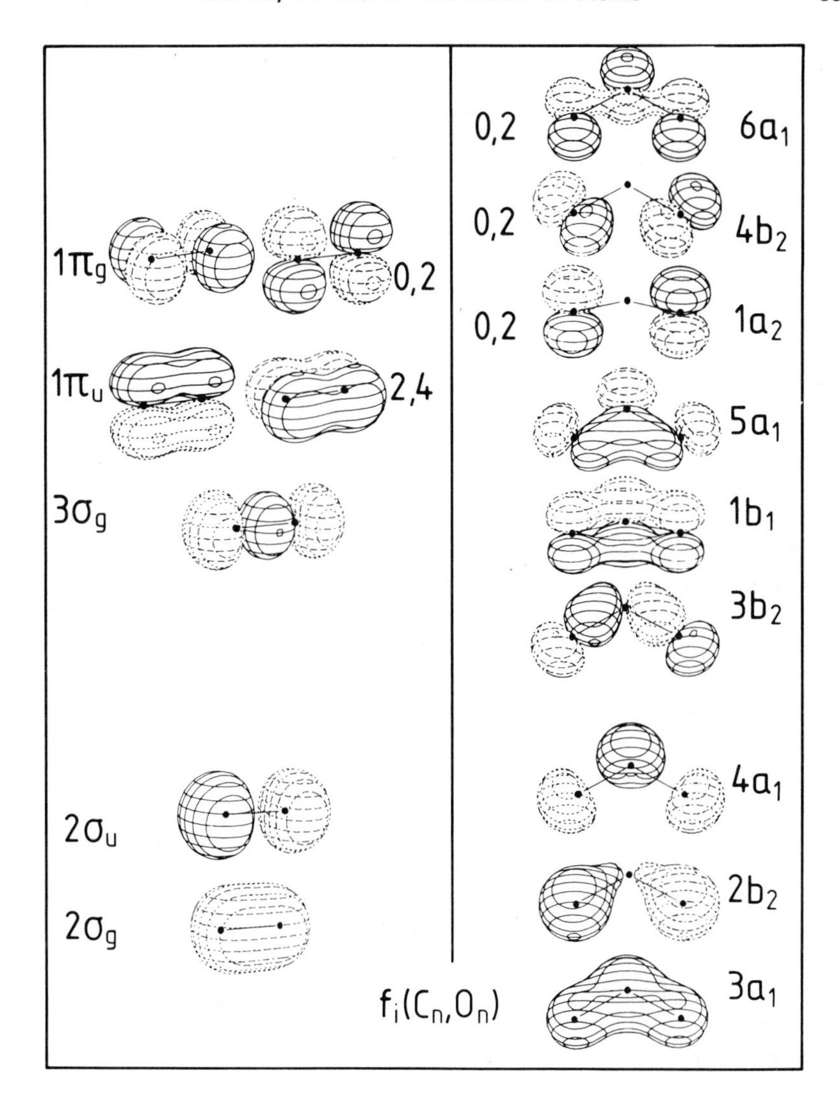


**Fig. 4.3** Molecular orbitals of $O_2$ (left) and $O_3$ (after ref. 27), with occupation numbers $f_i$ corresponding to $(C_2, O_2)$ and $(C_3, O_3)$. The other orbitals are doubly occupied (ref. 13).

parallels. In both the X–X and $X_2$–X bonds, the 2s-orbitals give rise to one bonding and one anti-bonding orbital, in the latter case with an additional non-bonding ($2b_2$) orbital. In both cases, an additional nodal plane is created, and our experience from the bonds in $Li_2$ suggests that the local density approximation will overestimate this contribution to the bond.

In $O_2$ and $O_3$, the 2p-derived shells are more than half-occupied and the occupancy of the $\pi_g$ and the $1a_2$-orbitals provide examples of occupancies of states with *additional* nodal planes. In view of the above discussion, it should not be surprising that the overestimate of the well-depth is larger in $O_3$ than in $C_3$. A further observation consistent with this picture comes from the low-lying states of $C_3$, which have linear geometries and for which the vertical excitation energies are known. The transition from the ground state to the (doubly degenerate) $^3\Pi_u$ and $^1\Pi_u$ states corresponds to the transfer of a $3\sigma_u$-electron to the $1\pi_g$-orbitals, which have an additional nodal plane. The LSD calculations[25] for the $^3\Pi_u$ and $^1\Pi_u$ excitation energies (1.8 and 2.6 eV, respectively) underestimate the measured values (2.10 and 3.06 eV).[13]

### 4.3.2 Correlation energies

The focus of our discussion of both atoms and small molecules has been the exchange energy. The difference between HF energies and experiment, the correlation energy $E_c$, has also been studied extensively by Gunnarsson and Jones,[13] particularly for atoms and ions. The results showed that the LSD approximation often provides a poor approximation of correlation energy differences where these are large, but the calculated energy differences are nevertheless closer to experiment than HF results. In general, the errors that result from a local density description of exchange tend to be reduced when correlation is included. A quantitative comparison shows, however, that the deviations are not negligible (up to $\sim 1$ eV) and the cancellation between the errors in the exchange and correlation contributions is neither complete nor systematic.

## 4.4 Discussion and concluding remarks

The above examples have been restricted in scope, but they illustrate general features of calculated energy differences. In cases where the atomic or molecular orbitals are occupied with the minimum number of nodal planes consistent with the exclusion principle, the trends in the interelectronic exchange energy are reproduced well by the local spin density approximation. In the systems discussed above, the absolute value is somewhat overestimated. On the other hand, energy transfer from such a state to one with *additional* nodal planes is often underestimated substantially by the LSD approximation. This is plausible, since the presence of nodal planes almost always results in cancellations in the integrand of the exchange integrals (eqn (4.10)).

In discussing extended systems, it is important to note that the Hartree–Fock approximation leads to unphysical results for properties related to the energy-eigenvalue spectrum, e.g. vanishing densities of states at the Fermi level, and substantial overestimates of bandwidths in metals and band gaps in semiconductors and insulators. In these cases, correlation effects are crucial and there are generally large differences between Hartree–Fock and local density results. In some cases, however, the above trends should be useful in estimating the errors to be expected. The properties of atomic clusters is a field of growing importance,[28] and cluster models are also useful in studying the chemisorption of single atoms on a surface or the presence of an isolated defect in an otherwise perfect system.[29] For such localized systems, we may anticipate errors similar to those in molecules. In the case of defects, it is sometimes of interest to study excitation energies of individual atoms. In such cases, one can expect local density approximations to result in errors comparable to those in excitation energies in atoms.

The results for energy differences in the iron-series atoms and ions have interesting and perhaps unexpected consequences for bulk systems. As noted above, the calculations of Gunnarsson and Jones[13] showed that the largest contribution to the errors in $\Delta_{sd}$ arose from the inability of local density approximations to describe energy differences arising from core–valence exchange, particularly the core–3d exchange energy. This is consistent with the smooth variation in the error with increasing atomic number, which is quite different from the situation for $\Delta_{sp}$ in first-row atoms and ions (Fig. 4.1). Changes in d-band occupancy, as occur in cohesion or in alloy formation, are likely to lead to errors in calculated formation energies. The error in core–valence exchange energies also suggests that calculations which replace the core by an effective potential (pseudopotential) will have difficulty in reproducing energy differences found in all-electron calculations.

I have made no attempt here to survey all the molecular calculations carried out with local density methods, and I have focused on atoms and small molecules where reliable data and calculations using other methods are available. The comparisons indicate that there are classes of problems for which local density approximations lead to predictions which must be treated with caution. In some cases, such as the sp-transfer in O and F or the sd-transfer in Mn, these arise in spite of there being small changes in both density and spin-density. Such problems provide ideal tests of improved non-local modifications to the LSD approximation. In spite of these difficulties, I believe that density functional methods will play an important role in molecular physics and chemistry, and I have discussed my reasons elsewhere.[30]

Density functional methods have found little resonance in the chemical community to date, and a contributory factor has been the over-optimistic assessment of the method by its protagonists, usually solid-state physicists. These methods are the basis of most parameter-free calculations in real solid-state materials, and it is perhaps understandable that solid-state physicists have tended to overstate the agreement between theory and experiment. However, the further development of density functional methods would be greatly aided by co-operation between solid-state and molecular physicists. It is important for us to identify the defects of the approximations we use, and to accept their limitations. I hope that this article will contribute to this development.

## References

1. Slater, J.C. (1951). *Phys. Rev.*, **81**, 385; **82**, 538.
2. Hohenberg, P. and Kohn, W. (1964). *Phys. Rev.*, **136**, B864.
3. Kohn, W. and Sham, L.J. (1965). *Phys. Rev.*, **140**, A1133.
4. See, for example, Slater, J.C. (1974), *Quantum Theory of Molecules and Solids*, Vol. IV, McGraw–Hill, New York.
5. Barth, U. von and Hedin, L. (1972). *J. Phys.*, **C5**, 2064.
6. For surveys of results of local density calculations, see articles in Lundqvist, S. and March, N.H. (eds) (1983). *Theory of the Inhomogeneous Electron Gas*, Plenum, New York.
7. Fu, C.L., Ohnishi, S., Wimmer, E. and Freeman, A.J. (1984). *Phys. Rev. Lett.*, **53**, 675.
8. Gunnarsson, O. and Jones, R.O. (1984). In *Local Density Approximations in Quantum Chemistry and Solid State Physics*, eds, Dahl T.P. and Avery J., Plenum, New York.
9. Gunnarsson, O. and Lundqvist, B.I. (1976). *Phys. Rev.*, **B13**, 4274.
10. Moore, C.E. (1949). *Atomic Energy Levels*, Natl Bur. Stand. (U.S.) Circ. 467, Vol. I; (1952). Vol. II; (1958). Vol. III.
11. Atomic Hartree–Fock energies are from Clementi, E. and Roetti, C. (1974). *At. Data Nucl. Data Tables*, **14**, 177 and Verhaegen, G. and Moser, C.M. (1970). *J. Phys.*, **B3**, 478.
12. Vosko, S.H., Wilk, L. and Nusair, M. (1980), *Can. J. Phys.*, **58**, 1200.
13. Jones, R.O. and Gunnarsson, O. (1985). *Phys. Rev. Lett.*, **55**, 107; Gunnarsson, O. and Jones, R.O. (1985). *Phys. Rev.* **B31**, 7588.
14. Slater, J.C. (1960). *Quantum Theory of Atomic Structure*, Vol. 2, Appendix 21, McGraw–Hill, New York.
15. Almbladh, C.-O., Ekenberg, U. and Pedroza, A.C. (1983. *Physica Scripta*, **38**, 389.
16. Almbladh, C.-O. and Pedroza, A.C. (1984), *Phys. Rev.* **A29**, 2322.
17. Almbladh, C.-O. and von Barth, U. (1985) *Phys. Rev.* **B31**, 3231.
18. Baerends, E.J. and Ros, P. (1978). *Int. J. Quant. Chem. (Symp.)*, **12**, 169.
19. Dunlap, B.I., Connolly, J.W.D. and Sabin, J.R. (1979). *J. Chem. Phys.*, **71**, 4993.
20. Jones, R.O. (1982). *J. Chem. Phys.*, **76**, 3098.
21. Painter, G.S. and Averill, F. (1982). *Phys. Rev.*, **B26**, 1781.
22. Gunnarsson, O., Harris, J. and Jones, R.O. (1977). *Phys. Rev.*, **B15**, 3027; (1977). *J. Chem. Phys.*, **67**, 3970.

23. Huber, K.P. and Herzberg, G. (1979). *Molecular Structure and Molecular Spectra IV. Constants of Diatomic Molecules*, Van Nostrand Reinhold, New York. The value for $Be_2$ is from Bondybey, V.E. and English, J.H. (1984). *J. Chem. Phys.*, **80**, 568.
24. The HF molecular energies (for the experimental equilibrium internuclear separation $r_e$) are taken from Cade, P.E. and Wahl, A.C. (1974). *At. Data Nucl. Data Tables*, **13**, 339. The atomic energies are taken from Clementi, E. and Roetti, C. (1974). *At. Data Nucl. Data Tables*, **14**, 177. The HF dissociation energies (for the HF value of the equilibrium separation) will be somewhat greater.
25. Jones, R.O. (1985). *J. Chem. Phys.*, **82**, 325.
26. Jones, R.O. (1985). *J. Chem. Phys.*, **82**, 5078.
27. Jorgensen, W.J. and Salem, L. (1973). *The Organic Chemist's Book of Orbitals*, Academic Press, New York.
28. See, for example, Martins, J.L., Buttet, J. and Car, R. (1984). *Phys. Rev. Lett.*, **53**, 655 (structural properties of Na clusters up to $Na_{13}$).
29. See, for example, Post, D. and Baerends, E.J. (1981). *Surf. Sci.*, **109**, 167 and (1982). *Ibid.*, **116**, 177 (CO adsorption on Li and Al clusters); Müller, J. and Harris, J. (1984). *Phys. Rev. Lett.*, **53**, 2493 ($H_2O$ adsorption on aluminium clusters).
30. Jones, R.O. (1987). In *Ab initio Methods in Quantum Chemistry*, Part I, ed. Lawley, K.P., Wiley, New York.

# 5 Solids: Defective and Perfect

N.H. MARCH

*Theoretical Chemistry Department, University of Oxford, 1 South Parks Road, Oxford OX13TG, England*



Single-Particle Density in Physics
and Chemistry ISBN 0-12-470518-9

## 5.1 Background

In the previous Chapter, the density functional method has been applied in detail to relatively small molecules, and also some conclusions have been drawn as to its value for solid-state physics. In this Chapter, and in the following one, the solid-state theory and applications are taken up in some detail.

The early part of the present Chapter has underlying it the Sommerfeld or jellium model of a metal. In this model, already discussed in Chapter 1, Section 1.8, the ion cores are smeared out to form a uniform background of positive charge in which the conduction electrons move, the overall system being electrically neutral. This model is already a useful starting point for simple s-p metals, Na and K being prime examples. However, in the jellium model, the electrons interact through the Coulomb potential energy $e^2/r_{ij}$ between electrons $i$ and $j$ at separation $r_{ij}$. This many-electron problem has been solved to high accuracy over a wide range of density and indeed the results have already been utilized in earlier Chapters to construct the local density approximation to the exchange-correlation potential $V_{xc}(\mathbf{r})$.

However, before summarizing the results of many-body theory applied to jellium, we first note that the Hohenberg–Kohn theorem proved in Chapter 1, Section 1.3, would merely tell us the (rather obvious) result in this case that the ground-state energy $E=E(r_s)$, where $r_s$ measures the mean interelectronic spacing through $\rho_0=3/4\pi r_s^3$, $\rho_0$ being the mean background density.

Therefore, for the jellium model to become interesting from the point of view of the single-particle density, it must be perturbed. Such a perturbation may be caused, physically, by an impurity atom, say an atom of Zn in Cu reflecting the excess positive charge $e$ on the (substitutional) Zn atom; or by a point defect, say a vacant lattice site in Al metal.

Thus, it will be convenient within the density functional framework to begin this Chapter with defective solids; in the primitive situations in which the defect can be treated as a perturbation on the Sommerfeld model.

Following this, we turn to perfect crystals where the lattice periodicity means that in non-magnetically ordered solids the total electron density $\rho(\mathbf{r})=\rho_\uparrow(\mathbf{r})+\rho_\downarrow(\mathbf{r})$ is periodic with the period of the lattice, the spin densities being such that $\rho_\uparrow(\mathbf{r})-\rho_\downarrow(\mathbf{r})$ is zero for all $\mathbf{r}$. The purpose of the discussion of periodic solids in this Chapter will be to assess in general terms the practical adequacy of density functional theory, with exchange-correlation potentials derived from the many-body theory of jellium (see also the following Chapter for fuller treatment of the many-body aspects). Instead of attempting to analyse the adequacy of such a local density approximation of say electronic band structure and cohesive energy calculations element by element, we shall present below some specific examples with characteristic properties and make some general comments, following Callaway and March.

Then, some attention will be paid to the ferromagnetic 3d metals, Fe, Ni and Co, which present, it will appear, greater difficulties for density functional theory. Some discussion of alloy properties will also be included, from however the point of view of cluster calculations.

In concluding the Chapter, we shall return to the simplest version of density functional theory, namely the Thomas–Fermi method, which remains valuable for treating matter under extreme conditions of temperature and pressure.

## 5.2 Screening of charges in metals

### 5.2.1 Test charge in a metallic medium

One of the areas in which the simplest density method, namely the Thomas–Fermi theory, and its extensions, has played a valuable role is in screening theory in simple metals. This theory makes quantitative the well-known qualitative consequence of electrostatics that long-range electric fields cannot exist in a conducting medium.

Suppose we place a test charge $ze$ at the origin in an originally uniform electron gas. Then the electrostatic potential $\chi(r)$ must tend to $ze/4\pi\varepsilon_0 r$ as $r$ tends to zero, $r$ measuring evidently the position from the test charge, and must fall off more rapidly than $1/r$ as $r$ tends to infinity. The potential energy felt by an electron is given by $-e\chi = V(\mathbf{r})$, and since $ze$ is a test charge we can use the linearized form of the Thomas–Fermi theory in eqn (1.9). Using units in which $4\pi\varepsilon_0 = 1$ for convenience, we then find, from Poisson's equation, the result

$$\nabla^2 V = q^2 V; \quad q^2 = 4k_F/\pi a_0, \tag{5.1}$$

$a_0$ being as usual the Bohr radius $h^2/4\pi^2 me^2$. Evidently the solution of eqn (5.1) with spherical symmetry appropriate to the single test charge at the origin and satisfying the boundary conditions discussed above, namely

$$V(r) \to -ze^2/r \text{ as } r \text{ tends to zero,}$$

$$V(r) \text{ tends to zero faster than } 1/r \text{ as } r \to \infty,$$

is given by

$$V(r) = \frac{-ze^2}{r} \exp(-qr). \tag{5.2}$$

Equation (5.1) and the solution (5.2) were first given by Mott[1]. It can be seen that according to the Thomas–Fermi theory, the Coulomb potential is screened out exponentially with distance $r$, with a screening length $q^{-1}$ which in a good metal is found from eqn (5.1) to be of the order of 1 Å.

While the Thomas–Fermi screening length $q$ derived above is of fundamental importance in treating charged perturbations in metals, there is one important feature which is missing from the above treatment. This omission is due to the semi-classical character of the Thomas–Fermi theory, and one really should be treating the scattering of electron waves from the test charge. Then the correct first-order treatment is given by combining the $\rho$–V relation of the fully wave-mechanical theory with Poisson's equation to obtain, following March and Murray[2] (see Appendix A 5.1),

$$\nabla^2 V = \frac{me^2}{\hbar^2}\frac{2k_F^2}{\pi^2}\int d\mathbf{r}'\frac{j_1(2k_F|\mathbf{r}-\mathbf{r}'|)}{|\mathbf{r}-\mathbf{r}'|^2}V(\mathbf{r}'), \tag{5.3}$$

which is the correct wave-mechanical generalization of the Mott equation (5.1), with $k_F$ the Fermi wavenumber.

Without solving eqn (5.3), it is easy to see that that the displaced charge can have a very different character at large $r$, depending on whether we use the wave-theory result (5.3) or the semi-classical result (5.1). Thus, consider a simple case when $V(\mathbf{r})$ is very short range, i.e. write $V(\mathbf{r})=\lambda\delta(\mathbf{r})$. Then the displaced charge according to eqn (5.1) is of similar short range to $V(\mathbf{r})$, while eqn (5.3) gives in contrast, if $\rho_0$ is as usual the unperturbed uniform density

$$\rho(\mathbf{r})-\rho_0 \propto \frac{j_1(2k_F r)}{r^2}; \quad \frac{k_F^3}{3\pi^2}=\rho_0; \quad j_1(x)=\frac{\sin x - x\cos x}{x^2}. \tag{5.4}$$

From the definition of the spherical Bessel function $j_1$ in eqn (5.4), it follows that at large $r$ the displaced charge decays as

$$\rho(r)-\rho_0 \propto \frac{\cos 2k_F r}{r^3}. \tag{5.5}$$

This is the new feature of the wave theory based on eqn (5.3) and cautions one to be careful to check that the conditions of validity of the Thomas–Fermi model are valid before using it in an application such as the present one. We shall see in Section 5.3.1 that the same theory, applied to ionized impurities in non-degenerate conditions in a semiconductor, is perfectly valid. That such "wiggles" as described by eqn (5.5) exist in the displaced charge round a given fixed perturbation in a Fermi gas was first pointed out by Blandin *et al.*[3,4]

We shall return to the above discussion below, when we treat the interaction energy between two charges in a Fermi gas. But we wish to study a little further the range of validity of the Thomas–Fermi solution in eqn (5.2). In order to do so, it proves important to introduce the dielectric function $\varepsilon(k)$, by working in Fourier transform.

*Wavenumber-dependent dielectric function*

We therefore define the Fourier components of the screened potential energy $V(\mathbf{r})$ treated above:

$$\tilde{V}(\mathbf{k})=\int d\mathbf{r}\, V(\mathbf{r}) \exp(i\mathbf{k}\cdot\mathbf{r}), \tag{5.6}$$

and if we use first the screened Coulomb potential (5.2) of the semi-classical Thomas–Fermi theory we then find

$$\tilde{V}_{\mathrm{TF}}(k)=-\frac{4\pi z e^2}{k^2+q^2}. \tag{5.7}$$

To see the limitations of this Thomas–Fermi description, let us return to the first-order wave-theory equation (5.3). This can be solved analytically in $\mathbf{k}$-space, using the properties of the Fourier transform of a convolution, to yield

$$\tilde{V}_{\text{wave theory}}=-\frac{4\pi z e^2}{k^2+(k_{\mathrm{F}}/\pi a_0)g(k|2k_{\mathrm{F}})}, \tag{5.8}$$

where the function $g$ is defined by

$$g(x)=2+\frac{x^2-1}{x}\ln\left|\frac{1-x}{1+x}\right|. \tag{5.9}$$

In the long-wavelength limit $k$ tends to zero, $g(x)$ tends to the value 4 and we find the same result

$$\tilde{V}(0)=-\frac{4\pi z e^2}{q^2} \tag{5.10}$$

from both the Thomas–Fermi approximation and from the correct first-order wave theory.

Often it proves valuable to express these results in terms of the dielectric function $\varepsilon(k)$ of the Fermi gas. This is best defined in the present context by writing

$$\tilde{V}(k)=-\frac{4\pi z e^2}{k^2\varepsilon(k)}. \tag{5.11}$$

It follows from eqn (5.7) that in the Thomas–Fermi approximation we have

$$\varepsilon_{\mathrm{TF}}(k) = \frac{k^2 + q^2}{k^2}, \tag{5.12}$$

while from the wave-theory result (5.8) we find

$$\varepsilon(k) = \frac{k^2 + (k_{\mathrm{F}}/\pi a_0) g(k/2k_{\mathrm{F}})}{k^2} \tag{5.13}$$

This latter expression is due to Lindhard;[5] it becomes equivalent to the Thomas–Fermi result (5.12) only when we replace $g(x)$ by its value $g(0) = 4$, which is easily seen by using the expression (5.9). Thus, the Thomas–Fermi approximation to the dielectric function is restricted in its range of validity to the long-wavelength regime. The Friedel wiggles displayed in eqn (5.5) arise from the kink in $\varepsilon(k)$ in eqn (5.13) at $k = 2k_{\mathrm{F}}$, the diameter of the Fermi sphere. This kink has interesting implications for electron–phonon interaction and is the origin of the Kohn anomaly[6] in the dispersion relation of lattice waves (phonons) in metals.[7]

### 5.2.2 Electrostatic model of interaction between charges in a Fermi gas

So far, we have been concerned with the screening out of the field of a single test charge. But in a number of important applications in condensed matter, we need to know the interaction energy between charges. We shall first use the semi-classical Thomas–Fermi theory below,[8] though the essential result obtained, namely the applicability of the so-called electrostatic model, is valid also in the wave theory discussed above.[9]

Let us suppose that ions 1 and 2, embedded in the bath of conduction electrons, are screened such that the potential due to ion 1 alone is $V_1$ and that due to ion 2 is $V_2$. Then in the linear approximation of eqn (5.1), it is immediately clear that the total potential $V$ of the two-centre problem is just the superposition potential given by

$$V = V_1 + V_2. \tag{5.14}$$

Hence each ion is surrounded by its own displaced charge and this is not affected by bringing up further ions. Incidentally, this is then the condition that the total potential energy of the ions can be written as a sum of pair potentials in the multi-centre case.[9]

The interaction energy between the ions, separated by a distance $R$ say, may now be obtained directly by calculating the difference between the total energy of the metal when the ions are separated by an infinite distance and when they are brought up to separation $R$. Clearly this is all to be done in the Fermi sea of constant density $\rho_0$ say and the result for the interaction energy will depend on the density $\rho_0$, or equivalently on the Fermi energy $E_F = p_F^2/2m \equiv \mu$.

Let us consider the changes in the kinetic and potential energy separately. The act of introducing an ion carrying a charge $ze$ into the Fermi gas changes the kinetic energy in the manner discussed in Appendix A 5.1. If $T_0$ denotes the total kinetic energy of the unperturbed Fermi sea, then in the Thomas–Fermi theory the result in the perturbed inhomogeneous gas, measured relative to the unperturbed value, is (cf. eqn (A 5.1.5)) given by

$$T - T_0 = E_F \int \Delta\rho \, d\mathbf{r} + \frac{E_F}{3\rho_0} \int d\mathbf{r} \, (\Delta\rho)^2 + O(\Delta\rho^3), \tag{5.15}$$

where $\Delta\rho$ denotes the displaced charge $\rho - \rho_0$. We now observe that the first term involves the normalization condition for the displaced charge, or in other words the condition that the ionic charge $ze$ is screened completely by the electron distribution. Clearly, this term will make no contribution to the energy difference between infinitely separated ions and the ion pair at separation $R$.

We are now in a position to calculate the changes in both kinetic and potential energy contributions when we bring the ions together to separation $R$ from infinity. We may write down the following contributions:[8]

(i) the interaction energy between the charge $ze$ of one "impurity" and the perturbing potential, say $V_2$, due to the other;

(ii) the interaction energy between the displaced charge $(q^2/4\pi e^2)V_1$ round the first ion and the potential $V_2$ due to the other; and

(iii) the change in kinetic energy.

These three terms are evidently given by

(i) $z^2 e^2 \exp(-qR)/R$; (5.16)

(ii) $-\dfrac{q^2}{4\pi e^2} \displaystyle\int d\mathbf{r} V_1 V_2$; and (5.17)

(iii) $$\left(-\frac{q^2}{4\pi e^2}\right)^2 \frac{E_F}{3\rho_0} \left[ \int d\mathbf{r} \{(V_1 + V_2)^2 - V_1^2 - V_2^2\} \right]$$

$$= \frac{q^2}{4\pi e^2} \int d\mathbf{r} \, V_1 V_2, \tag{5.18}$$

where (iii) follows from eqns (5.15) and (5.14). Thus it can be seen that contribution (ii) is precisely cancelled by the change in kinetic energy (iii) and we are left with the final result for the interaction energy $\Delta E$ as

$$\Delta E = z^2 e^2 \exp(-qR)/R. \tag{5.19}$$

But this is simply the electrostatic energy of an ion of charge $ze$ sitting in the electrostatic potential $(ze/R)\exp(-qR)$ of the second ion. This result, namely the electrostatic model, has been shown to hold also in the wave theory,[9] though now, of course, the correct screened potential with the Friedel wiggles must be used.

The above argument is very valuable for indicating the type of interaction energy between charged defects in a metal, as well as the nature of the ion–ion interaction in condensed metallic media. Of course we are usually dealing with ions of finite size, but if we can represent these by a bare pseudopotential $v_b(k)$ say in Fourier transform, then it can be shown that the interaction energy takes the form[10,7]

$$\Delta E(R)=\int\frac{k^2 v_b^2(k)}{\varepsilon(k)}\exp(\mathrm{i}\mathbf{k}\cdot\mathbf{R})\,\mathrm{d}\mathbf{k}. \qquad (5.20)$$

This reduces to the $r$-space form (5.19) for point ions, with $v_b(k)$ replaced by $4\pi ze^2/k^2$, if for $\varepsilon(k)$ we use the Thomas–Fermi dielectric constant. To obtain the Friedel oscillations in the long-range form of the interaction energy $\Delta E(R)$, we must use instead for $\varepsilon(k)$ the Lindhard dielectric function (5.13), or its refinements to include electron exchange and correlation (see Chapter 6; see also the summary in ref. 7).

These applications to interaction between charges caution us that while the Thomas–Fermi theory remains valuable for obtaining some important general results (for example the screening length $q^{-1}$ and the above derivation of the electrostatic model for the interaction between charges in a Fermi gas) its assumption that the screened potential varies slowly is too drastic in metals. The wave theory is therefore essential for quantitative work in this area. Nevertheless, there are two further interesting applications we wish to discuss; the problem of the vacancy formation energy in a simple close-packed metal and its relation to lattice properties; and secondly the important problem as to the way the Fermi level $E_F$ varies with concentration in a binary metallic alloy in which there is a valence difference between the components (e.g. a Cu–Zn or an Al–Mg alloy), in which the Thomas–Fermi screening theory remains valuable.

### 5.2.3 Relation of vacancy formation energy to Debye temperature in simple metals

The energy required to create a vacancy in a metal can be related to phonon properties, and in particular to the Debye temperature, through the linear screening theory discussed above. The argument goes as follows.[11] Let us consider first a potential energy $V(\mathbf{r})$ created by the vacancy screened by the charge it displaces. If we represent the vacancy in a simple metal of valency $Z$ by a negative charge $-Ze$ at the vacant site, $V(\mathbf{r})$ is given by the (now repulsive) screened Coulomb form (5.2) in the Thomas–Fermi approximation. Fortunately, in a linear theory, the change in the sum of the one-electron energies $\varepsilon$ from their free-electron values can be estimated without choosing such a detailed form of $V(\mathbf{r})$, but by merely asserting that it is sufficiently weak to allow the application of first-order perturbation theory. In the unperturbed free-electron metal the wave functions are the plane waves $\mathscr{V}^{-1/2}\exp(\mathrm{i}\mathbf{k}\cdot\mathbf{r})$, $\mathscr{V}$ being the total volume of the metal, and we can therefore write the change in the one-electron energy $\varepsilon_{\mathbf{k}}=\frac{1}{2}k^2$ as

$$\Delta\varepsilon_{\mathbf{k}}=\mathscr{V}^{-1}\int\exp(-\mathrm{i}\mathbf{k}\cdot\mathbf{r})V(\mathbf{r})\exp(\mathrm{i}\mathbf{k}\cdot\mathbf{r})\,\mathrm{d}\mathbf{r}=\mathscr{V}^{-1}\int V(\mathbf{r})\,\mathrm{d}\mathbf{r}, \qquad (5.21)$$

which is evidently independent of $\mathbf{k}$. Summing this over the $N$ electrons in the metal gives for the change $\Delta E_{\mathrm{s}}$ in the sum of the one-electron energies the result

$$\Delta E_{\mathrm{s}}=\rho_0\int V(\mathbf{r})\,\mathrm{d}\mathbf{r}, \qquad (5.22)$$

where $\rho_0$ as usual is the unperturbed electron density. But now, in linear theory, it is easy to show that this integral is determined merely by the requirement that the excess charge is perfectly screened, the result for the change in the one-electron energy sum being (from eqns (5.2) and (5.6))

$$\Delta E_{\mathrm{s}}=\tfrac{2}{3}ZE_{\mathrm{F}}, \qquad (5.23)$$

where $E_{\mathrm{F}}$ is the Fermi energy. If we neglect ionic relaxation, there is in the process of taking an atom from the bulk and placing it on the surface a reduction in kinetic energy of the Fermi electrons, owing to the increase of one atomic volume, which is readily calculated from eqn (1.12) to be[12]

$$\Delta T=\tfrac{2}{5}ZE_{\mathrm{F}}. \qquad (5.24)$$

This admittedly elementary theory of the vacancy formation energy $E_{\mathrm{v}}$ yields

$$E_{\mathrm{v}}\doteqdot\Delta E_{\mathrm{s}}-\Delta T=\tfrac{4}{15}ZE_{\mathrm{F}}. \qquad (5.25)$$

Though such a theory appears to be significant in the sense of extrapolating measured vacancy formation energies for $Z\geqslant1$ to the limit as $Z$ tends to zero, it turns out that $ZE_{\mathrm{F}}$ is not the appropriate unit in which to measure $E_{\mathrm{v}}$. That this is so was implicit in the work of Mukherjee,[13] who pointed out an empirical relation between $E_{\mathrm{v}}$ and the Debye temperature $\theta$; his results for some close-packed metals are summarized in Table 5.1.

**Table 5.1** Debye temperature $\theta$ related to vacancy formation energy $E_v$ in some close-packed metals (cf. eqn (5.33).

| *Metal* | *Debye temperature* (K) | *Vacancy formation energy* $E_v$ (eV) | $\theta/(E_v/M\Omega^{2/3})^{1/2}$ |
|---|---|---|---|
| Cu | 245 | 1.17 | 32 |
| Ag | 225 | 1.10 | 32 |
| Au | 165 | 0.95 | 34 |
| Mg | 406 | 0.89 | 34 |
| Al | 428 | 0.75 | 33 |
| Pb | 94.5 | 0.5 | 33 |
| Ni | 441 | 1.5 | 33 |
| Pt | 229 | 1.4 | 37 |

To see how such a relation arises, let us use the same linear theory of screening which led to eqn (5.25) to discuss the velocity of sound in a simple metal. This can be done by first noting that if the ions in a metal were embedded in a completely uniform electron gas, then they would vibrate with the ionic plasma frequency

$$\omega_p^{\text{ion}} = \left(\frac{4\pi n_i (Ze)^2}{M}\right)^{1/2} \tag{5.26}$$

determined by the ionic number density $n_i$, the ionic charge $Ze$ and mass $M$. Rewriting $n_i$ through the electron density $\rho_0 = Zn_i$ yields immediately

$$\omega_p^{\text{ion}} = \left(\frac{4\pi\rho_0 Ze^2}{M}\right)^{1/2}. \tag{5.27}$$

But this represents an optic mode and to obtain the desired acoustic mode of the form

$$\omega = v_s k \tag{5.28}$$

at small wavenumbers $k$, $v_s$ again being the velocity of sound, one must allow the electrons to pile up round the positive ions and screen them. This can be done by using the dielectric function of Section 5.2.1 and the consequence is that the bare Coulomb potential, written in $k$-space as $4\pi Ze^2/k^2$, must be screened such that

$$\frac{4\pi Ze^2}{k^2} \to \frac{4\pi Ze^2}{k^2 + q^2} \quad \text{or as } k \to 0,\ Z \to Zk^2/q^2. \tag{5.29}$$

Making the appropriate substitution (5.29) in eqn (5.27) leads to the so-called Bohm–Staver formula[14,15] for the velocity of sound

$$v_s = \left(\frac{Zm}{3M}\right)^{1/2} v_F, \qquad (5.30)$$

where $v_F$ is the Fermi velocity of the electrons. Between eqns (5.25) and (5.30) we can eliminate the quantity $ZE_F$, and with it at least some of the dependence of the argument on linear theory, to find

$$E_v = \text{const. } M v_s^2. \qquad (5.31)$$

Using in eqn (5.31) the usual (average) relation between $v_s$ and Debye temperature, namely

$$\theta = \frac{v_s}{\Omega^{1/3}} \left(\frac{3}{4\pi}\right)^{1/3} \frac{h}{k_B}, \qquad (5.32)$$

where $\Omega$ is the atomic volume and $k_B$ is Boltzmann's constant, leads to the Mukherjee relation

$$\theta = \frac{C E_v^{1/2}}{\Omega^{1/3} M^{1/2}}. \qquad (5.33)$$

The constant $C$ given in Table 5.1 would be correctly obtained by the above argument if, for example, the number $\frac{4}{15}$ in eqn (5.25) were replaced by $\frac{1}{6}$, so that the theory is already semi-quantitative for the constant $C$.

In fact, the theory of vacancy energy in the polyvalent metals Al and Pb is a non-linear problem and there is gross cancellation of energy terms as a fully non-linear calculation using the theory of the inhomogeneous electron gas shows.[16] The relation to the phonons in eqn (5.33) appears then to be a deeper way of tackling the problem of the vacancy formation energy. What the admittedly oversimplified linear theory given above shows is that, at this level, the problem of the response of the electron gas to the removal of an atom is related to that of introducing a phonon into the ionic lattice, via the dielectric function $\varepsilon(k)$. It should be pointed out that the argument presented above should not be used for open body-centred cubic metals and in particular the alkali metals, because relaxation of the ions around the vacant site now makes the dominant contribution. Nevertheless, it is remarkable that Mukherjee's relation (5.33) is again found to hold.[17]

### 5.2.4 Variation of Fermi level with concentration in a dilute binary metallic alloy

The question arises in an alloy such as Cu–Zn, with say a finite concentration of Zn atoms in a Cu matrix, as to the way in which the Fermi level $E_F$ varies with the impurity concentration $c$. The theory for binary metallic alloys was given by Friedel[18] in the case of a valence difference $Z$ between the constituent atoms. The elementary physical argument, which we shall make quantitative below, goes as follows. Whereas, in the so-called rigid band model, we would merely fill in the additional Zn electrons into the Cu density-of-states curve, at least at dilute concentration, and therefore the Fermi level would show a linear increase with concentration, in the screening theory presented in Section 5.2, each one-electron level would be lowered in energy by the attractive Zn impurities. It turns out that filling electrons in to the resulting density of states merely preserves the Fermi level intact to first order in the concentration.

To see how this result comes quantitatively from the Thomas–Fermi theory, and in the process to generalize it away from the dilute-concentration regime, we follow Friedel[18] in assigning to each impurity its own volume, which for simplicity one takes to be a sphere of radius $R$, this being related to the concentration of impurities $c$ by $c = 3/4\pi R^3$. We next use the Thomas–Fermi equation, but being careful to allow the Fermi level $E_F$ in the unperturbed metal to shift by an amount $\Delta E_F$. Then assuming the necessary condition $|\Delta E_F - V| \ll E_F$ for linearization we can write for the screened potential energy $V(r)$ around an impurity at the origin of its "own" sphere of radius $R$ the Poisson equation generalizing eqn (5.1):

$$\nabla^2 V = q^2(V - \Delta E_F). \qquad (5.34)$$

This has now to be solved subject to the boundary conditions

$$\left.\begin{aligned} &\left(\frac{dV}{dr}\right)_R = 0, \\ &V(R) = 0, \\ &V \to -\frac{Ze^2}{r}, \quad r \to 0 \end{aligned}\right\} \qquad (5.35)$$

the first condition expressing electrical neutrality of the cell through Gauss' theorem. The independent solutions of eqn (5.34) being of the form $(Ze^2/r)\exp(\pm qr)$ it is easy to obtain the solution satisfying the boundary condition that the electric field vanish at $R$:

$$V(r, R) - \Delta E_F = -\frac{Ze^2}{r}\,\frac{qR\cosh\{q(R-r)\} - \sinh\{q(R-r)\}}{qR\cosh qR - \sinh qR}. \qquad (5.36)$$

The second boundary condition (5.35) is then used to determine the shift in the Fermi level as

$$\Delta E_F = Ze^2 q/(qR\cosh qR - \sinh qR), \qquad (5.37)$$

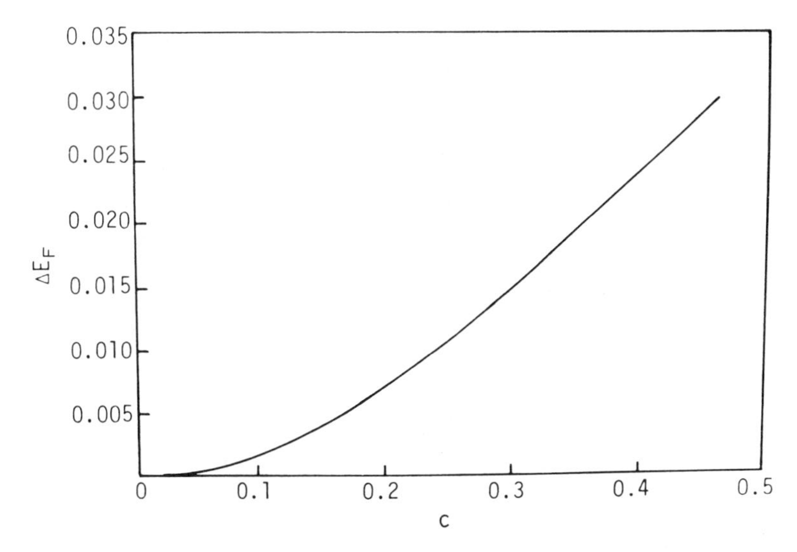


**Fig. 5.1** Shift in Fermi level of a dilute metallic alloy, such as Zn in Cu, with concentration $c$ of Zn.

and a plot of this is shown in Fig. 5.1. It will be seen that the slope at the origin $c=0$ is zero, and therefore that there is no term linear in the concentration. Experimental support for such a movement in the Fermi level is afforded by the optical experiments of Biondi and Rayne.[19]

Of course, such an argument as presented above is gross, and could only apply in simple metallic systems in at best a semi-quantitative way. Modern alloy theory, discussed in Chapter 10 of ref. 7, has of course now progressed far beyond such an approach in yielding results of quantitative accuracy. Nevertheless, the result that there is no shift in the Fermi level to $O(c)$ is an important qualitative point in alloy theory. As one example which can be cited, Bhatia and March[21] have used it in a discussion of the surface tension of liquid binary alloys, which generalizes the relation (10.9) of Chapter 10 for a pure metal to include the concentration fluctuations in a liquid binary metal alloy.

## 5.3 Screening in semiconductors

### 5.3.1 Ionized impurities in semiconductors

The theory of impurity scattering in semiconductors has a good deal of interest from a practical point of view, and can be tackled along rather similar lines to the corresponding problem in metals discussed in Section 5.2.1. However, it is evident that a number of modifications of that treatment must be made. For example, it is no longer permissible to suppose that in semiconductors we are dealing with a completely degenerate gas of free electrons. Indeed, in many cases the assumption that the electrons obey Maxwell–Boltzmann statistics is more appropriate. This latter case has been discussed by Debye and Conwell who give a formula for the mobility, derived by Herring. The generalization to arbitrary degeneracy has been considered independently by Dingle[22] and by Mansfield.[23]

In order to calculate the electric field round an ionized impurity centre, use is made of the generalized Thomas–Fermi theory, first formulated for low temperatures by Marshak and Bethe and for arbitrarily high temperatures by Sakai and independently Feynman *et al.* in the course of their work on equations of state of elements at high pressures and temperatures (see Section 5.8 below). In this generalized theory, the restriction made so far that the electron gas is completely degenerate is removed and the Fermi–Dirac distribution function must be introduced into the theory.

*Density–potential relation in generalized Thomas–Fermi theory*

The number of electrons with momenta of magnitude between $p$ and $p+\mathrm{d}p$ in a volume element $\mathrm{d}\tau$ must now be written as

$$4\pi p^2 \mathrm{d}p\, \mathrm{d}\tau \frac{2}{h^3}\left\{\exp\left[\left(\frac{p^2}{2m^*}+V\right)\Big/ k_{\mathrm{B}}T-\eta^*\right]+1\right\}^{-1}, \tag{5.38}$$

where $V$ is as usual the potential energy, $m^*$ is the effective electronic mass and $\eta^*$ is a constant, the reduced chemical potential of the electrons, which is to be determined as usual by normalization requirements. Then it follows that the electron density $\rho(r)$ is given by

$$\rho(r)=\int_0^\infty \frac{8\pi p^2}{h^3}\left\{\exp\left[\left(\frac{p^2}{2m^*}+V\right)\Big/ k_{\mathrm{B}}T-\eta^*\right]+1\right\}^{-1} \mathrm{d}p. \tag{5.39}$$

Equation (5.39) is the modified relation between density and potential in the generalized Thomas–Fermi approximation.

*Introduction of Fermi–Dirac functions*

It is convenient at this stage to introduce the Fermi–Dirac functions $I_k(\eta^*)$ defined by

$$I_k(\eta^*) = \int_0^\infty \frac{y^k \, \mathrm{d}y}{\exp(y - \eta^*) + 1}. \tag{5.40}$$

Then eqn (5.39) becomes

$$\rho(\mathbf{r}) = \frac{4\pi}{h^3} (2m^* k_B T)^{3/2} I_{1/2}\left(\eta^* - \frac{V}{k_B T}\right). \tag{5.41}$$

Next one wants to write down the self-consistent field equation to determine the screened potential $V(r)$ around an ionized impurity centre. In doing so, it is to be noted that if $\rho_0$ is the carrier density in the unperturbed lattice, then since $\rho_0$ is also the density of positive charge, Poisson's equation gives

$$\nabla^2 V = -\frac{4\pi e^2}{\varepsilon} (\rho - \rho_0), \tag{5.42}$$

$\varepsilon$ being the static dielectric constant. Assuming that $\eta^*$ is unchanged by the introduction of the impurity one has therefore

$$-\nabla^2 V = \frac{16\pi^2 e^2}{\varepsilon h^3} (2m^* k_B T)^{3/2} \left[ I_{1/2}\left(\eta^* - \frac{V}{k_B T}\right) - I_{1/2}(\eta^*) \right], \tag{5.43}$$

since $\rho_0$ is evidently given by

$$\rho_0 = \frac{4\pi}{h^3} (2m^* k_B T)^{3/2} I_{1/2}(\eta^*). \tag{5.44}$$

The boundary conditions appropriate to a single impurity centre carrying charge $e$, and taken as the origin of coordinates, are evidently, in the model of the dielectric medium adopted here:

$$\left.\begin{array}{lr} V \to -e^2/\varepsilon r & \text{as } r \to 0, \\ V \to 0 \quad \text{faster than } r^{-1} & \text{as } r \to \infty, \end{array}\right\} \tag{5.45}$$

*Linearized solution and screening length*

In order to obtain an approximate solution, Dingle[22] and Mansfield[23] assumed that

$$|\eta^* k_B T| \gg |V|;$$

eqn (5.43) can then be written precisely as in eqn (5.1), with modified screening length

$$q^2 = \frac{16\pi^2 e^2}{\varepsilon h^3} (2m^*)^{3/2} (k_B T)^{1/2} I'_{1/2}(\eta^*) \tag{5.46}$$

and explicit screened potential

$$V = -\frac{e^2}{\varepsilon r}\exp(-qr). \tag{5.47}$$

In fact the expression (5.46) for $q^2$ reduces in the case of complete degeneracy to the result in eqn (5.1) if one replaces $\varepsilon$ by unity and puts $m^* = m$. In the non-degenerate limit, on the other hand,

$$q^2_{\text{non-degenerate}} = \frac{4\pi e^2 \rho_0}{\varepsilon k_B T}, \tag{5.48}$$

and as an example to which eqn (5.48) is applicable, taking one impurity phosphorus atom for every $10^5$ Si atoms, the screening radius $q^{-1}$ is found to be about 60 Å at room temperature; i.e. almost two orders of magnitude longer than the screening length in a good metallic conductor.

Use will be made below in Section 5.8 of the generalized Thomas–Fermi theory to treat material under extreme conditions of temperature and pressure. But immediately below, related ideas will be employed (cf. also Section 5.2.1 for the case of metals) to obtain an expression for the dielectric function of a semiconductor.

### 5.3.2 Dielectric function of a semiconductor

As emphasized by Resta,[24] the wavenumber-dependent dielectric function of a semiconductor can be approximated usefully by means of the linearized Thomas–Fermi equation (cf. eqn (5.34)):

$$\nabla^2 V = q^2[V(r) - A], \quad q^2 = 4k_F/\pi a_0. \tag{5.49}$$

The boundary $R$ of the screening charge is now defined through

$$\rho(R) = \rho_0, \tag{5.50}$$

which yields for the constant $A$ in eqn (5.49)

$$A = V(R). \tag{5.51}$$

Beyond the screening radius $R$, the screened potential energy $V(r)$ of a point charge is given by

$$V(r) = -z/\varepsilon(0)r, \quad r \geqslant R, \tag{5.52}$$

where $\varepsilon(0)$ is the static dielectric constant. Inside the screening radius the potential is described by eqn (5.49) whose general solution may be written

$$V(r) = -(z/r)[\alpha \exp(qr) + \beta \exp(-qr)] + \mathrm{A}, \quad r < R. \tag{5.53}$$

By requiring continuity at $r=R$, using $rV(r)\to -z$ as $r\to 0$ and the condition (5.50), eqn (5.53) takes the form

$$V(r)=-\frac{z}{R}\frac{\sinh q(R-r)}{\sinh qR}-\frac{z}{\varepsilon(0)R}, \qquad r\leqslant R. \tag{5.54}$$

The screening radius $R$ is found by imposing continuity of the electric field at $r=R$, which yields

$$\sinh qR/qR=\varepsilon(0). \tag{5.55}$$

This equation produces a finite solution for $R$ for any $\varepsilon(0)>1$. The metallic situation, treated earlier in the Chapter, corresponds to $\varepsilon(0)=\infty$ and one then obtains $R=\infty$, in agreement with eqn (5.2).

If one introduces the spatial dielectric constant $\bar{\varepsilon}(r)$ defined by

$$V(r)=-z/\bar{\varepsilon}(r)r, \tag{5.56}$$

then one can write

$$\bar{\varepsilon}(r)=\begin{cases}\varepsilon(0)\,qR/[\sinh q(R-r)+qr], & r\leqslant R,\\ \varepsilon(0), & r>R\end{cases} \tag{5.57}$$

As in Section 5.2.1, it is often more useful to define the wavenumber-dependent dielectric function,

$$1/\varepsilon(k)=\tilde{V}(k)/V_0(k), \tag{5.58}$$

as the ratio of the Fourier transforms of screened to unscreened potentials. The result follows as

$$\varepsilon(k)=\frac{q^2+k^2}{[q^2/\varepsilon(0)]\sin kR/kR+k^2}. \tag{5.59}$$

The only input data needed to use eqn (5.59) are the static dielectric constant $\varepsilon(0)$ and the valence Fermi momentum $k_F$, as in the earlier Penn model.[25,26]

As mentioned above, the screening radius $R$ is found by solution of eqn (5.55): values for diamond, silicon and germanium are compared with the corresponding nearest-neighbour distances in the crystal in Table 5.2.

It is physically satisfactory that the values of $R$ thereby obtained are quite close to these near-neighbour distances. The results obtained for the dielectric function itself are quite close to those of, for example, Srinivasan,[27] using rather different arguments.

In concluding this section, we must note that a very refined density functional theory of the dielectric constant of silicon has been worked out by Baroni and Resta,[28] to whose paper the reader is referred for details.

**Table 5.2** Summary of relevant data for diamond, silicon and germanium (after Resta[24]).

| | *Diamond* | *Silicon* | *Germanium* |
|---|---|---|---|
| Nearest-neighbour distance (a.u.) | 2.91 | 4.44 | 4.63 |
| Valence Fermi wavenumber $k_F$ (a.u.) | 1.46 | 0.96 | 0.92 |
| Static dielectric constant $\varepsilon(0)$ | 5.7 | 11.9 | 16.0 |
| Screening radius $R$ from eqn (5.55) | 2.76 | 4.28 | 4.71 |

## 5.4 Exchange and correlation in the jellium model

In this section we summarize some of the more elementary aspects of the model of jellium, i.e. that model of a metal in which the positive ions are smeared out into a uniform background of positive charge which is just sufficient to neutralize the electronic negative charge (cf. Chapter 1, Section 1.8).

In the high-density limit of this model, the kinetic energy dominates over the potential energy. This is evident if we introduce the mean interelectronic spacing $r_s$, related to the mean electron density $\rho_0$ by

$$\rho_0 = 3/4\pi r_s^3. \tag{5.60}$$

Since the kinetic energy depends on the operator $\nabla^2$, it will scale like $r_s^{-2}$ per particle, while the potential energy per particle will be proportional to $r_s^{-1}$. Thus, in the high-density limit of very large $\rho_0$, or from eqn (5.60) $r_s \to 0$, the kinetic energy will dominate.

Therefore, the picture of the non-interacting electron gas used in Section 1.2 is the appropriate starting point. All states with momentum $p$ less than the Fermi momentum are occupied, and these states correspond to the plane waves exp $(i\mathbf{p}\cdot\mathbf{r})$. The total wave function is then an antisymmetrized product of these plane-wave states, or what is equivalent, a single Slater determinant of these free-particle wave functions.

### 5.4.1 First- and second-order density matrices

Now a property of such a single Slater determinant is that if the orbitals are denoted by $\Psi_i$ in general, the single-particle density matrix $\rho(\mathbf{r}, \mathbf{r}')$, defined precisely in eqn (5.63), is sufficient to determine the pair correlation function $\Gamma(\mathbf{r}_1, \mathbf{r}_2)$. In terms of the many-electron wave function $\Psi$, this pair correlation function is defined, apart from a normalization factor, by[29]

$$\Gamma(\mathbf{r}_1, \mathbf{r}_2) = \int \Psi^*(\mathbf{r}_1, \ldots, \mathbf{r}_N)\Psi(\mathbf{r}_1, \ldots, \mathbf{r}_N)\, d\mathbf{r}_3\, d\mathbf{r}_4 \cdots d\mathbf{r}_N. \tag{5.61}$$

With $\Psi$ approximated by a single Slater determinant, it can be shown from the definition (5.61) that

$$\Gamma(\mathbf{r}_1, \mathbf{r}_2) = \rho(\mathbf{r}_1)\rho(\mathbf{r}_2) - \tfrac{1}{2}\{\rho(\mathbf{r}_1, \mathbf{r}_2)\}^2, \tag{5.62}$$

where the general definition of the first-order density matrix $\rho$ from the many-electron wave function is (apart from normalization)

$$\rho(\mathbf{r}_1, \mathbf{r}_1') = \int \Psi^*(\mathbf{r}_1, \mathbf{r}_2, \ldots, \mathbf{r}_N)\, \Psi(\mathbf{r}_1', \mathbf{r}_2, \ldots, \mathbf{r}_N)\, d\mathbf{r}_2 \cdots d\mathbf{r}_N. \tag{5.63}$$

The form

$$\rho(\mathbf{r}_1, \mathbf{r}_1') = \sum \psi_i^*(\mathbf{r}_1)\psi_i(\mathbf{r}_1') \tag{5.64}$$

is again the special form following from the definition (5.63) when the many-electron wave function is a single Slater determinant built from orbitals $\psi_i$.

Calculating eqn (5.64) when the orbitals $\psi_i$ become the plane waves $\exp(i\mathbf{p}\cdot\mathbf{r})$ is a straightforward matter, the summation over occupied states $i$ being replaced by an integration over momenta through the occupied Fermi sphere. This is how the result (A 5.1.7) for free particles is calculated, as is readily verified. Turning to the pair function in the non-interacting high-density limit of jellium, one uses this result (A 5.1.7) in eqn (5.62), the density $\rho$ being simply the constant value $\rho_0$. In this way, the pair function in eqn (1.51) follows; the Fermi hole we discussed at some length in Chapter 1. This leads to the exchange energy per particle through eqn (1.52). Writing the results for the kinetic and exchange energy in terms of the interelectronic separation $r_s$ yields

$$\frac{E}{N} = \frac{2.21}{r_s^2} - \frac{0.916}{r_s}, \tag{5.65}$$

which is the Hartree–Fock energy for the jellium model. The correlation energy is defined as the lowering of the energy beyond this value, for a chosen $r_s$. In eqn (5.65) the energy is in Rydbergs if $r_s$ is in units of the Bohr radius $a_0$.

Many-body perturbation theory can be applied to calculate the correlation energy in the high-density limit (see ref. 99) discussed above, when eqn (5.65) becomes generalized to (cf. Section 1.8)

$$\frac{E}{N}=\frac{2.21}{r_s^2}-\frac{0.916}{r_s}+A\ln r_s+C+\cdots, \tag{5.66}$$

where $A=(2/\pi^2)(1-\ln 2)$ and $C=-0.096$.

### 5.4.2 Wigner electron crystal: low-density limit

Wigner[30] noted that although the Hartree–Fock theory based on plane waves is inappropriate in the strong coupling limit, nevertheless the inference from the different dependence of kinetic and potential energy on $r_s$ is that as $r_s$ tends to infinity the potential energy must dominate (cf. eqn (5.65)).

Therefore, one must choose a description of the electronic configuration that minimizes the potential energy and Wigner argued that this would be achieved if the electrons went onto a lattice, thereby avoiding one another optimally. The calculation of the potential energy is then equivalent to the calculation of the Madelung energy of a lattice of point electrons in a uniform positive background of neutralizing charge.[31] Of the lattices examined, the body-centred-cubic lattice turns out to have the lowest Madelung energy, the energy per electron $E/N$ in this extreme low-density limit being

$$\frac{E}{N}(r_s\to\infty)=-\frac{1.79}{r_s}. \tag{5.67}$$

This leads to a correlation energy per particle given by subtracting the exchange energy in eqn (5.65) from eqn (5.67), and this difference is embodied in the Wigner formula (1.56) in the limit $\rho\to 0$. As $r_s$ is reduced, the electrons vibrate about their (body-centred-cubic) lattice sites, and one can calculate the next term beyond that displayed in eqn (1.56) by essentially phonon theory applied to the Wigner electron crystal. The next term is found to be proportional to $r_s^{-3/2}$ but we shall refer the reader to a review on Wigner crystallization[32] for details.

We conclude by remarking that the Wigner electron crystal is an insulator at absolute zero, since the electrons are localized on lattice sites.

Obviously the high-density limit, in which the wave function is the Slater determinant of plane waves, is a delocalized state, and is metallic. For a long time it was a difficult matter to decide at what density the transition from the metal to the insulator occurred in the ground state. That matter has been settled by a Monte Carlo computer calculation (see Section 1.8) by Ceperley and Alder.[33] The critical value of $r_s$ is near to 100 $a_0$, which can be contrasted with the lowest density metal, Cs, with an $r_s$ of 5.5 $a_0$. A different approach, namely a one-body potential determination of the electron density in the Wigner lattice, is recorded in Appendix A 5.2.

The density functional theory of perfect crystals, to be discussed at some length in the following sections, leans heavily on the exchange and correlation energy calculation for the jellium model, summarized above.

## 5.5 Density functional theory of perfect crystals

### 5.5.1 Applications

In this and the following section we shall assess in a general way the practical adequacy of density functional theory as one presently understands how to use it; cf. the account of Callaway and March.[34] This means in the local density approximation with exchange and correlation potentials derived from calculations for these quantities in an electron liquid which is homogeneous (cf. Chapter 6). Instead of attempting to assess the successes and failures of band and cohesive energy calculations element by element we shall restrict the discussion to general remarks, bolstered by a few specific examples with characteristic properties. In this section, metals will be the primary focal point, with no long-range magnetic order, together with some comments on semiconductors and insulators. Then, because some problems arise with regard to band structures of ferromagnetic metals, this area will be treated separately in a following section.

One important point to make here is that the band structures produced by different exchange-correlation potentials for the same material are not greatly different. In fact, it turns out that quite reasonable band structures are often obtained in non-magnetic metals with a Slater-type exchange potential.[35] A clear example of this is the case of Cr, where Laurent *et al.*[36] have compared the results of calculations with the Slater–Kohn–Sham potential and the von Barth–Hedin potential.[37] Use of the exchange-correlation potential lowers the bands with respect to the exchange-only case by an amount which is roughly constant ($\sim$0.15 Ry) almost independently of symmetry. Changes in the relative positions of levels were found to be small; generally about 0.01 Ry. The more strongly attractive von Barth–Hedin potential produced a compression of the bands by about 0.008 Ry.

The reason for the small effect on the band shape appears to be the following. The correlation part of the exchange-correlation potential, as discussed by Callaway and March[34] (see also Chapter 6 below), is a slowly varying function of position, even in an atom where the electron density can vary rapidly. In particular, for large densities, the correlation energy per particle according to the homogeneous electron gas formula (5.66) of Gell-Mann and Brueckner varies as $\ln r_s$, where $r_s = (3/4\pi\rho)^{1/3}$, whereas the exchange energy per particle is proportional to $r_s^{-1}$. The slow variation of the correlation contribution of the exchange-correlation potential means that its main effect will be a lowering of the energy of electronic states rather than a relative shift. In contrast, if one tries to simulate the effect of correlation by multiplying the exchange potential by a parameter $\alpha$, as in the so-called X$\alpha$ method, there is a fair degree of sensitivity to this quantity, which multiplies $\rho^{1/3}$ (cf. eqn (1.60)).

### 5.5.2 Cohesive properties of metals

A highly successful application of density functional theory in the local density approximation has been the calculation of cohesive properties of metals. Moruzzi *et al.*[38] calculated the cohesive energy, equilibrium lattice constant and bulk modulus for 32 metallic elements, including metallic hydrogen, up to indium with $Z = 49$. In the course of their study, they obtained energy bands, electronic charge densities, the density of states and also the susceptibility enhancement. An exchange-correlation potential of the von Barth–Hedin form was employed, with however slightly modified parameters. Spin-polarized calculations were made for iron, nickel and cobalt which will be discussed later in this Chapter. The calculations were made using the Korringa–Kohn–Rostoker method, which is discussed, for example, in the book by Callaway.[39] They made a "muffin-tin" approximation to the electron density, in which this quantity is assumed to be spherically symmetric inside each of a set of non-overlapping spheres centred on nuclear sites and constant outside. Only body-centred-cubic (b.c.c.) and face-centred-cubic (f.c.c) lattice structures were considered, hexagonal close-packed structures being replaced by f.c.c. lattices.

It is not our purpose to review the conclusions of Moruzzi *et al.* with regard to the physical properties among the metallic elements. What is of concern here is the general level of accuracy achieved, since this bears on the degree of adequacy of current applications of the density functional theory to solids.

In general, these workers found good agreement between theory and experiment for all metals that do not have a partially filled 3d shell. The atomic volumes calculated for the alkali metals are about 10% smaller than the experimental results. Errors of 5% or less were found for Be, Mg, Al, Ni, Cu, Zn and all of the metals with an open 4d shell. The systems without 3d electrons yielding the least satisfactory results were the alkaline-earth metals

Sr and Ca, with errors of about 14%, and the alkali metals Li, Na, K and Rb, with errors of about 10%. The calculated bulk moduli were again satisfactory except for magnetically ordered 3d metals. Determination of the cohesive energy requires comparison of calculated total energies for both the solid and the free atom. In fact, the same pattern of general agreement is obtained as for the lattice constant and the bulk modulus; serious errors occur for materials with open 3d shells. While there is evidently considerable cancellation of errors, there is reason to believe that the most important difficulties are in the calculation of the total energy of the atom. This is to be expected in the sense that the charge density in an atom resembles that of a uniform electron assembly less than that in a solid, the atomic charge density decaying to zero exponentially far from the nucleus, as noted in Chapter 1, while that in a crystal is, of course, periodic; the conserved quantum numbers are different, etc. It seems that in the case of atoms with an unfilled 3d shell, application of straightforward local density theory frequently leads to a ground-state configuration with too few s electrons. Let us take the case of vanadium as an example. Local density theory predicts that the ground atomic state should have the configuration $(3d)^4(4s)^1$ whereas experimentally the $(3d)^3(4s)^2$ configuration is observed. Similar errors occur for Ti and Ni, whereas fractional occupancies of the 3d and 4s shells are predicted in Fe and Co, a situation which is possible when degeneracy exists at the Fermi level. The result of possibly more severe errors in the atom than in the solid is to find an overestimate of the cohesive energy. Moruzzi *et al.*[38] point out that it is not simply neglect of atomic multiplet structure which leads to difficulties because the same sort of errors are obtained in the case of Mn where the ground $^6S$ state can be represented by a single determinant in the Hartree–Fock approximation, as in the case of other open 3d-shell elements.

### 5.5.3 Example of lithium metal

In order to go into a little more detail to illustrate the state of calculations of this type, consider the calculations of Callaway *et al.*[40] on lithium. Numerical values taken from their work are collected in Table 5.3. A small contribution from the zero-point energy of the lattice vibrations has been included.

Figure 5.2 shows the total energy as a function of lattice constant. It is seen that in the neighbourhood of $a=11$, a ferromagnetic state lies lower than the non-magnetic state. There is a small region of unsaturated ferromagnetic metal near $a=11$, followed by a saturated ferromagnetic region, with the occupied 2s band becoming isolated from higher bands. In

**Table 5.3** Total energy $E_T$, equlibrium lattice space $a$, bulk modulus $B$ and cohesive energy $E_c$ for lithium.

| *Physical quantity* | *Experiment* | *Calculated* |
|---|---|---|
| $E_T$ (Ryd/atom) | −15.1 | −14.9 |
| $a$ (a.u.) | 6.60 | 6.52 |
| $B$ (Mbar) | 0.123 | 0.138 |
| $E_c$ (Ryd/atom) | 0.122 | 0.125 |

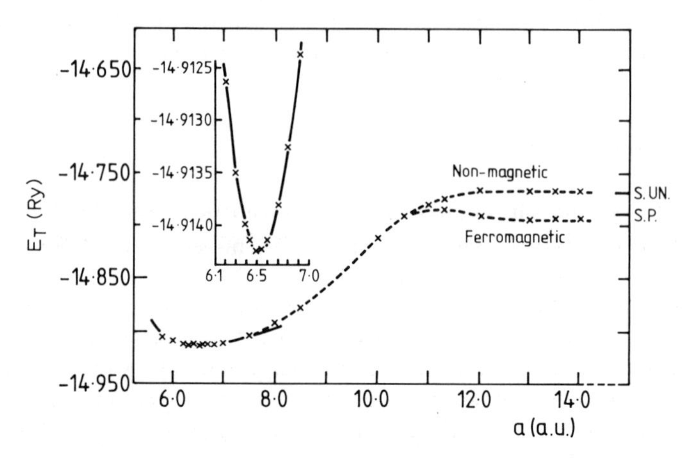


**Fig. 5.2** Total energy of lithium as a function of lattice constant $a$. Crosses indicate calculated energies. A ferromagnetic state has lower energy beyond $a \sim 11$. The inset shows the proximity to the equilibrium lattice constant.

this region, a ferromagnetic insulating state is obtained. It seems likely however that an antiferromagnetic insulating state, which was not explicitly calculated, would lie lower than the ferromagnetic state. However, it is to be noted that the curves of total energy versus lattice constant for both the ferromagnetic and non-magnetic states smoothly approach the atomic limits. This needs the small qualification that there is a minor discrepancy of 0.004 Ry between the apparent limit of the solid-state total-energy calculation and the atomic total-energy result. This is presumably a basis-set effect. If the solid limit were used, the cohesive energy quoted in Table 5.3 would be reduced to 0.121 Ry per atom. In contrast, in a Hartree–Fock calculation using a single determinant of Bloch functions, one would find that the total energy of the non-magnetic state approached too high a value owing to the presence of spurious ionic components in the wave function. The essential point to be stressed is that the local density calculation approaches a reasonable limit as the atomic separation is increased.

The results just discussed can depend to some extent on the particular exchange-correlation potential employed. The work of Vosko and Wilk[41] indicates that the total energy of lithium might be changed by 0.02 Ry and the cohesive energy by 0.01 Ry if a different potential $V_{xc}$ were used. These authors also found effects of the same order of magnitude (0.01 Ry) for the cohesive energies of the other alkali metals owing to differences in $V_{xc}$.

In this example, it should also be noted that the occupied bandwidth at the equilibrium lattice spacing is 3.55 eV. This is to be compared with an experimental value of 4.0 eV. In contrast, a Hartree–Fock calculation[42] gave 9.87 eV for this quantity. This exaggeration of bandwidths is a common fault of Hartree–Fock calculations in metals.

### 5.5.4 Fermi surfaces

It is also of interest to consider Fermi-surface predictions. In non-ferromagnetic metals, a body of evidence has built up that shows that such local density calculations are yielding Fermi surfaces that are at the very least qualitatively in agreement with experiment. Moreover, comparison with angularly resolved photoemission experiments demonstrates that the basic features of the band structure itself appear to be correct. No claim is made that agreement between theory and experiment is perfect, but major discrepancies are not found, as far as the writer is aware. But in ferromagnetic metals, covalently bonded semiconductors and insulators, the situation is a good deal less satisfactory at the time of writing.

These points will be illustrated by choosing the thoroughly studied case of copper, for which the Fermi surface has been carefully measured, and also photoemission has led to the determination of much of the band structure.

There are several calculations to date of the energy bands, Fermi surface and related properties of copper, using density functional theory.[43–46] If the band structure is found from self-consistent field calculations with an exchange-correlation potential; which particular one not seeming to be very significant, and the Fermi surface constructed, it is found that although the area of the (111) "belly" orbit is in satisfactory agreement with experiment, the area of the (111) "neck" where the Fermi surface intersects the zone face is almost 20% too low. The neck area can be brought into quantitative agreement with experiment without impairing the agreement with regard to the belly orbit if the band structure is calculated with an $X\alpha$-exchange-only potential with $\alpha = 0.77$. The belly orbit is not sensitive to the potential because its size is essentially determined by the required volume of the Fermi surface. However, the situation becomes less clear when the optical conductivity is considered. Thus Janak *et al.*[43,44] found it desirable to "expand" the bands calculated with $\alpha = 0.77$ by an empirical factor, in order to bring the position of the maxima in the calculated optical constants into better agreement with experiment. However, Bagayoko *et al.*[45] observed that although the energy at which interband transitions begin is slightly too low in comparison with experiment; in fact by about 0.3 eV, the experimental value being 2.15 eV, when an exchange-correlation potential is used the positions of major additional structure in the optical conductivity curve are given reasonably correctly without any expansion of the band structure. That the general behaviour of the band structure is more or less correct is seen from Figure 5.3 where the calculated band structure of Bagayoko *et al.*[45] is compared with the photoemission measurements of Thiry *et al.*[47] On the other hand, the d bands may be slightly too close to the Fermi energy, according to the optical conductivity results mentioned previously. Finally, it is also worth noting that the calculated Compton profile of Bagayoko *et al.*,[45] which reflects the electronic momentum distribution, is in general in accord with experiment, except possibly with regard to the directional anisotropy.

It can be concluded from all this that theory and experiment match quite

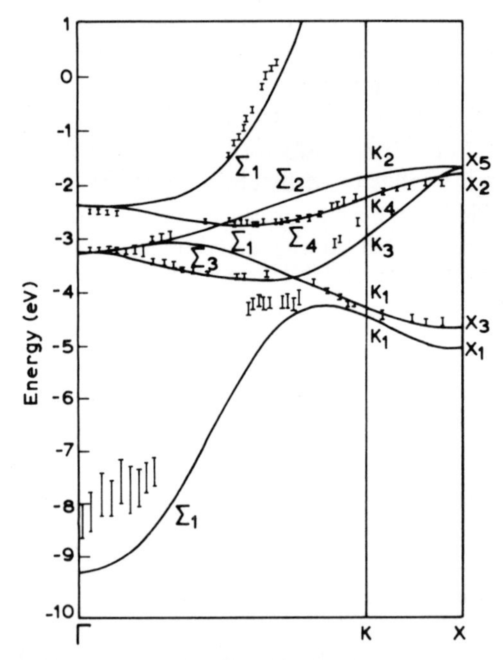


**Fig. 5.3** Calculated energy bands in Cu after Bagayoko et al. along the $\Gamma - K - X$ direction. Experimental points are taken from Thiry *et al.* $E = 0$ indicates the position of the Fermi level.

well, but that there are some specific aspects where improvement would be highly desirable, particularly with reference to the Fermi surface. Wang and Rasolt[48] have given arguments that there must be significant non-locality in the exchange-correlation potential that is ignored by the conventional approximations. Their calculations point to the possibility that a non-local modification of this potential could lead to greatly improved agreement with experiment.

### 5.5.5 Semiconductors: especially crystalline silicon

We shall here consider, fairly briefly, a semiconductor; and silicon, on which a good deal of work has been done, proves the best example to analyse. In general terms, the conclusions are the same as for the metals discussed above; namely, the cohesive properties can be calculated quite accurately, but some problems remain with the band structure, possibly more severe than for most (non-3d) metals.

With regard to cohesive properties, Yin and Cohen[49] combined a pseudopotential treatment of the valence electron–core interaction with a density functional calculation of the valence electrons. This amounts to a frozen-core approximation. The analysis for this approach was given by Ihm *et al.*,[50] while the pseudopotential employed was that constructed by Hamann *et al.*[51] The correlation contribution to the exchange-correlation potential was determined from Wigner's expression[30] for the correlation energy of jellium. The results of Yin and Cohen are collected in Table 5.4; it is plain that the approach gives impressive properties for this covalently bonded semiconductor.

Yin and Cohen were also able to show that the above approach gave the lowest total energy for the observed diamond structure. A transition to the $\beta$-tin structure occurs in silicon under high pressure and these workers were also able to calculate correctly the pressure at which this takes place. A severe difficulty is that the energy of the conduction band states are not placed correctly relative to the valence band. In particular, the calculated

**Table 5.4** Total energy $E_T$ for silicon (core electrons excluded), equilibrium lattice spacing $a$, bulk modulus $B$ and cohesive energy $E_c$.

| *Physical quantity* | *Experiment* | *Calculated* |
|---|---|---|
| $E_T$ (Ry/atom) | −7.919 | −7.909 |
| $a$ (Å) | 5.43 | 5.45 |
| $B$ (Mbar) | 0.99 | 0.98 |
| $E_c$ (ev/atom) | 4.63 | 4.67 |

band gap was 0.5 eV, rather than the observed 1.17 eV. We shall return to this point about the band gap later in this Chapter.

### 5.5.6 Metals with long-range magnetic order

The most serious discrepancies between band structure and cohesive energy calculations based on density functional theory and experiment occur for magnetically ordered 3d metals. The study of cohesive properties by Moruzzi *et al.*[38] revealed serious overestimates with regard to the cohesive energy, while the lattice constant was too small and the bulk modulus too large. Problems with the lattice constant and bulk modulus in their earlier calculations in which spin polarization (see Appendix A 5.4) was not included, were attributed to the existence of magnetization in the actual materials or to volume magnetostriction. Their subsequent spin-polarized calculations yielded markedly improved values for these quantities in the ferromagnetic metals Fe, Ni and Co. The basic idea here (cf. Appendix A 5.4) is that a ferromagnet can reduce the cost in kinetic energy of magnetic ordering by undergoing a lattice expansion. However, the cohesive energy remains a problem, and at the time of writing there still appear to be difficulties with the lattice constant and the bulk modulus of the antiferromagnetically ordered materials to which the proposed explanation for ferromagnets do not seem to apply.

On the other hand, the calculations of Moruzzi *et al.*[38] and of Callaway and Wang[52] for Fe, and of Wang and Callaway[53] for Ni do produce ferromagnetic states with the correct magnetic moment per atom. The charge, spin and momentum densities seem to be in generally reasonable agreement with regard to X-ray Bragg reflection intensities, neutron scattering form-factors and Compton line shape measurements. In addition, the calculation of hyperfine fields at the nuclear site, which are proportional to the value of the net spin density there, yields results of the correct sign (negative minority spins dominate) and of a magnitude which, although quantitatively too small, suggests that the major omission is that of relativistic effects (cf. Section 1.14). The calculated Fermi surfaces follow the pattern discussed above, which is that of general agreement without all quantitative detail being accurately reproduced by the theory. This is, in fact, a point of considerable significance; the agreement between theoretical and experimental Fermi surfaces indicates that the 3d electrons, though importantly correlated in their relative motions, are itinerant and not localized.

The first difficulty which bears directly on the question of the adequacy of spin-polarized exchange-correlation potentials is that of the exchange splitting of the d bands. It is curious that the disagreement between theory and experiment here exists for Ni but not for Fe. Co may also be a case where there is disagreement, but there are not, to the writer's knowledge at the time of writing, any reliable self-consistent spin-polarized local spin-density calculations for this element in its h.c.p. crystal structure, so attention will be restricted below to Ni and Fe. Calculations for Ni based on the Kohn–Sham exchange potential give an exchange splitting $\Delta E=0.88$ eV. Use of the potential constructed by von Barth and Hedin gives results which are significantly different from those using the Kohn–Sham potential; namely $\Delta E=0.63$ eV, while the experimental value is probably near to 0.3 eV.[54] This is one of the few features, it seems, which is sensitive to the difference between the Kohn–Sham and the von Barth–Hedin potential. The implication is that the latter potential unduly favours ferromagnetic alignment. However the absence of any discrepancy for Fe, where the exchange splitting is larger, remains puzzling.

The difficulties with bandwidth and band shape in Ni appear to be more serious. Reasonable agreement is again found for Fe. More information on this situation can be found from the work of Callaway[55] and of Himpsel *et al.*[56] In short, the total d-band width in Ni seems to be too small by an amount in the region between 12–25%, depending on the method of measurement. Certain levels at symmetry points (notably $L_2'$) are in unexpected positions. Moreover, there is some indication that the exchange splitting of bands based on d functions of $e_g$ symmetry is smaller by a factor of 2 or 3 than that of bands built from $t_{2g}$ symmetry. Attempts to resolve this discrepancy have, so far, focused on calculations of the self-energy (cf. Section 5.7 below) in second order with a Hubbard model Hamiltonian.

In view of the above problems with the Ni band structure, it is noteworthy that the calculation of the spin-wave stiffness $D$ gives a value[57] close to experiment. Specifically, the calculated $D=0.148$ Ry $a_0^2$, whereas the experimental value[58] is 0.146 Ry $a_0^2$. Although this close agreement might be to some extent fortuitous in the light of the questions raised above about the band structure, it appears that density functional theory is capable of giving a reasonable account of magnetic excitations near $T=0$.

Having treated perfect crystals at some length, we return in the next section to defective solids, treated now however using the density functional theory as the basis for cluster calculations.

## 5.6 Cluster calculations of electronic structure of transition-metal impurities in copper

The electronic structure of magnetic transition-metal impurities (Fe, Co, Ni) in Cu has been calculated by Blaha and Callaway[59] by means of a cluster approach and the local spin-density approximation. The wave functions were expanded into Gaussians and no shape approximation to the potential was made. These workers report the results of the energy level distribution in 13-atom f.c.c.-clusters $Cu_{13}$, $Cu_{12}Fe$ and in 19-atom clusters ($MCu_{12}Cu_6$, M = Cu, Fe, Co, Ni). The local cluster density of states (DOS) agrees very well with the respective bulk and surface DOS of Cu metal. For the Fe impurity they obtain a local moment of $3.0\,\mu_B$, which is in good agreement with experiment and Green function calculations. The spin densities show besides the strong localized Fe moment a negative polarization of the conduction electrons as indicated by a Mulliken population analysis or direct examination of the spin densities in the (100) plane. Previous observations of scattering in de Haas–van Alphen experiments, which showed that mainly spin-down states are involved, are in agreement with their local DOS. For the Co and Ni systems the impurity moment is reduced to 2.05 and $0.69\,\mu_B$; however, the Ni impurity might become non-magnetic by including more Cu shells or proper boundaries in the calculation. This is indicated through the level distribution as well as the high correlation of magnetism in both Cu shells.

Much experimental and theoretical research has been done with regard to dilute systems of transition-metal impurities alloyed in otherwise non-magnetic metals. These systems, sometimes also forming spin-glasses with increased concentration, are often called "Kondo-systems" and have been investigated by many authors. Common models describing the magnetic properties are (i) the virtual-impurity-state model and (ii) the impurity-ion crystal-field model.[60,61] Concepts like the "Kondo compensation cloud" or localized spin fluctuations are widely used.

The work of Blaha and Callaway was concerned with first-principles calculations of the electronic structure of such systems on the basis of density functional theory. Previous calculations have been of two types: calculations for finite clusters, and Green function calculations for impurities embedded in the bulk solid.

In particular, Green function impurity calculations have been made by a group in Jülich.[62] The latter authors argued that the cluster approach is unsuitable for the description of impurities in solids. However, in spite of major conceptual differences between bulk solids and small clusters (in contrast with a cluster, a bulk metal has a sharp Fermi surface, and excitations across it can occur with vanishingly small change in energy) there are reasons for continuing to study finite clusters as models of solids. Most particularly, a very high degree of self-consistency can be obtained in the electronic structure computations.

Calculations for Fe and Ni clusters were performed by Lee *et al.*[63] and generally good agreement with the bulk electroaic structure except for the spatial distribution of the spin density was achieved. Blaha and Callaway extend this approach to impurity systems, with the object of investigating the electronic structure of Fe, Co and Ni impurities in copper metal. Their results are compared with experiments and with other calculations.

### 5.6.1 Method

Only a brief description of the method employed in these calculations is given below. For a complete description, see, for example, ref. 64. The free-cluster calculations are performed on the basis of local spin-density functional theory (LSDF) by expanding the wave functions in an uncontracted Gaussian basis set of 14 s-, 9 p- and 5 d- type functions. Including angular dependences, 66 independent functions per atom are used; however, extensive use of the cubic symmetry of the cluster keeps the problem to a manageable size. The exponents of the orbital basis set are taken from the free-atom calculations performed by Wachters.[65]

In order to avoid the calculations of enormous numbers of two-electron integrals for the Coulomb matrix elements, Blaha and Callaway made an auxiliary fit to the charge density, using a separate Guassian basis set of 14 s and 9 p functions per atom. A variational fitting procedure was used, giving minimum errors in the electrostatic energy[66] instead of a least-squares fit to the charge density itself.

An exchange-correlation potential as parametrized by Rajagopal *et al.*[67] was used and the corresponding matrix elements were calculated by direct numerical integration. The cluster densities of states (CDOS) were obtained by broadening each eigenvalue with a Gaussian. Each state was decomposed by a Mulliken population analysis and the resulting *l*-like charges determine the width of the Gaussian used (0.6 eV for s and p types; 0.15 eV for d contributions) as well as the weights of the contribution to the CDOS.

### 5.6.2 Results

*$Cu_{13}$*. A starting point for considering impurities in metals should be the electronic structure of the pure metal. Therefore in Fig. 5.4 an energy-level diagram for a free $Cu_{13}$ cluster in f.c.c. geometry is shown. The nearest-neighbour distance of 4.83 a.u. corresponds to a Cu lattice constant of 6.83 a.u. It is found that the Fermi energy coincides with a state of $t_{2g}$ symmetry in accordance with previous calculations by Delley *et al.*[68] and Messmer *et al.*[69] using the discrete variation method (DVM) and the X$\alpha$-scattered wave method respectively. This indicates that the assumed geometry presumably would not be stable with respect to a Jahn–Teller

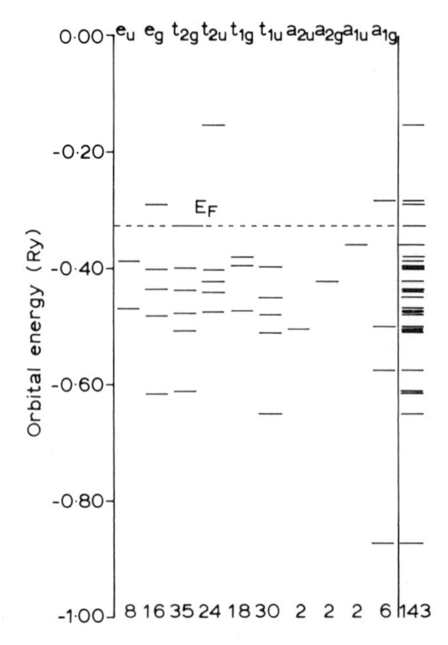


**Fig. 5.4** Energy level diagram for the $Cu_{13}$ cluster after Blaha and Callaway.[59] The symmetry (top) and occupancy (bottom) of these levels is also given. The dashed line denotes the Fermi level.

distortion. However, a detailed comparison with the level structure of Messmer *et al.* gives differences in both the ordering of levels of different symmetry (they find[59] for instance that the first $t_{1u}$ level is lower in energy that the first $t_{2g}$ and $e_g$ states, while the X$\alpha$ calculation predicts the $t_{2g}$ and $e_g$ states to be much lower than the $t_{1u}$ state), as well as the total width of the occupied states ($a_{1g} - E_F$:0.55 versus 0.48 Ry in X$\alpha$) or the distance of the highest d states from the Fermi energy (0.04 versus 0.12 Ry in X$\alpha$). A Hartree–Fock study of $Cu_{13}$ by Demuynck *et al.*[70] yields almost no overlap between the Cu 3d and 4s states and therefore predicts a completely different electronic structure from calculations based on local density functional theory.

A common way to compare cluster and bulk calculations is a comparison of the density of states. Figure 5.5 shows the CDOS of the $Cu_{13}$ cluster and for comparison the bulk f.c.c. Cu DOS calculated by Bagayoko.[71] The overall agreement between these two calculations is very good; in particular the four-peak structure of the bulk-metal DOS originating from the Cu 3d bands embedded in the Cu 4s band is similar to that of the CDOS, as well as the total d-band width in the cluster and the bulk solid. However, the lowest of these four peaks in the CDOS is separated from the others and the d band is shifted closer to the Fermi energy.

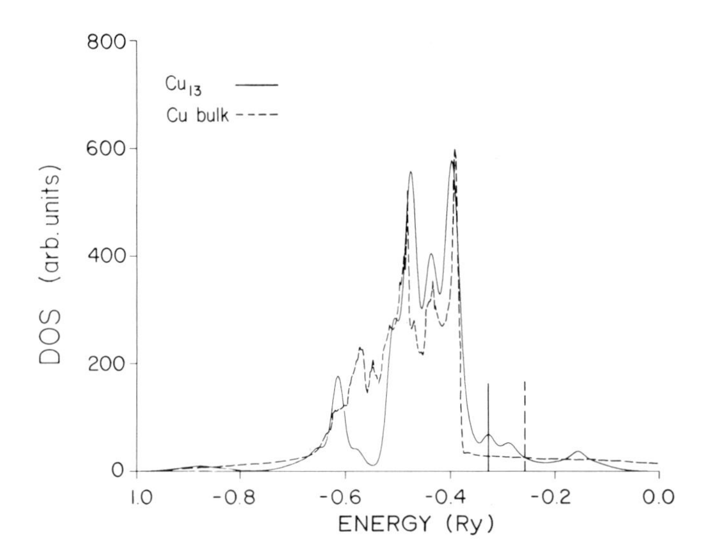


**Fig. 5.5** Cluster density of states (CDOS) for the $Cu_{13}$ cluster, after Blaha and Callaway.[59] In the CDOS, states are broadened and weighted according to their *l*-character in a Mulliken population analysis (0.6 eV for s and p, 0.15 eV for d). The Fermi energies are indicated by a dashed (bulk) and a solid (cluster) straight line. The bulk DOS is shifted in energy, so that the bulk and cluster 3d maxima coincide.

A Mulliken population analysis gives for the central atom a $3d^{9.66}4s^{2.0}$ configuration, and for the shell atoms, $3d^{9.76}4s^{1.17}$. It must be noted that an allocation of the very delocalized 4s electrons to the different Cu sites is somewhat ambiguous, as some small negative 4s populations (and also some greater than two) indicate.

The valence charge density is almost spherically symmetric around the nuclei as expected owing to the full d-shell, and agrees quite well with valence densities obtained from an APW bandstructure calculation.[59] The 3d maximum has a value of 3.86 e/a.u.$^3$ (3.79 e/a.u.$^3$ in APW) and is located 0.33 a.u. (0.32 a.u.) away from the nucleus. The $t_{2g}/e_g$ ratio of the d electrons of the central site is 1.48, a value which is near spherical symmetry (3 : 2).

*$Cu_{19}$*. It has been shown above that even a $Cu_{13}$ cluster can represent most of the electronic structure of bulk Cu. However, in order to get a more direct answer to how the replacement of the central Cu by a magnetic impurity atom affects the electronic structure, Blaha and Callaway made calculations for $Cu_{19}$ also. The energy-level diagram is shown in Fig. 5.6. Again there are major differences between their calculations and previous $X\alpha$-scattered wave[72] results. For instance they[59] find at the Fermi energy,

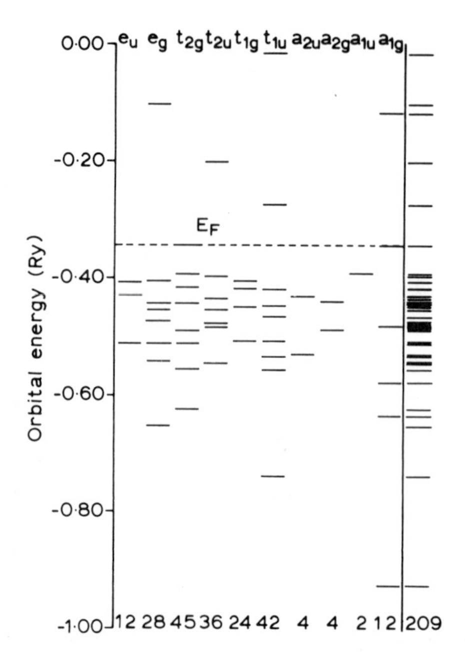


**Fig. 5.6** Energy level diagram for the $Cu_{19}$ cluster after Blaha and Callaway.[59].

two almost-degenerate levels of $a_{1g}$ and $t_{2g}$ symmetry, while X$\alpha$ puts the $t_{2g}$ clearly below the $a_{1g}$. Furthermore, they find the lowest $t_{1u}$ state below the first $t_{2g}$ and $e_g$ states in contrast to the X$\alpha$ results. This brings their d-band width into close agreement to bulk Cu.

As can be seen from Table 5.5 the addition of a second shell of Cu atoms leads to 3d populations, which are now largest at the central atom and not at the first shell as in $Cu_{13}$. Again, the CDOS of the centre plus the first-shell atoms (Fig. 7 of ref. 59) resembles the bulk-Cu DOS very well, while the $Cu_6$ atoms form a "surface DOS" showing a single peak at relatively high energy. This surface DOS was also shown by Delley *et al.*[68] on a $Cu_{79}$ cluster.

*$Cu_{12}Fe$.* When the central Cu atom is replaced by Fe and spin polarization is allowed, an energy-level diagram (Fig. 5.7) indicates that the pure Cu states show almost no spin splitting, but the Fe d states are split by 20–67 mRy (0.27–0.91 eV) at the bottom and top of the "Cu d band". This splitting causes an excess of two spin-up electrons in a $t_{2g}$ level. The exchange splitting of the Fe d states is substantially reduced compared with pure Fe clusters[63,64] (0.7–3.1 eV) or bulk Fe[52] (1.1–2.2 eV).

The spin density (cf. Fig. 5 of ref. 59) indicates the strong localization of the iron moment and the maxima point to the next Cu neighbours as

**Table 5.5** Mulliken population analysis from integrated cluster density of states (M stands for the central atom) (after Blaha and Callaway[59]).

| | $Cu_{12}Fe$ | $Cu_{18}Fe$ | $Cu_{18}Co$ | $Cu_{18}Ni$ | $Cu_{19}$ |
|---|---|---|---|---|---|
| M sp↑ | 1.00 | 1.15 | 1.09 | 1.04 | 1.14 |
| M sp↓ | 1.00 | 1.20 | 1.13 | 1.06 | |
| M d↑ | 4.45 | 4.87 | 4.89 | 4.89 | 4.94 |
| M d↓ | 2.20 | 1.82 | 2.84 | 4.20 | |
| M total | 8.65 | 9.01 | 9.95 | 11.19 | 12.16 |
| $Cu_{12}$sp↑ | 0.60 | 0.64 | 0.65 | 0.66 | 0.61 |
| $Cu_{12}$sp↓ | 0.61 | 0.62 | 0.63 | 0.65 | |
| $Cu_{12}$d↑ | 4.87 | 4.87 | 4.87 | 4.85 | 4.88 |
| $Cu_{12}$d↓ | 4.87 | 4.83 | 4.84 | 4.84 | |
| $Cu_{12}$total | 10.95 | 10.96 | 10.99 | 11.00 | 10.98 |
| $Cu_6$ sp↑ | | 0.58 | 0.58 | 0.57 | 0.54 |
| $Cu_6$ sp↓ | | 0.58 | 0.57 | 0.57 | |
| $Cu_6$ d↑ | | 4.87 | 4.86 | 4.84 | 4.89 |
| $Cu_6$ d↓ | | 4.87 | 4.86 | 4.83 | |
| $Cu_6$ total | | 10.90 | 10.87 | 10.81 | 10.86 |

expected from the level occupancy. The d electrons of Cu are also slightly positively spin-polarized owing to covalent interactions with Fe, but in the large "interstitial" region a small negative polarization occurs, indicating a dominance of spin-down 4s electrons. The spin densities at the nuclear sites of Fe and Cu are both negative (Table 5.6) and the large value at Cu

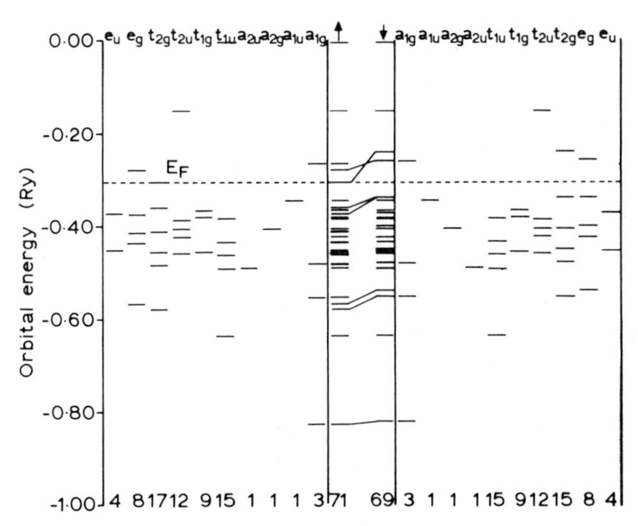


**Fig. 5.7** Energy level diagram for the $Cu_{12}Fe$ cluster (spin states separated) after Blaha and Callaway.[59]

indicates the strong magnetic interaction in contrast with bulk-copper-like behaviour. Note that in the valence electron density an $e_g$ dominance around the iron site is present, as one can see from the $t_{2g}/e_g = 1.2$ ratio.

The total magnetization of this cluster is $2\mu_B$ but a Mulliken analysis gives an Fe d moment of $2.2\mu_B$ which is partially screened by a negative polarization of 0.01 Cu 4s electrons (see Table 5.5). This could be interpreted as a spin-compensation cloud as suggested by many experiments or theoretical approaches[60,73,74].

*$Cu_{18}Fe$.* In order to get a more realistic model of a single Fe impurity in Cu a second shell of Cu atoms is now added to the cluster. This yields the level distribution shown in Fig. 5.8. The Fermi level coincides with a partially occupied $e_g$ spin-down level. This leads to a strong peak in the spin-down DOS (see Fig. 7 of ref. 59) and thus is consistent with de Haas–van Alphen measurements on dilute CuFe, where resonance scattering occurs mainly at the minority states.[75,76] The total magnetization is $4\mu_B$ and the exchange splitting for Fed states ranges from 20 mRy at the bottom to 90 mRy at the top (0.27–1.2 eV) of the occupied states.

The exchange splitting found by Blaha and Callaway is in good agreement with that found by the impurity calculations of Dederichs and coworkers.[62]

The total CDOS, which still resembles the bulk-Cu DOS, could be partitioned into spin-up and spin-down Fe, $Cu_{12}$ and $Cu_6$ contributions (Fig. 7 of ref. 59). The partial spin-up Fe CDOS is spread out over the entire

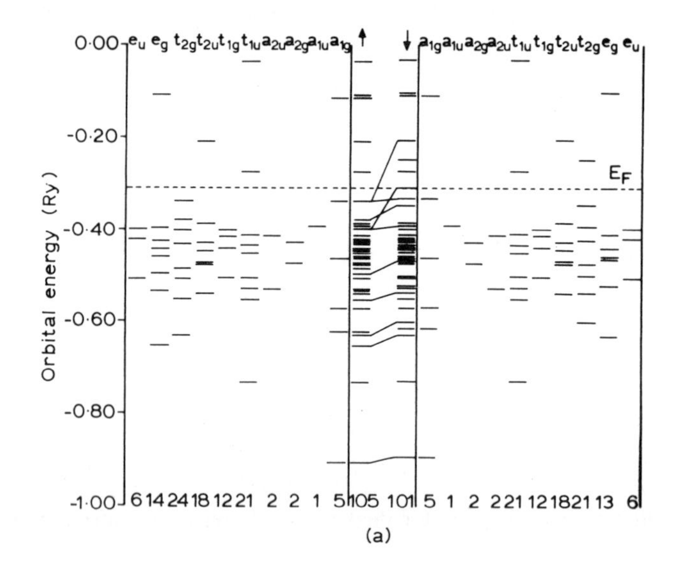


(a)

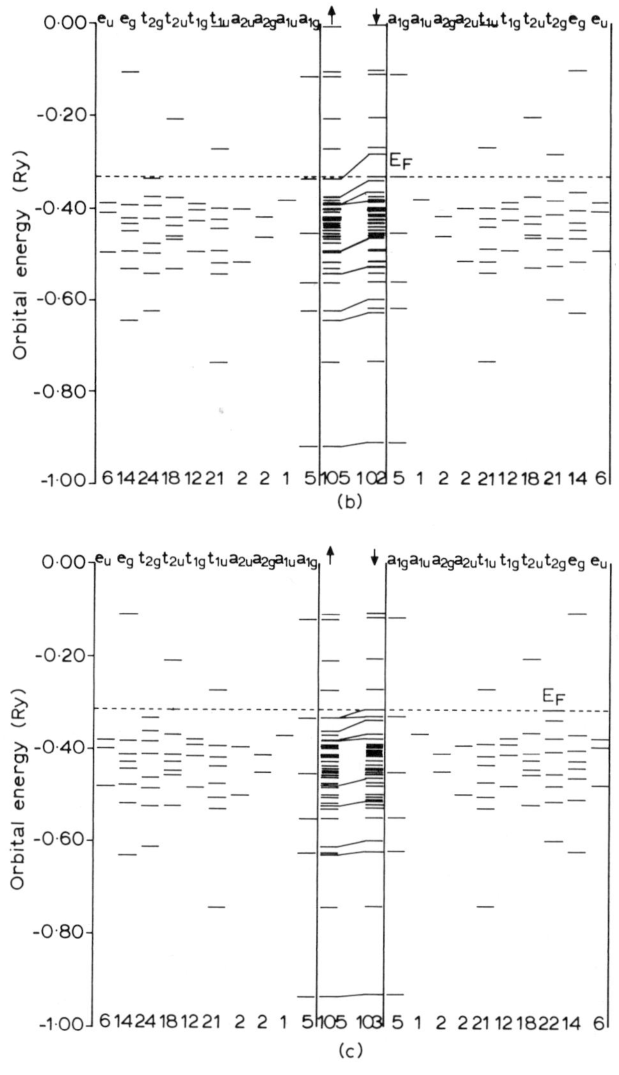


**Fig. 5.8** Energy level diagram for the 19-atom clusters after Blaha and Callaway:[59] (a) $Cu_{18}Fe$; (b) $Cu_{18}Co$; and (c) $Cu_{18}Ni$.

"Cu, d band" and is not restricted to a few eigenvalues. The spin-down Fe CDOS is strongly reduced, shifted to higher energies and shows a relative strong resonance at the Fermi energy. These facts are in good agreement with the impurity calculations mentioned above, although these authors find an additional well-defined sharp spin-up virtual bound state. Since the

Fe spin-up d states are much more hybridized with Cu d than the spin-down Fe d states, it is obvious that one cannot find such a resonance. Furthermore, such sharp resonances are not detected in CuFe by means of X-ray photoemission spectroscopy,[77] where only a smooth enhancement in the s,p region above the Cu d band was found. In addition, Cohen and Slichter[78] performed model calculations for 3d impurities in Cu, fitting experimental NMR satellite data. These authors find a peak only in the spin-down Fe DOS, which is located at the Fermi energy, but no spin-up resonance. The $Cu_{12}$ CDOS resembles bulk Cu, both in spin-up and spin-down, and the $Cu_6$ CDOS shows a single peak at relatively high energy corresponding to a "surface DOS".

One can see from Table 5.5 that the total moment of $4\,\mu_B$ is not located only at the Fe atom, but also the $Cu_{12}$ atoms show a slight ferromagnetic 3d polarization in agreement with the above cluster impurity calculation.[62] The pure Fe d moment of $3.05\,\mu_B$ is in good agreement with experimental data found by neutron diffraction measurements[74,79] or susceptibility measurements by Steiner *et al.*[80] and much larger than that of bulk Fe ($2.16\,\mu_B$). Again a small "spin-compensation cloud" can be seen (Fe 4s,p and Cu 4s,p contributions are indistinguishable) and the $Cu_6$ atoms are almost unpolarized.

The negative polarization in the spin density (Fig. 8 of ref. 59) is spatially reduced owing to the additional moment of the $Cu_{12}$ atoms, while on the $Cu_6$ atoms almost no polarization is present. On the iron nucleus there is a strong negative spin density of $-0.224$ e/a.u.$^3$, which could be compared to the bulk–Fe contact spin density[52] of $-0.406$ e/a.u.$^3$, while the Cu sites, in contrast with the $Cu_{12}$ Fe cluster, show only small and positive spin densities. In this $Cu_{18}$ Fe cluster no indication is found of oscillatory behaviour of the spin densities at the first and second shell of Cu atoms, as one would expect from Knight-shift measurements or model calculations. In the valence density (see Fig. 9 of ref. 59) the dominance of $e_g$ symmetry around Fe as well as the relatively strong deviation from spherical symmetry in the 3d density of the "surface atoms" $Cu_6$ is obvious. The latter indicates together with the partial CDOS of the $Cu_6$ atoms (Fig. 7 of ref. 59) a relative strong "cluster effect" on the electronic properties of atoms in such a cluster with free boundaries and suggests an increase of the cluster size (additional shells) or the use of proper boundary conditions in order to obtain results in better agreement with the solid.

*$Cu_{18}Co$.* The calculations of Blaha and Callaway for this cluster yield a total magnetic moment of $3\,\mu_B$ and a level structure shown in Fig. 5.8 (b). Again, the Fermi energy falls on a spin-down state ($a_{1g}$), but $t_{2g}$ and $a_{1g}$ states are very close together at $E_F$. The exchange splitting of the pre-

dominantly Co states is reduced in comparison with those of Fe in the $Cu_{18}Fe$ cluster to about 20–50 mRy in accordance with de Haas–van Alphen measurements. Single-impurity calculations by the Jülich group originally found Co to be non-magnetic. However, in a subsequent refinement of their method, now including not only the single impurity but also the first Cu neighbours in the self-consistent process, their Co cluster also becomes ferromagnetic.[62,59]

The partial CDOS is shown in Fig. 7 of ref. 59. It is found that there is no significant difference in the $Cu_{12}$ and $Cu_6$ partial CDOS between $Cu_{18}Fe$ and $Cu_{18}Co$. The Fermi energy falls now between the two characteristic impurity spin-down peaks ($a_{1g}$ level), whereas in the Fe cluster it was at the lower one.

The total magnetic moment of the cluster is $3\,\mu_B$, while the Co has a local moment of only $2\,\mu_B$. Again, a negative polarization of the conduction electrons can be seen and the neighbouring Cu atoms show a slight ferromagnetic polarization of $0.03\,\mu_B$ (Table 5.5). These facts are in qualitative agreement with ref. 62 but these authors find only $0.96\,\mu_B$ at the Co site. However, since the inclusion of the first Cu shell increased their magnetic moment from zero to nearly $1\,\mu_B$, further improvements may again change the absolute value, especially since the second-shell atoms become also slightly magnetic, in contrast with $Cu_{18}Fe$.

The spin density around Co shows an even stronger $t_{2g}$ symmetry than that around Fe, since now almost all spin-up excess comes from a $t_{2g}$ level. There is again a negative spin density between the local moment and the Cu neighbours and the $Cu_6$ atoms participate more strongly in the magnetic interaction than in the Fe case (see Fig. 10 of ref. 59.) The spin densities on the $Cu_{12}$ and $Cu_6$ sites show oscillatory and converging behaviour in contrast with the $Cu_{18}$ Fe results (Table 5.5).

*$Cu_{18}Ni$.* The experimental situation for Ni impurities in Cu is at the time of writing the most unclear one. It is generally believed that no strong localized magnetic moment exists. However, from NMR data[81,82] a small permanent moment can be deduced and many authors suggest consideration of spin fluctuations.[83]

The spin-polarized calculation of ref. 59 yields the electronic structure displayed in Fig. 5.8(c), along with a total magnetic moment of $2\,\mu_B$. The Fermi energy coincides with a singly occupied $t_{2g}$ down-state and it is obvious that a slight shift of the spin-up and spin-down $a_{1g}$ states to higher energies (20 mRy) would produce a non-magnetic cluster. However, the electronic structure in a non-spin-polarized scattered wave calculation[65] does not agree either with the results of Blaha and Callaway for the ordering of levels of different symmetry, or with regard to the total width of the occupied valence states ($E_F$–$a_{1g}$: 0.62 Ry versus 0.44 Ry in X$\alpha$). Their exchange splitting is further reduced to about 10–24 mRy.

The partial CDOS in Ni (see Fig. 7 of ref. 59) shows as well as the occupancies in Table 5.5 that almost all Ni states are now below $E_F$ and it is therefore reasonable that XPS measurements[77] found the strongest impurity resonances for Ni in the Fe, Co, Ni series.

In this highly correlated system, even the second nearest neighbour atoms $Cu_6$ show remarkably ferromagnetic aligned 3d moments. The local Ni moment is reduced to less than 0.7 $\mu_B$ and is quite comparable to that of bulk Ni[53] (0.57 $\mu_B$). For that purpose it is also reasonable that the contact spin density of $-0.100\,e/a.u.^3$ is close to the bulk value of $-0.110\,e/a.u.^3$. The spin density around the Ni site shows predominantly $t_{2g}$ character (see Fig. 11 of ref. 59). The spin density at the $Cu_{12}$ site is remarkably high (Table 5.6) and indicates again the high correlation of magnetic interactions in the CuNi system.

This example means that density functional calculations on clusters show promise in treating quantitatively transition-metal impurities in a variety of matrices.

## 5.7. Some problems remaining for density functional theory

Having discussed, in general terms, the kind of results that density functional calculations are yielding, we want now, following Pickett,[84] to highlight some problems and difficulties which remain at the foundations of the theory.

We saw in connection with the specific example of Si that the energy band gap was underestimated, and we take up this point in more general terms here. First, we note again that it is characteristic of current calculations of bands in the local density approximation that the energies of excited states in semiconductors and insulators are too low with respect to the valence band. In this connection, the work of Wang and Klein[85] may be cited. They made all-electron calculations of energy bands in six semiconductors including Si, by using an exchange-correlation potential, the correlation being of Wigner form.[30] The occupied bandwidth and X-ray form-factors compare favourably with experiment, but the fundamental band gap was only 0.65 eV. A similar problem was discussed by Khan and Callaway[86] with regard to the band structure of solid neon and argon. There, the calculated band gaps were found to be about one-half the experimental values, when calculated with an exchange-correlation potential. To obtain the correct band gap, it proved necessary to use an X$\alpha$ potential with $\alpha = 1.2$.

Below, the question of the foundations in density functional theory of determining the band gap will be exposed, following the survey of Pickett.[84] Within density functional theory, he notes, the description of single-particle excitation energies can be approached via the ground-state energy functional, as well as through a conventional description of excited states.

### 5.7.1 Excitations from ground-state energies

Let us consider a semiconducting system for which $N$ electrons exactly fill the valence bands. The corresponding total energy will be denoted by $E_N$. If one electron from the valence-band maximum is removed to infinity, the energy of the system is $E_{N-1}$, making the minimum energy necessary to excite an electron out of the crystal, namely the ionization potential,

$$I = E_{N-1} - E_N. \tag{5.68}$$

If an electron, originally at infinity, is introduced into the lowest conduction-band state, the change in energy is the electron affinity $A$:

$$A = E_N - E_{N+1}. \tag{5.69}$$

Then $I - A$, which is given by the difference $E_{N-1} + E_{N+1} - 2E_N$ in ground-state energies, is equal (neglecting electron–hole interaction effects) to the energy gap describing the lowest excitation of the semiconductor.

In contrast with the one-body potential form of density functional theory, with a local potential energy $V(\mathbf{r}) = V_{\text{Hartree}}(\mathbf{r}) + V_{\text{xc}}(\mathbf{r})$, in the formally exact theory of single-particle excitations (see ref. 84 for a simple example of the jellium model), the quasi-particle wave function $\phi$ is a solution of a similar Schrödinger equation, as discussed for instance by Pickett and Wang,[87] namely

$$[-\nabla^2 + V_{\text{ext}}(\mathbf{r}) + V_{\text{Hartree}}(\mathbf{r})]\phi(\mathbf{r}, E) + \int d\mathbf{r}'\, M(\mathbf{r}\mathbf{r}', E)\phi(\mathbf{r}', E) = E\phi(\mathbf{r}, E). \tag{5.70}$$

The main difference is the appearance in eqn (5.70) of a non-local energy-dependent potential $M$ which describes exchange-correlation effects and the replacement of discrete eigenvalues $\varepsilon_i$ with a continuous excitation energy $E$. Pickett draws attention to the fact that in a semiconductor, however, the low-energy solutions form undamped excitations that are very reminiscent of those calculated from the Kohn–Sham equations. It is natural then to identify the Kohn–Sham eigenvalues and eigenfunctions as approximations to actual single-particle excitation energies and wave functions. A similar point of view arose out of Slater's work,[35] and band-theory studies have implicitly assumed this connection.

The relationship between the gap $E_g = I - A$ and the Kohn–Sham eigenvalues $\varepsilon_i$ is derived by expressing the ground-state energy as

$$E = \sum_i \varepsilon_i \theta(\mu - \varepsilon_i) - \frac{1}{2}\int d\mathbf{r} \int d\mathbf{r}' \frac{\rho(\mathbf{r})\rho(\mathbf{r}')}{|\mathbf{r}-\mathbf{r}'|} + \int d\mathbf{r}\, \rho(\mathbf{r}) V_{xc}(\mathbf{r}, n) + E_{xc}[\rho]. \tag{5.71}$$

In this equation, $\Theta(x)$ is the Heaviside function; which is 1 for $x<0$, and zero otherwise. As pointed out by Perdew and Levy[88] and also by Sham and Schluter,[89] the gap can now be expressed in the form

$$E_g = \varepsilon_{N+1}(N) - \varepsilon_N(N) + V_{xc}(\mathbf{r};\rho)\Big|_{N-\delta}^{N+\delta}, \tag{5.72}$$

where $\varepsilon_i(J)$ is the $i$th eigenvalue in a system containing $J$ electrons and where the limits $N \pm \delta$ indicate that the exchange-correlation potential is to be evaluated for densities $\rho(\mathbf{r})$ containing fractionally more or less than $N$ electrons. In deriving this relation, one uses the notion that the densities of the $N+1$, $N$ and $N-1$ electron systems differ by order $N^{-1}$ and that $E_{xc}[\rho]$ is continuous for varying electron number. Making the reasonable assumption that $V_{xc}$ depends continuously on $N$, the last term in eqn (5.72) vanishes, and this relation equates the Kohn–Sham gap with the true gap.

But as we have discussed above, this is not found to be the case from density functional calculations, the available Kohn–Sham type calculations underestimating the band gap considerably.

This difficulty has been at least partially resolved by the realization that $V_{xc}$ is permitted to have a discontinuity at values of $N$ for which a complete set of bands is occupied. As a result, the Kohn–Sham gap is increased by the position-independent contribution

$$\Delta V_{xc} = V_{xc}(\mathbf{r};\rho)\Big|_{N-\delta}^{N+\delta}. \tag{5.73}$$

Pickett makes three observations about $\Delta V_{xc}$:

(i) all explicit proposals to date for $E_{xc}[\rho]$ lead to $\Delta V_{xc} = 0$;
(ii) it has not been proved as yet in any rigorous way that $\Delta V_{xc} \neq 0$, but several arguments suggest a non-vanishing value for this correction;
(iii) the only known procedure for calculating $\Delta V_{xc}$ is equivalent to the calculation of the single-particle self-energy discussed above.

Pickett also discusses density functionals for excitations. He refers to an approach which deals directly with the difference between the so-called "mass operator" $M$ in eqn (5.70) and $V_{xc}$. For metals, this difference is rather small; for semiconductors it is an order of magnitude or more smaller than $M$ itself, but we refer the reader to the article of Pickett for more details.

In concluding this Chapter, we shall consider solids under conditions of extreme temperature and pressure. While the latter variable has been already mentioned, and at $T=0$ such calculations are widespread using density functional theory, in the following Section we shall restrict ourselves to the simplest density description, namely the Thomas–Fermi theory and its generalizations, which was employed to discuss defects at the beginning of this Chapter. This will allow some asymptotically valid results to be established once and for all.

## 5.8 Matter under extreme conditions of pressure and temperature

It must be emphasized that the Thomas–Fermi theory is limited in its usefulness when we are dealing with perfect crystals under normal conditions since it cannot account for the Periodic Table effects. However, when pressures sufficient to obliterate the detailed influence of the outer electronic structure are applied, the results of the method should be valid for all elements. The information thus obtained on the behaviour of material at high pressures has proved of very considerable value in astrophysics and, to a lesser extent, in geophysics.

We shall confine ourselves here to two aspects:

(i) the equations of state of elements as given by the Thomas–Fermi method;
(ii) the generalization to the high temperatures often encountered in astrophysical problems.

The relation between the predictions of the theory and existing experimental information obtained from laboratory experiments (and from shock-wave measurements) is discussed in ref. 97.

### 5.8.1 Equations of state from Thomas–Fermi theory

In principle, the problem of describing the electronic density and the electric field in a crystal in the Thomas–Fermi theory can be reduced immediately to that of solving the Thomas–Fermi equation within the Wigner–Seitz cellular polyhedron surrounding a particular atom in the crystal, subject to the requirement of periodicity that the normal derivative of the potential should vanish over the surface of the atomic polyhedron. However, for many

**Table 5.6** Spin density at the nuclei (in $e$/a.u.$^3$) (after Blaha and Callaway[59]).

| | $Cu_{12}Fe$ | $Cu_{18}Fe$ | $Cu_{18}Co$ | $Cu_{18}Ni$ |
|---|---|---|---|---|
| M | −0.196 | −0.224 | −0.179 | −0.100 |
| $Cu_{12}$ | −0.205 | 0.035 | 0.092 | 0.133 |
| $Cu_6$ | — | 0.012 | −0.027 | −0.016 |

crystal structures, the Wigner–Seitz cell has high symmetry and it is a very useful approximation to replace the atomic polyhedron by a sphere of equal volume, as suggested by Slater and Krutter[90] in their pioneering paper on the Thomas–Fermi method for metals. The problem then reduces to solving the dimensionless Thomas–Fermi equation (1.74), already discussed fully for atoms and ions in Chapter 1, subject to the different boundary conditions that

$$\left.\begin{aligned} V(r) &\to -\frac{Ze^2}{r}, \quad r\to 0, \\ \left(\frac{\mathrm{d}V}{\mathrm{d}r}\right)_R &= 0, \end{aligned}\right\} \tag{5.74}$$

the latter condition being the analogue of the periodicity of $V(\mathbf{r})$ in the approximation of the spherical Wigner–Seitz cell of radius $R$.

In view of the fact that the Thomas–Fermi method uses uniform electron gas relations locally, it is physically appealing that the pressure $p$ turns out to be given by the same uniform-gas formula, provided the mean electron density is replaced by the cell boundary density $\rho(R)$. If we include both kinetic and exchange contributions (see Appendix A 5.4) in the expression for the pressure, which is equivalent to treating the free-electron gas by the Hartree–Fock approximation, then the pressure is given by

$$p = \tfrac{2}{3}c_k\{\rho(R)\}^{5/3} - \tfrac{1}{3}c_e\{\rho(R)\}^{4/3}, \tag{5.75}$$

where $\rho(R)$ is the boundary density calculated from the so-called Thomas–Fermi–Dirac approximation (see eqn (5.79) below). In a fully wave-mechanical treatment, there is no doubt about the importance of the boundary density in treatments of cohesion.

In the completely degenerate case, the equation of state given by the Thomas–Fermi theory in the first term of eqn (5.75) can conveniently be written in terms of the boundary value of the dimensionless description $\phi(x)$ (cf. eqn. (1.74)) of the self-consistent potential, at $x_0$ ($R = bx_0$), namely,

$$pv = \tfrac{2}{15}\,\frac{Z^2e^2}{b}\,x_0^{1/2}\,\phi(x_0)^{5/2}, \tag{5.76}$$

where $v$ is the atomic volume.

For small values of $v$, an expansion can be generated in which the first term is the free-electron gas pressure, leading to the equation of state[91]

$$pv = \frac{h^2}{5m}\left(\frac{3}{8\pi}\right)^{2/3}\frac{Z^{5/3}}{v^{2/3}}\left(1 - \frac{2\pi me^2}{h^2}(4Zv)^{1/3} + \cdots\right). \qquad (5.77)$$

A convenient representation of the Thomas–Fermi equation of state can be written, following Gilvarry[92] as

$$p^{2/5}\left[\sum_{n=2}^{6} A_n\left(\frac{3v}{4\pi b^3}\right)^{(n+2)/6}\right] = \left(\frac{Z^2e^2}{10\pi b^4}\right)^{2/5}, \qquad (5.78)$$

where the coefficients $A_n$ are recorded in Table 5.7. This fit should be sufficiently accurate for most purposes.

**Table 5.7** Coefficients $A_n$ in extreme high-pressure equation of state in form of eqn (5.78).

| | |
|---|---|
| $A_2$ | $4.8075 \times 10^{-1}$ |
| $A_3$ | 0 |
| $A_4$ | $6.934 \times 10^{-2}$ |
| $A_5$ | $9.700 \times 10^{-3}$ |
| $A_6$ | $3.3704 \times 10^{-3}$ |

### 5.8.2 Introduction of exchange in the equation of state

Equation (5.75) in principle defines the equation of state in the Thomas–Fermi–Dirac theory, and in terms of the boundary value $\phi(x_0)$ found by solving the Thomas–Fermi–Dirac equation,[91]

$$\frac{d^2\phi}{dx^2} = x\left[\alpha + \left(\frac{\phi}{x}\right)^{1/2}\right]^3; \quad \alpha = \frac{6^{1/3}}{4(\pi Z)^{2/3}}, \qquad (5.79)$$

we find

$$pv = \frac{2}{15}\frac{Z^2e^2}{b}x_0^3\left[\left(\frac{\phi(x_0)}{x_0}\right)^{1/2} + \alpha\right]^5\left[1 - \frac{\frac{5}{4}\alpha}{[\phi(x_0)/x_0]^{1/2} + \alpha}\right]. \qquad (5.80)$$

Equation (5.80) is the generalization of eqn (5.76) obtained by using eqn (1.74), $\phi$ and $x$ again being defined by eqns (1.72) and (5.79). Unfortunately the Thomas–Fermi–Dirac equation (5.79) must be solved separately for each atomic number $Z$. The analogue of the extreme high-pressure result

(5.77) can be shown to be[93]

$$pv = \frac{h^2}{5m}\left(\frac{3}{8\pi}\right)^{2/3}\frac{Z^{5/3}}{v^{2/3}}\left[1 - \frac{2\pi me^2}{h^2}(4Zv)^{1/3} - \frac{10\pi me^2}{3^{2/3}h^2}(4Zv)^{1/3}\alpha + \cdots\right]. \tag{5.81}$$

For many purposes, the graphical presentation of the equations of state by Feynman *et al.*[26] is adequate but we shall not reproduce their plots here.

### 5.8.3 Equations of state for the case of incomplete degeneracy

In this discussion of material under high pressure, it has so far been assumed that the electrons form a completely degenerate gas. However, in astrophysics one wants to relax this assumption and for the low-temperature case, first discussed by Marshak and Bethe,[91] results for the equations of state can again be expressed in a convenient analytical form applicable to all elements.[92,94,95] For this reason we shall consider these results briefly here, although the discussion in Section 5.8.4 is applicable to arbitrary temperatures and therefore in principle covers the case dealt with here.

If the pressure in the case of incomplete degeneracy is denoted by $P$, then this can be expressed in terms of the pressure $p$ in the completely degenerate case by the equation

$$P = p[1 + \tfrac{5}{2}(\sigma + 2\tau)\zeta(k_B T)^2]. \tag{5.82}$$

Here

$$\zeta = \pi^2 b^2 / 8Z^2 e^4, \tag{5.83}$$

while $\tau$ is defined in terms of the Thomas–Fermi boundary value $\phi(x_0)$ by

$$\tau\{\phi(x_0)\}^2 = x_0^2. \tag{5.84}$$

The quantity $\sigma$ can be represented approximately by

$$\sigma\phi(x_0) = \sum C_n x_0^n, \tag{5.85}$$

where in the summation, $n=3$, $n'$ and 5. The values of $n'$ and of the coefficients $C_n$ are recorded in Table 5.8, taken from the work of Gilvarry and Peebles.[94]

These results can be used whenever the temperature is low in comparison with the maximum kinetic energy of electrons near the boundary of the atomic sphere, or when the inequality

$$k_B T \ll \frac{Ze^2}{b}\left[\frac{\phi(x_0)}{x_0}\right] \tag{5.86}$$

**Table 5.8** Parametrization of equation of state for incomplete degeneracy by eqn (5.82). Coeficients $C_n$ in eqn (5.85) are given.

| $n$ | $C_n$ |
|---|---|
| 3 | $-3.205\times10^{-1}$ |
| $n'=4.215$ | $-2.331\times10^{-2}$ |
| 5 | $-2.519\times10^{-3}$ |

is satisfied. High temperatures can in fact fall within the range of this treatment in the limit as $x_0$ tends to zero.

### 5.8.4 Generalized Thomas–Fermi theory for arbitrary degeneracy

We have already outlined the argument showing how the relation between density and potential in the generalized Thomas–Fermi theory is derived. Combining this with Poisson's equation we find

$$\nabla^2 V = \frac{16\pi^2 e}{h^3}(2mk_B T)^{3/2} I_{1/2}\left(\eta^* - \frac{V}{k_B T}\right), \tag{5.87}$$

which is to be solved for the potential energy $V$. In the sphere approximation we again have the boundary conditions (5.74). The spherically symmetrical solutions of eqn (5.87) satisfying the boundary conditions (5.74) were first examined by Feynman *et al.*[91] and subsequently by Latter.[96] The pressure $P$ can be calculated either by considering the rate of transfer of momentum between the electrons and the surface of the atomic sphere or from the thermodynamic relation

$$P = -\left(\frac{\partial F}{\partial V}\right)_T, \tag{5.88}$$

where $F$ is thc Hclmholtz free energy. The result may be written

$$P = \frac{8\pi}{3h^3}(2mk_B T)^{3/2}(k_B T) I_{3/2}\left(\eta - \frac{V(R)}{k_B T}\right), \tag{5.89}$$

and once the self-consistent potential energy $V$ is known, the pressure may then be obtained from eqn (5.89) as a function of temperature $T$ and volume $v$. Results are given, for example, by Latter[96] for the pressure–volume relation for a series of temperatures.

Of course, the full machinery of energy band theory, within the density functional framework, is very valuable for treating matter at pressures much lower than those required to validate the Thomas–Fermi results discussed above, which, essentially, require all "Periodic Table" effects to be small. This general subject of matter at high pressures and also at high temperatures is reviewed by Ross.[97,98] While it would take us too far from the main theme to go into details here, suffice it to say that questions like those raised in Section 5.7, concerning in particular energy band gaps, may well be fruitfully studied with the aid of an additional physical variable, namely the lattice parameter or equivalently pressure.

However, in view of the fact that we considered atoms in high magnetic fields in Chapter 1, it seemed of interest to conclude the present Chapter by summarizing the density functional calculations of Jones[99] on crystals subject to intense magnetic fields.

Specifically, Jones has calculated the ground-state energies of rhombohedral and body-centred tetragonal atomic lattices in such fields. The exchange-correlation function (cf. Chapter 6 below) adopted is that used in his earlier calculations of atomic and molecular chain ground-state energies. This has the form, per electron,

$$\varepsilon_{xc} = 2\pi\rho \ln \rho + 13.7\,\rho - 37.8\,\rho^2 - (0.0096 \ln \rho + 0.122). \tag{5.90}$$

The field-dependent units of length and energy in this equation are

$$\left.\begin{aligned} (\hbar c/eB)^{1/2} &= 2.566 \times 10^{-10}\, B_{12}^{-1/2}\,\text{cm}, \\ e^2(\hbar c/eB)^{-1/2} &= 0.5612\, B_{12}^{1/2}\,\text{keV}. \end{aligned}\right\} \tag{5.91}$$

The advantage of these units is that atomic sizes expressed in terms of them are very slowly varying functions of $B_{12}$ (units of $10^{12}$ G).

Atomic cohesive energies and electron chemical potentials have been derived for atomic numbers and magnetic flux densities relevant to the stellar surface in radio pulsars. This work appears to have important implications for astrophysics, but the interested reader should refer to the paper by Jones[99] for further details.

## Appendix A. 5.1

### Kinetic energy in perturbative series in one-body potential

In Section 1.9, the density–potential relation of the Thomas–Fermi theory was shown to follow from a variation principle, in which the kinetic energy density was taken as that given by free-electron gas theory, in a local spatially varying theory. It is of interest to summarize here the results of the kinetic energy density to all orders in perturbation theory, and to see how the Thomas–Fermi theory can be regained.

The kinetic energy density is given in terms of the Dirac density matrix by

$$t = -\frac{\hbar^2}{2m}\sum \psi_i^* \nabla^2 \psi_i$$
$$= -\frac{\hbar^2}{2m}[\nabla_r^2 \rho(\mathbf{r},\mathbf{r}')]_{\mathbf{r}'=\mathbf{r}}. \quad \text{(A 5.1.1)}$$

Using the full perturbation expansion of the Dirac matrix in ref. 2 it can be shown that the kinetic energy density $\Delta t$ measured relative to the unperturbed Fermi gas is given by[100]

$$\Delta t = \sum_{j=1}^{\infty} \frac{j}{j+1} V(\mathbf{r})\rho_j(\mathbf{r}), \quad \text{(A 5.1.2)}$$

where

$$\rho(\mathbf{r}) = \sum_{i=0}^{\infty} \rho_i(\mathbf{r}). \quad \text{(A 5.1.3)}$$

Since it was shown in ref. 2 that $\rho_j(\mathbf{r}) = \text{const} \times (V)^j$ for slowly varying $V(\mathbf{r})$ it is easy to complete the summation in eqn (A 5.1.2) in terms of the one-body potential, to obtain

$$t = \text{const}\,(E_F - V(r))^{5/2}; \quad E_F = \tfrac{1}{2}k_F^2 \quad \text{(A 5.1.4)}$$

Using the density–potential relation one regains the Thomas–Fermi kinetic energy density given in eqn (1.12).

Returning to the full perturbation-theory results, it has been shown by Corless and March[9] that one can rewrite the kinetic energy $T$ measured relative to the Fermi-gas value $T_0$ as

$$T - T_0 = \frac{k_F^2}{2}\int d\mathbf{r}\,(\rho - \rho_0)$$
$$+ \frac{k_F^4}{\pi^3}\iint d\mathbf{r}_1\, d\mathbf{r}_2 \frac{V(\mathbf{r}_1)V(\mathbf{r}_2)j_1(2k_F|\mathbf{r}_2 - \mathbf{r}_1|)}{(2k_F|\mathbf{r}_2 - \mathbf{r}_1|)^2} + \cdots . \quad \text{(A 5.1.5)}$$

Using this expression variationally, one regains the first-order theory of the displaced charge $\rho - \rho_0$ given by March and Murray[2] as

$$\rho(\mathbf{r}) - \rho_0 = -\frac{k_F^2}{2\pi^3}\int d\mathbf{r}_1\, V(\mathbf{r}_1)\frac{j_1(2k_F|\mathbf{r} - \mathbf{r}_1|)}{|\mathbf{r} - \mathbf{r}_1|^2} + O(V^2). \quad \text{(A 5.1.6)}$$

The expression (A.5.1.5) generalizes the expression given by Alfred and March[8] and quoted in eqn (5.15) of the main text. The two expressions obviously will agree precisely only when $V(\mathbf{r})$ varies sufficiently slowly in space, namely by but a small fraction of itself over a characteristic electron de Broglie wavelength, in order to validate the Thomas–Fermi approximation.

To conclude this Appendix, it is worth recording here the results for the first-order density matrix $\rho(\mathbf{r},\mathbf{r}')$ for free particles $\rho_0(|\mathbf{r}-\mathbf{r}'|)$ and its correction $\rho_1(\mathbf{r},\mathbf{r}')$ to first order in the perturbing potential $V(\mathbf{r})$. The results are

$$\rho_0(|\mathbf{r}-\mathbf{r}'|)=\frac{k_F^2}{\pi^2}\frac{j_1(k_F|\mathbf{r}-\mathbf{r}'|)}{|\mathbf{r}-\mathbf{r}'|} \tag{A 5.1.7}$$

and[2]

$$\rho_1(\mathbf{r},\mathbf{r}')=\frac{-k_F^2}{2\pi^3}\int d\mathbf{r}_1\, V(\mathbf{r}_1)\frac{j_1(k_F|\mathbf{r}_1-\mathbf{r}'|+k_F|\mathbf{r}-\mathbf{r}_1|)}{|\mathbf{r}_1-\mathbf{r}'||\mathbf{r}-\mathbf{r}_1|}. \tag{A 5.1.8}$$

Equation (A 5.1.8) reduces on the diagonal ($\mathbf{r}=\mathbf{r}'$) to eqn (A 1.5.6). When eqn (A 5.1.6) is combined with Poisson's equation, the result (5.3) of the text is regained.

## Appendix A 5.2

### One-body potential generating ground-state electron density in Wigner electron crystal at low density

The density functional method has been applied to the Wigner electron crystal by Shore *et al.*[101] Their variational method led to a one-body potential $Ar^2/r_s^3$, with $r_s$ the mean inter-electronic spacing, and $A$ was found to be remarkably near to the value $A=1$ Ry given by the localized Einstein oscillator model.[32] In this Appendix, quantitative contact is made between these apparently very different potentials in the extreme low-density regime. The transition from a uniform electron liquid to a Wigner electron crystal has been known to occur for numerous decades. However, the density at which the transition occurs has now been quantitatively settled.[33] Before that, Shore *et al.*[101] had presented a density functional calculation for the Wigner crystal and, provided their method is restricted to densities lower than the transition density given in ref. 33, the electron densities thereby calculated should still be of value.

Shore *et al.* emphasized that one method, the variational determination of an effective one-body potential, was successful in leading to physically reasonable electron distributions; whereas employing a local density approximation, taken to self-consistency, was not appropriate for calculating the ground-state electron density in the Wigner electron crystal.

One point about their effective potential has remained perplexing. In a spherical-cell approximation which we also retain below, they adopted for their variational trial potential in the density functional scheme an isotropic three-dimensional harmonic oscillator potential of the form $A(r/r_s)^2 a_0/r_s$, and then $A$ was found variationally to be 0.9 Ry. Although Shore *et al.* did not comment on this value, it turns out to be extremely near to that given by the potential

$$\mathscr{V}(r)=\frac{e^2}{2a_0}\left(\frac{r}{r_s}\right)^2\frac{a_0}{r_s}, \tag{A 5.2.1}$$

which generates the Wigner oscillator orbital

$$\phi(r)=\left(\frac{\alpha}{\pi}\right)^{3/4}\exp\left(-\frac{\alpha r^2}{2}\right);\quad \alpha=r_s^{-3/2}a_0^{-1/2}. \tag{A 5.2.2}$$

The purpose here is to relate the one-body potential $V(r)$ of the density functional approach with the localized oscillator potential (A 5.2.1). This can be done by utilizing a result obtained recently by the present writer for the local potential needed to generate the square root of the ground-state electron density in an inhomogeneous electron assembly by solution of a Schrödinger equation.[102] This reads, with $\psi$ written for the square root of the electron density $\rho^{1/2}$:

$$\nabla^2\psi+\frac{2m}{\hbar^2}(\varepsilon-\mathscr{V}(r))\psi=0, \tag{A 5.2.3}$$

and the potential energy $\mathscr{V}(r)$ is given by

$$\mathscr{V}(r)=V(r)+V_{\text{Pauli}}(r), \tag{A 5.2.4}$$

where $V_{\text{Pauli}}$ is given to within an additive constant by

$$V_{\text{Pauli}}=\frac{\delta T_s}{\delta\rho(r)}+\frac{\hbar^2}{4m}\left[\frac{\nabla^2\rho}{\rho}-\frac{1}{2}\frac{(\nabla\rho)^2}{\rho^2}\right] \tag{A 5.2.5}$$

with $T_s$ the single-particle kinetic energy.

Regarding the Wigner crystal as developing from a uniform density as $r_s$ is lowered to the critical value at which a first-order liquid-crystal transition occurs, it seems appropriate to take the gradient expansion of $T_s$ to make contact between the potential $V(r)$ of Shore *et al.* and the Wigner oscillator potential, to be identified with $\mathscr{V}(r)$.

Then we can write, following Kirzhnitz, the equation

$$T_s=c_k\int\{\rho(r)\}^{5/3}\,d\mathbf{r}+\frac{\hbar^2}{72m}\int\frac{(\nabla\rho)^2}{\rho}\,d\mathbf{r};\quad c_k=\frac{3h^2}{10m}\left(\frac{3}{8\pi}\right)^{2/3}, \tag{A 5.2.6}$$

and we find after a simple calculation that the one-body potential to be put into conventional single-particle equations is given by

$$V(r)=\frac{1}{9}\frac{\frac{1}{2}e^2r^2}{r_s^3}-\frac{5}{3}c_k\{\rho(r)\}^{2/3}. \tag{A 5.2.7}$$

This equation can be readily evaluated with the Wigner oscillator density obtained by squaring the orbital $\phi$ in eqn (A 5.2.2). However, as a first orientation, let us make the comparison with the effective potential of Shore *et al.* by expanding eqn (A 5.2.7) to $O(r^2)$, when, apart from an additive constant we find

$$V(r)=\frac{e^2r^2}{2r_s^3}\left[\frac{1}{9}+2\left(\frac{\pi}{3}\right)^{1/3}\right]. \tag{A 5.2.8}$$

In this small-$r$ limit we obtain the Wigner oscillator potential $\mathcal{V}(r)$, multiplied though by a factor 2.1. This result is at least a partial justification from first principles of the trial form assumed by Shore *et al.* As an alternative procedure, we have calculated the expectation value of $T+V$ with a trial function $\exp(-\beta r^2)$, with $V$ given by eqn (A 5.2.7), and $\rho$ taken from the square of $\phi$ in eqn (A 5.2.2). Again, our conclusion is in favour of a factor between 2 and 3 multiplying the Wigner potential (A 5.2.1), by minimizing $E=T+V$ with respect to $\beta$.

The present approach seems also favourable for treating jellium in a weak external potential. The square root of the density, $\psi$, is then given by solving the Schrödinger equation with local potential $\mathcal{V}(r)$ given by

$$\mathcal{V}(\mathbf{r})=V(\mathbf{r})+\int F(|\mathbf{r}-\mathbf{r}'|)V(\mathbf{r}')\,d\mathbf{r}', \tag{A 5.2.9}$$

where $V(r)$ is the one-body potential of density functional theory. $F(x)$, the linear response function, is given explicitly in ref. 102. One of the points of interest for the future is that of determining how the $r_s$ critical value is reduced from that given in ref. 33 as a function of the strength of the external potential $V_{ext}$.

In summary, the difference is stressed between the density functional approach to the Wigner crystal, which obtains the inhomogeneous density by summing the squares of single-particle. wave functions up to the appropriate maximum energy, and the local-potential $\mathcal{V}(r)$ method which obtains the ground-state density as the square of a single wave function. From the Wigner–Einstein oscillator model, at sufficiently low density, one already has a good physical approximation to $\mathcal{V}(r)$ given by eqn (A 5.2.1). The precise relation between $\mathcal{V}(r)$ and the density functional potential $V(r)$ remains, of course, formal without an assumption as to the form of the single-particle kinetic energy functional $T_s$. Using the Kirzhnitz form, contact is established with the effective variational potential in the density functional approach of Shore *et al.*[101] It is possible that this form may lead to some improvement on their choice of variational potential, especially if the factor $\frac{1}{9}$ is replaced by a variational parameter, but we do not expect their main conclusions to be affected by such a refinement. The final comment is to reinforce their point about spin polarization being necessary in future refined treatments.

## Appendix 5.3

### Pair correlation function and exchange energy density in the infinite-barrier model of a metal surface

The purpose of this Appendix is to summarize the exact results which can be obtained for the Fermi hole, or pair correlation function, round an electron moving in a metal with a surface which is represented by the infinite-barrier model introduced by Bardeen.[103]

As Bardeen[103] showed in his original paper, the electron density profile $\rho(z)$ as a function of distance $z$ from the infinite barrier is given by

$$\rho(z)=\left\{\begin{array}{ll}\rho_0\left(1-\dfrac{3}{2}\dfrac{j_1(2k_F z)}{k_F z}\right) & \text{for } z>0, \\ 0 & \text{otherwise,}\end{array}\right\} \tag{A 5.3.1}$$

where $\rho_0$ is evidently the electron density in the bulk metal. The positive background density in the model, $\rho_b(z)$ say, is given

$$\rho_b(z)=-\rho_0\theta(z-\xi), \tag{A 5.3.2}$$

$\theta$ being the usual Heaviside function and $\xi$ being fixed by the condition of electrical neutrality, namely

$$\int dz\,[\rho(z)+\rho_b(z)]=0, \tag{A 5.3.3}$$

which gives $\xi=3\pi/8k_F$.

Though fully self-consistent density profiles through a metal surface are described in the article by Lang in ref. 104, which therefore avoid the drastic infinite barrier assumption of the Bardeen model, its merit is that in addition to the density $\rho(z)$ in eqn (A 5.3.1), the off-diagonal density matrix defined in eqn (5.64) can also be calculated exactly in this model. This, through the relation (5.62) given for a Slater determinant, allows the Fermi hole or pair correlation function to be obtained. The explicit form of this is given by

$$\Gamma(\mathbf{r},\mathbf{r}')=\rho(z)\rho(z')-\tfrac{1}{2}[\rho(\mathbf{r},\mathbf{r}')]^2$$

$$=\rho(z)\rho(z')-\frac{9}{2}\rho_0^2\left[\frac{j_1(2k_F|\mathbf{r}-\mathbf{r}'|)}{|\mathbf{r}-\mathbf{r}'|}-\frac{j_1(k_F\{|\mathbf{r}-\mathbf{r}'|^2+4zz'\}^{1/2})}{\{|\mathbf{r}-\mathbf{r}'|^2+4zz'\}^{1/2}}\right]^2, \quad \text{(A 5.3.4)}$$

which evidently also defines the first-order density matrix $\rho(\mathbf{r},\mathbf{r}')$ for this model. The expression (A 5.3.4) for the pair function $\Gamma(\mathbf{r},\mathbf{r}')$ is such that in the square bracket the first term alone would give back the uniform-gas result (1.51). The distance $|\mathbf{r}-\mathbf{r}'|$ appearing there is, in the second term in the square bracket, replaced by the distance between $\mathbf{r}$ and the image of the point $\mathbf{r}$ in the planar metal surface.[105]

If the total electron density, equal to $\rho(z)+\rho_b(z)$, is denoted by $\rho_t(z)$, then the electrostatic energy density is evidently given by

$$\tfrac{1}{2}e^2\int d\mathbf{r}'\,\rho_t(\mathbf{r})\rho_t(\mathbf{r}')/|\mathbf{r}-\mathbf{r}'|,$$

and this has been calculated analytically in ref. 105. Our interest here is the form of the exchange energy density $\varepsilon_x(z)$, defined in terms of the first-order density matrix, via the Fermi hole result as (cf. eqns (1.51) and (1.52) for the uniform electron gas)

$$\varepsilon_x(z)=\frac{-3e^2\rho_0 k_F}{4\pi}J(2k_F z). \quad \text{(A 5.3.5)}$$

The function $J$ is given by the rather complicated expression

$$J(2k_F z)=\left[\frac{2}{3x^2}+\frac{1}{6x^4}+\frac{4\cos x}{x^4}\right]\cos 2x+\left[\frac{1}{3x^3}+\frac{4\sin x}{x^4}\right]\sin 2x$$
$$+\left[-\frac{2}{15}+\frac{64}{15x^2}-\frac{36}{5x^4}+\frac{4}{x^4}f(x)+\frac{4\,\mathrm{si}\,(2x)}{x^3}\right]\cos x$$
$$+\left[-\frac{2}{15x}-\frac{56}{5x^3}+\frac{4}{x^3}f(x)-\frac{4\,\mathrm{si}\,(2x)}{x^4}\right]\sin x$$
$$+1+\frac{1}{3x^2}+\frac{91}{30x^4}-\frac{2x}{15}\mathrm{si}\,(x)+\frac{4}{3x}\mathrm{si}\,(2x), \quad \text{(A 5.3.6)}$$

with $x=2k_F z$. Here $f(x)$ is defined by

$$f(x)=C+\ln(2x)-\mathrm{ci}\,(2x), \quad \text{(A 5.3.7)}$$

$C$ is Euler's constant, ci (2x) is the cosine integral, and si $(x)$ is explicitly given by

$$\mathrm{si}\,(x)=-\int_x^\infty \frac{\sin t}{t}\,dt. \quad \text{(A 5.3.8)}$$

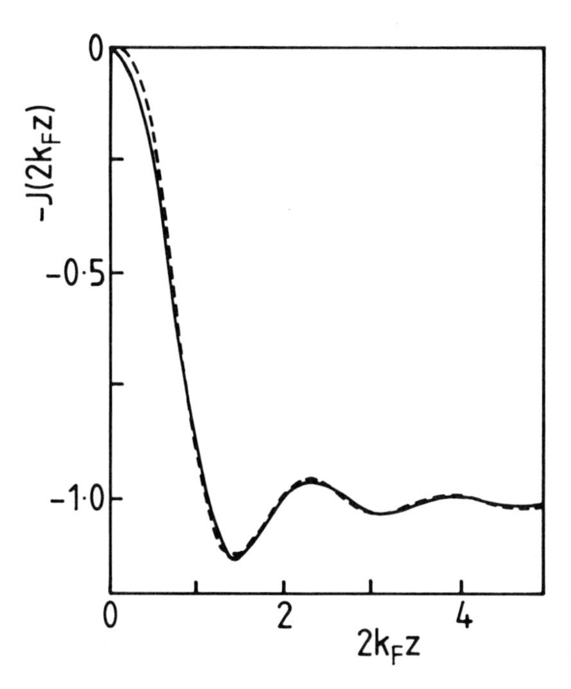


**Fig. 5.9** Local exchange energy density versus distance from barrier for the Bardeen model of a metal surface. The Dirac–Slater approximation (dashed curve) is seen to be an excellent fit to the exact calculation (solid curve).

The point to be emphasized is that here is an example of a truly inhomogeneous electron gas with a strong gradient in which we can test the approximate electron gas form (1.50) of the exchange energy density since we know this exactly from eqns (A 5.3.5) and (A 5.3.6), while $\rho(z)$ is available in eqn (A 5.3.1).

In Fig. 5.9 we show plots of the exchange energy density in units of $3e^2 \rho_0 k_F/4\pi$. The full line is the exact result of eqn (A 5.3.5), while the broken line is the local density approximation of eqn (1.50) calculated using the exact density (A 5.3.1). It can be seen that there is remarkable agreement between the exact result and the local density approximation for all $z$. Nevertheless, in spite of this, when the curves are integrated to obtain the total exchange energy, the exact and local theories differ by some 30%; the local density theory underestimating the magnitude of the total exchange energy. But there can be no doubt that this example is very encouraging for the local density theory of the exchange energy, even in the presence of a strongly varying density. In contrast, as Moore and March[106] demonstrate, the kinetic energy density approximated by the local form (1.13) in this same model is an approximation of much poorer quality than the exchange term discussed fully above.

## Appendix A 5.4

### Spin-density theory

The extension of the density functional method to include spin has been made by several authors. The essential point is that instead of just considering the particle density $\rho(\mathbf{r})$, one must now involve the spin-density vector $\mathbf{s}(\mathbf{r})$ as well. Similarly, instead of just considering a system subject to an external potential, there will be a static external magnetic field $B$, which in general is position-dependent, i.e. $B = B(\mathbf{r})$. The conclusion is, fairly clearly, that the ground-state energy is a unique functional of the charge and spin densities, which attains its minimum value when these quantities take their correct forms.

As in Chapter 1 a fully relativistic treatment incorporating the above points has been presented, we shall restrict this Appendix to a very simple account of the non-relativistic case. As the most elementary example of the charge density description developed in Chapter 1, generalized to treat spin density, consider the generalization of the Thomas–Fermi–Dirac theory for uniform electrons, described by plane waves. If $\rho_\uparrow$ and $\rho_\downarrow$ denote the densities, assumed constant, for the upward and downward spins, the single-particle kinetic energy of the Thomas–Fermi theory is readily generalized to read

$$t[\rho_\uparrow, \rho_\downarrow] = \text{const}\,(\rho_\uparrow^{5/3} + \rho_\downarrow^{5/3}). \tag{A 5.4.1}$$

Similarly the exchange energy density takes the form

$$\varepsilon_x[\rho_\uparrow, \rho_\downarrow] = \text{const}\,(\rho_\uparrow^{4/3} + \rho_\downarrow^{4/3}). \tag{A 5.4.2}$$

As Bloch was the first to show, by minimizing $t + \varepsilon_x$ with respect to $\rho_\uparrow$ and $\rho_\downarrow$, subject to $\rho_\uparrow + \rho_\downarrow = N/\Omega$ for $N$ electrons in volume $\Omega$, the ferromagnetic state becomes stable at sufficiently low densities, relative to the paramagnetic state. In terms of $\rho = \rho_\downarrow + \rho_\uparrow$ and the exchange constant $c_e$ in eqn (1.53), the condition for ferromagnetism in the ground state of this model is found to be

$$c_e \rho^{1/3}/E_F > \tfrac{3}{5}(2^{1/3} + 1), \tag{A 5.4.3}$$

where $E_F$ as usual denotes the Fermi energy.

This example is merely illustrative and we must caution the reader that the Hartree–Fock model overestimates the tendency to ferromagnetism because in this theory there are Fermi statistical correlations between all parallel-spin electrons but no correlation between antiparallel spins. In fact, in the uniform electron liquid, the introduction of correlation stabilizes magnetic phases relative to the paramagnetic phase only at very low densities.[33]

## Spin-dependent one-body potentials

We conclude this Appendix with a sketch of the generalization of one-body potential theory to describe spin densities in inhomogeneous electron clouds, leading to spin-dependent potentials.[107] Suppose a spin-dependent external potential $V_{\text{ext}}^{\sigma}$ is switched on, $\sigma$ being either $\uparrow$ or $\downarrow$. The ground-state energy, from the above discussion, can now be written formally as[107–110]

$$E[\rho_{\uparrow}(\mathbf{r}), \rho_{\downarrow}(\mathbf{r})] = E[\rho(\mathbf{r}), \mathrm{m}(\mathbf{r})], \tag{A 5.4.4}$$

where $m(\mathbf{r})$ is the resultant magnetization density $\rho_{\uparrow} - \rho_{\downarrow}$. This can be written a little more explicitly as

$$E = G[\rho, m] + \tfrac{1}{2}\int \rho V_{\mathrm{e}}\,\mathrm{d}\mathbf{r} + \int \rho V_{\text{ext}}(\mathbf{r})\,\mathrm{d}\mathbf{r} + E_{\text{field}}, \tag{A 5.4.5}$$

where $G$ subsumes kinetic, exchange and correlation energy densities as a functional of $\rho$ and $m$, and $E_{\text{field}}$ has been added to represent the interaction with a magnetic field $B$.

Performing the minimization of $E$ with respect to the spin densities, as a generalization of the charge-density argument of the main text, leads to two spin-dependent potentials

$$V_{\sigma}(\mathbf{r}) = V_{\text{ext}}(\mathbf{r}) + \int \frac{\rho(\mathbf{r}')\,\mathrm{d}\mathbf{r}'}{|\mathbf{r} - \mathbf{r}'|} + V_{\text{xc}}^{\sigma}(\mathbf{r}). \tag{A 5.4.6}$$

Again local density approximations are usually made for $V_{\text{xc}}^{\sigma}(\mathbf{r})$. Examples of this type of approach are given in the article by Ellis[111] and in the work of Langlinais and Callaway.[112] Later work by Hathaway *et al.*[113] contains numerous other references that the interested reader should consult.

## References

1. Mott, N.F. (1936). *Proc. Camb. Phil. Soc.*, **32**, 281.
2. March, N.H. and Murray, A.M. (1960). *Phys. Rev.*, **120**, 830; (1961). *Proc. Roy. Soc.*, **A261**, 119.
3. Blandin A., Daniel, E. and Friedel, J. (1959). *Phil. Mag.*, **4**, 180.
4. See also March, N.H. and Murray, A.M. (1960). *Proc. Roy. Soc.*, **A256**, 400.

5. Lindhard, J. (1954). *Kgl. Danske Mat. fys. Medd.*, **28**, 8.
6. Kohn, W. (1959). *Phys. Rev. Lett.*, **2**, 393.
7. Jones, W. and March, N.H. (1985). *Theoretical Solid-State Physics*, Vols. I and II, Dover (Reprints Series), New York.
8. Alfred, L.C.R. and March, N.H. (1957). *Phil. Mag.*, **2**, 985.
9. Corless, G.K. and March, N.H. (1961). *Phil. Mag.*, **6**, 1285.
10. See, for example, March, N.H. and Tosi, M.P. (1984). *Coulomb Liquids*, Academic Press, New York.
11. March, N.H. (1966). *Phys. Lett.*, **20**, 231.
12. Fumi, F.G. (1955). *Phil. Mag.*, **46**, 1007.
13. Mukherjee, K. (1965). *Phil. Mag.*, **12**, 915.
14. Bohm, D. and Staver, T. (1952). *Phys. Rev.*, **84**, 836.
15. See also Bardeen, J. and Pines, D. (1955). *Phys. Rev.*, **99**, 1140.
16. Stott, M.J., Baranovsky, S. and March, N.H. (1970). *Proc. Roy. Soc.*, **A316**, 201.
17. Flores, F. and March, N.H. (1981). *J. Phys. Chem. Solids*, **42**, 439.
18. Friedel, J. (1954). *Adv. Phys.*, **3**, 446.
19. Biondi, M.A. and Rayne, J.A. (1959). *Phys. Rev.*, **115**, 1522.
20. See, for example, ref. 7, Vol. II, Chap. 10.
21. Bhatia, A.B. and March, N.H. (1978). *J. Chem. Phys.*, **68**, 4651.
22. Dingle, R.B. (1955). *Phil. Mag.*, **46**, 831.
23. Mansfield, R. (1956). *Proc. Phys. Soc.*, **B69**, 76.
24. Resta, R. (1977). *Phys. Rev.*, **B16**, 2717.
25. Penn, D.R. (1962). *Phys. Rev.*, **128**, 2093.
26. See also the earlier work of Callaway, J. (1959). *Phys. Rev.*, **116**, 1368.
27. Srinivasan, G. (1969). *Phys. Rev.*, **178**, 1244.
28. Baroni, S. and Resta, R. (1986). *Phys. Rev.*, **B33**, 7017 and references therein.
29. See, for example, Löwdin, P.O. (1955). *Phys. Rev.*, **97**, 1474.
30. Wigner, E.P. (1934). *Phys. Rev.*, **46**, 1002; (1938). *Trans. Faraday Soc.*, **34**, 678.
31. See, for instance, Fuchs, K. (1935). *Proc. Roy. Soc.*, **A151**, 585.
32. Care, C.M. and March, N.H. (1975). *Adv. Phys.*, **24**, 101.
33. Ceperley, D.M. and Alder, B.J. (1980). *Phys. Rev. Lett.*, **45**, 566.
34. Callaway, J. and March, N.H. (1984). In *Solid State Physics*, Vol. 38, eds Ehrenreich, H. and Turnbull, D., Academic Press, New York, p. 135.
35. Slater, J.C. (1951). *Phys. Rev.*, **81**, 385.
36. Laurent, D.G., Callaway, J., Fry, J.L. and Brener, N.E. (1981). *Phys. Rev.*, **B23**, 4977.
37. von Barth, U. and Hedin, L. (1972). *J. Phys.*, **C5**, 1629.
38. Moruzzi, V.L., Janak, J.F. and Williams, A.R. (1978). *Calculated Electronic Properties of Metals*, Pergamon, Oxford.
39. Callaway, J. (1974). *Quantum Theory of the Solid State*, Academic Press New York.
40. Callaway, J., Zou, X. and Bagayoko, D. (1983). *Phys. Rev.*, **B27**, 631.
41. Vosko, S.H. and Wilk, L. (1980). *Phys. Rev.*, **B22**, 3812.
42. Pack, J.D., Monkhorst, H.J. and Freeman, D.L. (1979). *Solid St. Commun.*, **29**, 723.
43. Janak, J.F., Williams, A.R. and Moruzzi, V.L. (1972). *Phys. Rev.*, **B6**, 4367.
44. Janak, J.F., Williams, A.R. and Moruzzi, V.L. (1975). *Phys. Rev.*, **B11**, 1522; and ref. 38.
45. Bagayoko, D., Laurent, D.G., Singhal, S.P. and Callaway, J. (1980). *Phys. Lett.*, **76A**, 187.

46. Jepsen, O., Glötzel, D. and Mackintosh, A.R. (1981). *Phys. Rev.*, **B23**, 2684.
47. Thiry, P., Chandesris, D., Lecante, J., Guillot, C., Pinchaux, R. and Pétroff, Y. (1979). *Phys. Rev. Lett.*, **43**, 82.
48. Wang, J.S.Y. and Rasolt, M. (1977). *Phys. Rev.*, **B15**, 3714.
49. Yin, M.T. and Cohen, M.L. (1980). *Phys. Rev. Lett.*, **45**, 1004.
50. Ihm, J., Zunger, A. and Cohen, M.L. (1979). *J. Phys.*, **C12**, 4409.
51. Hamann, D.R., Schlüter, M. and Chiang, C. (1979). *Phys. Rev. Lett.*, **43**, 1494.
52. Callaway, J. and Wang, C.S. (1977). *Phys. Rev.*, **B16**, 2095.
53. Wang, C.S. and Callaway, J. (1977). *Phys. Rev.*, **B15**, 298.
54. Eastman, D., Himpsel, F.J. and Knapp, J.A. (1978). *Phys. Rev. Lett.*, **40**, 1514.
55. Callaway, J. (1980). *Inst. Phys. Conf. Ser.*, No. 55, p. 1.
56. Himpsel, F.J., Heimann, P. and Eastman, D.E. (1981). *J. Appl. Phys.*, **52**, 1658.
57. Callaway, J., Wang, C.S. and Laurent, D.G. (1981). *Phys. Rev.*, **B24**, 6491.
58. Mook, H.A., Lynn, J.W. and Nicklow, R.M. (1973). *Phys. Rev. Lett.*, **30**, 556.
59. Blaha, P. and Callaway, J. (1986). *Phys. Rev.*, **B33**, 1706.
60. Heeger, A.J. (1969). In *Solid State Physics*, Vol. 23, eds Ehrenreich, H., Seitz, F. and Turnbull, D., Academic Press, New York.
61. Kondo, J. (1969). In *Solid State Physics*, Vol. 23. eds Ehrenreich, H., Seitz, F. and Turnbull, D., Academic Press, New York.
62. Oswald, A., Zeller, R., Braspenning, P.J. and Dederichs, P.H. (1985). *J. Phys.* **F15**, 193 and references therein.
63. See, for instance, Lee, K., Callaway, J., Kwong, K., Tang, R., and Ziegler, A. (1985). *Phys. Rev.*, **B31**, 1796.
64. Lee, K., Callaway, J. and Dhar, S. (1984). *Phys. Rev.*, **B30**, 1724.
65. Wachters, A.J.H. (1970). *J. Chem. Phys.*, **52**, 1033.
66. Mintmire, J.W. and Dunlap, B.I. (1983). *Phys. Rev.*, **A25**, 88.
67. Rajagopal, A.K., Singhal, S.P. and Kimball, J.: cited by Rajagopal, A.K. (1979) in *Advances in Chemical Physics*, Vol. 41, eds Prigogine, I. and Rice, S.A., p. 59, Wiley, New York.
68. Delley, B., Ellis, D.E., Freeman, A.J., Baerends, E.J. and Post, D. (1983). *Phys. Rev.*, **B27**, 2132.
69. Messmer, R.P., Knudson, S.K., Johnson, K.H., Diamond, J.B. and Yang, C.Y. (1976). *Phys. Rev.*, **B13**, 1396.
70. Demuynck, J., Rohmer, M., Strich, A. and Veillard, A. (1981). *J. Chem. Phys.*, **75**, 3443.
71. Bagayoko, D. (1983). *Int. J. Quant. Chem.*, **17**, 527.
72. Johnson, K.H., Vvedensky, D.D. and Messmer, R.P. (1979). *Phys. Rev.*, **B19**, 1519.
73. Boyce, J. and Slichter, C.P. (1976). *Phys. Rev.*, **B13**, 379.
74. Dickens, M.H., Shull, C.G., Koehler, W.C. and Moon, R.M. (1975). *Phys. Rev. Lett.*, **35**, 595.
75. Coleridge, P.T., Scott, G.B. and Templeton, I.M. (1972). *Can. J. Phys.*, **50**, 1999.
76. Alles, H., Higgins, R.J. and Lowndes, D.H. (1973). *Phys. Rev. Lett.*, **30**, 705.
77. Hochst, H., Steiner, P. and Hüfner, S. (1980). *Z. Phys.*, **B38**, 201.
78. Cohen, J.D. and Slichter, C.P. (1980). *Phys. Rev.*, **B22**, 45.
79. Davis, J.R. and Hicks, T.J. (1979). *J. Phys.*, **F9**, L7.
80. Steiner, P., Hüfner, S. and Zdrojewski W.V. (1974). *Phys. Rev.*, **B10**, 4704.
81. Nevald, R. and Petersen, G. (1975). *J. Phys.* F. **5**, 1778.
82. Lo, D.C., Lang, D.V., Boyce, J.B. and Slichter C.P. (1973). *Phys. Rev.*, **B8**, 973.
83. Beaglehole, D. (1976). *Phys. Rev.*, **B14**, 341.

84. Pickett, W.E. (1985). *Comments Solid St. Phys.*, **12**, 1; (1986). *Ibid.*, **12**, 57.
85. Wang, C.S. and Klein, B.M. (1981). *Phys. Rev.*, **B24**, 3393.
86. Khan, M.A. and Callaway J. (1980). *Phys. Lett.*, **76A**, 441.
87. Pickett, W.E. and Wang, C.S. (1984). *Phys. Rev.*, **B30**, 4719.
88. Perdew, J.P. and Levy, M. (1983). *Phys. Rev. Lett.*, **51**, 1884.
89. Sham, L.J. and Schlüter, M. (1983), *Phys. Rev. Lett.*, **51** 1888.
90. Slater, J.C. and Krutter, H.M. (1935). *Phys. Rev.*, **47**, 559.
91. See the review by March, N.H. (1957). *Adv. Phys.*, **6**, 1.
92. Gilvarry, J.J. (1954). *Phys. Rev.*, **96**, 934.
93. March, N.H. (1955). *Proc. Phys. Soc.*, **A68**, 726.
94. Gilvarry, J.J. and Peebles, G.H. (1955). *Phys. Rev.*, **99**, 550.
95. March, N.H. (1955). *Proc. Phys. Soc.*, **A68**, 1145.
96. Latter, R. (1955). *Phys. Rev.*, **99**, 1854.
97. Ross, M. (1985). *Rep. Prog. Phys.*, **48**, 1.
98. See, for instance, McMahan, A.K., Albers, R.C. and Müller, J.E. (1985). *Phys. Rev.*, **B31**, 3435.
99. Jones, P.B. (1985). *Mon. Not. Roy. Astron. Soc.*, **216**, 503; (1985). *Phys. Rev. Lett.*, **55**, 1338; (1986). *Mon. Not. Roy. Astron. Soc.*, **218**, 477.
100. Stoddart, J.C. and March, N.H. (1967). *Proc. Roy. Soc.*, **A299**, 279.
101. Shore, H.B., Zaremba, E., Rose, J.H. and Sander, L. (1978). *Phys. Rev.*, **B18**, 6506.
102. March, N.H. (1986). *Phys. Lett.*, **113A**, 476.
103. Bardeen, J. (1936). *Phys. Rev.*, **49**, 653.
104. Lang, N.D. (1983). In *Theory of the Inhomogeneous Electron Gas*, eds Lundqvist, S. and March, N.H., Plenum, New York.
105. Miglio, L., Tosi, M.P. and March, N.H. (1981). *Surf. Sci.*, **111**, 119.
106. Moore, I.D. and March, N.H. (1976). *Ann. Phys. (N.Y.)*, **97**, 136.
107. Stoddart, J.C. and March, N.H. (1971). *Ann. Phys. (N.Y.)*, **64**, 174.
108. Pant, M.M. and Rajagopal, A.K. (1972). *Solid St. Commun.*, **10**, 1157.
109. von Barth, U. and Hedin, L. (1972). *J. Phys.*, **C5**, 1629.
110. Rajagopal, A.K. and Callaway, J. (1973). *Phys. Rev.*, **B7**, 1912.
111. Ellis, D.E. (1980). In *Electron and Magnetization Densities in Molecules and Crystals*, ed. Becker, P. (NATO Advanced Study Institutes Series, Physics, Vol. 48), p. 107, Plenum, New York.
112. Langlinais, J. and Callaway, J. (1972). *Phys. Rev.*, **B5**, 124.
113. Hathaway, K.B., Jansen, H.J.F. and Freeman, A.J. (1985). *Phys. Rev.*, **B31**, 7603 and references therein.

# 6 Exchange-Correlation Energy Functionals for Extended Systems

D.J.W. GELDART

*Department of Physics, Dalhousie University, Halifax, Nova Scotia, Canada B3H3J5*

and

M. RASOLT

*Solid State Division, Oak Ridge National Laboratory, Oak Ridge, Tennessee 37831, U.S.A.*



## 6.1 Introduction

The study of homogeneous many-particle systems of *uniform* density has been very useful in developing our understanding of the average or bulk properties of condensed matter. These studies have proceeded by a wide variety of analytic and computational methods. For the past few decades,

Single-Particle Density in Physics
and Chemistry ISBN 0-12-470518-9

there has been greatly renewed interest in inhomogeneous systems of *non-uniform* density. This renewed interest has partly been due to the need to meet the challenges in understanding the properties of materials and devices of technological importance. Additional major stimulus for the study of inhomogeneous systems was provided by the rapid development of density functional theory (DFT) which is a far-reaching generalization of Thomas–Fermi theory.[1–4] From the initial work of Hohenberg, Kohn and Sham,[5–7] DFT has developed extensively both a general theoretical framework and a variety of approximate procedures for the quantitative description of many-particle systems.

In its simplest form, DFT states that the ground-state energy of a system of particles interacting via a two-body potential $v$ (which is Coulombic or short range for many-fermion systems) and moving in an external potential $V_{\text{ext}}$ is a unique functional $E[\rho]$ of the particle density $\rho(\mathbf{r})$, which attains its minimum value for the ground-state energy, when the density is the ground-state density of particles, $\rho(\mathbf{r})$. This statement of DFT for the ground-state energy of a single-component system can be extended considerably; for example, generalizations to finite temperature[8–12] and to multi-component systems have been given.[13–32] Relativistic extensions have also been given.[33–36]

It is convenient to write the various contributions to $E[\rho]$ explicitly via[5]

$$E[\rho] = \int d^3r\, V_{\text{ext}}(\mathbf{r})\rho(\mathbf{r}) + F[\rho], \tag{6.1}$$

where $F[\rho]$ is a universal functional of the density only:

$$F[\rho] = \frac{1}{2}\int d^3r \int d^3r'\, \rho(\mathbf{r})v(\mathbf{r}-\mathbf{r}')\rho(\mathbf{r}') + G[\rho], \tag{6.2}$$

which separates the long-range Hartree energy from

$$G[\rho] = T[\rho] + E_{\text{xc}}[\rho], \tag{6.3}$$

in which $T[\rho]$ is the kinetic-energy functional for a system of non-interacting particles of density $\rho(\mathbf{r})$ and the remainder is the exchange-correlation energy functional, $E_{\text{xc}}[\rho]$. It is obviously the latter which is difficult to construct, as full knowledge of $E_{\text{xc}}[\rho]$ is tantamount to having an exact solution for the energy of any arbitrary interacting many-particle system.

The development of adequate approximate methods to describe the universal functional $E_{\text{xc}}[\rho]$ has proved to be a major challenge, as expected. Some important observations concerning its structure were already provided in the seminal paper of Hohenberg and Kohn.[5] $E_{\text{xc}}[\rho]$ is universal and applies equally to all systems, whether small (e.g. atoms) or large (e.g.

a solid); whether the density variation is large or small; and so on. The capacity of a single functional to describe in a universal way all systems implies a considerable complexity of structure which must be very non-trivial. Consequently, in order to gain some limited understanding, it is useful to examine simple limiting cases. Some of these simple cases are reviewed in Section 6.2. These simple cases provide motivation for a number of approximation schemes. Their major features are discussed and their relative advantages and disadvantages are indicated.

## 6.2 General procedures for approximations to $E_{xc}[\rho]$

In this section, we consider general procedures for generating tractable approximations to the exchange-correlation energy functional in various circumstances. It is convenient to distinguish three situations.

(1) If the particle density is nearly uniform throughout the system, the external potential responsible for the small non-uniformity in the particle density is small. Thus, straightforward perturbation theory is applicable, and within the usual context of a perturbation expansion, the properties of the weakly inhomogeneous system can be rigorously calculated.
(2) If the particle density is slowly varying in space, the system can be regarded as locally uniform. This constitutes the local density approximation (LDA) as a first approximation. Corrections to this lowest-order approximation can be generated in the form of a series expansion in powers of various spatial gradients of the particle density.
(3) Formally exact expressions for $E_{xc}[\rho]$ in terms of the density–density correlation function are known. Some of these are expressed in terms of the exchange-correlation hole. Approximations to the latter quantity, based largely on experience in uniform systems, can be generated.

In the following subsections, we review briefly the formulation of these procedures. Their relative merits as well as their weaknesses are indicated.

### 6.2.1 Almost uniform density

Following Hohenberg and Kohn, we first examine the case of almost uniform density, where

$$\rho(\mathbf{r}) = \rho_0 + \tilde{\rho}(\mathbf{r}) \tag{6.4}$$

with $|\tilde{\rho}(\mathbf{r})| \ll \rho_0$, where $\rho_0$ is the uniform particle density in the absence of the very weak external potential $V_{\text{ext}}(\mathbf{r})$, which induces the small density variation. This situation is amenable to a straightforward (in principle) perturbation expansion. Since $\tilde{\rho}(\mathbf{r})$ is a small quantity, it is a suitable parameter for a functional series expansion of well-behaved quantities. For example, we must have

$$G[\rho] = G[\rho_0] + \int d^3r\, K_1(\mathbf{r})\tilde{\rho}(\mathbf{r})$$

$$+ \frac{1}{2}\int d^3r \int d^3r'\, K_2(\mathbf{r},\mathbf{r}')\tilde{\rho}(\mathbf{r})\tilde{\rho}(\mathbf{r}') + \cdots \tag{6.5}$$

The expansion coefficients of such a functional Taylor expansion are appropriate derivatives of $G[\rho]$ evaluated at uniform density. Thus

$$K_1(\mathbf{r}) = \left[\frac{\delta G}{\delta\rho(r)}\right]_{\rho=\rho_0} \tag{6.6}$$

and

$$K_2(\mathbf{r},\mathbf{r}') = \left[\frac{\delta^2 G}{\delta\rho(\mathbf{r})\delta\rho(\mathbf{r}')}\right]_{\rho=\rho_0}. \tag{6.7}$$

The fact that these derivatives are evaluated for a system of uniform density implies translational invariance. Thus $K_1(\mathbf{r}) = \text{const}$ (independent of position). Then the second term on the right-hand side of eqn (6.5) vanishes, owing to

$$\int d^3r\, \tilde{\rho}(\mathbf{r}) = 0, \tag{6.8}$$

because turning on an external potential does not change the number of particles in the system. Also, the implication of translational invariance for $K_2(\mathbf{r},\mathbf{r}')$ is

$$K_2(\mathbf{r},\mathbf{r}') = K(\mathbf{r}-\mathbf{r}'). \tag{6.9}$$

The first non-trivial effects of inhomogeneity of the system thus appear at second order and eqn (6.5) becomes

$$G[\rho] = G[\rho_0] + \frac{1}{2}\int d^3r \int d^3r'\, K(\mathbf{r}-\mathbf{r}')\tilde{\rho}(\mathbf{r})\tilde{\rho}(\mathbf{r}') + \cdots, \tag{6.10}$$

for a system of total volume $\Omega$. In order to identify $K(\mathbf{r}-\mathbf{r}')$, we use standard perturbation theory to obtain the shift in the energy and in the particle density induced by $V_{\text{ext}}$,[5,37]

$$\Delta E = \frac{1}{2\Omega}\sum_{\mathbf{k}} \chi(\mathbf{k})|V_{\text{ext}}(\mathbf{k})|^2 + \cdots \tag{6.11}$$

and

$$\tilde{n}(\mathbf{k}) = \chi(\mathbf{k}) V_{\text{ext}}(\mathbf{k}). \tag{6.12}$$

The density–density correlation function (also known as the linear density response function) $\chi(\mathbf{k})$ can be written as

$$\chi(\mathbf{k}) = -\pi(\mathbf{k})/\varepsilon(\mathbf{k}), \tag{6.13}$$

where the dielectric function,

$$\varepsilon(\mathbf{k}) = 1 + v(\mathbf{k})\pi(\mathbf{k}), \tag{6.14}$$

is expressed in terms of the interparticle interaction, $v(\mathbf{k}) = 4\pi e^2/\mathbf{k}^2$ in the present case of interest, and the irreducible polarization (screening) function $\pi(\mathbf{k})$. The energy shift includes, of course, the direct coupling of the particle density to the external potential, and the Hartree energy. Using eqns (6.1) and (6.2) and (6.11)–(6.14), these contributions to $\Delta E$ can be isolated so that eqn (6.11) becomes

$$\Delta E = \int \mathrm{d}^3 r\, V_{\text{ext}}(\mathbf{r})\tilde{\rho}(\mathbf{r}) + \frac{1}{2}\int \mathrm{d}^3 r \int \mathrm{d}^3 r'\, \tilde{\rho}(\mathbf{r}) v(\mathbf{r}-\mathbf{r}')\tilde{\rho}(\mathbf{r}') + \Delta G, \tag{6.15}$$

where

$$\Delta G = \frac{1}{2\Omega}\sum_{\mathbf{k}} |\tilde{\rho}(\mathbf{k})|^2/\pi(\mathbf{k}), \tag{6.16}$$

which immediately allows the identification of the kernel $K$ in eqn (6.10) via

$$K(\mathbf{k}) = 1/\pi(\mathbf{k}). \tag{6.17}$$

If the same procedures are followed for a system having the same, almost uniform, density but without two-body interactions, the resulting kernel is

$$K_0(\mathbf{k}) = 1/\pi_0(\mathbf{k}), \tag{6.18}$$

where $\pi_0(\mathbf{k})$ is the well-known Lindhard function. According to eqn (6.3), we can thus subtract the kinetic energy contribution to $G[\rho]$ to obtain the explicit exchange-correlation contribution:

$$\Delta E_{\text{xc}}[\rho] = \frac{1}{2\Omega}\sum_{\mathbf{k}} |\tilde{\rho}(\mathbf{k})|^2 K_{\text{xc}}(\mathbf{k}), \tag{6.19}$$

where

$$K_{\text{xc}}(\mathbf{k}) = 1/\pi(\mathbf{k}) - 1/\pi_0(\mathbf{k}). \tag{6.20}$$

We emphasize that this result is a rigorous determination of the inhomogeneity contribution to $E_{\text{xc}}$ to second order in the external potential and is therefore very valuable. However, there are also some disadvantages in this approach. First, $\pi(\mathbf{k})$ is still rather incompletely known, although a good deal of progress has been made.[38-44] However, a more serious disadvantage is that even if $\pi(\mathbf{k})$ were known exactly, eqn (6.19) is valid only to second order in $V_{\text{ext}}$ so is not of practical utility, in itself, in common situations where the external potential generating the particle-density non-uniformity is strong. Thus, the above procedure is incomplete as regards practical applications but, in several ways which are discussed in the following, it does serve as a valuable benchmark and as a guide in constructing and evaluating the potential of more ambitious approximation schemes.

### 6.2.2 Slowly varying density

We consider first the situation where the density is not only slowly varying but is almost uniform in the sense that $\tilde{n}(\mathbf{k})$ has only small-wavenumber components. Then only small-$\mathbf{k}$ terms contribute to eqn (6.19), which permits simplifications. By rotational invariance, $K_{\text{xc}}$ is a function of $k=(\mathbf{k}^2)^{1/2}$ only. We assume that $K_{\text{xc}}(\mathbf{k})$ can be expanded in powers of $k$ to give

$$E_{\text{xc}}[\rho]=E_{\text{xc}}[\rho_0]+\frac{1}{2\Omega}\sum_{\mathbf{k}}|\tilde{\rho}(\mathbf{k})|^2[K_{\text{xc}}(0)+\tfrac{1}{2}k^2K''_{\text{xc}}(0)+\cdots]$$

$$=E_{\text{xc}}[\rho_0]+\int d^3r\,[\tfrac{1}{2}K_{\text{xc}}(0)(\tilde{\rho}(\mathbf{r}))^2+\tfrac{1}{4}K''_{\text{xc}}(0)(\nabla\tilde{\rho}(\mathbf{r}))^2+\cdots]. \quad (6.21)$$

By virtue of $\partial E/\partial N=\mu$ and the Ward identity, leading to

$$\partial\mu/\partial\rho_0=1/\pi(0), \quad (6.22)$$

the first two terms on the right-hand side of eqn (6.21) are recognized as the first two terms in the expansion of the LDA in powers of $\tilde{\rho}$; that is,

$$E_{\text{xc}}^{\text{LDA}}[\rho]=\int d^3r\,A_{\text{xc}}(\rho(\mathbf{r})), \quad (6.23)$$

where $A_{\text{xc}}(\rho(\mathbf{r}))$ is the exchange-correlation energy per unit volume of a uniform system of the local density $\rho(\mathbf{r})=\rho_0+\tilde{\rho}(\mathbf{r})$. Expansion of eqn (6.23) to second order in $\tilde{\rho}(\mathbf{r})$ yields the above-mentioned terms of eqn (6.21).

The third term on the right-hand side of eqn (6.21) goes beyond viewing the system as quasi-uniform and yields the lowest-order gradient correction to the LDA. We must emphasize that while $A_{\text{xc}}$ is given fully by the density dependence of the energy of a uniform system, $K''_{\text{xc}}(k=0)$ is not obtainable from this knowledge of the energy of a uniform system but requires information concerning the wavenumber dependence of the density fluctuation correlation function of the uniform system.

There are several ways that eqn (5.21) could be extended in order to generate a possibly useful computational tool within DFT. Higher-order terms in powers of $\tilde{\rho}$ could be included beyond the second-order perturbation expansion in $V_{\text{ext}}$. This would introduce higher-order response functions, involving triplet, quadruplet, etc. correlation functions of the interacting uniform system depending on a correspondingly increased number of independent wave-vectors. If the simple structure associated with gradient corrections is desired, we could expand all of these new terms in powers of their various wave-vector dependences to obtain contributions involving

$$\int d^3r[B_{xc}(\rho_0)(\nabla\tilde{\rho}(\mathbf{r}))^2+C_{xc}(\tilde{\rho})(\nabla^2\rho_0(\mathbf{r}))^2$$
$$+D_{xc}(\rho_0)(\nabla^2\tilde{\rho}(\mathbf{r}))(\nabla\tilde{\rho}(\mathbf{r}))^2+E_{xc}(\rho_0)(\nabla\tilde{\rho}(\mathbf{r}))^4+\cdots]. \qquad (6.24)$$

Note that the last three terms are of second, third and fourth order, respectively, in $V_{\text{ext}}$, just by counting factors of $\tilde{\rho}$. There will also be terms involving zero, two, four, etc., spatial derivatives each with various factors of $\tilde{\rho}(\mathbf{r})$. When all results are collected, the net effect is that all of the gradient coefficients $A_{xc}(\rho_0)$, $B_{xc}(\rho_0)$, etc., and their various derivatives with respect to $\rho_0$ will enter with precisely the right series of powers of $\tilde{\rho}(\mathbf{r})$ so that each gradient coefficient is evaluated at the local density $\rho(\mathbf{r})=\rho_0+\tilde{\rho}(\mathbf{r})$. Since $\nabla\tilde{\rho}(\mathbf{r})=\nabla\rho(\mathbf{r})$ is trivially true, it follows that this particular partial resummation of the perturbation expansion leads to

$$E_{xc}[\rho]=E_{xc}^{\text{LDA}}[\rho]+E_{xc}^{g}[\rho], \qquad (6.25)$$

where the LDA is indeed given by eqn (6.23) to all orders in the perturbation expansion and the gradient corrections are similarly given by

$$E_{xc}^{g}[\rho]=\int d^3r\,[B_{xc}(\rho(\mathbf{r}))(\nabla\rho(\mathbf{r}))^2+C_{xc}(\rho(\mathbf{r}))(\nabla^2\rho(\mathbf{r}))^2$$
$$+D_{xc}(\rho(\mathbf{r}))(\nabla^2\rho(\mathbf{r}))(\nabla\rho(\mathbf{r}))^2+E_{xc}(\rho(\mathbf{r}))(\nabla\rho(\mathbf{r}))^4+\cdots]. \qquad (6.26)$$

At this point, we shall summarize some of the advantages and disadvantages of this approach to inhomogeneous systems.

The LDA is an extension of the original Thomas–Fermi theory to include interactions. Corrections to this quasi-uniform picture are provided by the series of gradient corrections. The resulting energy functional representation, which is sometimes referred to as extended Thomas–Fermi (ETF), has some strong positive points, at least as regards basic principles and simplicity. We note that the initial reference density $\rho_0$ is no longer present in $E_{xc}[\rho]$ so this representation is indeed a universal functional as desired in DFT. The form of $E_{xc}[\rho]$ is very simple in structure so that it is widely applicable to a variety of geometries and situations and is easily generalized as required (for example, to finite temperature). In addition, $E_{xc}[\rho]$ has a sound foundation in that it is a systematic expansion based on a rigorous starting point.

The disadvantages of the gradient expansion for $E_{xc}[\rho]$ are also significant. Only the first gradient-expansion coefficient $B_{xc}(\rho)$ is well known at present. The other coefficients $C_{xc}(\rho)$, etc., have never been calculated. Even if we assume that they all exist, when appropriately formulated in a physical problem, the convergence of the expansion in powers of gradients is presumably only asymptotic. This is due, at least in part, to the initial expansion of $K_{xc}(\mathbf{k})$ and all higher-order response functions in powers of small wavenumbers. Such an expansion misses non-analytic structures associated with the sharpness of the Fermi surface (at zero temperature, $T=0$). The convergence properties of the gradient expansion for $E_{xc}$ will also be dependent on the class of density profiles under consideration. A detailed study of this question must await adequate information concerning the expansion coefficients.

It may be of interest to note here that in the case of non-interacting particles, the corresponding gradient expansion through fourth order in gradients has been carried out by the above perturbative resummation method. The structure of eqn (6.26), which is expected by symmetry, is also found and all of the expansion coefficients (the counterparts of the above $A$, $B$, $C, \ldots$), after resummation, agree with other methods of calculation for the (much simpler) non-interacting system as required by symmetry considerations.[45,46] Sixth-order gradient-expansion coefficients have also been given.[47] We note that the question of whether the kinetic energy for non-interacting particles is adequately accounted for by its gradient expansion has been studied quite extensively and answered in the affirmative when this expansion is truncated after including the fourth-order gradient correction provided unphysical (too rapidly varying) density profiles are avoided. Further discussion of this point is given in Section 6.4.

In the absence of more detailed information concerning the gradient-expansion coefficients in $E_{xc}[\rho]$, we have two alternatives. We may apply eqn (6.26) truncated at second order and certainly hope to exhibit correct trends and orders of magnitude. The precise question of how well this

$$E_{xc}^{(2)}[\rho]=\int d^3r\, B_{xc}(\rho(\mathbf{r}))(\boldsymbol{\nabla}\rho(\mathbf{r}))^2 \tag{6.27}$$

accounts for the true exchange-correlation energy in a class of physical problems must then be determined by comparison, preferably, with the results of a highly accurate and carefully tested benchmark calculation. This possibility and its conclusions concerning eqn (6.27) will be discussed in Section 6.4.

As an alternative to the simple truncation at second order, and indeed to any explicit appearance of gradient corrections, we might attempt to organize the partial resummation of the basic perturbation expansion in different ways. As one example, we return to eqn (6.10) where we now make explicit the translational symmetry and the dependence of the kernel on the density $\rho_0$ of the reference system:

$$K(\mathbf{r},\mathbf{r}')=K(\mathbf{r}-\mathbf{r}';\rho_0). \tag{6.28}$$

Then the exchange-correlation contribution to $G[\rho]$ is approximated by

$$E_{xc}[\rho]=E_{xc}^{\mathrm{LDA}}[\rho]-\frac{1}{4}\int \mathrm{d}^3r\int \mathrm{d}^3r'\,K_{xc}(\mathbf{r}-\mathbf{r}';\rho_{\mathrm{AV}})(\rho(\mathbf{r})-\rho(\mathbf{r}'))^2, \tag{6.29}$$

where the non-universal reference density $\rho_0$ has been eliminated in favour of an average density

$$\rho_{\mathrm{AV}}=\rho_{\mathrm{AV}}(\mathbf{r},\mathbf{r}'), \tag{6.30}$$

which requires specification and might be taken as $\frac{1}{2}(\rho(\mathbf{r})+\rho(\mathbf{r}'))$ or $\rho(\frac{1}{2}(\mathbf{r}+\mathbf{r}')$ or various other forms.[48] More complicated non-local specifications of this average density can also be made. We emphasize that the expansion in powers of $\tilde{\rho}$ of eqn (6.29) will always reproduce exactly the second-order results of eqn (6.10). Of course, the expansion of eqn (6.30) will also produce an infinite series of higher-order terms in powers of $\tilde{\rho}$ and, possibly, in powers of gradients of the density. This series of higher-order terms is extremely sensitive to the choice of the relatively unrestricted $\rho_{\mathrm{AV}}$, of course, and it is by no means clear that the non-local structure of the high-order terms in the perturbation expansion will be correctly captured by a given prescription for this quantity. In fact, this will be seen to be a major problem when attempting to devise approximations. The question of the adequacy of terms like eqn (6.29) is therefore *a priori* very difficult and must also be settled by comparison with benchmark calculations of known accuracy or by other appropriate means. Approximations of this type have been applied in atoms and in metallic surface problems, for example. In some cases there are very serious difficulties such as completely unphysical divergent contributions to the energy. These divergence problems are due to incorrect specification of the non-local structure of the energy. This point emphasizes the extremely delicate non-local structure of the exchange-correlation energy for an inhomogeneous system. It is therefore clear that there is great potential for serious errors when attempting to construct approximate exchange-correlation functionals. The essential point is that the non-local structure of $E_{xc}[\rho]$, when applied to inhomogeneous systems, must be rather delicate. This structure is unknown *a priori* and we cannot expect to discover and understand it without careful investigation of new features of specifically non-uniform systems. Thus eqn (6.29) is unsuitable in spite of its attractive features, chief among which is perhaps the fact that the long-wavelength expansion of eqn (6.21) is avoided so that quantum density oscillations are incorporated.

### 6.2.3 Exchange-correlation hole methods

An interesting class of approximate exchange-correlation energy functionals has been generated by the exact result[49,50]

$$E_{xc}[\rho]=\frac{1}{2}\int d^3r\int d^3r'\,v(\mathbf{r}-\mathbf{r}')\int_0^1 d\lambda\,\rho(\mathbf{r})[g_\lambda(\mathbf{r},\mathbf{r}')-1]\rho(\mathbf{r}'), \qquad (6.31)$$

where $g_\lambda(\mathbf{r},\mathbf{r}')$ is the pair-correlation function of a non-uniform system of the correct physical density $\rho(\mathbf{r})$ but with the two-body interaction scaled by a factor $\lambda$. It is common to express this result in terms of an exchange-correlation hole defined by

$$\rho_{xc}(\mathbf{r},\mathbf{r}')=\rho(\mathbf{r}')\int_0^1 d\lambda\,[g_\lambda(\mathbf{r},\mathbf{r}')-1]. \qquad (6.32)$$

Although detailed knowledge of the pair correlations in a general non-uniform system is not available, various approximation schemes may be motivated by replacing the unknown pair-correlation function in eqn (6.32) by the corresponding pair-correlation function $g^h_\lambda$ for a homogeneous translationally invariant system of appropriate density. This then necessitates the introduction of one or more quantities like $\rho_{AV}$, of Section 6.2.2. This approach has been developed extensively primarily by Gunnarsson *et al.*[51,52] Their local density (LD) approximation is defined by

$$\rho^{LD}_{xc}(\mathbf{r},\mathbf{r}')=\rho(\mathbf{r})\int_0^1 d\lambda\,[g^h_\lambda(\mathbf{r}-\mathbf{r}';\,\rho(\mathbf{r}))-1]. \qquad (6.33)$$

In the average density (AD) approximation, an explicit attempt is made to take account of non-locality by prescribing both the prefactor in eqn (6.32) and $\rho_{AV}$ to be given by a weighted average

$$\bar{\rho}(\mathbf{r})=\int d^3r'\,W(\mathbf{r}-\mathbf{r}';\;\acute{\rho}(\mathbf{r}'))\bar{\rho}(\mathbf{r}'), \qquad (6.34)$$

where $W$ is a specified density-dependent weighting factor. This provides the approximation

$$\rho^{AD}_{xc}(\mathbf{r},\mathbf{r}')=\bar{\rho}(\mathbf{r})\int_0^1 d\lambda\,[g^h_\lambda(\mathbf{r}-\mathbf{r}';\,\bar{\rho}(\mathbf{r}))-\mathbf{1}]. \qquad (6.35)$$

The motivation for these approximations stems from the fact that they both satisfy

$$\int d^3 r' \, \rho_{xc}(\mathbf{r}, \mathbf{r}') = -1. \tag{6.36}$$

This sum rule is well known for homogeneous systems and therefore applies to the LD and AD approximations by virtue of the role played by $g^h$ in their definitions and it is also applicable to inhomogeneous systems as will be discussed briefly in Section 6.2.4.

It is also possible to suggest similar approximation schemes in which the sum rule is explicitly satisfied locally; that is, eqn (6.36) holds at each point **r** of the system. One such approximation is the weighted density (WD) approximation for which

$$\rho_{xc}^{WD}(\mathbf{r}, \mathbf{r}') = \rho(\mathbf{r}') \int_0^1 d\lambda \, [g_\lambda^h(\mathbf{r} - \mathbf{r}'; \bar{\rho}(\mathbf{r})) - 1], \tag{6.37}$$

where the weighted density $\bar{\rho}(\mathbf{r})$ is now to be determined so as to satisfy

$$\int d\mathbf{r}' \int_0^1 d\lambda \, [g_\lambda^h(\mathbf{r} - \mathbf{r}'; \bar{\rho}(\mathbf{r})) - 1] \, \rho(\mathbf{r}') = -1. \tag{6.38}$$

It is clear that a number of other approximations to $E_{xc}$ via the exchange-correlation hole could be given.

Extensive applications of the LD, AD and WD approximations have been made to various systems, including atoms and metallic surfaces. We refer to Section 6.4 for a detailed discussion of numerical results and we wish to summarize here only the major overall features. We first consider atomic systems which are good test-cases since their exchange energies and correlation energies can be obtained accurately by other means. Errors in the atomic exchange energies tend to be of order 5%, 1% and 10% for the LD, AD and WD approximations, respectively. The correlation energies in the LD and WD approximations are a factor of two or more too large, whereas the AD approximation gives correlation energies which are too small and even of the wrong sign in some cases. These results are disappointing, particularly since the rather sophisticated WD approximation scheme does not yield any improvement over the cruder approximations, contrary to expectations. Results using these approximations schemes for the jellium model of a semi-infinite metal within the infinite-barrier model have been obtained and they are also disappointing. Unphysical divergent integrals appear as a result of inappropriate specification of quantities such as $\rho_{AV}$, just as in Section 6.2.2. The exchange-correlation surface energy is too large (perhaps even infinite) in the AD approximation. For the WD approximation, the exchange and correlation energies taken separately are also divergent, which is also wrong. The total exchange-correlation energy is finite, as required, but is about a factor of two smaller than the numerical results suggested by more direct calculations for this model. Thus, both the AD and WD approximations are inferior to the relatively crude LD approximation, in fact.[51,52]

It is important to appreciate why these approximations give such disappointing results, in spite of efforts to incorporate non-locality beyond local density pictures, for both atoms and metal surfaces. The central guiding principle was the imposition of the sum rule, eqn (6.36), which requires the exchange-correlation hole to contain exactly one particle. Although this sum rule may be regarded as generally valid, it is nevertheless true that it carries very limited implications for the important spatial dependences of the exchange-correlation hole $\rho_{xc}(\mathbf{r}, \mathbf{r}')$. This question is taken up in some detail in Section 6.2.4. In addition to this question, there is still the fundamental problem of how to specify appropriate average densities in $\rho_{xc}(\mathbf{r}, \mathbf{r}')$. On the basis of the above results, it is to be concluded that the suggested approximation schemes seriously misrepresent the exchange-correlation hole for both atoms and metallic surfaces and that the detailed essentially non-local structure of $E_{xc}[\rho]$ has not been captured by such procedures.

Another class of approximations based on the exchange-correlation hole has been suggested by Langreth and Perdew.[50,53-55] In this scheme, a method is proposed to decompose $E_{xc}[\rho]$ into different wavenumber contributions. The sum rule, eqn (6.36), and other considerations are invoked to gain insight into the structure of the decomposed $E_{xc}[\rho; \mathbf{q}]$. Although the original suggestion has been substantially modified, all versions of this type of scheme suffer from the lack of adequate fundamental guides to specify the non-local structure of $E_{xc}[\rho]$. These procedures have also been applied to numerical studies of the exchange-correlation energy of atoms and metal surfaces. These results are discussed in Section 6.4 after we describe recent work based on a direct variational calculation of the jellium surface energy.

In Section 6.2.4 we discuss the structure factor of general inhomogeneous systems and indicate the limited relevance of the sum rule, eqn (6.36), in providing a fundamental guide to the full structure of the exchange-correlation hole.

### 6.2.4 Structure factor for inhomogeneous systems

In this subsection, we consider various correlation functions, with emphasis on the properties of the structure factor, for inhomogeneous systems. We are interested in the wavenumber variation of the static structure factor in the limiting region of long wavelength, in particular, and we address the questions raised in the previous subsection. We shall see that it is of crucial importance to distinguish essentially finite systems, such as atoms, from extended non-uniform systems, such as semi-infinite metals, for which the thermodynamic limit has been taken. In the latter case, the static structure factor generates a discontinuity at zero wave-vector, in general, and the $\mathbf{q}\to\mathbf{0}$ limiting value of $S(\mathbf{q},\mathbf{q})$ is not universal.

The primary correlation function relating to a pair of particles is

$$G(\mathbf{r},\mathbf{r}')=\sum_{\substack{ij\\ i\neq j}}\langle\delta(\hat{\mathbf{x}}_i-\mathbf{r})\delta(\hat{\mathbf{x}}_j-\mathbf{r}')\rangle, \tag{6.39}$$

where the average is to be taken for a system of $N$ particles in a volume $\Omega$ having density

$$\rho(\mathbf{r})=\sum_i\langle\delta(\hat{\mathbf{x}}_i-\mathbf{r})\rangle, \tag{6.40}$$

and the particles interact via a two-body interaction $v(\mathbf{r}-\mathbf{r}')$. This pair-correlation function is written in a number of other equivalent forms, such as

$$\begin{aligned}G(\mathbf{r},\mathbf{r}')&=\langle\hat{\rho}(\mathbf{r})\hat{\rho}(\mathbf{r}')\rangle-\rho(\mathbf{r})\delta(\mathbf{r}-\mathbf{r}')\\&=\langle\delta\hat{\rho}(\mathbf{r})\delta\hat{\rho}(\mathbf{r}')\rangle+\rho(\mathbf{r})\rho(\mathbf{r}')-\rho(\mathbf{r})\delta(\mathbf{r}-\mathbf{r}')\\&=\rho(\mathbf{r})g(\mathbf{r},\mathbf{r}')\rho(\mathbf{r}'),\end{aligned} \tag{6.41}$$

where

$$\delta\hat{\rho}(\mathbf{r})=\hat{\rho}(\mathbf{r})-\rho(\mathbf{r}) \tag{6.42}$$

is the operator for density fluctuations, or in terms of the static structure factor

$$NS(\mathbf{r},\mathbf{r}')=\langle\delta\hat{\rho}(\mathbf{r})\delta\hat{\rho}(\mathbf{r}')\rangle. \tag{6.43}$$

These quantities are physically observable. The correlation functions at intermediate coupling constant required in the construction of $E_{\mathrm{xc}}[\rho]$, eqn (6.31), are indicated by a subscript $\lambda$. The exchange-correlation hole is thus expressed by[49,50]

$$\rho_{\mathrm{xc}}(\mathbf{r},\mathbf{r}')=\int_0^1 \mathrm{d}\lambda\,[G_\lambda(\mathbf{r},\mathbf{r}')-\rho(\mathbf{r})\rho(\mathbf{r}')]/\rho(\mathbf{r}). \tag{6.44}$$

For the purposes of discussing general features of correlation functions, the coupling-constant integration is inessential. Consequently, we begin by considering the sum rules associated with the physical correlation function at $\lambda=1$ and the integration can be reintroduced when required.

To explore the question of sum rules, first construct

$$\int_{\Omega} d^3r \int_{\Omega} d^3r' \, S(\mathbf{r},\mathbf{r}')=0, \tag{6.45}$$

which follows trivially from the definitions, eqns (6.41) and (6.43) provided only the total number of particles is really fixed. This is certainly the case for strictly finite systems, such as atoms, or even for macroscopic systems if described by a canonical ensemble. On the other hand, if the system is treated by a grand canonical ensemble which allows number fluctuations with only the average particle number $\bar{N}$ fixed at $N$,[56]

$$\int_{\Omega} d^3r \int_{\Omega} d^3r' \, S(\mathbf{r},\mathbf{r}')=\langle(\delta\hat{n})^2\rangle/N. \tag{6.46}$$

We are not concerned here with ensemble-dependent details such as the differences, for general temperature and geometry, which exist between (6.45) and (6.46). Similar statements can be made for the exchange-correlation hole. From eqns (6.41) and (6.44), we find

$$\int_{\Omega} d^3r' \, n_{xc}(\mathbf{r},\mathbf{r}')=-1 \tag{6.47}$$

provided we are concerned with systems with fixed particle number. Our attention will be restricted in the following to such situations. There is then no doubt that global sum rules, reflecting particle conservation, exist and the important question is the extent to which they are useful in providing adequate information for constructing approximations to $E_{xc}[\rho]$.

The exchange-correlation energy involves the integral of a correlation function weighted by the interaction. Fourier-transforming the interaction

$$v(\mathbf{r},\mathbf{r}')=\int d^3q \frac{\exp[i\mathbf{q}\cdot(\mathbf{r}-\mathbf{r}')]}{(2\pi)^3} v(\mathbf{q}), \tag{6.48}$$

we rewrite eqn (6.31) as[50,57]

$$E_{xc}[\rho]=\frac{N}{2}\int \frac{d^3q}{(2\pi)^3} v(\mathbf{q}) \int_0^1 d\lambda \, [S_\lambda(\mathbf{q},\mathbf{q})-1], \tag{6.49}$$

where

$$S(\mathbf{q},\mathbf{q})=\int_{\Omega} d^3r \int_{\Omega} d^3r' \exp[i\mathbf{q}\cdot(\mathbf{r}-\mathbf{r}')] \, S(\mathbf{r},\mathbf{r}'). \tag{6.50}$$

The sum rule, eqn (6.45), expresses particle conservation only and relates only to the $\mathbf{q}=\mathbf{0}$ value of $S(\mathbf{q},\mathbf{q})$. To evaluate $E_{xc}$, we require knowledge of $S(\mathbf{q},\mathbf{q})$ for all $\mathbf{q}$. What determines the scale of variations in $S(\mathbf{q},\mathbf{q})$? For a finite system, such as an atom for example, the scale of variations in $S(\mathbf{q},\mathbf{q})$

will be $\sim 1/L_{at}$ where $L_{at}$ is the characteristic dimension of the atom. Thus, for an atom or molecule, $S(\mathbf{q},\mathbf{q})$ increases from zero to its maximum in a range $\sim 1/L_{at}$ and ultimately approaches unity at large wavenumber. Thus, from a knowledge of length scales and the sum rule,

$$S(\mathbf{q}=\mathbf{0},\mathbf{q}=0)=0, \tag{6.51}$$

a plausible picture of the structure factor can be suggested. This can certainly be helpful indeed.

We next consider the case of a block of metal whose volume $\Omega=L^3$, although possibly large, is finite. The characteristic length scales for this system include $L$, the Fermi wavelength $k_F^{-1}$ and the screening distance $k_{TF}^{-1}$. Consequently, there will necessarily be structure in $S(\mathbf{q},\mathbf{q})$ at the corresponding wave-vectors $q_L \sim L^{-1}, k_F$ and $k_{TF}$. We know that $S(\mathbf{q},\mathbf{q})$ is zero for $\mathbf{q}$ identically zero and it is very important to realize that $S(\mathbf{q},\mathbf{q})$ can rise from zero to a value of order unity over a very small range of $\mathbf{q}$ in the case of a large system. To appreciate this fact, consider again the definition, eqn (6.50). For $\mathbf{q}=\mathbf{0}$, the value of zero requires very precise cancellation of contributions when integrating $S(\mathbf{r},\mathbf{r}')$ over the entire volume. When the oscillatory phase factor $\exp[i\mathbf{q}\cdot(\mathbf{r}-\mathbf{r}')]$ is included in the integrand, this precise cancellation is relatively unaffected only if $q \ll 1/L$. On the other hand, for $q \gg 1/L$, this detailed cancellation of contributions is completely removed by the oscillations owing to the exponential phase factor. We conclude that the influence of total particle number conservation on $S(\mathbf{q},\mathbf{q})$ extends only over a region of order $1/L$ which is insignificant for a macroscopic system. In fact, for any theoretical description in which the usual thermodynamic limit ($N,\Omega \to \infty$ with the density appropriately specified) has been taken, this region disappears totally. As a result, we see that the $\mathbf{q} \to \mathbf{0}$ limit and the thermodynamic limit for macroscopic systems are not uniformly interchangeable in general. In this sense, the structure factor develops a discontinuity at $\mathbf{q}=\mathbf{0}$ in the thermodynamic limit and the sum rule has no implications for any finite $\mathbf{q}$. We emphasize that while

$$\lim_{q\to\infty} \lim_{\Omega\to\infty} S(\mathbf{q},\mathbf{q}) \neq \lim_{\Omega\to\infty} S(\mathbf{q}=0,\ \mathbf{q}=0)=0 \tag{6.52}$$

is expected in general, this does not forbid cases where the limiting value of the static structure factor vanishes. The point is that the limiting value of the long-wavelength structure factor is not universal and cannot be completely specified for all systems by purely general arguments. A specific real calculation is required and the result will be dependent upon details of the system including the nature of the inhomogeneity, the density profile structure and the interparticle interaction.

A detailed discussion of the static structure factor for non-uniform macroscopic systems in the thermodynamic limit will be given in the following section. For purposes of comparison, it will be instructive to recall the corresponding properties of a uniform system. To define a uniform translationally invariant system, we should take the limit of $\Omega \to \infty$ in the previously discussed $\rho(\mathbf{r})$ and $S(\mathbf{r}, \mathbf{r}')$ in such a way that both $\mathbf{r}$ and $\mathbf{r}'$ become infinitely distant from the receding walls. Then all boundary effects disappear and $\rho(\mathbf{r}) \to \rho_0 = \text{const}$ while $S(\mathbf{r}, \mathbf{r}') \to S^{\mathrm{h}}(\mathbf{r} - \mathbf{r}')$ which is a function of the difference in coordinates (and the density). The resulting density and pair correlations are thus typical of the bulk (interior) properties and can be conveniently evaluated in a grand canonical ensemble with periodic boundary conditions. This limiting process applied to eqn (6.46) leads to[56]

$$\int \mathrm{d}^3 R\, S^{\mathrm{h}}(\mathbf{R}) = k_{\mathrm{B}} T \frac{\partial \rho_0}{\partial \mu} \xrightarrow[T \to 0]{} 0. \tag{6.53}$$

For the homogeneous system, assuming "reasonable" properties of pair correlations at large separation, it is highly plausible but non-trivial that, at zero temperature ($T = 0$),

$$\lim_{q \to 0} S^{\mathrm{h}}(\mathbf{q}) = S^{\mathrm{h}}(\mathbf{q} = \mathbf{0}). \tag{6.54}$$

We emphasize that large separation here does not refer to the structure of the previous $S(\mathbf{r}, \mathbf{r}')$ of the inhomogeneous system as this structure was removed by the way the uniform system was defined in the thermodynamic limit.

Our primary interest in this review lies in many-electron systems interacting by Coulomb interactions. A proof of eqn (6.54) has been given for the case of a translationally invariant many-electron system (jellium) by Pines and Nozières[58] (PN). It is useful to recall the main points of their proof and to indicate where an attempt to generalize it to inhomogeneous systems would falter. We therefore very briefly review the general argument of PN which leads to the following small-$q$ limit for the uniform-electron-gas structure factor $S^{\mathrm{h}}(q)$

$$\lim_{q \to 0} S^{\mathrm{h}}(\mathrm{q}) = \frac{q^2}{2m\omega_{\mathrm{p}}}, \tag{6.55}$$

where $\omega_{\mathrm{p}} = (4\pi\rho_0 e^2/m)^{1/2}$ and we occasionally set $\hbar = 1$. Pines and Nozières consider the dynamic structure factor $S^{\mathrm{h}}(q, \omega)$ and separate it into a plasmon-dominated part $S^{\mathrm{h}}_{\mathrm{pl}}(q, \omega)$ and a particle–hole excitation $S^{\mathrm{h}}_{\mathrm{inc}}(q, \omega)$,

$$S^{\mathrm{h}}(q, \omega) = S^{\mathrm{h}}_{\mathrm{pl}}(q, \omega) + S^{\mathrm{h}}_{\mathrm{inc}}(q, \omega), \tag{6.56}$$

which determine the static structure factor by

$$\rho_0 S^{\mathrm{h}}(q) = \frac{1}{2\pi} \int_0^{\infty} \mathrm{d}\omega\, S^{\mathrm{h}}(q, \omega). \tag{6.57}$$

The plasmon part must be of the form

$$S^{\mathrm{h}}_{\mathrm{pl}}(q,\omega)=A_q\delta(\omega-\omega_{\mathrm{p}}). \tag{6.58}$$

Now the spectral decomposition of $S^{\mathrm{h}}(q,\omega)$ is

$$S^{\mathrm{h}}(q,\omega)=2\pi\sum_n \langle 0|\delta\hat{n}(\mathrm{q})|n\rangle\langle n|\delta\hat{n}(q)|0\rangle\ \delta(\omega-\omega_{n0}) \tag{6.59}$$

where $|n\rangle$ are the eigenstates and $\omega_{n0}$ the eigenvalues for the interacting uniform electron gas.

A crucial step now enters the argument. One can show, for the uniform case, that the matrix element for the particle–hole pair excitations can be written as[58]

$$\langle 0|\delta\hat{n}(q)|n\rangle\propto\frac{\langle 0|\hat{n}^1_q|n\rangle}{\varepsilon(q)}, \tag{6.60}$$

where $\langle 0|\hat{n}^1_q|n\rangle$ is the corresponding matrix element for a non-interacting system and is clearly of order unity. $\varepsilon(q)$ is the static dielectric function, which for small $q$ is $\varepsilon(q)=1+k^2_{\mathrm{sc}}/q^2$ where $k^2_{\mathrm{sc}}=4\pi e^2(\mathrm{d}\rho_0/\mathrm{d}\mu)$. This leads to *perfect* screening of the bare matrix element $\langle 0|\hat{n}^1_q|n\rangle$. We emphasize that this perfect screening is a product of translational property of a *uniform* system. This will not be true in a *non-uniform* many-electron system! Therefore, from eqn (6.60)

$$\langle 0|\delta\hat{n}(q)|n\rangle\propto q^2, \tag{6.61}$$

and from eqns (6.59) and (6.61) the contribution from the particle–hole pair excitations for $S^{\mathrm{h}}(q)$ goes like $q^5$. The final fifth power of $q$ is a result of the sum over the intermediate states $|n\rangle$ which are restricted to a $q$ phase space owing to momentum conservation. One can also show[58] from the same momentum-conservation considerations that multi-pair excitation contributes a $q^4$ term to $S^{\mathrm{h}}(q)$.

We can now introduce eqn (6.56) into the f sum rule[58]

$$\int_0^\infty \mathrm{d}\omega\,\omega S^{\mathrm{h}}(q,\omega)=\frac{q^2\rho_0}{2m} \tag{6.62}$$

to conclude that particle–hole pair excitations do not contribute at small $q$ and also to determine the value of $A_q$ in eqn (6.58):

$$A_{\mathrm{q}}=\frac{\rho_0 q^2}{2m\omega_{\mathrm{p}}}, \tag{6.63}$$

and from eqn (6.57),

$$S^{\mathrm{h}}(q)=\frac{q^2}{2m\omega_{\mathrm{p}}}. \tag{6.64}$$

Two crucial steps have been introduced in the derivation of eqn (6.64) which are manifestly properties of the uniform but *not* the non-uniform many-electron system. First, the perfect screening of $\delta n^1(q)$ is not true for the inhomogeneous system. The momentum conservation which adds higher powers of $q$ to $S^h(q)$ is also not true in non-uniform systems (this was demonstrated explicitly in an *exchange-only* calculation of $S(\mathbf{q}, \mathbf{q})$ in a jellium surface;[57] see Section 6.3.5). We therefore have absolutely no reason to expect explicit calculations of $S(\mathbf{q}, \mathbf{q})$ in non-uniform systems to lead to behaviour for $S(\mathbf{q}, \mathbf{q})$ at small $\mathbf{q}$ which is the same or even similar to that found for $S^h(q)$; this will indeed be demonstrated in Section 6.3.5.

## 6.3 Formulation of gradient expansions

Our approach then is to consider a non-uniform interacting electron gas which is complex enough to expose many of the subtleties encountered in going from the uniform to the non-uniform electron gas, yet simple enough to be handled *exactly* in certain limits. As we shall see, our results will provide the basis for making some important, more general, predictions concerning such non-uniform corrections to $E_{xc}(\rho)$ and to the structure factor $S(\mathbf{r}, \mathbf{r}')$ (see Sections 6.1 and 6.2) in imhomogeneous systems.

For the system we have in mind, we consider a uniform electron gas of density $\rho_0$ on which we superimpose an arbitrary potential

$$V_{ext}(\mathbf{r}) = \frac{1}{\Omega} \sum_{\mathbf{k}} V_{ext}(\mathbf{k}) \exp(i\mathbf{k} \cdot \mathbf{r}). \tag{6.65}$$

Here $\Omega$ is the volume of the system, which we always take to be macroscopic. Such a weakly non-uniform electron gas in the high-density limit (HDL) was considered first by Ma and Brueckner[37] (MB) and by Sham.[59] Here we review important new results based on further extensions of such a model.

The first is the extension of the model to metallic densities, second is its extension to multi-component systems, and third are the implications of this model to the nature of the structure factor $S(\mathbf{r}, \mathbf{r}')$ in non-uniform systems.

The expression presented in Section 6.2 for the exchange and correlation energy (eqn (6.19)) is appropriate to the weakly varying non-uniform electron gas. It is already too complex for simple analytical treatment. We need to make one further requirement and force the external potential $V_{ext}(\mathbf{r})$ to be slowly varying. The Fourier wave-vector components $\mathbf{k}$, in eqn (6.65), must therefore satisfy $|\mathbf{k}| \ll k_F$. We then can expand the irreducible screening function $\pi(\mathbf{k})$ (eqn (6.65)) in a power expansion

$$\pi(\mathbf{k}) = a^{-1} - bk^2 - ck^4 - dk^6 \tag{6.66}$$

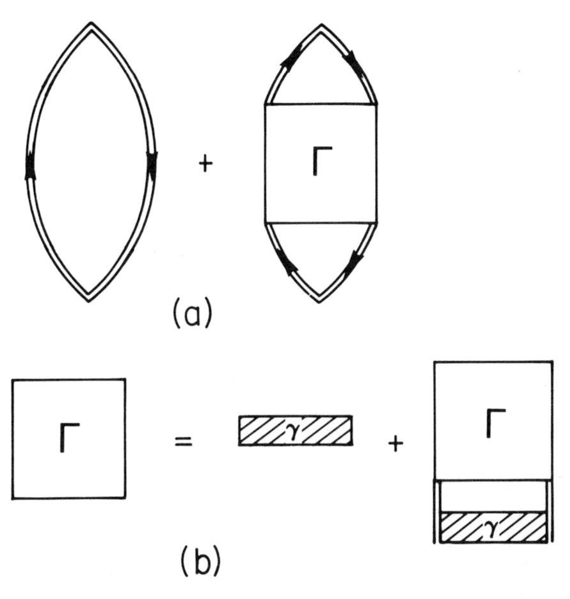


**Fig. 6.1** Series of graphs for $\pi(k)$ in terms of the exact propagators and the scattering functions. The reducible scattering function $\Gamma$ satisfies an integral equation with the irreducible scattering frunction given by $\gamma$.

plus logarithmic corrections, such as $k^4 \ln k^2$, and stop at second order in $k$. The coefficient $a^{-1}$ in the expansion is given by the compressibility sum rule (see eqn (6.22)) as $a = \mathrm{d}\mu/\mathrm{d}\rho_0$. It has been extensively studied by many previous authors. The second coefficient, $b$, provides the first signature of the non-uniformity imposed on the uniform system. It is clearly much more difficult to calculate. In fact, to calculate $b$ at arbitrary uniform density by directly expanding $\pi(\mathbf{k})$ is almost intractable. One can, however, derive an exact closed-form expression for $b$ which turns out to be very useful.[60, 61] To see this, consider the irreducible graphs for $\pi(\mathbf{k})$ (Fig. 6.1(a)) written in terms of the reducible scattering function $\Gamma(\mathbf{k})$ (Fig. 6.1(b)). $\Gamma(k)$ satisfies the matrix equations

$$\Gamma(\mathbf{k}) = \gamma(\mathbf{k}) + \gamma(\mathbf{k})R(\mathbf{k})\Gamma(\mathbf{k}), \tag{6.67}$$

$$\Gamma(\mathbf{k}) = \gamma(\mathbf{k}) + \Gamma(\mathbf{k})R(\mathbf{k})\gamma(\mathbf{k}), \tag{6.68}$$

where $\gamma(\mathbf{k})$ is the irreducible scattering function (Fig. 6.1(b)) and the matrix elements of $R(\mathbf{k})$ are given by

$$R_{p,p'}(\mathbf{k}) = G(p + \tfrac{1}{2}\mathbf{k})G(p - \tfrac{1}{2}\mathbf{k})\delta_{p,p'}, \tag{6.69}$$

with $G$ denoting the exact one-electron propagator and $p = (\mathbf{p}, p_0)$. Now since $b$ corresponds to the $\mathbf{k}^2$ term of $\pi(\mathbf{k})$ we next expand $\gamma(\mathbf{k})$ and $R(\mathbf{k})$ to that order, i.e.

$$\gamma(\mathbf{k})=\gamma(\mathbf{0})+\mathbf{k}^2\gamma^{(2)}+\mathrm{O}(\mathbf{k})^4 \tag{6.70}$$

and

$$R(\mathbf{k})=R(\mathbf{0})+\mathbf{k}^2R^{(2)}+\mathrm{O}(\mathbf{k})^4. \tag{6.71}$$

Some matrix algebra using eqns (6.67)–(6.71) yields

$$\Gamma(\mathbf{k})-\Gamma(\mathbf{0})=\Gamma(\mathbf{0})\mathbf{k}^2R^{(2)}\Gamma(\mathbf{0})+[1+\Gamma(\mathbf{0})R(\mathbf{0})]\,\mathbf{k}^2\gamma^{(2)}[1+R(\mathbf{0})\Gamma(\mathbf{0})]. \tag{6.72}$$

Next we define a vertex function $\Lambda(\mathbf{k})$ given by

$$\Lambda(\mathbf{k})=\lambda+\Gamma(\mathbf{k})R(\mathbf{k})\lambda, \tag{6.73}$$

where $\lambda$ is a column vector with components $\lambda_p=1$. In terms of $\Lambda(\mathbf{k})$, $\pi(\mathbf{k})$ is given by (see Fig. 6.1)

$$\pi(\mathbf{k})=-2\tilde{\lambda}R(\mathbf{k})\Lambda(\mathbf{k}), \tag{6.74}$$

where $\tilde{\lambda}$ is the transpose of $\lambda$. Some matrix algebra using eqns (6.72) – (6.74) yields

$$\begin{aligned}\mathbf{k}^2b=2\tilde{\Lambda}(\mathbf{0})\mathbf{k}^2R^{(2)}\Lambda(\mathbf{0})+2\tilde{\lambda}R(\mathbf{0})[1+\Gamma(\mathbf{0})R(\mathbf{0})]\\ \times\mathbf{k}^2\gamma^{(2)}[1+R(\mathbf{0})\Gamma(\mathbf{0})]R(\mathbf{0})\lambda,\end{aligned} \tag{6.75}$$

where $\tilde{\Lambda}(\mathbf{0})$ is the transpose of $\Lambda(\mathbf{0})$. Noting that

$$-\frac{\mathrm{d}G(p)}{\mathrm{d}\mu}=[1+R(\mathbf{0})\Gamma(\mathbf{0})]R(\mathbf{0})\lambda, \tag{6.76}$$

where

$$\frac{\mathrm{d}G(p)}{\mathrm{d}\mu}=\frac{\partial G(p)}{\partial p_0}+\frac{\partial G(p)}{\partial\mu},$$

we get the final expression for $b$:

$$b=2\,\mathrm{tr}_p\,\Lambda_p(\mathbf{0})R_p^{(2)}\Lambda_p(\mathbf{0})+2\,\mathrm{tr}_p\,\mathrm{tr}_{p'}\left[\frac{\mathrm{d}G(p)}{\mathrm{d}\mu}\gamma_{p,p'}^{(2)}\frac{\mathrm{d}G(p')}{\mathrm{d}\mu}\right], \tag{6.77}$$

where

$$\mathrm{tr}_p[\cdots]=\int\frac{\mathrm{d}^4p}{(2\pi)^4\mathrm{i}}[\cdots], \tag{6.78}$$

and where the vertex function at $\mathbf{k}=\mathbf{0}$ is given by the Ward identity

$$\Lambda_p(0)=\frac{\mathrm{d}G^{-1}(p)}{\mathrm{d}\mu}. \tag{6.79}$$

Equation (6.77) for $b$ has been expressed as far as possible in terms of one-electron propagators and its derivatives.

### 6.3.1 Exchange only

As a first application of eqn (6.77) to the single-component non-uniform interacting electron gas, we calculate $b$ with exchange alone.[59,62-64] $\gamma(\mathbf{k})$ in the Hartree–Fock (HF) approximation is given by

$$\gamma_{p,p'}(\mathbf{k}) = \lim_{\lambda \to 0} \frac{-4\pi e^2}{|\mathbf{p}-\mathbf{p}'|^2+\lambda^2}. \tag{6.80}$$

Since $\gamma(\mathbf{k})$ in the HF approximation is independent of $\mathbf{k}$, the last term in eqn (6.77) vanishes. All we need to know, then, is the (frequency-independent) HF self-energy given by

$$\Sigma(p) = -\int \frac{d^3p'}{(2\pi)^3} v(\mathbf{p}-\mathbf{p}') f(E(\mathbf{p}')), \tag{6.81}$$

where $f(E(\mathbf{p}))$ is the usual Fermi function and the particle dispersion law is

$$E(\mathbf{p}) = \frac{\hbar^2\mathbf{p}^2}{2m} + \Sigma(\mathbf{p}). \tag{6.82}$$

$R_p^{(2)}$ in eqn (6.77) is the $\mathbf{k}^2$ expansion of eqn (6.69) after integration over the frequency $p_0$, i.e.

$$R_{\mathbf{p}}(\mathbf{k}) = \int_{-\infty}^{+\infty} \frac{dp_0}{2\pi i} R_p(\mathbf{k}) = \frac{f(E(\mathbf{p}+\frac{1}{2}\mathbf{k})) - f(E(\mathbf{p}-\frac{1}{2}\mathbf{k}))}{E(\mathbf{p}+\frac{1}{2}\mathbf{k}) - E(\mathbf{p}-\frac{1}{2}\mathbf{k})}, \tag{6.83}$$

and the $\mathbf{k}^2$ expansion gives

$$\mathbf{k}^2 R_{\mathbf{p}}^{(2)} = \tfrac{1}{8} f''(E(p))[(\mathbf{k}\cdot\nabla_{\mathbf{p}})^2 E(p)] + \tfrac{1}{24} f'''(E(p))[\mathbf{k}\cdot\nabla_{\mathbf{p}} E(p)]^2.$$

Using this final form in eqn (6.77) we get the following for $b_{\mathrm{ex}}$:

$$b_{\mathrm{ex}} = \frac{1}{24\pi^2} \int_0^\infty dp\, p^2 [\Lambda(p)]^2 \left\{ \left[ E''(p) + \frac{2}{p} E''(p) \right] \right.$$

$$\left. \times f''(E(p)) + \frac{1}{3} [E'(p)]^2 f'''(E(p)) \right\}. \tag{6.84}$$

Here the primes denote differentiation with respect to the indicated argument. The integration over $\mathbf{p}$ is simple to perform, and all the entries are known (i.e. $E(\mathbf{p})$ is given in eqns (6.81) and (6.82) and

$\Lambda(p)=1-\partial\Sigma(\mathbf{p})/\partial\mu$. The final result for $b_{ex}$ is[57,59]

$$b_{ex}=\frac{e^2}{72\pi^3[\mu'(k_F)]^2}(A+B), \tag{6.85}$$

where

$$A=5+\left(\frac{k_F^2-2\lambda^2}{k_F^2}\right)\ln\left(\frac{4k_F^2+\lambda^2}{\lambda^2}\right)-\frac{64k_F^4-4k_F^2\lambda^2-3\lambda^4}{(4k_F^2+\lambda^2)^2}-8\pi k_F a_0 \tag{6.86}$$

and

$$B=\frac{\{2\pi k_F a_0+2+2\lambda^2/(4k_F^2+\lambda^2)-(\lambda^2/k_F^2)\ln[(4k_F^2+\lambda^2)/\lambda^2]\}^2}{2\pi k_F a_0-2+[(2k_F^2+\lambda^2)/2k_F^2]\ln[(4k_F^2+\lambda^2)/\lambda^2]}, \tag{6.87}$$

$$\mu'(k_F)=\frac{\hbar^2k_F}{m}-\frac{e^2}{\pi}\left[1-\frac{\lambda^2}{4k_F^2}\ln\left(\frac{4k_F^2+\lambda^2}{\lambda^2}\right)\right], \tag{6.88}$$

and where $a_0=\hbar^2/me^2$ is the usual Bohr radius. Of course, the leading coefficient $a^{-1}$, in eqn (6.66), can also be calculated exactly in the HF approximation. It is

$$a=\frac{d\mu}{d\rho_0}=\frac{\hbar^2\pi^2}{mk_F}-e^2\frac{\pi}{k_F^2}\left[1-\frac{\lambda^2}{4k_F^2}\ln\left(\frac{4k_F^2+\lambda^2}{\lambda^2}\right)\right]. \tag{6.89}$$

If we carefully examine eqns (6.86)–(6.89) we discover that for fixed $e^2$,

$$\lim 4\pi e^2 b_{ex}\to-\frac{\ln(\lambda^2)}{18(\pi k_F a_0-1)^2}, \tag{6.90}$$

whereas for fixed $\lambda$ the expansion in $e^2$ gives

$$4\pi e^2 b\to-\frac{1}{18}\left[\frac{6}{\pi k_F a_0}+\frac{5}{(\pi k_F a_0)^2}-\frac{(\ln\lambda^2)^2}{2(\pi k_F a_0)^3}\right]. \tag{6.91}$$

Even at this level of approximation (i.e. in the slowly varying HF approximation), the non-uniform electron gas has already revealed some unexpected and subtle features when we consider the exchange and correlation energy beyond the LDA. This we see by considering again the form given in eqn (6.20). For the weakly perturbed uniform electron gas the following connection between $B_{xc}(\rho_0)$ and $b$ is readily established:

$$B_{xc}(\rho_0)=\tfrac{1}{2}(a^2b-a_0^2b_0), \tag{6.92}$$

where $a_0$ and $b_0$ are the expansion coefficients of eqn (6.66) for the corresponding non-interacting electron gas; $a_0=\hbar^2\pi^2/mk_F$ and $b_0=m/12\pi^2k_F\hbar^2$. As we define the case of Coulomb interactions by taking the limit of $\lambda\to 0$ in the exact HF result in eqn (6.90), we see that the corresponding $B_x(\rho_0)$ becomes infinite. On the other hand, if we prefer to examine $B_x(\rho_0)$ order by order in the inter-particle coupling $e^2$ but still with $\lambda \to 0$, then to order $e^2$ the contribution of the graphs of (Fig. 6.1) are well defined. However, we encounter again a divergent contribution to $b_{ex}$ (or $B_x(\rho_0)$) in order $e^4$ of the coupling constant, as can be seen from the last term of eqn (6.91). In short, imposing a non-uniformity on the uniform system has made the separation of the exchange and correlation in non-uniform systems within DFT much more subtle. Presumably the gradient series, eqn (6.26), is infinite for exchange and correlation separately, but remains finite from the sum of the two. One thing already clear at the HF level of approximation for the non-uniform electron gas is that the separation of the exchange and correlation in DFT is not an obvious matter and probably not even a properly defined approach. This is the first result which our simple but exact model has already demonstrated.

### 6.3.2 Correlations in the high-density limit

Again we turn to the expansion in eqn (6.66) and to the coefficients $a^{-1}$ and $b$. In the high-density limit (HDL), $a$ is known exactly. It is given by

$$a = \frac{\hbar^2 \pi^2}{m k_F} - \frac{e^2 \pi}{k_F^2} + O(e^4 \ln e^2). \tag{6.93}$$

The first two terms on the right-hand side of eqn (6.93) are the exchange contributions (see eqn (6.89) with $\lambda \to 0$). Correlation introduces a higher-order connection in $e^2$ and gives the usual contribution associated with correlation in the uniform electron gas in the HDL, which varies as $e^4 \ln e^2$ (ref. 65). The HDL contributions to $b$ are associated with the $\mathbf{k}^2$ expansion of the terms of $\pi(\mathbf{k})$ shown in Fig. 6.2. The contributions can be written at once using eqn (6.77). We only need to consider the appropriate form for $\gamma_{p,p}(\mathbf{k})$ (see Fig. 6.3), the relation between $\Lambda_p(\mathbf{0})$ and the self-energy $\Sigma(p)$ (as given by the Ward identity in eqn (6.79)), and the propagator $G$ to order $e^4$. The results are the ones first presented by MB,[37] and these are

$$b = b_0 + b^{(1)} + b^{(2)} + b_1^{(3)} + b_2^{(3)}. \tag{6.94}$$

$b^{(1)}$ gives the self-energy contribution (see Fig. 6.2(a) and the first two terms in Fig. 6.2(b)):

$$b^{(1)}\mathbf{k}^2 = \hbar^2 \frac{\mathbf{k}^2}{m} \operatorname{tr}_p \left[ \Sigma(\mathrm{p}) \left( -\frac{1}{3} \frac{\partial^3}{\partial \mu_0^3} G_0(p) + \frac{1}{9} \frac{\partial^4}{\partial \mu_0^4} G_0(\mathrm{p}) \right) \right] - \frac{\mathbf{k}^2 \Sigma(k_F, 0) m^2}{12\pi^2 \hbar^4 k_F^3}, \tag{6.95}$$

where

$$\Sigma(p) = -\operatorname{tr}_{p'} [V(p') G_0(p + p')], \tag{6.96}$$

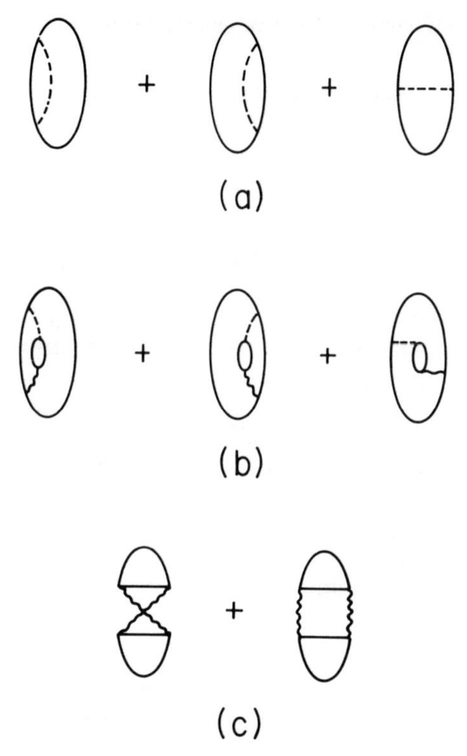


**Fig. 6.2** Lowest-order contributions to $\pi$ in the RPA. (a) Lowest-order exchange Hartree–Fock graphs; the dotted line represents the bare Coulomb interaction. (b) Lowest-order dymanic corrections to the exchange graphs; the wavy line represents the dynamically screened Coulomb interaction. (c) Lowest-order members of the non-Hartree–Fock class of graphs.

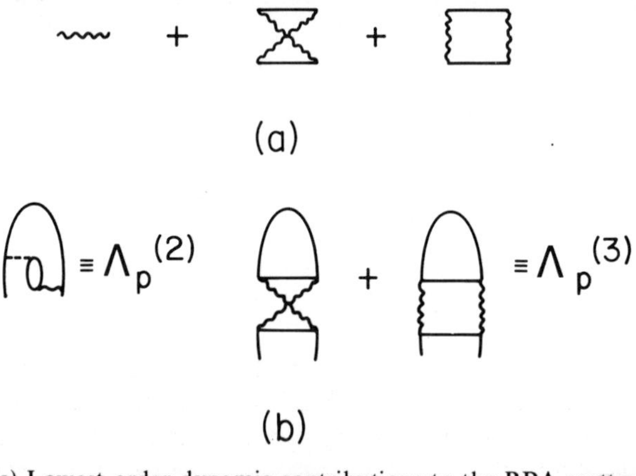

(a)

$\equiv \Lambda_p^{(2)}$ $\quad$ + $\quad$ $\equiv \Lambda_p^{(3)}$

(b)

**Fig. 6.3** (a) Lowest-order dynamic contributions to the RPA scattering function, $\gamma$. (b) Lowest-order correlation corrections to the bare vertex function.

with

$$V(p')=\frac{4\pi e^2}{\mathbf{p}'^2}\Big/\varepsilon(p')$$

and

$$G_0(p)=[p_0-\varepsilon_{\mathbf{p}}+\mu_0+\mathrm{i}\delta\,\mathrm{sgn}\,(\varepsilon_{\mathbf{p}}-\mu_0)]^{-1}; \tag{6.97}$$

$\varepsilon_{\mathbf{p}}=\hbar^2\mathbf{p}^2/2m$, $\mu_0=\hbar^2k_{\mathrm{F}}^2/2m$ and $\varepsilon(p')$ is the dynamic dielectric function in the random-phase approximation (RPA): $\varepsilon(p')=1+(4\pi e^2/|\mathbf{p}'|^2)\pi^0(p')$.

$b^{(2)}$ gives the vertex contribution (see Fig. 6.2(b), last term):

$$b^{(2)}\mathbf{k}^2=\frac{\hbar^2\mathbf{k}^2}{m}\,\mathrm{tr}_{\mathrm{p}}\left[\Lambda_p^{(2)}\left(\frac{1}{2}\frac{\partial^2}{\partial\mu_0^2}G_0(p)-\frac{1}{9}\varepsilon_{\mathbf{p}}\frac{\partial^3}{\partial\mu_0^3}G_0(p)\right)\right], \tag{6.98}$$

where $\Lambda_{\mathrm{p}}^{(2)}$ is the vertex function of Fig. 6.3(b). $b_2^{(3)}$ gives the vertex contribution of $\Lambda_{\mathrm{p}}^{(3)}$ (see Fig. 6.3(b)):

$$b_2^{(3)}\mathbf{k}^2=\frac{\hbar^2\mathbf{k}^2}{m}\,\mathrm{tr}_{\mathrm{p}}\left[\Lambda_{\mathrm{p}}^{(3)}\left(\frac{1}{2}\frac{\partial^2}{\partial\mu_0^2}G_0(p)-\frac{1}{9}\varepsilon_{\mathbf{p}}\frac{\partial^3}{\partial\mu_0^3}G_0(p)\right)\right]. \tag{6.99}$$

Finally, $b_1^{(3)}$ corresponds to the last contribution of eqn (6.77):

$$b_1^{(3)}\mathbf{k}^2=\frac{-\hbar^2}{24}\mathbf{k}^2\mathrm{tr}_{\mathrm{p}}\left[\left(\frac{\partial}{\partial\mu_0}\pi^0(p)\right)^2[V(p)\nabla_{\mathbf{p}}^2V(p)-(\nabla_{\mathbf{p}}V(p))^2]\right]. \tag{6.100}$$

We give a brief sketch of the calculation of these four contributions in Appendix A6.1. The forms listed in eqns (A6.10)–(A6.14) of the Appendix are now used in the HDL. First we can extract from eqns (16.10) and (A6.11) the three HF contributions (see Fig. 2(a)) already discussed in eqn (6.91). After careful integration over the frequency variable $y$ in eqns (A6.10) and (A6.11) when the dynamic interaction $V(q,y)$ is replaced by $4\pi e^2/(2k_{\mathrm{F}}q)^2$, we get (note from the Appendix that $q=p'/2k_{\mathrm{F}}$ is now a dimensionless variable)

$$b_{\mathrm{ex}}=b_0-\frac{m^2}{3\pi^2}\int\frac{\mathrm{d}^3q}{(2\pi)^3}v(\mathbf{q})\left[-q\theta(1-q)+\frac{11}{6}k_{\mathrm{F}}\delta(2k_{\mathrm{F}}(q-1))+\frac{k_{\mathrm{F}}^2}{3}\delta'(2k_{\mathrm{F}}(q-1))\right]. \tag{6.101}$$

If we perform the integration over $q$ we get at once the second term on the right-hand side of eqn (6.91). The HDL correlation contribution is now given by taking the $q\to 0$ limit of eqns (A 6.10) and (A 6.11) after we have subtracted the $4\pi e^2/(2k_{\mathrm{F}}q)^2$ in the appropriate terms of $V(q,y)$. We group our results into two parts.

We define

$$b_c = \frac{e^4 m^3}{(2\pi)^3 k_F^4}(Z_1 + Z_2), \tag{6.102}$$

where

$$Z_1 = \frac{8k_F}{3\pi^2}\int_0^\infty dq\, q \int_0^\infty dy \frac{1}{(y^2+1)^2}\frac{1}{[\tilde{\varepsilon}(q,y)]^2}\left\{\left(\frac{\frac{4}{3}y^2+\frac{1}{2}}{y^2+1}+\frac{2}{3}\frac{y^2(2+y^2)}{3(y^2+1)^2}\right)\right.$$
$$+\frac{1}{k_F a_0 4\pi\tilde{\varepsilon}(q,y)}\left(-4+\frac{7y^2}{y^2+1}-\frac{y^2(3y^2+1)}{(y^2+1)^2}+6R(y)\right)$$
$$+\frac{1}{(k_F a_0)^2 4\pi^2}\frac{1}{[\tilde{\varepsilon}(q,y)]^2}$$
$$\left.\times\left[-1+\frac{y^2(y^2+2)}{(y^2+1)^2}-3R(y)\left(\frac{2y^2}{y^2+1}-2+3R(y)\right)\right]\right\}, \tag{6.103}$$

and

$$Z_2 = \frac{8k_F^2 a_0}{3\pi}\int_0^\infty \frac{dq}{q}\int_0^\infty dy\left(\frac{-q^2}{\tilde{\varepsilon}(q,y)}+1\right)$$
$$\times\left(\frac{1}{2}\frac{1}{y^2+1}+\frac{\frac{7}{2}-\frac{7}{6}y^2}{(y^2+1)^2}+\frac{4y^2-2+\frac{2}{3}y^4}{(y^2+1)^3}-\frac{2}{3}\frac{17y^2+8y^4+3}{(y^2+1)^4}\right), \tag{6.104}$$

and where $\tilde{\varepsilon}(q,y) = q^2 + 1/(k_F a_0 \pi)R(y)$ and $R(y) = 1 - y\tan^{-1}(1/y)$. These representations of the HDL results, as functions of $q$ and $y$, are different from the forms of MB although the final contribution to the HDL when both $q$ and $y$ integrals are performed reduce identically to the MG HDL, i.e.

$$b_c = \frac{e^2 m^2}{(2\pi)^3 k_F^2}(1.9756), \tag{6.105}$$

where the final factor was obtained by numerical evaluation of the above. We shall see, however, in Section 6.3.5, that the results as presented in eqns (6.102) to (6.104) (see Appendix A 6.1) can be directly related to the structure factor $\mathbf{S}(\mathbf{r}, \mathbf{r}')$ of this weakly perturbed and slowly varying non-uniform electron gas.

The results of eqn (6.105) already exhibit something special in the correlation energy of the non-uniform electron gas which is absent in the uniform case. In the uniform electron gas, correlation contributions are *always* of higher order in the coupling $e^2$ than exchange contributions. These terms vary as $e^4 \ln e^2$ (see eqn (6.93)). For the first non-uniform correction represented by the $b$ coefficient, this is no longer the case. Both exchange (see second term on the right-hand side of eqn (6.91)) and correlation (eqn (6.105)) are of the same order in $e^2$. We shall see in Section 6.3.5 that this is related to non-universal long-range behaviour in the exchange-correlation hole (see Section 6.3.5 for what we mean by "non-universal") of non-uniform systems. This structure does not exist in the uniform counterpart. A correct treatment of these features of non-uniform systems is fundamental to a meaningful extension of exchange and correlation forms beyond the LDA (see Section 6.2).

For a final comment of the structure of $b$ in the HDL, we return to the form of eqn (6.94) and calculate $b_1^{(3)}$ and $b_2^{(3)}$ separately. These contributions to the screening graphs correspond to Fig. 6.2(c). From eqns (6.99) and (6.100) the results in the HDL are[61]

$$b_2^{(3)}=\frac{e^2m^2}{(2\pi)^3k_{\mathrm{F}}^2}\frac{2}{9}\int_0^{\infty}\frac{\mathrm{d}y}{\pi}\frac{y^2(13+9y^2)}{R(y)(y^2+1)^4}=\frac{e^2m^2}{(2\pi)^3k_{\mathrm{F}}^2}(1.4556), \qquad (6.106)$$

and

$$b_1^{(3)}=\frac{e^2m^2}{(2\pi)^3k_{\mathrm{F}}^2}\frac{2}{9}\int_0^{\infty}\frac{\mathrm{d}y}{\pi}\left(2+\frac{1}{1+y^2)}\right)\frac{1}{(1+y^2)^2R(y)}$$

$$=\frac{e^2m^2}{(2\pi)^3k_{\mathrm{F}}^2}(0.5913). \qquad (6.107)$$

From eqn (6.105) we see that the terms of Fig. 6.2(c) capture almost all of the HDL contribution to $b_c$. What this says is that any attempts to include correlation contribution by screening the HF contributions could never be right. The *dynamic* screening in the inter-particle interaction going beyond HF-like graphs (see Figs. 6.2(a) and (b)) is essential to the correlation contributions of non-uniform systems.

### 6.3.3 Metallic densities

The above results in the HDL are indicative of the new features to be found in non-uniform systems. In order, however, to construct the partial sum for $B_{xc}(\rho(\mathbf{r}))$ discussed in Section 6.2 (eqn (6.2.6)) we need to extend the HDL results to metallic densities. This was carried out in refs 60 and 61 (see also ref. 66) and here we review this calculation. We start with the coefficient $a$. A calculation of $a$ at metallic densities cannot be terminated at lowest order in $e^2$, as was done in the HDL or eqn (6.73). Since $a^{-1}$ is proportional to the compressibility, which is directly related to the second density derivative of the energy of the uniform system, we can obtain $a^{-1}$ from the available electron-gas ground-state energies. An essential point to note is that $a^{-1}$ is a very strong function of the density[38] and, in fact, becomes singular at $r_s = r_{s0} \gtrsim 5$ ($r_s$ is related to the density $\rho_0$ of the uniform electron gas by $r_s = (3/4\pi\rho_0)^{1/3}$, and is given in atomic units).

To extend $b$ to the metallic range we return to eqn (6.77). It is immediately apparent that $b$ depends quadratically on the vertex function $\Lambda_p(\mathbf{0})$ (which is the origin of the density-dependent enhancement effect in $a^{-1}$ resulting in a singularity at $r_s = r_{s0}$). Hence $b$ is expected to also be a strong function of $r_s$. To include these higher-order effects in $b$ we shall be closely guided by eqn (6.77). We focus on the first term on the right-hand side of eqn (6.77), which we denote by $b_{\mathrm{I}}$, i.e

$$b_{\mathrm{I}} = 2\,\mathrm{tr}_p[\Lambda_p(\mathbf{0}) R_p^{(2)} \Lambda_p(\mathbf{0})]. \tag{6.108}$$

We write

$$\Lambda_p(\mathbf{0}) = 1 + \tilde{\Lambda}_p(\mathbf{0}); \tag{6.109}$$

then

$$b_{\mathrm{I}} = 2\,\mathrm{tr}_p\,\{1 + 2\tilde{\Lambda}_p(\mathbf{0}) + [\tilde{\Lambda}_p(\mathbf{0})]^2\}\, R_p^{(2)}. \tag{6.110}$$

We next make the following two approximations. (1) We treat the coupling *of* $R_p^{(2)}$ to the first-order part of $\tilde{\Lambda}_p(\mathbf{0})$ *exactly*, while the additional higher-order contributions from $\tilde{\Lambda}_p(\mathbf{0})$ are kept in an average way in the spirit of the Hubbard[67] approximation. (2) For the moment we replace the exact $R_p^{(2)}$ with that evaluated from the free propagators (i.e. $G \to G_0$). The first-order coupling of $\tilde{\Lambda}_p(\mathbf{0})$ and $R_p^{(2)}$ is nothing more than the functions $b'$, $b''$ and $b'''$ discussed in Appendix 6.1. The fact that in eqns (A 6.10)–(A 6.14) no small expansion in $q$ was made, gives us at once the full form of $b'$, $b''$ and $b'''$ at all $r_s$. The higher-order contributions to $\tilde{\Lambda}_p(\mathbf{0})$ enter in an average way by a factor $(1-\eta)^{-1}$. The three terms on the right-hand side of eqn (6.110) then become

$$b_{\mathrm{I}} = b_0 + b_{\mathrm{I}}^{(1)} \frac{1}{(1-\eta)} + b_{\mathrm{I}}^{(1)} \frac{\frac{1}{2}\eta}{(1-\eta)^2}, \tag{6.111}$$

where

$$b_{\mathrm{I}}^{(1)} = b_{\mathrm{ex}} + b' + b'' - \frac{m^2}{12\pi^2 k_{\mathrm{F}}^3} \Sigma(k_{\mathrm{F}}, 0) \tag{6.112}$$

and $b'$ and $b''$ are given in Appendix A6.1. Note that in $b_{\mathrm{I}}^{(1)}$ we have also included the first-order contribution from the self-energies of the full propagators in $R_p^{(2)}$. The way this happens is discussed in ref. 57.

We next turn to the second term in eqn (6.77) which we denote by $b_{\mathrm{II}}$. To maintain a consistent analysis we again replace the full propagator by

$G \to G_0$ and include higher-order effects by treating the vertex function in an average way, i.e.

$$\frac{dG}{d\mu} = -R(\mathbf{0})\Lambda(\mathbf{0}) = \frac{1}{(1-\eta)}\frac{\partial G_0}{\partial \mu_0}; \tag{6.113}$$

then

$$b_{II} = b'''/(1-\eta)^2 \tag{6.114}$$

and finally

$$b_I = b_0 + b_I^{(1)}\frac{1-\frac{1}{2}\eta}{(1-\eta)^2} + \frac{b'''}{(1-\eta)^2}, \tag{6.115}$$

where $b'''$ is again given in Appendix A6.1. To get $B_{xc}$ we simply substitute eqn (6.115) in eqn (6.92). The crucial thing to observe (from Fig. 6.2) is that since $a^{-1}$ corresponds to $\pi(\mathbf{k}=0)$ it can be expressed again in terms of $\tilde{\Lambda}_p(\mathbf{0})$ and in fact provides the definition for the enhancement factor $\eta$, i.e.

$$a^{-1} = a_0^{-1}/(1-\eta). \tag{6.116}$$

Therefore the singularity at $\eta = 1$ is *precisely* cancelled by $a$ in eqn (6.92) and the final expression for $B_{xc}$ is left finite for all $r_s$, as expected.

The two-dimensional integrals for $b'$, $b''$ and $b'''$ have been calculated numerically for a range of values of $r_s$ in the metallic density range. The required values for $\Sigma(k_F, 0)$ and $\eta$ were taken from the review article of Hedin and Lundquist.[68] It is convenient to express $B_{xc}$ in the form[61]

$$B_{xc} = \frac{e^2 C_{xc}(r_s)}{\rho_0^{4/3}}. \tag{6.117}$$

The results for the dimensionless $C_{xc}(r_s)$, as a function of $r_s$, are plotted in Fig. 6.4. An interpolation that accounts accurately for the required values of $C_{xc}(r_s)$ is[69]

$$C_{xc}(r_s) = 10^{-3}\frac{(2.568 + ar_s + br_s^2)}{1 + cr_s + dr_s^2 + 10br_s^3}, \tag{6.118}$$

with $a = 23.266$, $b = 7.389 \times 10^{-3}$, $c = 8.723$ and $d = 0.472$. Application to metallic surface energy calculations will be discussed in Section 6.4.

### 6.3.4 The multi-component weakly and slowly varying non-uniform electron gas

As already mentioned in Section 6.2, the forms for $E_{xc}$ can be readily generalized to multi-component systems.[13–32] Such generalizations are not

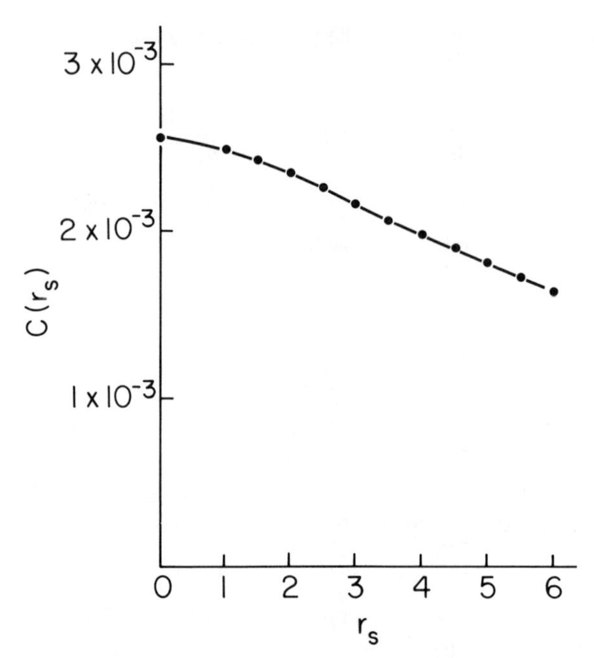


**Fig. 6.4** Calculated $r_s$ dependence of $C(r_s)$ which is related to the gradient coefficient $B_{xc}$ by $B_{xc} = C(r_s)e^2/\rho_0^{4/3}$.

just a formal exercise. Itinerant ferromagnets and the electron–hole droplet (EHD) are just two examples where $E_{xc}(\rho_1, \ldots, \rho_\alpha)$ is a functional of all $\alpha$ components. The expansion in eqn (6.27) to second order in $\nabla\rho_i(r)$ now takes the form

$$E_{xc}^{(2)} = \int d^3r \sum_{i,j=1} [B_{ij}^{xc}(\rho_1(\mathbf{r}), \ldots, \rho_\alpha(\mathbf{r}))\nabla\rho_i(\mathbf{r}) \cdot \nabla\rho_j(\mathbf{r})] \tag{6.119}$$

We can again relate these coefficients to the response of a uniform system where each component has its own uniform density $\rho_{\alpha 0}$. We perturb the system with an external potential $V_{ext}^i(\mathbf{k})$ which couples independently to each of the $\alpha$ components. The energy shift is now

$$\Delta E = \frac{1}{2\Omega} \sum_{\substack{\mathbf{k} \\ i,j=1}}^{\alpha} \chi^{ij}(\mathbf{k}) V_{ext}^i(\mathbf{k})^* V_{ext}^j(\mathbf{k}). \tag{6.120}$$

*Two-component itinerant ferromagnets*

We can follow the same rearrangement of terms as for the one-component case. The connection between $\chi_{ij}(\mathbf{k})$ and the irreducible screening func-

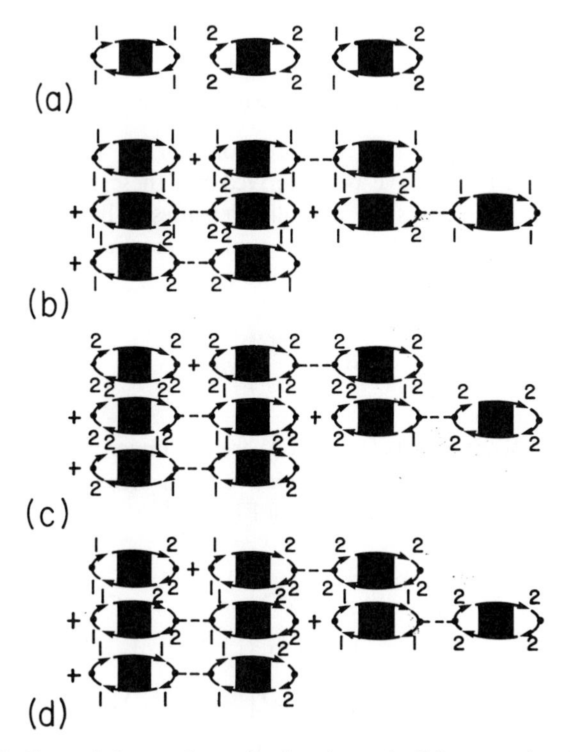


**Fig. 6.5** (a) General form of graphs for the reducible screening function: The lowest-order terms in the decomposition of the reducible screening functions into irreducible screening functions for (b) $\chi_{11}$, (c) $\chi_{22}$ and (d) $\chi_{12}$. The dashed line represents the Coulomb interaction which is repulsive for electron–electron interactions and attractive for electron–hole interactions.

however, is now a bit more complex (see Fig. 6.5). The final relations between $B^{ij}_{\text{xc}}$ and the $\pi_{ij}(\mathbf{k})$ are now obtained from[27] the terms of order $\mathbf{k}^2$ in the expansions of

$$\frac{1}{2}\left(\frac{\pi_{\downarrow\downarrow}(\mathbf{k})}{\Delta(\mathbf{k})}\right)-\frac{1}{\pi^0_{\uparrow\uparrow}(\mathbf{k})}=\text{const}+B^{\text{xc}}_{\uparrow\uparrow}\mathbf{k}^2+\cdots, \tag{6.121}$$

$$\frac{1}{2}\left(\frac{\pi_{\uparrow\uparrow}(\mathbf{k})}{\Delta(\mathbf{k})}-\frac{1}{\pi^0_{\downarrow\downarrow}(\mathbf{k})}\right)=\text{const}+B^{\text{xc}}_{\downarrow\downarrow}\mathbf{k}^2+\cdots, \tag{6.122}$$

$$\frac{\pi_{\uparrow\downarrow}(\mathbf{k})}{\Delta(\mathbf{k})}=\text{const}+B^{\text{xc}}_{\uparrow\downarrow}\mathbf{k}^2, \tag{6.123}$$

where

$$\Delta(\mathbf{k})=\pi_{\uparrow\uparrow}(\mathbf{k})\pi_{\downarrow\downarrow}(\mathbf{k})-(\pi_{\uparrow\downarrow}(\mathbf{k}))^2. \tag{6.124}$$

(We have $i=1=\uparrow$ and $i=2=\downarrow$.) Expressions for the $b_{\uparrow\uparrow}$, $b_{\downarrow\downarrow}$ and $b_{\uparrow\downarrow}$ of the kind given in eqn (6.77) have been derived but these have not as yet been used to extend the results to metallic densities. The HDL calculations, however, have been carried out. The appropriate Feynman graphs in the HDL are illustrated in Figs. 6.6–6.8. If we write

$$B_{ij}^{\mathrm{xc}}=\frac{e^2 C_{ij}^{\mathrm{xc}}}{\rho_i^{2/3}\rho_j^{2/3}}, \tag{6.125}$$

we find that the HDL results for the $C_{ij}^{\mathrm{xc}}$ are no longer a single constant, but rather a strong function of the relative spin densities parameter $\beta=(\rho_\uparrow/\rho_\downarrow)^{1/3}$. These results are presented in Fig. 6.9.[27]

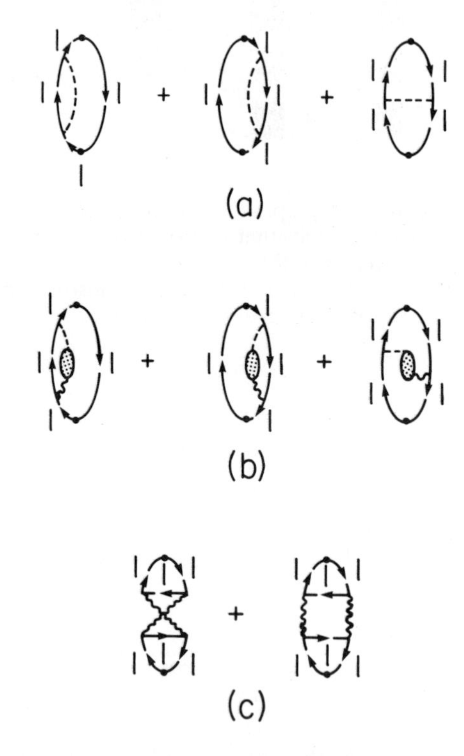


**Fig. 6.6** Lowest-order irreducible contributions to $\pi_{11}$ in the HDL: (a) bare exchange, (b) screening of exchange graphs, and (c) pure correlation graphs not of Hartree–Fock origin.

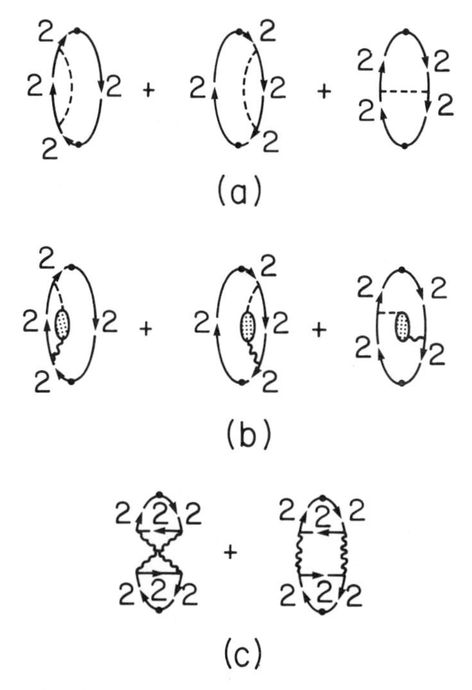


**Fig. 6.7** Lowest-order irreducible contributions to $\pi_{22}$ in the HDL: (a) bare exchange, (b) screening of exchange graphs, and (c) pure correlation graphs not of Hartree–Fock origin.

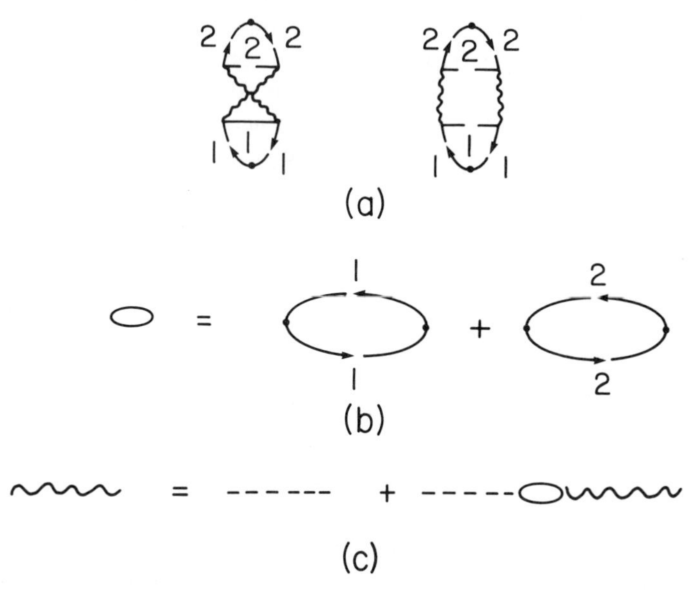


**Fig. 6.8** (a) Lowest-order irreducible contributions to $\pi_{12}$ in HDL, (b) the Lindhard screening function, and (c) the RPA screened interaction.

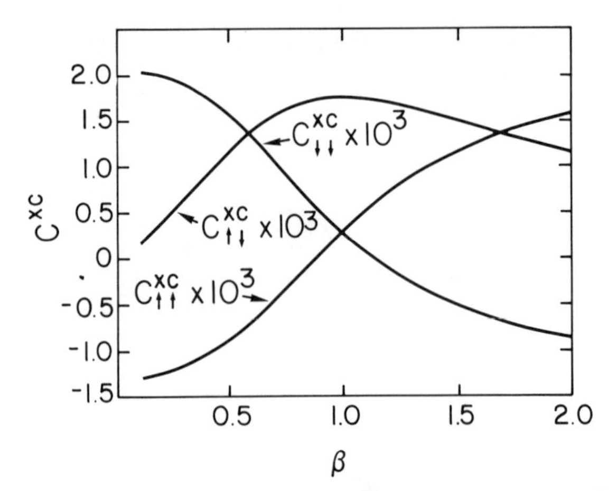


**Fig. 6.9** (a) Calculated HDL contributions to the exchange-correlation gradient coefficients for a spin-polarized electron gas as a function of the spin-up to spin-down densities ratio via $\beta=(n_\uparrow/n_\downarrow)^{1/3}$. Note that these results are consistent with eqn (6.125), while in ref. 27, the $C^{xc}_{\uparrow\downarrow}$ was mislabelled and should be a factor of larger.

*The many-component electron–hole droplet*

To proceed from a two-component system to one of an arbitrary number is relatively straightforward. A minor change arising from the fact that the electrons and holes do not carry the same mass is readily accounted for. The electron–hole droplet (EHD) in Ge is generally treated as a five-component system of four equivalent electron pockets plus one hole pocket (the light and heavy hole pockets are generally combined into one). Similarly the EHD of Si is treated as six equivalent electron pockets and one hole pocket. In all cases, the dispersion relations are taken to be isotropic.

Using the various symmetries for the $B_{ij}$ coefficients in eqn (6.119), $E^2_{xc}$ can be written for Ge as

$$E^{(2)}_{xc}=\int d^3r\,[4^{1/3}(B^{xc}_{11}+3B^{xc}_{12})(\nabla\rho_e(\mathbf{r}))^2 + B^{xc}_{55}(\nabla\rho_h(\mathbf{r}))^2+2(4^{2/3})B^{xc}_{15}(\nabla\rho_e(\mathbf{r}))\cdot(\nabla\rho_h(\mathbf{r}))], \tag{6.126}$$

where $i=1$ to 4 are the indices for the electron pockets and $i=5$ is the index for the hole pocket. For Si, there are seven components, so

$$E^{(2)}_{xc}=\int d^3r[6^{1/3}(B^{xc}_{11}+5B^{xc}_{12})(\nabla\rho_e(\mathbf{r}))^2 + B^{xc}_{77}(\nabla\rho_h(\mathbf{r}))^2+2(6^{2/3})B^{xc}_{17}(\nabla\rho_e(\mathbf{r}))\cdot(\nabla\rho_h(\mathbf{r}))] \tag{6.127}$$

where $i=1$ to 6 denotes the electron pockets and $i=7$ the hole pocket.

The $B_{ij}$ have been listed in detail in refs 25 and 26. Here we give an illustration taking the case of local charge neutrality in which case $\rho_e(\mathbf{r})=\rho_h(\mathbf{r})=\rho(\mathbf{r})$. We can then reduce eqns (6.126) and (6.127) to the form

$$E_{xc}^{(2)}=\int d^3r\, B_{eff}(\rho(\mathbf{r}))(\nabla\rho(\mathbf{r}))^2, \tag{6.128}$$

where

$$B_{eff}(\rho)=\frac{e^2C_{eff}}{\rho^{4/3}} \tag{6.129}$$

and

$$C_{eff}^{Ge}=4^{1/3}(C_{11}+3C_{12})+C_{55}+2(4^{2/3})C_{15}, \tag{6.130}$$

$$C_{eff}^{Si}=6^{1/3}(C_{11}+5C_{12})+C_{77}+2(6^{2/3})C_{17}. \tag{6.131}$$

From the values of these $C_{ij}$ we find that[26]

$$C_{eff}^{Ge}=6.643\times10^{-3},\quad C_{eff}^{Si}=7.227\times10^{-3}. \tag{6.132}$$

If we compare these results with the HDL of a single component ($C_{xc}=2.568\times10^{-3}$; see eqn (6.118) at $r_s=0$), we see that the non-uniform exchange and correlation-energy contribution rapidly increases as the number of fermion components increases.

### 6.3.5 The structure factor in the non-uniform electron gas

As we saw in Section 6.2 much of the work on non-uniform exchange and correlation energy contributions does not follow the systematic development as presented in the previous Sections. Rather, many of the other approaches rely more on intuitive arguments for the exchange and correlation hole (or, equivalently, structure factor $S(\mathbf{r},\mathbf{r}')$) for non-uniform systems. Such schemes invariably assume that $S(\mathbf{r},\mathbf{r}')$, in non-uniform systems, does not deviate in form significantly from its uniform counterpart. The idea then is almost always to force this homogeneous structure factor to accommodate features of the non-uniformity, while leaving the general form for $S(\mathbf{r},\mathbf{r}')$ to be that of the uniform system. We already saw in Section 6.3.2 that the HDL result for $b_c$ (eqn (6.105)) was totally different than one would expect from the uniform $S^h(\mathbf{r}-\mathbf{r}')$. Here we examine further such intuitive approaches to $S(\mathbf{r},\mathbf{r}')$.

Again we turn to the weakly non-uniform electron gas whose exchange and correlation energy is given exactly to second order in $V_{\text{ext}}(\mathbf{k})$ in eqn (6.10). No further approximation is required. Now the connection with the structure factor $S(\mathbf{q}, \mathbf{q})$ follows from eqn (6.30), i.e.

$$\Omega A_{\text{xc}}(\rho_0)+\frac{1}{2\Omega}\sum_{\mathbf{k}}\tilde{\rho}^2(\mathbf{k})\left(\frac{1}{\pi(\mathbf{k})}-\frac{1}{\pi_0(\mathbf{k})}\right)$$
$$=\frac{N}{2}(2k_{\text{F}})^3\int\frac{\mathrm{d}^3q}{(2\pi)^3}\,v(q)\int_0^1 \mathrm{d}_\lambda[S\lambda(\mathbf{q},\mathbf{q})^{-1}]. \tag{6.133}$$

To unravel the screening function $\pi(\mathbf{k})$ in terms of its individual $\mathbf{q}$-components one avoids the integral over $\mathbf{q}$ in the wiggly lines of Fig. 6.10. The relation to $S(\mathbf{q}, \mathbf{q})$ is then clear. (We should mention in this connection that some complications do arise from the coupling-constant integration $\lambda$, but these do not affect the points we wish to make here.) We should emphasize at the outset that we expose $\pi(\mathbf{k})$ in terms of its individual $\mathbf{q}$-vectors only to gain insight into the structure of $S(\mathbf{q}, \mathbf{q})$. For the purpose of constructing $B_{\text{xc}}$, this procedure is totally unnecessary.

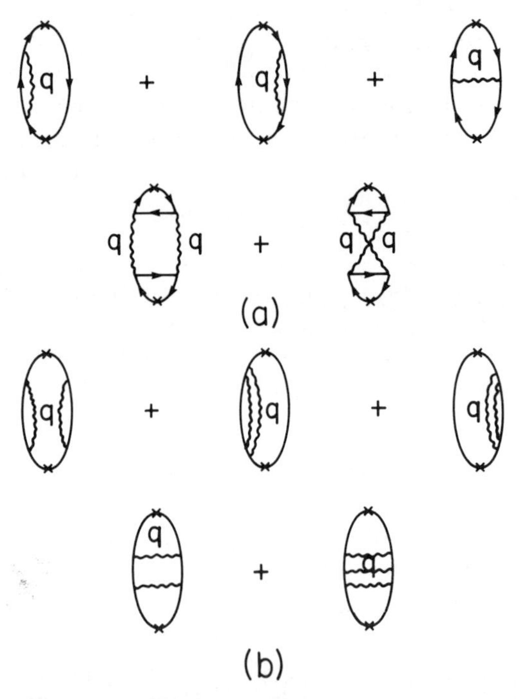


**Fig. 6.10** (a) Lowest-order RPA non-local contributions to $S(\mathbf{q}, \mathbf{q})$ to second order in $V_{\text{ext}}$ which is indicated by crosses at vertices. (b) Some examples of higher-order contributions to $S(\mathbf{q}, \mathbf{q})$ again to second order in $V_{\text{ext}}$.

The relation to the small-$\mathbf{k}$ expansion coefficient $b$, of $\pi(\mathbf{k})$, is also obvious. We shall ignore the higher-order contributions to $\pi(\mathbf{k})$ (see Fig. 6.10) which in the coefficient $b$ were handled via the enhancement factor $\eta$ (see Section 6.3.3). This leaves us with only the exchange and correlation terms in lowest-order RPA (see Fig. 6.10). The results we present are rigorous therefore only in the HDL.

$S(\mathbf{q}, \mathbf{q})$, accordingly, is proportional to the coefficient $b$ before integration over $q$; for example, in lowest-order exchange we can use eqn (6.101) and

$$S_{\text{ex}}(\mathbf{q}, \mathbf{q}) \propto q\theta(1-q) - \frac{11}{6} k_{\text{F}}\, \delta(2k_{\text{F}}(q-1)) - \frac{k_{\text{F}}^2}{3} \delta'(2k_{\text{F}}(q-1)) . \quad (6.134)$$

(We neglect the uniform contribution $S_{\text{h}}^{\text{ex}}(q)$.) The additional delta function and its derivative (in eqn (6.134)) have no equivalents in the uniform case. Such functions, transformed to real space, introduce terms in the structure factor of the weakly and slowly varying non-uniform electron gas which go like $\sin(2k_{\text{F}}R)/R$ and $\cos(2k_{\text{F}}R)$ where $R = |\mathbf{r} - \mathbf{r}'|$. No simple modification of the homogeneous-electron-gas structure factor can account for such important details which are clearly an essential input to the structure of $E_{\text{xc}}^{\text{g}}$. We furthermore note that these forms cannot be cancelled when correlation is included; at finite $q$, correlation must enter $S(\mathbf{q}, \mathbf{q})$ with higher powers of $e^2$. This can be verified directly by looking at similar delta-function contributions to $b_{\text{c}}$ (see eqn (A 6.13)). These contributions are listed in ref. 70; they are

$$S_{\text{c}}(\mathbf{q}, \mathbf{q}) \propto A\delta(2k_{\text{F}}(q-1)) + B\delta'(2k_{\text{F}}(q-1)), \quad (6.136)$$

where $A$ and $B$ are constants independent of $q$ and are of order $e^2$.

We next turn to the small-$q$ limit. According to Section 6.2.4 the sum rule for $S^{\text{h}}(q)$ at small $q$ cannot be naively extended to the non-uniform $S(\mathbf{q}, \mathbf{q})$. To look at the leading small-$q$ behaviour of $S(\mathbf{q}, \mathbf{q})$, we turn to the results of eqns (6.103)–(6.105). These results are exact in the HDL. The small-$q$ limit of $S(\mathbf{q}, \mathbf{q})$ is

$$\lim_{q \to 0} S(\mathbf{q}, \mathbf{q}) \propto e^4 (Z_1 + Z_2), \quad (6.136)$$

where in $z_1$ and $z_2$ only the $y$-integration is carried out. It is not difficult to do this integration and we find

$$\lim_{q \to 0} S(\mathbf{q}, \mathbf{q}) \approx \text{const.} \quad (6.137)$$

This result is very important. It implies some very important features which are unique to the non-uniform system. A similar calculation was carried out for the uniform system (in the RPA) by Nozières and Pines.[71] They found, in contrast with $S(\mathbf{q}, \mathbf{q})$, that $S^{\text{h}}(q)$ varies as $q^2$ at small $q$, in agreement with other general considerations (see also Section 6.2.4). Therefore, in the uniform system, the exchange-correlation hole is sufficiently localized so that

$$\lim_{q\to 0} S^{h}(q) = S^{h}(0) = 0.$$

As discussed in Section 6.2.4, the last equality comes from conservation of particle number and is always true for any canonical ensemble at zero or finite temperatures. It is equally true for small systems (like atoms) or for large systems in the thermodynamic limit. However, it clearly contains no information about the range of the structure factor. (The range could be a few angströms or a few metres; it does not matter.) The $q\to 0$ limit of $S^{h}(q)$, however, does! For example, the $q^2$ dependence of $S^{h}(q)$ implies that the real-space $S^{h}(R)$ falls off faster than $1/R^5$ at large $R$. For the uniform system in the exchange approximation alone, $S^{h}(q) \propto q$ as $q\to 0$. This implies that $S^{h}(R) \propto 1/R^{3+\varepsilon}$ $(0<\varepsilon\leqslant 2)$ at large $R$. (A direct calculation[58] leads to $1/R^4$ behaviour at large $R$.) For the non-uniform system, eqn (6.137) implies that $S(\mathbf{r}, \mathbf{r}')$ falls off like $1/R^3$ or slower. In fact, as already discussed in Section 6.3.2, the behaviour at small $q$ (in eqn (6.137)) relates closely to the surprising dependence of the MB $B_c$, which goes like $e^2$ in the HDL (i.e. the same as $B_{ex}$). Therefore, the expectation that the conservation of particle number sum rule (eqn (6.65)) is a good guide for $S(\mathbf{r}, \mathbf{r}')$, in non-uniform systems of macroscopic dimensions, is not true. No twisting of the homogeneous structure factor can accommodate these important changes as we go from the uniform to the non-uniform systems. Such features can only follow direct calculations in a non-uniform background.

It turns out that the constant in eqn (6.137) is also extremely sensitive to the type of non-uniformity,[72] in particular, to how slowly varying is $V_{ext}(\mathbf{k})$ (eqn (6.65)) (i.e. the weight in the small-$\mathbf{k}/k_F$ spectrum of $V_{ext}(\mathbf{k})$). It simply means that the size (or long range) of the exchange-correlation hole in non-uniform systems is not universally governed by intrinsic length scales of the uniform system (like $k_F$ or $k_{TF}$). It is largely decided by the length scale $|\mathbf{k}|$ and is therefore not universal and, in particular, is not characterized by properties of the uniform electron gas (see Section 6.3.2).

It is interesting to note one more non-uniform system which supports the contention that $S(\mathbf{q}, \mathbf{q})$ is drastically different from its uniform counterpart. As we saw, the exchange hole in a uniform system is sufficiently localized so that $S^{h}(q) \propto q$ as $q\to 0$. One can do an equivalent *exchange-only* calculation for a jellium surface. One finds that $\lim_{q\to 0} S^{ex}(\mathbf{q}, \mathbf{q}) = \text{const.}$ (ref. 57). Again, non-uniformity changes entirely the range of the uniform exchange hole. For short-range inter-particle interaction $v(\mathbf{q})$, this result will not be modified by terms beyond exchange.

We should emphasize that the above strongly non-universal behaviour of the structure factor at small $\mathbf{q}$ is seen also at finite temperature. For example, it is well known that for a uniform system with short-range interactions treated in a canonical ensemble, the limiting value as $\mathbf{q}\rightarrow\mathbf{0}$ of $S^{h}(\mathbf{q})$ (ref. 73) at $T\neq0$ is not $S^{h}(\mathbf{q}=\mathbf{0})=0$ in the thermodynamic limit. Conservation of particle number is irrelevant to this limit and the limiting value of $S^{h}(\mathbf{q})$ as $\mathbf{q}\rightarrow\mathbf{0}$ is related instead to the compressibility in the usual way. The corresponding non-zero limiting values of $S(\mathbf{q},\mathbf{q})$, which persist even at $T=0$ in the absence of thermal fluctuations, are the reflection of purely quantum fluctuations of the particle density relative to its non-uniform average value.

As a final observation, we turn to the multi-component electron gas. Here we find that the individual structure factors $S_{ij}(\mathbf{q},\mathbf{q})$, i.e.

$$S_{ij}(\mathbf{q},\mathbf{q})=\frac{1}{N}\langle 0|(\hat{n}_i(\mathbf{q})-n_i(\mathbf{q}))(\hat{n}_j(\mathbf{q})-n_j(\mathbf{q}))|0\rangle, \tag{6.138}$$

even for the uniform case, are no longer perfectly screened[16,74] (i.e. $S^{h}_{ij}(q)\propto q$ at small $q$). We expect, therefore, $S_{ij}(\mathbf{q},\mathbf{q})$ to be even more sensitive to any density non-uniformity. This similarly should be true for the corresponding $E^{g}_{xc}$ (eqn. (6.119)).

## 6.4 Applications

Applications of DFT, and in particular of the first term in the expansion of eqn (6.23) (i.e. LDA), have been very extensive. These forms have been widely used not only in ground-state energy calculations (where DFT applies rigorously) but even more so in band structure calculations which presumably describe the quasi-particle excitation spectrum.[75-81] We shall review only some of the calculations which reflect strongly on the non-local structures and in particular on the gradient form of eqn (6.25). It is, however, almost without exception the case that when any of the non-local forms (see Section 6.2) are applied to the bulk-excitation spectrum, such forms all make very small corrections to the results of the LDA.[17,32,82,83]

We turn then to examples of ground-state energy calculations where we are confident that eqn (6.25) should apply. One such example would certainly be the bulk cohesive energy of simple metals but, unfortunately, the conduction-band electron density varies relatively slowly and the non-local connections are therefore again very small. An electron gas confined (or bounded) by a barrier potential (like the surface of a simple metal or the surface of the EHD) or attracted by an impurity (like a proton in a simple metal) does involve a significant density variation so that the non-local corrections do make a significant difference. The density variations in such systems, however, do not satisfy the criteria $|\nabla\rho|/\rho \ll k_F$ and $|\nabla^2\rho|/|\nabla\rho| \ll k_F$ (equivalently $|\mathbf{k}|k_F \ll 1$, see eqn (6.66)) for eqn (6.25) to apply *strictly*. Nevertheless, as already discussed in Section 6.2, the convergence of the expansion is not simple to judge. First eqn (6.25) includes a partial *infinite* sum of terms when $\rho_0$ is replaced by $\rho(\mathbf{r})$; this partial sum must play a crucial role. Secondly, the gradient series is an asymptotic series and its convergence can only be judged by properly terminating the series at a certain level of gradients and applying it to a specific density variation; it is to be hoped, a physical variation! The convergence of an asymptotic series does not necessarily get better by adding additional terms.

We can make some definite statements by looking first at the kinetic-energy gradient expansion alone. In this case, the gradient expansion coefficients are known to sixth order and extensive studies for systems of present interest have been made through fourth order. In addition, the full kinetic energy for both the jellium surface energy and the energy of a proton in jellium can be calculated exactly in terms of a set of phase shifts[86] and we can compare this exact value with the gradient expansion predictions. In Table 6.1, we present such a comparison for both the energy of a jellium surface and for a proton in jellium.[87–90] We see that if we stop at fourth order, provided we include *all* fourth-order gradient contributions $\Delta T_4^{(2)}$, $\Delta T_4^{(3)}$ and $\Delta T_4^{(4)}$ to $\Delta T_4$, then the expansion can be accurately terminated at this order in both the planar and spherical geometries. The correction to the LDA for the kinetic energy is of order 10%.[91] We also note that any partial sum of gradients of the type given by eqn (6.25) would overlook two of the

**Table 6.1** The surface kinetic energy (in erg cm$^{-2}$) for jellium at $r_s=2$. $\Delta T$ is the exact result of Lang and Kohn.[100,86] $\Delta T_0$, $\Delta T_2$ and $\Delta T_4$ are the local and the first two gradient corrections to the kinetic energy corresponding to eqns (6.24) and (6.25). The linear and the two non-linear contributions to $\Delta T_4$ are explicitly indicated. For a proton in jellium at $r_s=2.07$, $\Delta T$(in Ry) is given by Perrot and Resolt[88] as are the local and first two gradient corrections to the kinetic energy.

| | | | | | |
|---|---|---|---|---|---|
| Surface energy in erg cm$^{-2}$ of jellium at $r_s=2$ | $\Delta T=-5600$ | $\Delta T_0=-6211$ | $\Delta T_2=482$ | $\Delta T_4=85$ | $\Delta T^g=-5644.$ |
| Contributions to $\Delta T_4$ at $r_s=2$ in erg cm$^{-2}$ | $\Delta T_4=85$ | $\Delta T_4^{(2)}=302$ | $\Delta T_4^{(3)}=-331$ | $\Delta T_4^{(4)}=114$ | |
| Proton in jellium at $r_s=2.07$ in Ry | $\Delta T=1.593$ | $\Delta T_0=1.436$ | $\Delta T_2=0.085$ | $\Delta T_4=0.027$ | $\Delta T^g=1.548.$ |

fourth-order gradient contributions ($\Delta T_4^{(3)}$ and $\Delta T_4^{(4)}$); such a partial sum is therefore not very useful. It is also clear from the form of $\Delta T_4$[12,45,46] that no amount of guesswork can generate the correct structure of this expansion; it requires a direct calculation of the gradient coefficients $B$, $C$, $D$ and $E$ (see eqn (6.24)) in the presence of a non-uniform background. Similar conclusions are found for the kinetic energy of atoms[92] and for the exchange-only energy of jellium surfaces.[93–96]

The exchange-correlation problem is more difficult than the kinetic energy for two reasons. First, there exists no *exact* benchmark calculation to compare with, and, secondly, $B_{xc}(\rho(\mathbf{r}))$ is not known exactly, while the higher-order exchange-correlation gradient coefficients are not as yet known at all. However, the convergence of $E_{xc}^{g}$ should not be entirely different from the kinetic-energy gradient expansion and we may expect the forms of eqn (6.25) to add about 10% to 20% correction to the surface energy in the LDA. In other words, the truncated $E_{xc}^{g}$ accounts for about 70% to 80% of the non-local corrections to the surface energy. Some of these speculations can be at least partially substantiated by recent calculations of the surface energy of jellium by Sun *et al.*[97] and Krotscheck *et al.*[98,99] Both groups avoid DFT and solve for the surface energy using a Jastrow-like many-body wave function. We present these results along with the LDA and gradient correction (eqn (6.25)) in Table 6.2. If we subtract the results of Kohn and Lang[100] in the column labelled LK from the last column labelled GE, we get the non-local correction as given by eqn (6.27). We see then that $E_{xc}^{g}$ makes a 15% correction to the LDA exchange-correlation surface energy[100] and does account for about 80% of the difference between LK and FHNC||0; this is not entirely different from the

**Table 6.2** Comparison of the total surface energies (in erg cm$^{-2}$) obtained for the jellium model at various densities by Lang and Kohn[100] in column 3, by Sun, Li, Farjam and Woo[97] in column 4, by Krotscheck, Kohn and Qian[98] in column 5 and by the LDA plus gradient corrections eqn (6.26) for the non-local exchange-correlation energy in column 6. Note that the results of LK correspond to the Wigner[101] form for the LDA. More recent forms[102] would tend to increase the LK results at $r_s = 2.07$ by approximately 100 erg cm$^{-2}$.

| $r_s$ | *Metal* | *LK* | *SFW* | *FHNC‖0* | *GE* |
|---|---|---|---|---|---|
| 2.07 | Al | −730 | 102 | −220 | −280 |
| 2.30 | Pb | −130 | 278 | 181 | 170 |
| 2.66 | Mg | 110 | 309 | 383 | 305 |
| 3.28 | Li | 210 | 363 | 360 | 305 |
| 3.99 | Na | 160 | 204 | 261 | 210 |
| 4.96 | K | 100 | 94 | 159 | 125 |
| 5.23 | Rb | 85 | 76 | 105 | 105 |

situation in the kinetic-energy expansion (Table 6.1). On the other hand, methods based on the exchange-correlation hole sum rule[50,53–55] via "modifications" of the gradient expansion, account for only about 10%–20% of this non-local component of $E_{xc}$ in sharp contrast to $E^{g}_{xc}$ which accounts for almost all of the non-local exchange-correlation energy. This comparison with accurate surface-energy calculations emphasizes clearly the failure of the above exchange-correlation hole procedures to capture the essence of the non-local structure of $S(\mathbf{q}, \mathbf{q})$. The same expansion was also applied to real systems;[84] however, because of the ionic contributions, which introduce great uncertainty in the theoretical predictions and also the uncertainty in the experimental results, we cannot draw any accurate conclusions about such corrections. Benchmark model calculations are therefore extremely valuable.

In multi-component systems such as the EHD in Si and Ge[16–30] (see Section 6.3.4), the non-local contribution to the surface energy also amounts to about 15% of the LDA.[25,26] More specifically, the LDA contribution in Ge is $\sim 5 \times 10^{-4}$ erg cm$^{-2}$ and the gradient corrections give $\sim 0.76 \times 10^{-4}$ erg cm$^{-2}$.

What about the extension of eqn (6.25) to microscopic systems, such as atoms or molecules?[22,48,103–107] Here we agree with the conclusion of MB that microscopic systems with their discrete spectrum of bound states are *fundamentally* different from extended metallic systems with the corresponding continuous spectrum of single-particle excitations. The local correlation structure of atoms is *not* closely related to electrons in a uniform system! Because of the continuum of electron–hole excitations possible in the uniform electron gas, correlation contributions must be much larger in extended systems than in atoms. Note that this is precisely what is found for both the LDA and gradient corrections.[37] This can be further corroborated, in an entirely different way, by looking at the structure factor $S(\mathbf{q}, \mathbf{q})$ for the weakly non-uniform electron gas (i.e. the $\mathbf{q}$ dependence of $B_{xc}(\mathbf{q})$; see Section 6.3.5). Its form is sketched in Fig. 6.11 along with the form for two finite-size systems of length $L_1$ and $L_2$. For a *microscopic* system, the particle conservation sum rule, eqn (6.51), and the $\mathbf{q} \to \mathbf{0}$ limit of $S(\mathbf{q}, \mathbf{q})$ must be the same and equal to zero. From Fig. 6.11, we see why predictions based on forms for extended systems *must* overestimate the non-local correlation contribution of finite systems. At this point, we can suggest a plausible reason why the LDA does even as well as it does for atoms. The exchange-correlation hole for a uniform system is localized with a length scale not entirely dissimilar to atomic dimensions. Consequently, $E^{\mathrm{LDA}}_{xc}$ is of the correct magnitude and is also relatively insensitive to boundary effects of atoms. To improve upon this situation in a systematic way is not trivial.

Perhaps a functional of the gradients which sums higher-order gradient

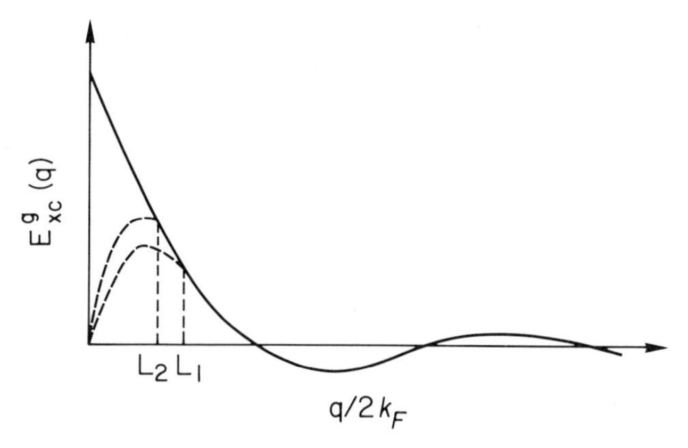


**Fig. 6.11** Sketch of the expected behaviour of the non-local contribution to the exchange and correlation energy (given by the dashed curves) in two microscopic systems of lengths $L_1$ and $L_2$ (where $L_2 > L_1$). The solid curve gives the behaviour for a macroscopic non-uniform electron gas. This small-$q$ region has been shown to be very sensitive to the non-uniform perturbation and not to be determined by any intrinsic length scales (like $k_F$ or $k_{TF}$) of the uniform electron gas. The graph is therefore to be regarded as only a sketch of this small-$q$ region. Clearly, the extended results much overestimate the correlation energy in microscopic systems, and the amount overestimated is very sensitive to the system (e.g. $L_1$ or $L_2$).

contributions could improve matters in small systems. Such a form must account for *all* order of gradients available.[37,103] Even then it is not clear whether, for localized systems, the appropriate starting point is the uniform electron gas. Such questions remain for future investigations and at this stage it is clear that an accurate universal functional for the exchange and correlation energy which encompasses both extended and localized systems is not available.

## 6.5 Conclusions

Our objective has been to review relevant aspects of the problems encountered in attempts to construct approximate exchange-correlation energy functionals for use in applications of DFT with particular emphasis on inhomogeneous many-electron systems in their ground states. The central quantity is the static structure factor $S(\mathbf{q}, \mathbf{q})$ of eqn (6.50) which contains all required information concerning inter-particle interactions, geometrical restrictions, inhomogeneities and external potentials. It is of the utmost importance to distinguish strictly finite systems, such as atoms, from truly macroscopic systems. For finite systems, the sum rule, $S(\mathbf{q}=\mathbf{0}, \mathbf{q}=\mathbf{0})=0$, which is related to particle-number conservation, at least at zero temperature, can be a useful guide. However, the range of $\mathbf{q}$ in which this sum rule plays a role is $\sim 1/L$, where $L$ is the characteristic linear dimension of the system. This vanishes completely for a macroscopic system in the thermodynamic limit and has the consequence that the limiting value of $S(\mathbf{q}, \mathbf{q})$ as $\mathbf{q} \to \mathbf{0}$ need not then be zero and, in fact, is non-universal depending on interactions, density profiles and details of the non-uniform system. For this reason, approaches to $E_{xc}$ based on forcing the exchange-correlation hole to obey the sum rule are not correct for extended systems in the thermodynamic limit. Adequate information concerning $S(\mathbf{q}, \mathbf{q})$ can only be obtained by detailed calculations.

The non-local structure of $E_{xc}$ is certainly very delicate and unknown, *a priori*. In the simplest approach to this problem, the formally exact perturbation expansion of the ground-state energy is rearranged and partially resummed to yield $E_{xc}$ as the local density approximation plus a presumably asymptotic series of gradient corrections $E_{xc}^{g}$. In practice, $E_{xc}^{g}$ is estimated from the lowest-order term alone. The accuracy of this approximation can be judged by comparing with recent numerical work on the surface energy of jellium which shows that this $E_{xc}^{g}$ accounts for about 80% of the total corrections to the LDA. In contrast, methods based on the sum rule underestimate these corrections by a wide margin (e.g. by a factor of 5 for jellium of Al density). The gradient expansion also provides a direct evaluation of $S(\mathbf{q}, \mathbf{q})$ and explicitly exhibits its structure which is much more complex for inhomogeneous systems than for uniform systems in the thermodynamic limit. In particular, the non-universal behaviour for $\mathbf{q} \to \mathbf{0}$ is obtained.

We summarize the major points of this discussion.

(1) The non-local aspects of the structure factor of a non-uniform many-electron system are subtle and are easily misrepresented or even totally missed by methods which are heavily based on too limited and incorrect extensions of results from uniform systems.

(2) The structure factor of an inhomogeneous many-electron system in the thermodynamic limit does not have universal behaviour as $\mathbf{q} \to \mathbf{0}$ and arguments based on particle conservation and perfect screening as for uniform systems have no relevance.

(3) The gradient expansion does remarkably well in accounting for corrections to the LDA when applied to the standard reference point of the surface energy of jellium. Computational schemes based on the LDA plus gradient corrections are thus reasonably accurate, as well as extremely simple, and constitute a feasible approach to calculating properties of those inhomogeneous extended, i.e. macroscopic, many-electron systems which are closely related, by virtue of their energy

spectrum, to the uniform electron gas. Note also that for purposes of calculating $E_{xc}$, it is neither necessary nor desirable to introduce the complications of $S(\mathbf{q}, \mathbf{q})$ which has non-universal $\mathbf{q}$ dependence; all relevant information is contained in the gradient coefficients, such as $B_{xc}$, which are universal functions of the density. The utility of such methods is clear in situations such as those of low symmetry, particularly where simple results may serve as input to the more accurate but computationally arduous calculations that will ultimately be required.

(4) Correlations in finite systems with a discrete energy spectrum, such as atoms, are quite different from correlations in extended systems which have a continuous energy spectrum of excitation. Procedures based on the uniform electron gas should not be used.

## Acknowledgments

This work was supported (D.J.W.G.) by the Natural Science and Engineering Research Council of Canada and (M.R.) by the Division of Materials Science, U.S. Department of Energy under Contract DE-AC05-840R21400 with Martin Marieta Energy Systems, Inc. We wish to acknowledge the help of Lisa Rasolt during the preparation of the manuscript and to thank Bridget Trim for her typing of the manuscript. We are also grateful to the editors, Professor N.H. March and B.M. Deb, for their patience.

## Appendix A 6.1

In this Appendix we give a brief sketch of the calculation of $b$ in eqn (6.94). It turns out to be more convenient to combine the results of $b^{(1)}+b^{(2)}+b_1^{(3)}+b_2^{(3)}$ into three different functions using the Ward identity

$$\Lambda_p^{(2)}+\Lambda_p^{(3)}=-\frac{\partial \Sigma(p)}{\partial \mu_0}. \tag{A 6.1}$$

Then according to MB,

$$b=b_0+b'+b''+b'''-\frac{m}{12\pi^2 k_F^3}\Sigma(k_F, 0),$$

where

$$b'=-\frac{\hbar^2}{m}\frac{\partial}{\partial \mu_0}\mathrm{tr}_p\left[\Sigma(p)\left(\frac{1}{2}\frac{\partial^2}{\partial \mu_0^2}G_0(p)-\frac{1}{9}\varepsilon_p\frac{\partial^3}{\partial \mu_0^3}G_0(p)\right)\right], \tag{A 6.2}$$

$$b'' = \frac{\hbar^2}{m}\frac{1}{6}\operatorname{tr}_{\mathrm{p}}\left[\Sigma(p)\frac{\partial^3}{\partial\mu_0^3}G_0(p)\right] \tag{A 6.3}$$

and

$$b''' = b_1^{(3)}. \tag{A 6.4}$$

Next, using eqn (6.96) for $\Sigma(p)$ and $V(p') = 4\pi e^2/\mathbf{p}'^2\varepsilon(p')$, we can rewrite $b'$ as

$$b' = \frac{\hbar^2}{m}\operatorname{tr}_{p'}\left\{V(p')\operatorname{tr}_p\left[\frac{1}{2}\frac{\partial}{\partial\mu_0}\left(G_0(p+p')\frac{\partial^2}{\partial\mu_0^2}G_0(p)\right)\right.\right.$$
$$\left.\left.-\frac{1}{9}\frac{\partial}{\partial\mu_0}\left(G_0(p+p')\varepsilon_{\mathbf{p}}\frac{\partial^3}{\partial\mu_0^3}G_0(p)\right)\right]\right\}$$
$$+\frac{\hbar^2}{m}\operatorname{tr}_{p'}\left\{\left(\frac{\partial}{\partial\mu_0}V(p')\right)\operatorname{tr}_p\left[\frac{1}{2}G_0(p+p')\frac{\partial^2}{\partial\mu_0^2}G_0(p)\right.\right.$$
$$\left.\left.-\frac{1}{9}G_0(p+p')\varepsilon_p\frac{\partial^3}{\partial\mu_0^3}G_0(p)\right]\right\} \tag{A 6.5}$$

and we can also rewrite $b''$ as

$$b'' = -\frac{\hbar^2}{m}\frac{1}{6}\operatorname{tr}_{p'}\left\{V(p')\operatorname{tr}_p\left[G_0(p+p')\frac{\partial^3}{\partial\mu^3}G_0(p)\right]\right\}. \tag{A 6.6}$$

Following MB we next define the functions

$$I_1(p') = \operatorname{tr}_p\left[G_0(p+p')\frac{\partial^2}{\partial\mu_0^2}G_0(p)\right], \tag{A 6.7}$$

$$I_2(p') = \operatorname{tr}_p\left[G_0(p+p')\frac{\partial^3}{\partial\mu_0^3}G_0(p)\right] \tag{A 6.8}$$

and

$$I_3(p') = \operatorname{tr}_p\left[G_0(p+p')\varepsilon_{\mathbf{p}}\frac{\partial^3}{\partial\mu_0^3}G_0(p')\right]. \tag{A 6.9}$$

Then in terms of these functions

$$b' = \frac{\hbar^2}{m}\operatorname{tr}_{p'}\left[\frac{\partial V(p')}{\partial\mu_0}\left\{\frac{1}{2}I_1(p') - \frac{1}{9}I_3(p')\right\}\right]$$
$$+\frac{\hbar^2}{m}\operatorname{tr}_{p'}\left[V(p')\left\{\frac{1}{2}\frac{\partial}{\partial\mu_0}I_1(p') - \frac{1}{9}\frac{\partial}{\partial\mu_0}I_3(p')\right\}\right] \tag{A.6.10}$$

and

$$b'' = -\frac{\hbar^2}{m}\frac{1}{6}\,\mathrm{tr}_{p'}[V(p')I_2(p')], \qquad \text{(A 6.11)}$$

and $b'''$ is still given in eqn (6.100).

To calculate the terms $I_1(p')$, $I_2(p')$ and $I_3(p')$ takes some algebra, but the calculation is fairly straightforward. The final integration in $\mathrm{tr}_{p'}[\ldots]$ is a two-dimensional integral over $|\mathbf{p}'|$ and $p_0'$. It is preferable to avoid the rapid structure one encounters along the real $p_0'$-axis, by rotating $p_0'$ to the imaginary axis. With the substitutions $p_0' \to 2k_F^2 qyi/m$ and $|\mathbf{p}'| \to q2k_F$, we get the following forms for the $I$s:

$$I_1(q) = \frac{m^3}{8k_F^3\pi^2 q^2}\left\{\frac{(1-2q^2)(-y^2+q^2-1)-2y^2(2q^2+1)}{(y^2+1-q^2)^2+4y^2q^2} + \frac{1}{4}\frac{1}{q}\ln\left[\frac{y^2+(q+1)^2}{y^2+(q-1)^2}\right]\right\}, \qquad \text{(A 6.12)}$$

$$I_2(q) = \frac{m^4}{8\pi^2 k_F^5 q^2}\left\{\frac{(3+2q^2)(-y^2+q^2-1)+4q^2y^2}{(y^2+1-q^2)^2+4y^2q^2} + \frac{(3-4q^2)[(y^2+1-q^2)^2-4q^2y^2]+16y^2(y^2+1-q^2)(\frac{3}{4}+q^2)}{[(y^2+1-q^2)^2-4y^2q^2]^2+16y^2q^2(y^2+1-q^2)^2} + \frac{im^3}{32\pi k_F^3}\delta(p_0')\,\delta(q-1)\right\}, \qquad \text{(A 6.13)}$$

$$I_3(q) = \frac{m^3}{4\pi^2 k_F^3 q^2}\left\{\frac{(\frac{3}{4}-2q^2)(-y^2+q^2-1)-2y^2(\frac{5}{4}+2q^2)}{(y^2+1-q^2)^2+4y^2q^2} + \frac{y^2[4y^2q^2-(y^2+1-q^2)^2+2(y^2+1-q^2)(y^2-q^2)]}{[(y^2+1-q^2)^2-4y^2q^2]^2+16y^2q^2(y^2+1-q^2)^2} + \frac{3}{16}\frac{1}{q}\ln\left[\frac{y^2+(q+1)^2}{y^2+(q-1)^2}\right]\right\} + \mu_0 I_2(q), \qquad \text{(A 6.14)}$$

and the integrals over $q$ and $y$ in (A 6.10) and (A 6.11) (and eqn (6.100) as well) are carried out numerically.

Our results in eqns (A 6.10)–(A 6.14) differ from MB in two important ways. First we did not proceed, at the outset, to take the small-$q$ limit of these equations. The small-$q$ limit is adequate for the HDL, but the more general forms, for arbitrary $q$, are needed for the extension to metallic densities (see Section 6.3.3). Secondly, MB simplify the calculation of $b'$ by integrating eqn (A 6.2) over $p'$ first, and then differentiating with respect to

$\mu_0$. This is sufficient for the total value of $b'$ but as we see in Section 6.3.5 it does not give the correct form for the structure factor of this weakly and slowly varying non-uniform electron gas; a form and its implications we wish to consider in Section 6.3.5.

## References


1. Thomas, L.H. (1927). *Proc. Camb. Phil. Soc.*, **23**, 542.
2. Fermi, E. (1927). *Rend. Acad. Naz. Lincei*, **6**, 602.
3. March, N.H. (1975). *Self-Consistent Fields in Atoms*, Pergamon, New York.
4. See also the series of review articles in *Theory of the Inhomogeneous Electron Gas*, eds Lundqvist, S. and March, N.H. Plenum, New York, 1983.
5. Hohenberg, P. and Kohn, W. (1964). *Phys. Rev.*, **136**, B864.
6. Kohn, W. and Sham, L.J. (1965). *Phys. Rev.*, **140**, A1133.
7. Sham, L.J. and Kohn, W. (1966). *Phys. Rev.*, **145**, 561.
8. Mermin, N.D. (1965). *Phys. Rev.*, **137**, A1441.
9. Perrot, F. (1979). *Phys. Rev.*, **A20**, 586.
10. Messer, J. (1981). *Temperature Dependent Thomas–Fermi Theory*, Lecture Notes in Physics Vol. 147, Springer, Berlin.
11. Brack, M. (1984). *Phys. Rev. Lett.*, **53**, 119.
12. Geldart, D.J.W. and Sommer, E. (1985). *Phys. Rev.*, **B32**, 7694.
13. Brueckner, K.A., Buchler, J.R., Clark, R.C. and Lombard, R.J. (1969). *Phys. Rev.*, **181**, 1543.
14. Baym, G., Bethe, H.A. and Pethick, C.J. (1971). *Nucl. Phys.*, **A175**, 225.
15. Brueckner, K.A., Chirico, J.H. and Meldner, H.W. (1971). *Phys. Rev.*, **C4**, 732.
16. Combescot, M. and Nozières, P. (1972). *J. Phys.*, **C5**, 2369.
17. von Barth, U. and Hedin, L. (1972). *J. Phys.*, **C5**, 1629.
18. Sander, L.M., Shore, H.B. and Sham, L.J. (1973). *Phys. Rev. Lett.*, **31**, 533.
19. Bütner, H. and Gerlach, E. (1973). *J. Phys.*, **C6**, L433.
20. Brinkman, W.F. and Rice, T.M. (1973). *Phys. Rev.*, **B7**, 1508.
21. Reinecke, T.L. and Ying, S.C. (1974). *Solid St. Commun.*, **14**, 381.
22. Gunnarsson, O., Lundqvist, B.I. and Wilkens, J.W. (1974). *Phys. Rev.*, **B10**, 1319.
23. Vosko, S.H., Perdew, J.P. and MacDonald, A.H. (1975). *Phys. Rev. Lett.*, **35**, 1725.
24. Gunnarsson, O. and Lundqvist, B.I. (1976). *Phys. Rev.*, **B13**, 4274.
25. Rasolt, M. and Geldart, D.J.W. (1977). *Phys. Rev.*, **B15**, 979.
26. Rasolt, M. and Geldart, D.J.W. (1977). *Phys. Rev.*, **B15**, 4804.
27. Rasolt, M. (1977). *Phys. Rev.*, **B16**, 3234; note that the quantities $dS_i/d\mu_j$ in eqns (42) and (44) of this reference should be replaced by the appropriate vertex functions as is seen from the preceding eqns (39)–(41).
28. Rice, T.M. (1977). In *Solid State Physics*, Vol. 32, eds Seitz, F., Turnbull, D. and Ehrenreich, H., p. 1, Academic Press, New York.
29. Rose, J.H. and Shore, H.B. (1978). *Phys. Rev.*, **B17**, 1884.
30. Kalia, R.K. and Vashista, P. (1978). *Phys. Rev.*, **B17**, 2655.
31. Vosko, S.H., Wilk, L. and Nusair, M. (1980). *Can. J. Phys.*, **58**, 1200.
32. Rasolt, M. and Davis, H.L. (1981). *Phys. Lett.*, **86A**, 45.
33. Rajagopal, A.K. and Callaway, J. (1973). *Phys. Rev.*, **B7**, 1912.

34. Rajagopal, A.K. (1978). *J. Phys.*, **C11**, L943.
35. MacDonald, A.H. and Vosko, S.H. (1979). *J. Phys.*, **C12**, 2977.
36. Rajagopal, A.K. (1980). In *Advances in Chemical Physics*, Vol. 41, eds Prigogine, I. and Rice, S.A., p. 59, Wiley, New York.
37. Ma, S.-K. and Brueckner, K.A. (1968). *Phys. Rev.*, **165**, 18.
38. Geldart, D.J.W. and Vosko, S.H. (1966). *Can. J. Phys.*, **44**, 2137.
39. Geldart, D.J.W. (1967). *Can. J. Phys.*, **45**, 3139.
40. Geldart, D.J.W. and Taylor, R. (1970). *Can. J. Phys.*, **48**, 155.
41. Geldart, D.J.W. and Taylor, R. (1970). *Can. J. Phys.*, **48**, 167.
42. Singwi, K.S., Sjolander, S., Tosi, M.P. and Land, R.H. (1970). *Phys. Rev.*, **B1**, 1044.
43. Dagens, L. (1971). *J. Phys., Paris*, **32**, 719.
44. Vashista, P. and Singwi, K.S. (1972). *Phys. Rev.*, **B6**, 4883.
45. Kirzhnits, D.A. (1957). *Zh. Eskp. Teor. Fiz.*, **32**, 115 (*Sov. Phys. JETP*, **5**, 64).
46. Hodges, C.H. (1973). *Can. J. Phys.*, **51**, 1428.
47. Murphy, D.R. (1981). *Phys. Rev.*, **A24**, 1682.
48. Sham, L.J. (1973). *Phys. Rev.*, **B7**, 4357.
49. Harris, J. and Jones, R.O. (1974). *J. Phys.*, **F4**, 1170.
50. Langreth, D.C. and Perdew, J.P. (1977). *Phys. Rev.*, **B15**, 2884.
51. Gunnarsson, O., Jonson, M. and Lundqvist, B.I. (1979). *Phys. Rev.*, **B20**, 3136.
52. Gunnarsson, O. and Jones, R.O. (1980). *Physica Scripta*, **21**, 394.
53. Langreth, D.C. and Perdew, J.P. (1980). *Phys. Rev.*, **B21**, 5469.
54. Langreth, D.C. and Perdew, J.P. (1982). *Phys. Rev.*, **B26**, 2810.
55. Langreth, D.C. and Mehl, M.J. (1983). *Phys. Rev.*, **B28**, 1809; (1984). *Ibid.*, **B29**, 2310.
56. Landau, L.D. and Lifshitz, E.M. (1980). *Statistical Physics*, Part 1, Pergamon, New York.
57. Rasolt, M., Malmström, G. and Geldart, D.J.W. (1979). *Phys. Rev.*, **B20**, 3012.
58. Pines, D. and Nozières, P. (1966). *The Theory of Quantum Liquids*, Benjamin, New York.
59. Sham, L.J. (1971). In *Computational Methods in Band Theory*, eds Marcus, P.J., Janak, J.F. and Williams, A.R., p. 458, Plenum, New York.
60. Rasolt, M. and Geldart, D.J.W. (1975). *Phys. Rev. Lett.*, **35**, 1234.
61. Geldart, D.J.W. and Rasolt, M. (1976). *Phys. Rev.*, **B13**, 1477.
62. Geldart, D.J.W., Rasolt, M. and Ambladh, C.O. (1975). *Solid St. Commun.*, **16**, 243.
63. Kleinman, L. (1974). *Phys. Rev.*, **B10**, 2221.
64. Beattie, A.M., Stoddart, J.C. and March, N.H. (1971). *Proc. Roy. Soc.*, **A326**, 97.
65. Gell-Mann, M. and Brueckner, K.A. (1957). *Phys. Rev.*, **106**, 364.
66. Peuckert, V. (1976). *J. Phys.*, **C9**, 4173.
67. Hubbard, J. (1957). *Proc. Roy. Soc.*, **A243**, 336.
68. Hedin, L. and Lundqvist, S. (1969). In *Solid State Physics*, Vol. 23, eds Seitz, F., Turnbull, D. and Ehrenreich, H., p. 1, Academic Press, New York.
69. Rasolt, M. and Geldart, D.J.W. (1986). *Phys. Rev.*, **B34**, 1325.
70. Rasolt, M. and Geldart, D.J.W. (1980). *Phys. Rev.*, **B21**, 3158.
71. Nozières, P. and Pines, D. (1958). *Phys. Rev.*, **111**, 442.
72. Rasolt, M. and Geldart, D.J.W. (1982). *Phys. Rev.*, **B25**, 5133.
73. Engelstaff, P.A. (1967). *An Introduction to the Liquid State*, Academic Press, New York.

74 Rasolt, M. (1983). *Phys. Rev.*, **27**, 5653.
75. Janak, J.F., Williams, A.R. and Moruzzi, V. L. (1975). *Phys. Rev.*, **B11**, 1522.
76. Ihm, J., Zunger, A. and Cohen, M.L. (1979). *J. Phys.*, **C12**, 4409.
77. Wang, C.S. and Freeman, A.J. (1980). *Phys. Rev.*, **B21**, 4585.
78. Schlüter, M. and Sham, L.J. (1982). *Phys. Today*, **35**, No. 2, p. 36.
79. Rasolt, M., Nickerson, S.B. and Vosko, S.H. (1975). *Solid St. Commun.*, **16**, 827.
80. Hybertsen, M.S. and Louie, S.G. (1985). *Phys. Rev. Lett.*, **55**, 1418.
81. Perdew, J.P. and Levy M. (1983). *Phys. Rev. Lett.*, **51**, 1884; Sham, L.J. and Schlüter, M. (1983). *Phys. Rev. Lett.*, **51**, 1888.
82. Hamann, D.R. (1979). *Phys. Rev. Lett.*, **42**, 662.
83. Car, R. and von Barth, U. (1984). In *Many-Body Phenomena at Surfaces*, eds Langreth, D.C. and Suhl, H., p. 31, Academic Press, New York.
84. Rose, J.H., Shore, H.B., Geldart, D.J.W. and Rasolt, M. (1976). *Solid St. Commun.*, **19**, 619.
85. Ferrantee, J. and Smith, J.R. (1979). *Phys. Rev.*, **B19**, 3911.
86. Lang, N.D. (1973). In *Solid State Physics*, Vol. 28, eds Seitz, F., Turnbull, D. and Ehrenreich, H., p. 225, Academic Press, New York.
87. Wang, J.S.-Y. and Rasolt, M. (1976). *Phys. Rev.*, **B13**, 5330.
88. Perrot, F. and Rasolt, M. (1983). *Phys. Rev.*, **B27**, 3273.
89. Jena P., Fradin, S.Y. and Ellis, D.E. (1979). *Phys. Rev.*, **B20**, 3543.
90. Popovic, D., Stott, M.J., Carbotte, J.P. and Piercey, G.R. (1978). *Phys. Rev.*, **B18**, 590.
91. Ma, C.Q. and Sahni, V. (1977). *Phys. Rev.*, **B16**, 4249.
92. Plumer, M.L. and Geldart, D.J.W. (1983). *J. Phys.*, **C16**, 677.
93. Harris, J. and Jones, R.O. (1970). *J. Phys.*, **F4**, 1170.
94. Mahan, G.D. (1975). *Phys. Rev.*, **B2**, 5585.
95. Rasolt, M., Wang, J.S.-Y. and Kahn, L.M. (1977). *Phys. Rev.*, **B15**, 580.
96. Sahni, V., Gruenebaum, J. and Perdew, J.P. (1982). *Phys. Rev.*, **B26**, 4371.
97. Sun, X., Li, T., Farjam, M. and Woo, C.W. (1983). *Phys. Rev.*, **B27**, 3913.
98. Krotscheck, E., Kohn, W. and Qian, G.-X. (1985). *Phys. Rev.*, **B32**, 5693.
99. Krotscheck, E. and Kohn, W. (1986). *Phys. Rev. Lett.*, **57**, 862.
100. Lang, N.D. and Kohn, W. (1970). *Phys. Rev.*, **B1**, 4555.
101. Wigner, E.P. (1938). *Trans. Faraday Soc.*, **34**, 678.
102. Ceperley, D.M. (1978). *Phys. Rev.*, **B18**, 3126.
103. Slater, J.C. (1951). *Phys. Rev.*, **81**, 385.
104. Herman, F., van Dyke, J.P. and Ortenburger, I.B. (1969). *Phys. Rev. Lett.*, **22**, 807.
105. Brueckner, K.A. (1969). In *Mathematical Methods in Solid State and Superfluid Theory*, eds Clark, R.C. and Derrick, G.H. p. 235, Oliver & Boyd, Edinburgh.
106. Tong, B.Y. (1971). *Phys. Rev.*, **A4**, 1375.
107. Gunnarsson, O. and Jones, R.O. (1985). *Phys. Rev. Lett.*, **55**, 107.

# 7 Density Functional Theory for Excited States

A.K. THEOPHILOU

*"Demokritos" Scientific Research Centre, Aghia, Paraskevi Attikis, Greece*



## 7.1 Introduction

The attractiveness of energy density functional theory derives from the fact that one can get rid of the many variables of the Schrödinger equations and study a many-particle system as a problem in one position variable. Although this is done at the expense of linearity, it makes the problem numerically manageable. The success of the Hohenberg–Kohn (HK) theory[1] for the ground state, and the elegance of the proof of the relevant theorems, presented a challenge to develop a theory for the excited states. Further, the Kohn–Sham (KS) theory[2] presented an intuitive argument or, rather, a secret hope that the energies of the one-particle wave functions might have some relation to the excitation spectrum of a many-particle system.

The first step for a KS theory for excited states came from Slater and co-workers[3–5] with transition-state theory.

The equations of Slater's transition-state theory were exactly like the KS equations. However, the density through which the self-consistent exchange and correlation potential was calculated involved fractional occupation numbers. This is a strange feature indeed, as a Slater determinant representing the state of a many-fermion system cannot have fractionally occupied orbitals. The arguments of Slater about the statistical nature of the fractional occupation number were not convincing for most physicists although the application of this theory to atoms, molecules and solids gave results that compare well with the experimental data.[3–5]

Single-Particle Density in Physics
and Chemistry ISBN 0-12-470518-9

The rigorous justification of Slater's transition-state theory came in 1979 when the present author introduced the subspace theory for excited states.[6] The fractional occupation numbers are derived rigorously in this theory. The density, however, is not the density corresponding to a single many-particle system, but to the sum of the $m$ lowest-energy eigenstate densities. The same holds for the energy and all other measured quantities. Thus the information about the $m$th eigenstate is derived by solving two sets of KS equations, i.e. the set for the $m$ and $m-1$ lowest-energy eigenstate densities.

In practice one usually performs the calculations with increasing order, so this is not a disadvantage of the theory. The disadvantage of this subspace theory is that one has to find expressions for the new density functionals for the exchange and correlation energy and this is a hard job indeed. Stoddart and Davis[7] have derived such expressions and have shown that the exchange and correlation energy for the sum of the ground and first excited state densities is similar in form to that of the ground state.

A new step in excited-state functional theory was made by Gunnarsson and Lundqvist,[8] who showed that if one restricts the space of wave functions to those characterized by a set of quantum numbers, then one can find a one-to-one correspondence between density and the lowest-energy eigenstate. This approach has the advantage of dealing with a single-state theory. However, for the higher excited state of each representation, one has to establish a subspace theory. Von Barth[9] made some improvements on the Gunnarsson–Lundqvist approach which are based on intuition and lack a rigorous mathematical foundation. In Section 7.4 I shall show that this approach is correct.

The major breakthrough in energy density functional theory came with the introduction, by Levy[10–12] and independently by Lieb,[13–14] of the density functional of the internal energy of a fermion system. The Levy–Lieb functional is not restricted to the ground state but also involves a large class of excited states. The special features of this class and its extent have not yet been investigated. This class may include all excited states which can be expressed as real-valued functions, but a proof of this conjecture is not yet available. The Levy–Lieb functional made possible rigourous formulations of the HK and KS theories for the ground state and was directly extended to subspaces, thus making possible the rigorous formulation of Theophilou's excited-state theory.[14–17]

Excited-state functionals were introduced by Valone and Capitani.[18] This theory has the advantage of using single-state functionals, in contrast with the Theophilou subspace functionals. However the functionals derived involve not only the density but also the potential. Thus, they do not lead to HK and KS theories for excited states.

A detailed mathematical investigation of the above theories has recently being initiated by Lieb, Levy and Perdew, English and English, Hadjisavvas and others.[14–26] These investigations involve concepts such as tangent functionals, Frechet derivatives, lower semicontinuities and, in general, obscure concepts from functional analysis, far beyond the present mathematical background, not only of an experimentalist, but also of most theorists. Although many would argue that these authors prove what is obvious for a physicist, the investigation of minor details reveals features that have far-reaching repercussions for the development of density functional theory as well as finding more accurate expressions for the density functionals. The mathematical physicist's approach is reliable but slow. A lot of theoretical physics however is based on intuition rather than rigorous mathematical proofs. The comparison of theoretical results with the experimental data is the main criterion of correctness. The KS approach is a good example of a theory which lacked a sound mathematical foundation. However, physicists never stopped using it to calculate the ground-state properties of atoms, molecules and solids as their results compared well with experiment. The rigorous mathematical foundation came some twenty years later.

## 7.2 The subspace functionals

As I believe that new ideas in science do not come out of the blue, I think it is worthwhile devoting a few words to the chain of thoughts which led me to define functionals and variational principles for subspaces. The idea came after trying to give a geometrical interpretation to the eigenvalue equation $H|\psi\rangle = E|\psi\rangle$. From the geometrical point of view this equation means that we search for a one-dimensional subspace of the Hilbert space, which is invariant under the operator $H$. In the same way one can search for subspaces $S$ of two or more dimensions which are invariant under $H$, i.e. a point in $S$ is mapped by $H$ to another point in $S$. This gives more freedom and the problem becomes easier as we do not care where exacly the point is mapped; it is sufficient to know that it is mapped into the subspace $S$. The next step is learn how to establish variational principles equivalent to that of eigenstates. In order to do this, one has to search for quantities which depend only on the subspace and not on the particular choice of eigenstates (points). Such quantities are, for example, the determinant of an operator when restricted to $S$, and its trace. In fact the trace in $S$ can be defined in terms of a determinant, but we shall not bother about tiresome mathematical definitions here. Thus we consider an orthonormal basis in our subspace $S$ and define the quantity

$$G_H(S) = \sum_{i=1}^{M} \langle\psi_i|H|\psi_i\rangle, \quad \langle\psi_i|\psi_j\rangle = \delta_{ij}, \quad \psi_i \in S. \tag{7.1}$$

The quantity $G_H(S)$ depends only on $S$ as one can go from one basis to another again in $S$, by a unitary transformation. It is easy then to prove that such transformations preserve trace. We can then search for the subspace of dimension $M$ that gives the minimum $G_H$. Let us call this $S_{\min}$. Then we have a new function

$$K_H(S)=\min_{S'} G_H(S')=G_H(S_{\min}), \tag{7.2}$$

the value of which is the sum of the $M$ lowest eigenstates of $H$.

Once we have the minimizing space $S$ we can diagonalize the matrix with elements $\langle\phi_i|H|\phi_j\rangle$, with $|\phi_i\rangle$ being any orhonormal basis in $S$, and obtain the $M$ lowest-energy eigenstates $E_i$, $E_i \leqslant E_{i+1}$. The matrix representing $H$ in $S$ is always diagonalizable since $S$ is of finite dimension.

In order to proceed to the HK theory, we have to show that for the minimizing subspace $S_V$ of $G_{H_V}(S)$,

$$H_V = H_0 + \int \hat{\rho}(\mathbf{r})V(\mathbf{r})\,\mathrm{d}\mathbf{r} \tag{7.3}$$

is not identical with that of $H_{V'}$. After subtracting $H_V$ from $H_{V'}$ we have

$$H_{V'} - H_V = \int \hat{\rho}(\mathbf{r})\,(V'(\mathbf{r}) - V(\mathbf{r}))\,\mathrm{d}\mathbf{r}. \tag{7.4}$$

Then if $S_V = S_{V'}$, $S_V$ is an invariant subspace of a potential operator

$$\hat{U} = \int \hat{\rho}(\mathbf{r})(V'(\mathbf{r}) - V(\mathbf{r}))\,\mathrm{d}\mathbf{r}.$$

If the difference of the two potentials is a constant then the two potentials are equivalent. So we have to examine the case $\hat{U}=CI$. If $\hat{U}$ has an invariant, the subspace of finite dimension also has eigenstates. The corresponding eigenfunctions of $U$, $\psi(\mathbf{r}_1, \ldots, \mathbf{r}_N)$, vanish in a finite volume:

$$\left\{\left(\sum_i U(\mathbf{r}_i)\right) - \lambda\right\}\psi(\mathbf{r}_1, \ldots, \mathbf{r}_N) = 0. \tag{7.5}$$

But such eigenstates are not in the set where we search for the eigenstates of $H_V$. Hence $S_V \neq S_{V'}$.

A mathematician would ask about the exact definition of the space where we search for the eigenstates. This is a legitimate question, as is the question as to whether the minimizing subspace really belongs to the set of subspaces of a Hilbert space. These questions have recently been answered by Lieb,[14] English and English,[17] and Hadjisavvas and Theophilou.[15,16]

We now proceed to give our fundamental theorems.

For the subspace density we use the notation $\rho(\mathbf{r}, S) = G_{\hat{\rho}(\mathbf{r})}(S)$.

*Theorem* 1
If $S_V$ is a minimizing subspace for $H_V$ and $S_{V'}$ that of $H_{V'}$ then the corresponding subspace densities are different.

*Proof*
From the minimum property of $G_{\mathrm{H}}(S_V)$ and $G_H(S_{V'})$, defined by eqn (7.2), and the fact that $S \neq S_{V'}$ we have

$$G_{H_0}(S_V) + \int \rho\,(\mathbf{r}, S_V)V(\mathbf{r})\,\mathrm{d}\mathbf{r} < G_{H_0}(S_{V'}) + \int \rho(\mathbf{r}, S_{V'})V(\mathbf{r})\,\mathrm{d}\mathbf{r}, \tag{7.6a}$$

$$G_{H0}(S_{V'}) + \int \rho(\mathbf{r}, S_{V'})V'(\mathbf{r})\,\mathrm{d}\mathbf{r} < G_{H0}(S_V) + \int \rho\,(\mathbf{r}, S_V)V'(\mathbf{r})\,\mathrm{d}\mathbf{r}. \tag{7.6b}$$

By adding and rearranging terms we get the inequality

$$\int (\rho(\mathbf{r}, S_V) - \rho(\mathbf{r}, S_{V'}))(V(\mathbf{r}) - V'(\mathbf{r}))\,\mathrm{d}\mathbf{r} < 0. \tag{7.7}$$

Thus it follows that the two densities are identical only if $V(\mathbf{r}) - V'(\mathbf{r}) = C$. □

*Theorem* 2
If $S_V$ is non-degenerate, i.e. if the minimizing subspace in eqn (7.2) is uniquely defined, then there is a one-to-one correspondence between the density and the subspace.

This theorem follows directly from Theorem 1.

A result of the above theorem is that all subspace functionals $G_A(S_V)$ can be expressed in terms of $\rho$. For these functionals we shall use the notation

$$A(\rho_V) = G_A(S_V), \quad S_V \rightrightarrows \rho_V. \tag{7.8}$$

By $S \rightrightarrows \rho$ we mean a subspace with density $\rho$.

Non-degenerate subspaces may occur when the highest eigenvalue $E_i$ of $H_V$ in $S$ is degenerate, and at least one of the degenerate eigenstates is not in the minimizing subspace. Degeneracy does not always affect the one-to-one correspondence $S \rightleftarrows \rho$, as one can easily verify by considering the one-particle case. For the definition of the energy as a functional of the density the $S \rightleftarrows \rho$ relation is not necessary, because different $S$ corresponding to the same $\rho$ may give the same energy. The density $\rho$ need not be unique either. Thus, as noted by English and English the minimum principle for the energy-density functional holds for degenerate states as well.[17]
The variations in the subspaces can be separated into two categories:

(a) the $S$ with the same $\rho$, i.e. $S \rightrightarrows \rho$; and
(b) those with different densities.

Then

$$E[V]=\min_{\rho}\left\{\min_{S\rightarrow\rho}\{G_{H_0}[S]+\int\rho(\mathbf{r})V(\mathbf{r})\,\mathrm{d}\mathbf{r}\}\right\}. \tag{7.9}$$

Thus one can define a new functional for subspaces that is similar to the Levy–Lieb functional for the single states, i.e.

$$H_0[\rho]=\min_{S\rightrightarrows\rho}\{G_{H_0}[S]\}. \tag{7.10}$$

Then

$$E[V]=\min_{\rho}\left\{H_0[\rho]+\int\rho V\right\}. \tag{7.11}$$

Note that the $H_0[\rho]$ defined above is not restricted to densities corresponding to a subspace $S_V$ but to all densities. This is an advantage since one can establish a minimum principle with variations $\delta\rho$ not restricted to the set of ground-state densities. Thus from eqn (7.11) we have

$$\delta H_0[\rho]+\int V\,\delta\rho(\mathbf{r})\,\mathrm{d}\mathbf{r}=\mu_N\int\delta\rho(\mathbf{r})\,\mathrm{d}\mathbf{r}, \tag{7.12}$$

where $\mu_N$ is a Lagrange multiplier arising from the condition $\int\rho(\mathbf{r})\,\mathrm{d}\mathbf{r}=MN$, $N$ being the number of particles and $M$, as always in this Chapter, the dimension of the subspace. Since the variations are arbitrary we have

$$\frac{\delta H_0(\rho)}{\delta\rho(\mathbf{r})}=-V(\mathbf{r})+\mu_N. \tag{7.13}$$

Thus we have the HK equation for subspace densities. It is to be noted that eqn (7.13) gives not only minimizing densities, but other stationary subspace densities as well. The one to be chosen is that which gives the minimum $E(V)$, as defined by eqn (7.11); then

$$H_0(\rho_M)+\int\rho_M(\mathbf{r})\,V(\mathbf{r})\,\mathrm{d}\mathbf{r}=\sum_{i=1}^{M}E_i, \tag{7.14}$$

and from the corresponding $\rho_{M-1}$ density of the subspace $S_{M-1}$ one can find the eigenvalue $E_M$:

$$E_M=E_V(\rho_M)-E_V(\rho_{M-1}), \tag{7.15}$$

and all properties of the eigenstate $\psi_M$, i.e. for any operator $A$,

$$\langle\psi_M|A|\psi_M\rangle=A(\rho_M)-A(\rho_{M-1}). \tag{7.16}$$

It is to be noted that the functionals $E_V(\rho_M)$ and $A(\rho_M)$ are different for subspaces of different dimension.

### 7.2.1 The equivalent of the KS theory for excited states

As the kinetic-energy functional is not known with accuracy, it is useful to develop a KS theory for excited states. This is done by assigning to each subspace of the interacting system $S$ a subspace of the "non-interacting" system $R$, in such a way that the two subspaces have the same density. By "non-interacting" system we mean here a system whose energy eigenstates are single Slater determinants.

In the following, we denote by $G_A(S)$ the functionals as defined earlier divided by $M$, the dimension of the subspace. This definition will make the comparison with the ground-state KS theory easier, as the new density $\rho_R(\mathbf{r}) = G_{\rho(\mathbf{r})}(R)$ corresponds to $N$ particles. Similarly the new definition of the other functionals correspond to "average values" in the subspace.

We next define the following functionals for every non-interacting subspace $R$. Note that the internal-energy operator of the non-interacting system is $H_0 = T$, the kinetic-energy operator; whereas for the interacting system,

$$H_0 = T + H_{\text{int}}, \tag{7.17a}$$

where $H_{\text{int}}$ is the electron–electron interaction operator:

$$H_{\text{int}} = \frac{1}{2}\sum_{i \neq j} \frac{e^2}{|\mathbf{r}_i - \mathbf{r}_j|}; \tag{7.17b}$$

$$G_T(R'_R) = \min\{G_T(R')\colon \dim R' = M, R' \rightrightarrows \rho_R\}, \tag{7.18}$$

$$G_{H_0}(S_R) = \min\{G_{H_0}(S)\colon \dim S = M, S \rightrightarrows \rho_R\}, \tag{7.19}$$

$$\Delta T(R) = G_T(S_R) - G_T(R'_R), \tag{7.20}$$

$$E_{\text{xc}}(R) = G_{H_{\text{int}}}(S_R) - \frac{1}{2}\int \frac{\rho_R(\mathbf{r})\rho_R(\mathbf{r}')}{|\mathbf{r} - \mathbf{r}'|}\, d\mathbf{r}\, d\mathbf{r}', \tag{7.21}$$

$$H[R] = G_T[R] + \Delta T[R] + \int V(\mathbf{r})\rho_R(\mathbf{r})\, d\mathbf{r} + E_{\text{xc}}[R] + \frac{1}{2}\int \frac{\rho_R(\mathbf{r})\rho_R(\mathbf{r}')}{|\mathbf{r} - \mathbf{r}'|}\, d\mathbf{r}. \tag{7.22}$$

The physical meaning of these functionals can be seen as follows. Suppose that $R$ is the eigensubspace corresponding to the first $M$ eigenvalues of the Hamiltonian of a non-interacting system for some external potential. Suppose further that the corresponding density is also the density of an eigensubspace corresponding to the first $M$ eigenvalues of an interacting system, for some other external potential. Then it follows easily that the subspaces $R'_R$ and $S_R$ defined by relations (7.18) and (7.19) are nothing but the subspaces $R$ and $S$ respectively. Thus, $\Delta T[R]$ is the difference of the kinetic energy of the interacting and the non-interacting system, and $G_T[R]$ and $E_{xc}[R]$ are, respectively, the kinetic energy and the exchange and correlation energy of the interacting system, expressed as functionals of $R$. However, the fundamental advantage of the definitions (7.18)–(7.22) is that they have meaning for any $R$, eigenspace or not, and no assumptions are made on the nature of the density of $R$.

One can next prove the following theorem which establishes a variational principle.

*Theorem* 3

The minimum of $H[R]$ over all non-interacting subspaces $R$ is equal to the subspace energy of the exact Hamiltonian and the density corresponding to the minimizing $R$ is equal to the exact subspace density, i.e.

$$\min_R H[R] = \frac{1}{M} \sum_{i=1}^{M} E_i, \tag{7.23a}$$

$$\rho_R(\mathbf{r}) = \frac{1}{M} \sum_{i=1}^{M} \rho_i(\mathbf{r}), \tag{7.23b}$$

where the $\rho_i(\mathbf{r})$ are the densities corresponding to the eigenstates $|\psi_i\rangle$ of the exact Hamiltonian.

The proof of the above theorem is given in ref. 16.

Let us now write the explicit form of the Slater determinants of the subspace $R$. In order that $\langle\Phi_i|\Phi_j\rangle = 0$ for $i \neq j$, $|\Phi_i\rangle$ and $|\Phi_j\rangle$ have to differ by at least one orbital:

$$\left.\begin{aligned}
|\Phi_1\rangle &= \{\varphi_1, \varphi_2, \ldots, \varphi_{N-1}, \varphi_N\}, \\
|\Phi_2\rangle &= \{\varphi_1, \varphi_2, \ldots, \varphi_{N-1}, \varphi_{N+1}\}, \\
|\Phi_3\rangle &= \{\varphi_1, \varphi_2, \ldots, \varphi_{N-2}, \varphi_{N+1}, \varphi_N\}, \\
|\Phi_4\rangle &= \{\varphi_1, \varphi_2, \ldots, \varphi_{N-2}, \varphi_{N-1}, \varphi_{N+2}\}.
\end{aligned}\right\} \tag{7.24}$$

The subspace density and kinetic energy expressions are

$$\rho(\mathbf{r}) = \sum_i \frac{n_i}{N} |\varphi_i(\mathbf{r})|^2, \tag{7.25a}$$

$$T(\rho) = \sum_i \frac{n_i}{N} |\nabla\varphi_i(\mathbf{r})|^2. \tag{7.25b}$$

As $n_i/N \leqslant 1$, fractional occupation numbers appear in the density as in the Slater transition-state theory. The $\varphi_i$, being orbitals of a "non-interacting" system, obey a system of equations exactly like those of Kohn and Sham, i.e.

$$-\tfrac{1}{2}\nabla^2\varphi_i+(V_{xc}[\rho]+V+V_{es}(\mathbf{r}))\varphi_i=\varepsilon_i|\varphi_i\rangle, \tag{7.26a}$$

where

$$V_{xc}=\frac{\delta E_{xc}[\rho]}{\delta\rho(\mathbf{r})}, \tag{7.26b}$$

$$V_{es}(\mathbf{r})=\int\frac{\rho(\mathbf{r}')}{|\mathbf{r}'-\mathbf{r}|}\,d\mathbf{r}'. \tag{7.26c}$$

Here $V$ is the external potential. For the two-dimensional subspace it is clear how $|\Phi_1\rangle$ and $|\Phi_2\rangle$ are chosen. For three-dimensional subspaces one has to choose $|\Phi_3\rangle$ in different ways, i.e. by replacing $\varphi_{N-1}$ in $|\Phi_1\rangle$ by $\varphi_{N+1}$, or $\varphi_N$ by $\varphi_{N+2}$, and search for the minimum values of the energy for the self-consistent density. For higher dimensions one has to try more combinations. It is clear that inner-shell excitations are allowed for $M>2$.

In practice one uses as an input to the calculation a given density, and by solving the KS equations one gets the eigenstates. With the output eigenstates one can try various combinations of fractional occupation numbers for the higher-energy orbitals and search for a self-consistent density, taking into account that the occupation numbers allowed are those which allow the construction of an orthogonal system of Slater determinants. In this way one gets upper bounds for the energy. By modifying the occupation numbers one can get new densities and lower energy bounds until one reaches the ground state. This can be used as the minimization procedure when only the single ground-state energy and density are of interest. Calculational methods based on the subspace theory have been developed by Bendt and Zunger.[23]

For many-particle systems such as solids one does not expect a significant change of the density for the low-energy excitation spectrum, as a change of the highest occupied orbitals will introduce only a small perturbation to the density. One can easily show that the new effective potential calculated with the perturbed density differs from the ground-state potential by $C/N$. Thus the self-consistent potential of the Kohn–Sham theory is practically self-consistent for the lowest-energy eigenstates. For this reason energy band theory gives information about the low-lying excitation spectrum of solids. For the high-energy excitation spectrum the use of the orbitals of the KS theory will give upper bounds. Therefore, exact self-consistency calculations for densities with non-integral occupation numbers are needed. The only calculations so far carried out are those of Slater and co-workers.[3–5] Although their results agree reasonably with experimental data, more accurate results can be derived if better approximations for the exchange and correlation potential for subspaces are used.

For two-dimensional subspaces an expression for $V_{xc}[\rho]$ has been derived by Stoddart and Davis:[7]

$$V_{xc}^{SD}(\rho)=C(N)\rho^{1/3}(r). \tag{7.27a}$$

The coefficient $C$ depends on the number of particles, and is different from that of the ground state:

$$C(N)=\frac{(3\pi^2)^{4/3}}{12\pi^3}\left\{1+\frac{1}{2}\frac{N-1}{N}\left[\left(\frac{N-1}{N}\right)^{1/3}+\frac{8}{3}\left(\frac{N+1}{N}\right)^{1/3}-\frac{8}{3}\right]\right.$$
$$\left.+\frac{1}{2}\left(\frac{N+1}{N}-\frac{8}{3}\right)\left(\frac{N+1}{N}\right)^{1/3}\right\}. \tag{7.27b}$$

Their method was based on a Hartee–Fock approach for subspaces and the procedure used was that developed by Stoddart, Beattie and March.[29]

## 7.3 Other results derivable from the subspace theory

From the inequality for the ground state,

$$\int(\rho_1(\mathbf{r})-\rho_1'(\mathbf{r}))(V(\mathbf{r})-V'(\mathbf{r}))\,d\mathbf{r}<0, \tag{7.28a}$$

and that for the two-dimensional subspace densities (eqn (7.7)),

$$\int(\rho_1(\mathbf{r})-\rho_1'(\mathbf{r}))(V(\mathbf{r})-V'(\mathbf{r}))+(\rho_2(\mathbf{r})-\rho_2^{1}(\mathbf{r}))(V(\mathbf{r})-V'(\mathbf{r}))\,d\mathbf{r}, \tag{7.28b}$$

weaker inequalities result that establish a connection with the ensemble densities.[12,14,34] These inequalities are obtained by multiplying (7.28a) and (7.28b) by positive constants $c_1$ and $c_2$ and adding. Then one can easily prove that convex combination of densities of the ground and excited state uniquely determine the potential, i.e. if

$$\rho(\mathbf{r},\lambda_1,\lambda_2)=\lambda_1\rho_1(\mathbf{r})+\lambda_2\rho_2(\mathbf{r}),\quad \lambda_1+\lambda_2=1,\quad 0<\lambda_2<\lambda_1\leqslant 1, \tag{7.29}$$

then

$$\int(\rho(\mathbf{r},\lambda_1,\lambda_2)-\rho'(\mathbf{r},\lambda_1,\lambda_2))(V(\mathbf{r})-V'(\mathbf{r}))\,d\mathbf{r}<0. \tag{7.30}$$

Thus one can establish variational principles and functionals for pairs of orthogonal states. The generalization for a convex combination of "densities" belonging to mutually orthogonal states is straightforward. The variational principle is

$$E(V, \lambda_1, \ldots, \lambda_m) = \min_{\rho(\lambda_1, \ldots, \lambda_m)} \left\{ \min \left\{ \sum_{i=1}^{M} \lambda_i \langle \psi_i | H_V | \Psi_i \rangle : \quad \rho(\mathbf{r}, \lambda_1, \ldots, \lambda_m) = \right.\right.$$

$$\sum_i \lambda_i \langle \Psi_i | \hat{\rho}(\mathbf{r}) | \Psi_i \rangle, \lambda_i > 0, \sum_i \lambda_i = 1,$$

$$\left.\left. \langle \Psi_i | \Psi_j \rangle = \delta_{ij} \right\}\right\}. \tag{7.31}$$

Note that in the ordering $\lambda_i \leqslant \lambda_{i+1}$ the energy eigenvalues will occur in the same order. The difficulty is that one has to define new functionals for each $n$-tuple of the parameters $\lambda_i$. The other disadvantage is that the higher the excited state the lower the contribution Thus for $M=2$, say, one has a smaller contribution from the excited state, about which information is sought, than from the ground state for which the information can be extracted from the HK theory.

Instead of the $n$-tuple of mutually orthogonal states, one can define a state

$$\left. \begin{aligned} |\Psi(C_1, \ldots, C_M)\rangle = \sum_{i=1}^{M} C_i | \Psi_i \rangle, \quad \langle \Psi_i | \Psi_j \rangle = \delta_{ij}, \\ C_i > 0, \quad C_i \text{ real}, \sum_i C_i^2 = 1, \end{aligned} \right\} \tag{7.32}$$

and the functional

$$\hat{E}_V(C_1, \ldots, C_M) = \min \langle \Psi(C_1, \ldots, C_M) | H_V | \Psi(C_1, \ldots, C_M) \rangle \tag{7.33}$$

or

$$\hat{E}(V, C_1, \ldots, C_M) = \min \left\{ \sum_i |C_i|^2 \langle \Psi_i | H_V | \Psi_i \rangle + \sum_{i \neq j} C_i C_j \langle \Psi_i | H_V | \Psi_j \rangle, \right.$$

$$\left. \langle \Psi_i | \Psi_j \rangle = \delta_{ij} \right\}. \tag{7.34}$$

Then by using the method of Lagrange multipliers for the orthogonality constraint exactly as in ref. 6 we find that the $|\Psi_i\rangle$ have to satisfy the following system of equations:

$$|C_i|^2 H | \Psi_i \rangle + \sum_{i \neq j} (C_i C_j - \mu_{ij}) | \Psi_j \rangle = |C_i|^2 E_i | \Psi_i \rangle, \quad i = 1, 2, \ldots, M. \tag{7.35}$$

By using the orthogonality relation for the eigenstates one can easily show that eqns (7.35) reduce to

$$H_V | \Psi_i \rangle = E_i | \Psi_i \rangle \tag{7.36}$$

and the minimum is obtained for the $M$ lowest-energy eigenstates of $H_V$. Then by a procedure similar to that of HK we can establish a one-to-one correspondence between $\rho(C_1, \ldots, C_M)$ and $|\Psi(C_1, \ldots, C_M)\rangle$. Note that

$$E(\lambda_1, \lambda_2, \ldots, \lambda_\mu) = \hat{E}(\lambda_1^{1/2}, \lambda_2^{1/2}, \ldots, \lambda_\mu^{1/2}). \tag{7.37}$$

A special case of the functional $\hat{E}(C_1, \ldots, C_M)$ is the functional defined by von Barth[9] in dealing with states of different symmetries. This functional will be discussed below.

## 7.4 Density functionals and symmetries

If one restricts the space of wave functions to a certain infinite-dimensional subspace of a Hilbert space then it is possible to derive density functionals restricted to this space. This was first realized by Gunarssan and Lundquist who generalized the HK theorem for the lowest-energy states of each symmetry.[8] Thus consider a linear set of operators $Q_i$ which commute with the Hamiltonian of a system, i.e.
Hamiltonian of a system, i.e.

$$[Q_i, H] = 0, \tag{7.38a}$$

$$Q_i |\Psi(q_1, \ldots, q_n)\rangle = q_i |\Psi(q_1, \ldots, q_n)\rangle. \tag{7.38b}$$

Then since the operators $Q_i$ are linear, a linear combination of such $\Psi$ will be an eigenstate of $Q_i$. Thus these functionals form a linear space that is a subspace of the Hilbert space. Then by following a procedure like that of HK one can develop a functional theory for the lowest energy of each symmetry.[8]

As an example, one can take the space of states $|\Psi_m^l\rangle$:

$$\mathbf{L}^2 |\Psi_m^l\rangle = l(l+1) |\Psi_m^l\rangle, \tag{7.39a}$$

$$L_z |\Psi_m^l\rangle = m |\Psi_m^l\rangle. \tag{7.39b}$$

Then one can establish a one-to-one correspondence between $\Psi_m^l$ and $\rho_m^l$. Note that the density $\rho_m^l$ is not an eigenstate of $\mathbf{L}^2$ and $L_z$. Another application is Katriel's alternative proof of Theorems 1 and 2, by searching for the antisymmetric lowest-energy state of a super-Hamiltonian in the product space of eigenstates.[26]

Instead of restriction to certain quantum numbers, one can consider the states transforming according to the irreducible representation of a group $G$, i.e. if $g$ are the group elements of $G$ and $D^l(g)$ the representation matrix, then by the definition of a representation,

$$g |\Psi_m^l\rangle = \sum_{m'} D_{m'm}^l(g) |\Psi_{m'}^l\rangle. \tag{7.40}$$

As the states with fixed $l$ but different $m$ form a subspace, one can show that the subspace density

$$\rho^l(\mathbf{r})=\sum_m \langle \Psi^l|\hat{\rho}(\mathbf{r})|\Psi^l_m\rangle \qquad (7.41)$$

transforms according to the identity representations of the group.[30] The advantage is that the effective potential corresponding to $\rho^l(r)$ has the same symmetry as the external potential. This is not obviously the case in the single-state theory of Gunnarsson and Lundqvist.

For higher excited states belonging to the same irreducible representation, one has to consider the direct sum of subspaces and define

$$H_0[\rho]=\min\sum_i G_{H_0}(S_i), \quad \frac{1}{m}\sum_i \rho(\mathbf{r}, S_i)=\rho(\mathbf{r}). \qquad (7.42)$$

The main problem in the HK and the KS theories is to find reasonable approximations for the functionals. So, for each definition of a functional its explicit functional form, i.e. its explicit expression, is needed. However it is much easier to work out functionals for high-symmetry potentials, such as spherically symmetric potentials, than it is to develop a general theory. These cases may assist in finding more accurate functional forms of the general theory. Note that most of the presently known approximate functionals for $E_{xc}(\rho)$ derive from the electron gas, which is characterized by a highly symmetric Hamiltonian. For spherically symmetric systems, the features of the eigenstate densities of atoms found by Parr and his collaborators[31,32] offer an excellent opportunity for insight and inspiration for the study of density functionals and functional forms.

Gunnarson and Lundqvist[8] in deriving the lower-energy states of atoms of each symmetry used fractional occupation numbers in order to construct densities having the same symmetry as that of the external potential. This was the right choice according to subspace theory, even though the single Slater determinant coorespondence to the non-interacting system is violated.

In order to overcome the above difficulties von Barth[9] developed the theory of mixed symmetry, i.e. instead of trying to associate to the state $|\Psi\rangle$ a Slater determinant $|\Phi\rangle$ having the same symmetry, he constructed states from linear combinations of eigenstates of the exact Hamiltonian which have different symmetries, for example

$$|\Psi(l_1, l_2; a_1, a_2)\rangle = a_1|\Psi_{l_1}\rangle + a_2|\Psi_{l_2}\rangle, \quad a_1^2+a^2=1, \qquad (7.43)$$

where $l$ denotes a set of quantum numbers that characterizes different irreducible representations. Then he showed that for $a_1$ and $a_2$ fixed, the minimum is attained for a definite $|\Psi\rangle$ which is a functional of the density.

The coefficients $a_i$ are chosen so that to the mixed-symmetry state corresponds a Slater determinant of the non-interacting system with the same density. However it is not clear how this correspondence is made as one can find many determinants with the same density. Neither is it possible to find a single determinant which corresponds to the lowest-energy state of an irreducible representation because the choice of the exact state was from a reduced subspace. Neither is the density $\rho(\mathbf{r})$ symmetric. However, the energy levels that von Barth derived for atoms[9,33] indicate that a clear and rigorous formulation is possible.

Thus let us consider the lowest-energy states of a non-interacting system corresponding to a potential which is invariant under a group of transformations. Instead of referring to any group we shall for convenience refer to the familiar group of rotations, i.e. the potential considered is a spherically symmetric potential. The energy levels are degenerate and they form a subspace $S$. This subspace is reducible, i.e. one can find lower-dimensional subspaces of $S$, the $S^l$ irreducible subspaces, which are invariant under rotations. Assume now that we switch on an electron–electron interaction having a small coupling constant $g$ and apply perturbation theory for the degenerate states. The energy levels split and the corresponding eigenstates belong to the irreducible subspaces $S^l$of the rotation group. The new states form a basis in $S$.

Although in $S$ one could find an orthogonal basis consisting of single determinants, the $\Psi^l_m$ are linear combinations of determinants. However, linear combinations of determinants cannot always be expressed as single determinants.

Let us now express the initial basis in $S$, i.e. the determinants $D^m_i$, in terms of the new basis. (We use in our notation only two indices for $D^m_i$, although more are usually needed.) Then

$$|D^m_i\rangle = \sum_l C^l_{im} |\Psi^l_m\rangle; \tag{7.44}$$

that is, the determinants of the basis have definite transformation properties although they do not belong to a unique irreducible representation of the rotation group. Instead of the approximate energy eigenstates, let us consider any states $|\Phi^l_m\rangle$ of the Hilbert space and construct from them a set of states $|\Phi^m_i\rangle$ having coefficients $C^l_{im}$ as above. The linear combinations of these states form an infinite-dimensional subspace of the Hilbert space which is invariant under rotations. The expectation value of $H_V$ for a state $\Psi_i(\Phi^{l_1}, \ldots, \Phi^l_1)$ is a functional

$$\begin{aligned} G(\Phi^{l_1}, \ldots, \Phi^{l_n}, C^l_{im}) &= \langle \Psi_i(\Phi^{l_1}, \Phi^{l_2}, \ldots, \Phi^{l_n}) | H | \Psi_i(\Phi^{l_1}, \Phi^{l_2}, \ldots, \Phi^{l_n}) \rangle \\ &= \sum_l |C^l_{im}|^2 \langle \Phi^l_m | H_0 | \Phi^l_m \rangle + \int \sum_l |C^l_{im}|^2 \langle \Phi^l_m | \rho(r) | \Phi^l_m \rangle \times V(\mathbf{r})\, d\mathbf{r}. \end{aligned} \tag{7.45}$$

The minima of $G$, as von Barth has shown, establish a variational principle and an energy density functional for the minimizing state $|\Psi_i\rangle$. This density is not, however, spherically symmetric as it involves off-diagonal terms of the energy eigenstates $|\Psi^l_m\rangle$.

So far we have established a variational principle as von Barth did, but with a definite choice of the coefficients $C^l_{im}$. The next step is to derive an HK theory. This can be done by assigning to each $|\Phi^m_i\rangle$ a Slater determinant $D^m_i$ having the same density as the state $|\Phi^m_i\rangle$.

The establishment of the non-interacting state minimization principle can be obtained by following a procedure similar to that used by Hadjisavvas and Theophilou for a rigorous formulation of the KS theory. Thus, provided that the special theorems relating to the requirement for special transformation properties of the determinants $D^m_l$ are proved, a rigorous formulation of the von Barth theory is possible. Further, by the procedure followed in Section 7.3 for deriving variational principles for convex combinations of densities, one can derive an equivalent variational principle for fixed coefficients $|C^l_{im}|^2$. The advantage of the latter formalism is that the density so derived is spherically symmetric. This however is not a single-state formalism, but rather a formalism for an $n$-tuple of states.

## 7.5 The Valone–Capitani functional

If $\lambda$ is a real parameter then

$$F(\Psi;\lambda)=\langle\Psi|(H_V-\lambda I)^2|\Psi\rangle\geqslant 0. \tag{7.46}$$

The minimum of $F(\Psi;\lambda)$ for fixed $\lambda$ occurs for the eigenstate $|\Psi_i\rangle$ of $H$ whose eigensvalue $E_i$ is closest to $\lambda$.[35] Thus one can define the functional

$$G_{VG}(V,\lambda)=\min_{\|\Psi\|=1}\langle\Psi|(H-\lambda I)^2|\Psi\rangle. \tag{7.47}$$

Next one can define the density functional

$$G_{VG}(\rho;\mathbf{V},\lambda)=\min_{\rho}\left\{\min_{\Psi\rightrightarrows\rho}\langle\Psi(H-\lambda)^2|\Psi\rangle\right\} \tag{7.48}$$

The above functional was introduced by Valone and Capitani[18] and is a functional which gives excited-state energies. They also defined similar functionals for even powers of $H-\lambda I$.

The advantage of the Valone–Capitani functional is that it is a single-state functional, whereas the Theophilou functional is defined on $n$-tuples of mutually orthogonal states. However, theirs is a potential-dependent functional and cannot be used to express the internal energy of a system as a functional of the density. Thus it does not lead to HK and KS theories.

## 7.6 Conclusions

The KS equations can give the excitation spectra of many-electron systems, provided that fractional occupation numbers are allowed for the density. The allowed orbitals are those that allow the construction of mutually orthogonal Slater determinants. The densities involved in the construction of the EDF theories are in general composite densities of mutually orthogonal states. Alternative proofs of the fundamental theorems have been presented which allow more physical, and mathematical, insight. However, little work has been done on the explicit expressions of the density functionals. The rich experience in ground-state functionals[29, 36–62] can be used in this new field. In particular, approximate expressions for the exchange and correlation potential will offer the possibility of carrying out calculations for the excitation spectra and other physical properties of atoms, molecules and solids.

The discovery that a class of excited states are minimizing states for the internal energy of a many-electron system under the density constraint has given a new impetus to the excited-state theory.[63, 64] The determination of further mathematical and physical features of this class is still an open problem. Potential-dependent functional theories have also been developed.[18] Long-forgotten simple mathematical relations such as those of orthogonal polynomials[65] have been retrieved because of their importance in energy density functional theory, while the use of modern mathematical weaponry has started only recently.

As the field of excited-state density functional theory is still in its infancy, it is an open field not only for mathematically minded physicists, but also for the intuitive and pragmatic who have more interest in calculations and their interpreation.

## Acknowledgments

The author would like to thank Professor N.H. March and Dr N. Hadjisavvas for useful discussions.

## References

1. Hohenberg, P.C. and Kohn, W. (1964). *Phys. Rev.*, **B136**, 846.
2. Kohn, W. and Sham, L.J. (1965). *Phys. Rev.*, **A140**, 1133.
3. Slater, J.C. (1974). *The Self-Consistent Field for Molecules and Solids*, McGraw-Hill, New York.
4. Slater, J.C. and Wood, J. (1971). *Int. J. Quant. Chem.*, **4**, 3.

5. Slater, J.C., Mann, J.B., Wilson, T.M. and Wood, J.H. (1969). *Phys. Rev.*, **184**, 672.
6. Theophilou, A. (1979). *J. Phys.*, **C12**, 5419.
7. Stoddart, J.C. and Davis, K. (1982). *Solid St. Commun.*, **42** 147.
8. Gunnarsson, O. and Lundqvist, B.J. (1976). *Phys. Rev.*, **B13**, 4274.
9. Von Barth, U. (1979). *Phys. Rev.*, **A20**, 1693.
10. Levy, M. (1979). *Proc. Natl Acad. Sci. U.S.A.*, **76**, 6062.
11. Hohenberg, P. and Kohn, W. (1964). *Phys. Rev.*, **B136**, 846.
12. Levy, M. (1979). *Bull. Am. Phys. Soc.*, **24**, 626.
13. Lieb, E. (1983). *Int. J. Quant. Chem.*, **24**, 243.
14. Lieb, E. (1985). In *Density Functional Methods in Physics*, eds Dreizler, R.M. and da Providencia, X., Plenum, New York.
15. Hadjisavvas, N. and Theophilou, A. (1984). *Phys. Rev.*, **A30**, 2183.
16. Hadjisavvas, N. and Theophilou, A. (1985). *Phys. Rev.*, **A32**, 720.
17. English, H. and English, R. (1983). *Physica*, **121A**, 253.
18. Valone, S.M. and Capitani, J.F. (1981). *Phys. Rev.*, **A23**, 2127.
19. Nalewajski, R. and Carlton, T.S. (1983). *J. Chem. Phys.*, **78**, 1616.
20. Levy, M. and Perdew, J.P. (1985). In *Density Functional Methods in Physics*, eds Dreizler, R.M. and da Providencia, K., Plenum, New York.
21. English, H. and English R. (1984). *Phys. Stat. Sol.* (b), **123**, 711; **124**, 373.
22. Perdew, J. P. and Zunger, A. (1981). *Phys. Rev.*, **B23**, 5048.
23. Pathak, R. (1984). *Phys. Rev.*, **A29**, 978.
24. Percus, J.K. (1978). *Int. J. Quant. Chem.*, **13**, 89.
25. Freed, K.F. and Levy, M. (1982). *J. Chem. Phys.*, **77**, 396.
26. Katriel, J. (1980). *J. Phys.*, **C13**, L375.
27. Harriman, J.E. (1981). *Phys. Rev.*, **A24**, 680.
28. Bendt, P. and Zunger, A. (1982). *Phys. Rev.*, **B26**, 3114.
29. Stoddart, J.C., Beattie, A.M. and March, N.H. (1971). *Int. J. Quant. Chem.* **4**, 35.
30. Theophilou, A. (1981). In *Recent Developments in Condensed Matter Physics*, Vol. 2, eds DeVreese, J.T., Lemmens, L.F., Van Doren V.E. and Van Royen, J., Plenum, New York.
31. Parr, R. (1985). In *Density Functional Methods in Physics*, eds Dreizler R.M. and da Providencia, X., Plenum, New York.
32. Wang, W.P. and Parr, R. (1977). *Phys. Rev.*, **A16**, 91.
33. Von Barth, U. (1985). In *Density Functional Methods in Physics*, eds Dreizler, R.M. and da Providencia, X., Plenum, New York.
34. Valone, S.M. (1980). *J. Chem. Phys.*, **73**, 1344; 4653.
35. McDonald, A.H. (1979). *J. Phys.*, **C12**, 2977.
36. Brosens, F. (1975). *Phys. Stat. Sol.* (b), **68**, 645.
37. Deb, B.M. and Ghosh, S.K. (1983). *Int. J. Quant. Chem.*, **23**, 1.
38. Hodges, C.H. (1973). *Can. J. Phys.*, **51**, 1428.
39. Stoddart, J.C. and March, N.H. (1967). *Proc. Roy. Soc.*, **A299**, 279.
40. Antoniewicz, P.R. and Kleinman, L. (1985). *Phys. Rev.*, **B31**, 6779.
41. Kleinman, L. (1974). *Phys. Rev.*, **B10**, 222.
42. Singwi, K.S., Tosti, M.P. and Land, R.H. (1968). *Phys. Rev.*, **176**, 589.
43. Singwi, K.S., Sjolander, A., Tosi, M.P. and Land, R.H. (1970). *Phys. Rev.*, **B1**, 1044.
44. Chakravarty, S. and Chia-Wei Woo, (1976). *Phys. Rev.*, **B13**, 4815.
45. Pathak, R.K. and Bartolotti, L.J. (1985). *Phys. Rev.*, **A31**, 3557.
46. Huntington, H.B. (1979). *Phys. Rev.*, **B20**, 3165.

47. Mohammed, A.-R. E. and Sahni, U. (1985). *Phys. Rev.*, **B31**, 4879.
48. Mohammed, A.-R. E. and Sahmi, U. (1984). *Phys. Rev.*, **B29**, 3687.
49. Geldart, D.J.W. and Rasolt, M. (1976). *Phys. Rev.*, **B13**, 1477.
50. Rasolt, M. and Geldart, D.J.W. (1975). *Phys. Rev. Lett.*, **35**, 1234.
51. Rasolt, M. and Vosko, S.H. (1974). *Phys. Rev.*, **B10**, 4195.
52. Nickersan, S.B. and Vosko, S.H. (1976). *Phys. Rev.*, **B14**, 4399.
53. Rau, A.R.P. and Rajagopal, A.K. (1975). *Phys. Rev.*, **B11**, 3604.
54. Rajagopal, A.K. (1978). *Phys. Rev.*, **B17**, 2980.
55. Vinter, B. (1978). *Phys. Rev.*, **B17**, 2429.
56. Almbbladh, C.O. and von Barth, U. (1985). *Phys. Rev.*, **B31**, 3231.
57. Alonso, J.A. and Girifalco, L.A. (1978). *Phys. Rev.*, **B17**, 3735.
58. Csavinsky, P. and Vosman, F. (1983). *Int. J. Quant. Chem.*, **23**, 1973.
59. Ossicini, S. and Bertoni, C.M. (1985). *Phys. Rev.*, **A31**, 3550.
60. Plumer, M.L. and Stott, M.J. (1985). *J. Phys.*, **C18**, 143.
61. Ludena, E.V. (1983). *Int. J. Quant. Chem.*, **23**, 127.
62. Santos, E. and Leal, M. (1983). *Phys. Rev.*, **A28**, 459.
63. Perdew, J.P. and Levy, M. (1983). *Phys. Rev. Lett.* **51**, 1884.
64. Perdew, J.P. and Levy, M. (1985). *Phys. Rev.*, **B31**, 6264.
65. Dawson, K.A. and March, N.H. (1984). *Phys. Lett.* **106A**, 161.

# 8 Quantum Fluid Dynamical Significance of the Single-Particle Density

B.M. DEB

*Theoretical Chemistry Group, Department of Chemistry, Panjab University, Chandigarh – 160 014, India*

and

S.K. GHOSH

*Heavy Water Division, Bhabha Atomic Research Centre, Bombay – 400 085, India*



Single-Particle Density in Physics and Chemistry ISBN 0-12-470518-9

## 8.1 Introduction: analogy between quantum systems and classical fluids

The previous Chapters have essentially dealt with time-independent aspects of the single-particle density. The development of a general time-dependent density functional theory, which is still incomplete, is intimately related to developments in the general area of quantum fluid dynamics (QFD),[1,2] which views the electron cloud in a many-electron system as a "classical" fluid moving under the action of classical Coulomb forces augmented by forces of quantum origin. In this Chapter, we discuss various fundamental aspects of QFD, its links with density functional theory (DFT) and certain applications to many-electron systems.

Since Schrödinger's early formulation of quantum mechanics in terms of a complex wave function having no direct physical significance, attempts* have been made to interpret quantum phenomena within the framework of classical mechanics. Although the last two decades have seen an upsurge of interest in the hydrodynamical analogy of quantum mechanics, this analogy was first established by Madelung[4] who transformed the Schrödinger single-particle equation into a pair of hydrodynamical equations. Thus, even at such an early stage, a "classical" language could be employed, in principle, to describe quantum processes. De Broglie's immediate attempt to prescribe a deeper signifance to this picture did not find many followers and interest in his theory of double solution[5,6] or the pilot wave theory has grown only recently.[7] Among other approaches to classical interpretations of quantum mechanics, there is the important stochastic analogy,[8-11] where a particle-like inhomogeneity is thought to undergo random fluctuations (Brownian motion) in a background fluid. Its link with QFD will also be discussed in this Chapter.

As a result of pioneering work by Bohm[12] and Takabayashi,[13] the hydrodynamical analogy has now attracted a great deal of attention and has also found many applications[1,2] in atomic, molecular, solid-state and nuclear physics, especially the latter. The associated theoretical framework has also been broadened and developed further. Among the various routes leading to QFD, the most popular, and historically the earliest, is to substitute the polar form of the wave function into the Schrödinger equation.[4] However, this is restricted to pure quantum states. An alternative approach,[14] using density matrices, is applicable to mixed states as well. The ensemble interpretation of quantum mechanics, whose foundations were laid by formalisms in phase space,[15-18] provides another route to QFD. Both the density matrix and the phase-space formalisms are

*For an excellent summary of various interpretations of quantum mechanics in an historical perpective, see ref. 3.

suitable to arrive at the hydrodynamics corresponding to macroscopic quantum systems. QFD is also employed at a phenomenological level[19] to study, for example, problems in solid-state physics. All these developments of QFD are surveyed in this Chapter, emphasizing the interconnections between the various approaches.

The development of the stochastic interpretation of quantum mechanics[10,20,21] lends additional significance to QFD. Associated with the concepts of an internal local pressure[22-24] and local temperature,[25] there is a "thermodynamic" picture[26] of the many-electron fluid leading to various interpretive aspects and several schemes of potential utility.[27,28] Using DFT, there follows a more rigorous formulation of certain commonly used chemical concepts.[29,30] Density functional theory is formally equivalent to QFD in three-dimensional (3D) space,[31] whose basic variables are the 3D charge density $\rho(\mathbf{r})$ and current density $\mathbf{j}(\mathbf{r})$ of the electron fluid. Therefore, through QFD, DFT can go not only into time-dependent (TD) situations,[32,33] but into excited states (see Chapter 6) as well. This promises to be of considerable benefit in studying atomic and molecular collisions since here one is concerned with both TD situations and excited states. In fact, a joint attack on high-energy ion–atom collisions through DFT and QFD* leads to an interesting non-linear Schrödinger equation[34] through which one can, in principle, follow a collisional process from start tó finish. Besides, quantities such as frequency-dependent multipole polarizabilities[35] and hyperpolarizabilities,[36] which are difficult to calculate by any other method, are particularly amenable to a QFDFT approach.

## 8.2 Transcription of non-relativistic quantum mechanics (pure states) into fluid dynamics

### 8.2.1 Quantum hydrodynamics corresponding to a single-particle time-dependent Schrödinger equation

Consider a one-electron system characterized by a time-dependent external potential $V(\mathbf{r}, t)$ described by the time-dependent Schrödinger equation:

$$-\frac{\hbar^2}{2m}\nabla^2\psi + V\psi = \mathrm{i}\hbar\frac{\partial\psi}{\partial t}. \tag{8.1}$$

Substitution of the polar form of the wave function, i.e.

$$\psi(\mathbf{r}, t) = R(\mathbf{r}, t)\exp\left[\mathrm{i}S(\mathbf{r}, t)/\hbar\right], \tag{8.2}$$

* We shall designate any approach based jointly on QFD and DFT as QFDFT.

leads to the pair of fluid dynamical equations, namely the continuity equation

$$\frac{\partial \rho}{\partial t} + \nabla \cdot \mathbf{j} = 0, \tag{8.3}$$

and the Euler-type equation of motion

$$m\rho \frac{\mathrm{d}\mathbf{v}}{\mathrm{d}t} = -\rho \nabla (V + V_{\mathrm{q}}), \tag{8.4}$$

where

$$\frac{\mathrm{d}\mathbf{v}}{\mathrm{d}t} = \frac{\partial \mathbf{v}}{\partial t} + (\mathbf{v} \cdot \nabla)\mathbf{v}; \tag{8.5}$$

the density $\rho = R^2$, the velocity field $\mathbf{v} = (1/m)\nabla S$, the quantum potential $V_{\mathrm{q}} = (-\hbar^2/2m)\nabla^2 R/R$, and $\mathbf{j}$ ($\equiv \rho \mathbf{v}$) is the current density. The analogy of quantum mechanics with classical fluid dynamics is then clear; the electron fluid is subjected to a force of quantum origin, in addition to the classical Coulomb force. The charge and current densities $\rho$ and $\mathbf{j}$ respectively are the quantities of interest here and eqns (8.3) and (8.4) form the basis of QFD.

The Euler equation (8.4) can also be recast into the form of a Navier–Stokes equation:[1]

$$m \frac{\partial}{\partial t}(\rho v_\mu) = \sum_\nu \nabla_\nu \{ m\rho v_\mu v_\nu + \tilde{p}^{(\mathrm{q})}_{\mu\nu} \} - \rho \nabla_\mu V, \tag{8.6}$$

where $v_\mu$ is the $\mu$th component of the velocity vector $\mathbf{v}$, $\nabla_\mu$ represents the $\mu$th gradient component and $\tilde{p}^{(\mathrm{q})}_{\mu\nu}$ is the quantum stress tensor given by

$$\tilde{p}^{(\mathrm{q})}_{\mu\nu} = -\frac{\hbar^2}{4m} \nabla^2 \rho \delta_{\mu\nu} + \frac{\hbar^2}{4m} \frac{\nabla_\mu \rho \nabla_\nu \rho}{\rho}. \tag{8.7}$$

The viscosity coefficients of the electron fluid can easily be identified[23] from the stress tensor.

The velocity field $\mathbf{v}$ appearing in the fluid dynamical equations is in fact simply the real part of a more general quantum mechanical "local" velocity field (ref. 37) $\mathbf{u}$, defined through the momentum operator as

$$\mathbf{u} = -\mathrm{i} \frac{\hbar}{m} \frac{\nabla \psi}{\psi} = \mathbf{v} + \mathrm{i}\mathbf{v}_{\mathrm{i}} \tag{8.8}$$

where the real part $\mathbf{v}$ ($= \nabla S/m$) and the imaginary velocity $\mathbf{v}_{\mathrm{i}}$ is given by

$$\mathbf{v}_{\mathrm{i}} = -\frac{\hbar}{2m} \nabla \ln \rho. \tag{8.9}$$

Both these velocity fields are, however, irrotational in nature, i.e. $\nabla \times \mathbf{v}_i = \mathbf{0}$. The quantum stress tensor of eqn (8.7) differs from that of classical hydrodynamics in that it depends on the derivatives of density rather than velocity. This can be rationalized by considering this dependence to be on the imaginary velocity field. Several other aspects of the velocity field and of the single-particle Schrödinger fluid have been given by Kan and Griffin.[38]

Although the imaginary velocity field or the closely related stress tensor appear in the local differential form of the equation of motion, they disappear on integration, leaving behind the simple form

$$m\frac{d\langle \mathbf{v} \rangle}{dt} = -\langle \nabla V \rangle \tag{8.10a}$$

or

$$m\frac{d}{dt}\int \mathbf{j}(\mathbf{r})\, d\mathbf{r} = -\int \rho(\mathbf{r}) \nabla V(\mathbf{r})\, d\mathbf{r}, \tag{8.10b}$$

Thus, the net quantum force on the system is always zero for isolated quantum systems. (It may be noted that this is not true[39] if the integration is over a molecular subsystem[40] or if one considers an enclosed system.[41]) Thus, it is purely an inner force. For a static stationary state with zero velocity field, the left-hand side of eqn (8.10) vanishes, indicating that the integrated classical force density also vanishes. For a single-particle system, $V$ is only the external potential and this result is merely a special case of the more general result of Levy and Perdew,[42] who have proved that the integrated force density due to the external potential is zero even for a many-electron system. Additional interesting consequences for a many-electron system will be discussed in the next Section.

Another result that might be of interest in this connection is that the quantum forces do not have a contribution to the resultant moment of force, namely

$$\langle \mathbf{r} \times \nabla V_q \rangle = \mathbf{0}, \tag{8.11}$$

thus implying that the change in angular momentum is caused solely by the moment of the classical force.

The quantum fluid dynamical equations are described in terms of the charge and current densities; their time-independent solutions correspond to stationary states. The physical interpretation associated with the static stationary state ($\mathbf{v} = \mathbf{0}$) is that the classical electrostatic force acting on the electron fluid is exactly balanced by the quantum force arising from the Bohm potential, thus giving an explanation of the stability of matter. The quantum mechanical kinetic energy is merely manifested as a quantum potential energy. For the dynamical stationary state ($\mathbf{v} \neq \mathbf{0}$), the uncompensated force leads to an additional macroscopic kinetic energy ($\frac{1}{2}mv^2$) of the electron fluid. Thus, it is interesting to note that the total kinetic energy is properly represented in terms of the complex velocity field $\mathbf{u}$ i.e. as $\frac{1}{2}mu^2$, where $\mathbf{u} = \mathbf{v} + \mathbf{v}_i$.

The velocity fields, on integration, yield the trajectories and streamlines for the electron fluid. The origin of the angular momentum in stationary states can thereby be interpreted and detailed insight into collision phenomena is possible. Various consequences of the vorticity arising from the velocity field and their relations with the nodes of the wave function have been discussed elsewhere.[1]

QFD can be generalized to include the effect of the magnetic vector potential. Thus, if the quantum particle is acted on by additional external time-dependent electric and magnetic fields, characterized by the scalar potential $\phi(\mathbf{r}, t)$ and vector potential $\mathbf{A}(\hat{\mathbf{r}}, t)$ respectively, the corresponding Schrödinger equation (taking electronic charge as $-e$),

$$\frac{1}{2m}\left(-i\hbar\nabla + \frac{e}{c}\mathbf{A}\right)^2 \psi + V\psi - e\phi\psi = i\hbar\frac{\partial \psi}{\partial t}, \tag{8.12}$$

would then be equivalent to the continuity equation

$$\frac{\partial \rho}{\partial t} + \nabla \cdot (\rho \mathbf{v}) = 0, \tag{8.13a}$$

and an equation of motion

$$m\rho\frac{\mathrm{d}\mathbf{v}}{\mathrm{d}t} + \rho\left(e\mathbf{E} + \frac{e}{c}\mathbf{v} \times \mathbf{B}\right) = -\rho\nabla(V + V_q), \tag{8.13b}$$

where the Coulomb gauge $\nabla \cdot \mathbf{A} = 0$ has been employed for the vector potential $\mathbf{A}$ and the electric and magnetic field strengths are given by

$$\mathbf{E} = -\nabla\phi - \frac{1}{c}\frac{\partial \mathbf{A}}{\partial t}, \tag{8.14a}$$

$$\mathbf{B} = \nabla \times \mathbf{A}. \tag{8.14b}$$

The new velocity field is

$$\mathbf{v} = \frac{1}{m}\left(\nabla S + \frac{e}{c}\mathbf{A}\right). \tag{8.14c}$$

Several interesting consequences of this extension and the effect of taking electron spin into account have been summarized by Ghosh and Deb.[1]

### 8.2.2 Quantum hydrodynamics for a many-electron system

The above formalism is restricted to a single-particle system. Consider now the more general case of an $N$-electron system characterized by an external potential $V(\mathbf{r})$ and subjected to external electric as well as magnetic fields described by the scalar and vector potentials $\phi(\mathbf{r},t)$ and $A(r,t)$ respectively. The corresponding time-dependent Schrödinger equation is (electronic charge is taken as $-e$)

$$\mathscr{H}\psi = \mathrm{i}\hbar\frac{\partial\psi}{\partial t}, \tag{8.15a}$$

where

$$\mathscr{H} = \sum_j \frac{1}{2m}\left(\hat{\mathbf{p}}_j + \frac{e}{c}\mathbf{A}_j\right)^2 + V' - \sum_j e\phi_j, \tag{8.15b}$$

with $\mathbf{A}_j \equiv \mathbf{A}(\mathbf{r}_j, t)$; $\phi_j \equiv \phi(\mathbf{r}_j, t)$ and $V'$ denoting the total potential energy, i.e. $V' = \Sigma_j V(\mathbf{r}_j) + \frac{1}{2}\Sigma_{j\neq k}|\mathbf{r}_{jk}|^{-1}$. Substituting the polar form of the wave function, i.e. $\psi = R(\mathbf{r}_1, \ldots, \mathbf{r}_N; t)\exp[\mathrm{i}S(\mathbf{r}_1, \ldots, \mathbf{r}_N; t)/\hbar]$, one recovers[31] the hydrodynamical-continuity and the Euler-type equations, namely

$$\frac{\partial\rho_N}{\partial t} + \sum_j \nabla_j \cdot (\rho_N \mathbf{v}_j) = 0 \tag{8.16a}$$

and

$$m\frac{\partial\mathbf{v}_k}{\partial t} + m\sum_j (\mathbf{v}_j \cdot \nabla_k)\mathbf{v}_j + m\sum_j (1 - \delta_{jk})\mathbf{v}_j \times (\nabla_k \times \mathbf{v}_j)$$

$$+ \left(e\mathbf{E}_k + \frac{e}{c}\mathbf{v}_k \times \mathbf{B}_k\right) = -\nabla_k(Q + V'), \tag{8.16b}$$

where the velocity field for the $j$th electron is given by

$$\mathbf{v}_j = \frac{1}{m}\left(\nabla_j S + \frac{e}{c}\mathbf{A}_j\right) \tag{8.16c}$$

and the quantum potential is

$$Q = \sum_j Q_j = \sum_j -\frac{\hbar^2}{2m}\frac{\nabla_j^2 R}{R}. \tag{8.16d}$$

The QFD equations (8.16) are, however, coupled complicated equations in configuration space and therefore it is quite difficult to interpret them physically. Unless the many-electron Schrödinger fluid is projected to

3D space, it loses physical clarity, thereby becoming essentially a device to supplement the mathematical formalism. The continuity equation (8.16a) can easily be transformed to 3D space by integrating over $N-1$ electron coordinates, i.e.

$$\frac{\partial \rho}{\partial t}+\nabla \cdot \mathbf{j}=0, \tag{8.17}$$

where $\rho$ and $\mathbf{j}$ are the 3D charge and current densities respectively, given by

$$\left.\begin{aligned} \rho(\mathbf{r}) &= N\int \rho_N \, d\mathbf{r}_2 \cdots d\mathbf{r}_N, \\ \mathbf{j}(\mathbf{r}) &= \frac{1}{m} N \int_{\mathbf{r}_1=\mathbf{r}} \rho_N \nabla_1 S(\mathbf{r}_1, \ldots, \mathbf{r}_N)\, d\mathbf{r}_2 \cdots d\mathbf{r}_N. \end{aligned}\right\} \tag{8.18}$$

However, one cannot directly obtain an Euler-type equation of motion in 3D space, although the QFD will be appealing only if local quantities $\rho$ and $\mathbf{j}$ in 3D space are employed as the basic dynamical variables. This has been repeatedly emphasized in our works (see e.g. ref. 2).

Success has, however, been achieved in formulating QFD in 3D space for many-electron systems within the framework of single-particle self-consistent theories which employ an orbital partitioning of the charge density. Thus, Takabayashi[13] has outlined the 3D QFD corresponding to a many-electron system within Hartree theory, while the same has been obtained by Wong[43] for Hartree–Fock theory. An "exact" description is possible using the natural-orbital Hamiltonian.[31] Another formally "exact" theory suitable for this description is the time-dependent density functional theory[32] (TDDFT) involving the 3D charge and current densities. TDDFT is now acquiring increasing importance[32–35, 44–50] and is briefly discussed below. Also available for this purpose (reduction to 3D space) are the dynamical Thomas–Fermi theory and its variants;[51] these are discussed later.

All the self-consistent theories mentioned above involve a set of single-particle equations of the form

$$\left(-\frac{\hbar^2}{2m}\nabla^2+V_{\text{ext}}+V_k^{\text{SCF}}\right)\psi_k=i\hbar\frac{\partial \psi_k}{\partial t}, \tag{8.19}$$

where the effective potential $V_{\mathbf{k}}^{\text{SCF}}$ is determined self-consistently from the solutions $\{\psi_k\}$. The density $\rho(\mathbf{r}, t)$ and the current density $j(\mathbf{r}, t)$ are obtained from

$$\rho(\mathbf{r}, t)=\sum_k |\psi_k(\mathbf{r}, t)|^2 \tag{8.20a}$$

and

$$\mathbf{j}(\mathbf{r}, t) = \frac{i\hbar}{2m} \sum_k (\psi_k^* \nabla \psi_k - \psi_k \nabla \psi_k^*). \tag{8.20b}$$

The 3D QFD corresponding to eqn (8.19) is easily obtained by again employing the polar form of the single-particle orbitals, i.e. $\psi_k = R_k \exp[iS_k/\hbar]$. Thus, one obtains

$$\frac{\partial \rho_k}{\partial t} + \nabla \cdot \mathbf{j}_k = 0 \tag{8.21a}$$

and

$$m \frac{\partial \mathbf{j}_k}{\partial t} + m\nabla \cdot \left( \frac{\mathbf{j}_k \mathbf{j}_k}{\rho_k} \right) = -\rho_k \nabla (U_k + V_k) \tag{8.21b}$$

as the continuity and Euler-type equations respectively in simple forms. The effective potential $V_k^{\mathrm{SCF}}$ is a local one depending on the density ($\rho_k \equiv R_k^2$) as is the case with the time-dependent version[32] of Kohn–Sham[52] density functional theory. Here, the $k$th orbital current density is given by $\mathbf{j}_k = (1/m)\rho_k \nabla S_k$, the $k$th orbital quantum potential by $U_k = -(\hbar^2/2m)\nabla^2 R_k/R_k$ and $V_k = V_{\mathrm{ext}} + V_k^{\mathrm{SCF}}$. In the presence of external electric and magnetic fields, an additional force term given by $-e\rho_k(\mathbf{E} + (1/c)\mathbf{j}_k \times \mathbf{B})$ appears on the right-hand side of eqn (8.21b) and the current density is given by

$$\mathbf{j}_k = \frac{1}{m} \rho_k \left( \nabla S_k + \frac{e}{c} \mathbf{A} \right), \tag{8.21c}$$

where $\mathbf{A}$ is the vector potential giving rise to a magnetic field $\mathbf{B}$.

In the above approach, since each orbital corresponds to a pair of hydrodynamical equations, eqns (8.21a) and (8.21b) are in 3D space. The continuity equation (8.21a) may be summed over $k$ (i.e. all the orbitals) to yield a continuity equation in terms of the net charge density and current density. Similarly, eqn (8.21b) can also be summed over $k$. The net charge and current densities are respectively given by

$$\rho = \sum_k \eta_k \rho_k, \tag{8.22a}$$

$$\mathbf{j} = \sum_k \eta_k \mathbf{j}_k, \tag{8.22b}$$

where $\{\eta_k\}$ are the occupation numbers of the orbitals. It is also worthwhile to note[31] that although each individual velocity field $\mathbf{v}_k$ is irrotational, the net velocity field $\mathbf{v}(=\mathbf{j}/\rho)$ is rotational and not additive over the individual $\mathbf{v}_k$ s. Quantum fluid dynamics of this type in 3D space have been reported by

Takabayashi[13] for Hartree theory, Wong[43] for Hartree–Fock theory, Ghosh and Deb[31] for natural-orbital theory, as well as Deb and Ghosh[32] for density functional theory. The work of Ghosh and Deb[31] includes certain interesting consequences arising from nuclear motion and electron spin. These have been discussed elsewhere.[53]

If the effective potential $V_{\mathbf{k}}^{\mathrm{SCF}}$ in eqn (8.19) is non-local, the corresponding QFD is somewhat modified. In the case of DFT, this potential is the same for all $k$ and is local for the time-independent case.[54] For the Hartree–Fock theory, however, the exchange potential which is a part of $V_k^{\mathrm{SCF}}$ is non-local. Thus, the earlier reported QFD for the Hartree–Fock case[43] may be expressed in a slightly different form. Following Kreuzer,[55] one can express the net current density in terms of a local and a non-local contribution. Thus, when the self-consistent equation contains a non-local potential of the form

$$V_{\mathrm{NL}}^{\mathrm{SCF}}\psi_k=\int V_k(\mathbf{r},\mathbf{r}')\psi_k(\mathbf{r}')\,\mathrm{d}\mathbf{r}', \tag{8.23}$$

the current density is given by

$$\mathbf{j}=\mathbf{j}^{\mathrm{L}}+\mathbf{j}^{\mathrm{NL}},$$

where $\mathbf{j}^{\mathrm{L}}$ is defined through eqn (8.20b) and $\mathbf{j}^{\mathrm{NL}}$ is formally defined by[55]

$$\nabla\cdot\mathbf{j}_{\mathrm{k}}^{\mathrm{NL}}=\frac{\mathrm{i}}{\hbar}\int V_k(\mathbf{r},\mathbf{r}')\,[\psi_k^*(\mathbf{r},t)\psi_k(\mathbf{r}',t)-\psi_k^*(\mathbf{r}',t)\,\psi_k(\mathbf{r},t)]\,\mathrm{d}\mathbf{r}'. \tag{8.24}$$

Further, it may be noted that, for non-local potentials, the definitions of the velocity as $\dot{\mathbf{r}}$ and $\mathbf{p}/m$ do not coincide.[56] Consequently, a non-local contribution arises as follows:

$$\begin{aligned}\left(\psi_k^*\frac{\hat{\mathbf{p}}}{m}\psi_k\right)&=(\psi_k^*\dot{\mathbf{r}}\psi_k)-\mathrm{i}(\psi_k^*[V(\mathbf{r},\mathbf{r}'),\mathbf{r}]\psi_k)\\&=(\psi_k^*\dot{\mathbf{r}}\psi_k)-\mathrm{i}\psi_k^*\int V(\mathbf{r},\mathbf{r}')\,(\mathbf{r}'-\mathbf{r})\psi_k(\mathbf{r}')\,\mathrm{d}\mathbf{r}'.\end{aligned} \tag{8.25}$$

The conventional velocity field is defined as

$$\mathbf{v}_k=\mathrm{Re}\left\{\psi_k^*\frac{\mathbf{p}}{m}\psi_k\right\}\Big/\{\psi_k^*\psi_k\}. \tag{8.26}$$

Possible consequences in QFD which follow from the above difference caused by non-locality may well be worthy of further study.

### 8.2.3 Applications

We shall now briefly discuss a few applications* arising out of the QFDFT viewpoint (other applications are discussed in Chapter 9). All these applications depend on the continuing efforts to find (i) a satisfactory kinetic energy density (KED) functional and (ii) a single equation for the direct calculation[57] of charge and/or current density. The first problem of designing a KED functional with proper local and global behaviour as well as a proper functional derivative has been a crucial one for both time-independent and time-dependent DFT. Various attempts to solve this problem have been reviewed by Chattaraj and Deb.[58] This is also intimately linked with problem (ii). A full solution of these two problems remains a matter for future work.

In an effort to understand problem (i) better, one has frequently gone back to the roots of DFT, namely the Thomas–Fermi method[59] (see Chapter 1). Thus, Chattaraj *et al.*[60] re-examined the vintage open-ended problem of $Z$-dependence ($Z$ is the nuclear charge) of the energies of neutral atoms. They simulated the Hartree–Fock ground-state atomic energy as

$$-E_{\mathrm{HF}}(Z)=C_7Z^{7/3}+C_6Z^2+C_5Z^{5/3}+C_4Z^{4/3}+C_3Z, \tag{8.27}$$

where $C_4$ $(=0.0129783\,\text{a.u.})$ and $C_3$ $(=0.00126492\,\text{a.u.})$ are exchange corrections; $C_7$ $(=0.76874512\,\text{a.u.})$, $C_6$ $(=-0.5\,\text{a.u.})$ and $C_5$ $(=0.2699\,\text{a.u.})$ had been determined by other workers (see ref. 60). For $2\leqslant Z\leqslant 110$, this leads to excellent agreement with Hartree–Fock energies, which generally lie lower than the simulated energies. For atoms, the Hartree–Fock KED may be approximated as[61]

$$\begin{aligned} t_{\mathrm{HF}}^{\mathrm{app}}(\mathbf{r};\rho) &= t_{\mathrm{TF}}(\rho)+t_r(\mathbf{r};\rho) \\ &= \frac{3}{10}(3\pi^2)^{2/3}\rho^{5/3}-\frac{1}{40}\frac{\mathbf{r}\cdot\nabla\rho}{r^2}, \end{aligned} \tag{8.28}$$

in which an unusual "first-gradient" connection has been added to the Thomas–Fermi term. Using Hartree–Fock atomic densities, eqn (8.28) shows better local behaviour than $t_{\mathrm{TF}}+t_2$ and better global behaviour than $T_{\mathrm{TF}}+T_2+T_4$. If one includes the exchange energy density functional, then the corresponding Euler–Lagrange equation turns out to be a quadratic non-differential Thomas–Fermi–Dirac (TFD)-type equation, which is easier to solve and yields both atomic density and energy that are as good as other TFD calculations.[62] However, the atomic shell structure is not obtained. As suggested by Bader and Essen,[63] the shell structure can be

*Atomic units are employed in this Section.

recovered if one incorporates the $-\nabla^2\rho$ term in eqn (8.28), by appropriately modifying the coefficients.[64] However, using an uncertainty-corrected form[65] of the Thomas–Fermi kinetic energy ($N$ is the number of electrons),

$$T_f = \frac{3}{10}(3\pi^2)^{2/3}\int \rho^{5/3}\,d\mathbf{r}$$

$$+\frac{N^{\pi/e}}{6\pi}\frac{1}{8N^2}\left\{N\int d\mathbf{r}\frac{\nabla\rho\cdot\nabla\rho}{\rho}+\left(\int d\mathbf{r}\frac{\mathbf{r}\cdot\nabla\rho}{r}\right)^2\right\}, \tag{8.29}$$

in conjunction with Hartree–Fock atomic densities, gives probably the best overall agreement so far with Hartree–Fock atomic energies. Note that eqn (8.29) includes the Weizsäcker correction* $T_w = 9T_2 = \frac{1}{8}\int d\mathbf{r}(\nabla\rho\cdot\nabla\rho)\rho^{-1}$.

In yet another development, Deb and Ghosh[67] proposed a KED of the form

$$t(\mathbf{r};\rho) = -\frac{1}{4}\nabla^2\rho + \frac{1}{8}\frac{\nabla\rho\cdot\nabla\rho}{\rho} + \frac{3}{10}(3\pi^2)^{2/3} f(\mathbf{r})\rho^{5/3}, \tag{8.30}$$

where the last term is a modulated Thomas–Fermi term, $f(\mathbf{r})$ being the modulating function. The corresponding Euler–Lagrange equation is a *non-linear* differential equation for the direct calculation of the electron density:

$$[-\tfrac{1}{2}\nabla^2 + V_{\text{eff}}(\mathbf{r};\rho)]\phi(\mathbf{r}) = \mu\phi(\mathbf{r}), \tag{8.31}$$

where $\rho = |\phi(\mathbf{r})|^2$. Equation (8.31) could be solved for noble-gas atoms only in a model potential framework. The results, however, were quite gratifying.[67] Extending this idea of working in terms of a single equation to time-dependent situations such as ion–atom high-energy collisions,† Deb and Chattaraj[34] employed a QFDFT approach to obtain a time-dependent Kohn–Sham-type equation in 3D space. The equation appears to be a new non-linear Schrödinger equation in which "time" is explicitly embedded as a parameter. Its solution describes the time evolution of the electron density of the interacting system and an effective potential. The current density can also be obtained. One can thus directly follow the interactions process from start to finish within the novel framework of a (time-dependent) *pulsating potential* surface. Even though one has to solve a highly non-linear integro-differential equation, this approach may still result in computational economy because it deals with only *three* space variables, apart from "time".

*It is worthwhile noting that there is no linear correlation between $T_w$ and the binding energies of diatomic molecules[66] (see also Section 1.12).

†For diatomic molecules and two-centre collision problems, one adds a suitable $R$-dependent term to the atomic KED, $R$ being the internuclear distance.

The equation requires a new algorithm for its numerical solution and a new stability analysis.[68] The extension of this work to explore probable non-linear features in gas-surface dynamics[69] is a future possibility.

One of the most useful applications of QFDFT is to calculate the dynamic (frequency-dependent) polarizabilities[35,70-72] and hyperpolarizabilities[73] (see also Chapter 9). A time-dependent version of the single density equation (8.31) may be written as[35]

$$[-\tfrac{1}{2}\nabla^2 + V_{\text{eff}}(\mathbf{r}, t; \rho)]\psi = \mathrm{i}\frac{\partial \psi}{\partial t}. \tag{8.32}$$

Obviously, eqn (8.32) would also lead to a continuity equation and an Euler-type equation of motion in terms of the (time-dependent) charge density $\rho = |\psi|^2$ and current density $\mathbf{j} = \mathrm{Re}\,(\psi^*(\hat{\mathbf{p}}/m)\psi)$. The Euler-type equation has an additional force term[74] arising from the gradient of the potential corresponding to the modulated Thomas–Fermi term that has been absorbed in $V_{\text{eff}}$ in eqn (8.32).

In order to calculate frequency-dependent $2^L$-pole ($L = 1, 2, \ldots$) polarizabilities, the weak perturbing potential is taken as[35]

$$v_{\text{ext},L}(\mathbf{r}, t) = r^L P_L(\cos\theta) F_L^0[\exp(\mathrm{i}\omega t) + \exp(-\mathrm{i}\omega t)]. \tag{8.33}$$

Here the frequency $\omega$ is not large enough to cause excitation, and $F_L^0$ denotes the magnitude of the electric field for $L = 1$ (dipole), the field gradient for $L = 2$ (quadrupole), the second derivative of the field for $L = 3$ (octupole) and so on. A perturbative treatment in which time dependence is transformed into frequency dependence then leads to a pair of equations, involving both positive and negative frequencies, for calculating the first-order correction to the density. The numerical solution of these equations leads to the calculation of $\alpha_{2^L}(\omega)$, the frequency-dependent multipole polarizability. The formalism has been applied[35] to the noble-gas atoms, in a model-potential framework, using Hartree–Fock densities as the unperturbed functions. The *static* polarizabilities agree very well with coupled Hartree–Fock results (see ref. 35 for details). This is probably the first, relatively painless, calculation of $\alpha_{2^L}(\omega)$, which do not appear to have been calculated or experimentally determined before for the systems examined, because of difficulties associated with their evaluation. Further, in this approach an increase in the value of $L$ does not increase computational labour significantly. Thus, the treatment of linear response (see also ref. 72) is particularly successful within QFDFT, although the situation regarding non-linear response[73] requires further study.

Before we leave this Section, a word is needed about the current status of TDDFT. A TD version of the Hohenberg–Kohn theorem (see Chapters 1 and 2) can be formally established provided that the map between the TD potential and the TD density is invertible. If the TD potential consists of a time-independent part and a TD part, then invertibility is satisfied when the latter is in the neighbourhood of zero.[75] However, the TD situation is not always so restrictive. Thus, Deb and Ghosh,[34] and subsequently Bartolotti,[44,45] developed a TDKS equation for TD harmonic potentials with frequencies not large enough to cause excitation (see above). Runge and Gross[33] have then shown that the above invertibility is valid for the case of arbitrary TD density (pure states) provided certain conditions regarding the initial density and the time-evolution of the potential are satisfied. Xu and Rajagopal[48] have argued that the Runge–Gross proof is invalid for a uniform system under a TD perturbation and that, more justifiably, an invertible mapping exists between the TD potential and the TD current density (rather than the charge density). However, Dhara and Ghosh[50] are of the opinion that these criticisms of the Runge–Gross proof do not stand, provided the TD potentials are well-behaved at large distances. The Runge–Gross work has been extended by Kohl and Dreizler[49] through the construction of a Levy-type functional[76] (see Chapter 2) and by Li *et al.*[46,47] to TD ensembles. With all these, it is reasonable to assume that the QFDFT approach is likely to emerge as a more powerful and versatile approach than DFT itself.

## 8.3 Quantum kinetic equations and QFD (mixed states)

Quantum kinetic equations are equations of motion of reduced density matrices[77] or the corresponding distribution functions in phase space. For classical systems, the formal development of kinetic equations and their relations to fluid dynamical equations are well established.[78] Analogously, QFD equations can also be obtained from the quantum kinetic equations, either in terms of the reduced density matrices[14] or the phase-space distribution functions,[79] both of which are discussed in this Section. This approach has the additional advantage of being able to deal with mixed states (linear combinations of pure states), systems at a finite temperature as well as quantum transitions.

### 8.3.1 Quantum hydrodynamics through reduced density matrices

Quantum kinetic equations for reduced density matrices have been used, in deriving hydrodynamical equations, by Fröhlich,[14,80] Kreuzer,[81,82] Wong *et al.*,[83,84] and many others. Work has been reported on electron correlation;[85] density fluctuations, consequent plasmon propagation in periodic lattices and their dispersion;[86] and superfluid hydrodynamics.[87]

The first step is to derive the equation of motion for the reduced density matrix. For this purpose, one can employ the Schrödinger equation to obtain[88] an equation for the $N$-particle density matrix which can be subsequently integrated over several degrees of freedom to reduce it to a few-particle density matrix equation. Alternatively, one can[14] start with the Heisenberg equation of motion for the single-particle density matrix, which gives the required kinetic equation directly. In any case, the equation of motion for the single-particle density matrix $\rho(\mathbf{r}',\mathbf{r}'')$ involves the two-particle density matrix $\rho_2$ for a system in which there is a two-particle interaction and is given by*

$$i\hbar\frac{\partial\rho(\mathbf{r}',\mathbf{r}'')}{\partial t}=-\frac{\hbar^2}{2m}(\nabla'^2-\nabla''^2)\rho(\mathbf{r}',\mathbf{r}'')$$
$$+\int d\mathbf{r}_2\,[V(\mathbf{r}'-\mathbf{r}_2)-V(\mathbf{r}''-\mathbf{r}_2)]\rho_2(\mathbf{r}',\mathbf{r}_2;\mathbf{r}'',\mathbf{r}_2)$$
$$+(v(\mathbf{r}')-v(\mathbf{r}''))\rho(\mathbf{r}',\mathbf{r}''), \tag{8.34}$$

where $v(\mathbf{r})$ is the external potential (which may be time-dependent) and $V(\mathbf{r}'-\mathbf{r}'')$ is the two-particle interaction term. Equation (8.34) is the required quantum kinetic equation; similar equations were employed by Born and Green[89] in their work on fluid dynamical equations for macroscopic systems. Within the Hartree–Fock approximation, of course, eqn (8.34) involves only the single-particle density matrix.

Although the kinetic equation can be written for higher-order density matrices and one can obtain fluid dynamical equations corresponding to higher-order distribution functions, our interest is mainly in the QFD in 3D space and therefore eqn (8.34) is sufficient for our purpose.

One can express the density matrix in polar form as

$$\rho(\mathbf{r}',\mathbf{r}'')=\sigma(\mathbf{r}',\mathbf{r}'')\exp[i\chi(\mathbf{r}',\mathbf{r}'')/\hbar], \tag{8.35}$$

where the amplitude function $\sigma(\mathbf{r}',\mathbf{r}'')=\sigma(\mathbf{r}'',\mathbf{r}')$ is symmetric, whereas the phase function $\chi(\mathbf{r}',\mathbf{r}'')=-\chi(\mathbf{r}'',\mathbf{r}')$ is antisymmetric (note that $\chi(\mathbf{r},\mathbf{r})=0$). The fluid dynamical equation can be obtained from eqn (8.34) by substituting eqn (8.35) and taking suitable limits. For pure states, this becomes analogous to the approach of Madelung discussed in Section 8.1. However, for general mixed states or transition densities, this approach provides certain new insights. One can also start directly with eqn (8.34) which in the

*For simplicity, in Section 8.3, the time dependences of various functions and matrices are not shown explicitly although it would be clear to the reader that these, in general, may be time-dependent.

limit $\mathbf{r}' \rightarrow \mathbf{r}''$ gives the continuity equation

$$\frac{\partial \rho}{\partial t} + \nabla \cdot \mathbf{j} = 0, \tag{8.36}$$

where the density $\rho$ and the current density $\mathbf{j}$ are obtained as

$$\rho(\mathbf{r}) = \lim_{\mathbf{r}' \rightarrow \mathbf{r}''} \rho(\mathbf{r}', \mathbf{r}'') = \lim_{\mathbf{s} \rightarrow 0} \rho(\mathbf{r} + \tfrac{1}{2}\mathbf{s},\ \mathbf{r} - \tfrac{1}{2}\mathbf{s}) = \sigma(\mathbf{r}, \mathbf{r}) \tag{8.37a}$$

and

$$\mathbf{j}(\mathbf{r}) = \frac{\hbar}{2mi} \lim_{\mathbf{r}' \rightarrow \mathbf{r}''} (\nabla' - \nabla'')\rho(\mathbf{r}', \mathbf{r}'') = \frac{\hbar}{im} \lim_{\mathbf{s} \rightarrow 0} \nabla_s \rho(\mathbf{r} + \tfrac{1}{2}\mathbf{s}, \mathbf{r} - \tfrac{1}{2}\mathbf{s}). \tag{8.37b}$$

Here $\nabla' = \nabla_{\mathbf{r}}$; the new coordinates $\mathbf{r}$ and $\mathbf{s}$ represent the familiar centre-of-mass and difference coordinates defined by $\mathbf{r} = \frac{1}{2}(\mathbf{r}' + \mathbf{r}'')$ and $\mathbf{s} = \mathbf{r}' - \mathbf{r}''$, respectively. One has the following relations between the two coordinate systems:

$$\left.\begin{aligned} &\nabla_{\mathbf{r}} = \nabla' + \nabla''; \quad \nabla_s = \tfrac{1}{2}(\nabla' - \nabla''); \\ &\nabla' = \tfrac{1}{2}\nabla_r + \nabla_s; \quad \nabla'' = \tfrac{1}{2}\nabla_r - \nabla_s. \end{aligned}\right\} \tag{8.38}$$

In deriving eqn (8.36), we have first expressed eqn (8.34) in the form

$$\begin{aligned} i\hbar \frac{\partial}{\partial t} &\rho(\mathbf{r} + \tfrac{1}{2}\mathbf{s}, \mathbf{r} - \tfrac{1}{2}\mathbf{s}) \\ &= -\frac{\hbar^2}{\mathrm{m}} \nabla_r \cdot \nabla_s \rho(\mathbf{r} + \tfrac{1}{2}\mathbf{s},\ \mathbf{r} - \tfrac{1}{2}\mathbf{s}) + W(\mathbf{r} + \tfrac{1}{2}\mathbf{s}, \mathbf{r} - \tfrac{1}{2}\mathbf{s}), \end{aligned} \tag{8.39}$$

where

$$\begin{aligned} W(\mathbf{r}', \mathbf{r}'') = &\int d\mathbf{r}_2 \, [V(\mathbf{r}' - \mathbf{r}_2) - V(\mathbf{r}'' - \mathbf{r}_2)] \rho_2(\mathbf{r}', \mathbf{r}_2; \mathbf{r}'', \mathbf{r}_2) \\ &+ (v(\mathbf{r}') - v(\mathbf{r}''))\rho(\mathbf{r}', \mathbf{r}''). \end{aligned} \tag{8.40}$$

Employing the definition of current density given by eqn (8.37b), clearly eqn (8.39) reduces to the continuity equation (8.36) by taking the limit $s \rightarrow 0$ since $W(\mathbf{r}, \mathbf{r}) = 0$. By making use of the polar form in eqn (8.35), it is easy to show from eqn (8.37b) that

$$\mathbf{j}(\mathbf{r}) = \rho(\mathbf{r})\ \mathbf{v}(\mathbf{r}), \tag{8.41a}$$

where the velocity field $\mathbf{v}(\mathbf{r})$ is given by

$$\mathbf{v}(\mathbf{r}) = \frac{1}{m} \lim_{s \rightarrow 0} \nabla_s \chi(\mathbf{r} + \tfrac{1}{2}\mathbf{s},\ \mathbf{r} - \tfrac{1}{2}\mathbf{s}). \tag{8.41b}$$

The Euler equation of motion can be derived[82] by taking the derivative of the kinetic equation (8.39) with respect to the non-local coordinate **s** and then taking the limit $s \to 0$. Thus, one obtains from the last term of the right-hand side of eqn (8.39),

$$\lim_{s\to 0} \nabla_s W(\mathbf{r}+\tfrac{1}{2}\mathbf{s},\mathbf{r}-\tfrac{1}{2}\mathbf{s}) = -\int d\mathbf{r}_2\, \nabla_r V(\mathbf{r}-\mathbf{r}_2)\rho_2(\mathbf{r},\mathbf{r}_2;\mathbf{r},\mathbf{r}_2) - \rho(\mathbf{r})\nabla V(\mathbf{r}). \qquad (8.42)$$

In order to evaluate the contribution from the other term, we first evaluate

$$-\frac{\hbar^2}{m}\lim_{s\to 0}\frac{\partial}{\partial s_k}\frac{\partial}{\partial s_l}\rho(\mathbf{r}+\tfrac{1}{2}\mathbf{s},\,\mathbf{r}-\tfrac{1}{2}\mathbf{s}) = -\frac{\hbar^2}{m}\lim_{s\to 0}\exp\left(\frac{i\chi}{\hbar}\right)\left[\frac{\partial^2\sigma}{\partial s_k\,\partial s_l} + \frac{i}{\hbar}\frac{\partial\chi}{\partial s_k}\frac{\partial\sigma}{\partial s_l} + \frac{i}{\hbar}\frac{\partial\sigma}{\partial s_k}\frac{\partial\chi}{\partial s_l} + \frac{i\sigma}{\hbar}\frac{\partial^2\chi}{\partial s_k\,\partial s_l} - \frac{1}{\hbar^2}\sigma\frac{\partial\chi}{\partial s_k}\frac{\partial\chi}{\partial s_l}\right] \qquad (8.43)$$

and note that only the first and the last terms on the right-hand side survive in the limit $s \to 0$ since $\sigma$ and $\chi$ of eqn (8.35) obey the symmetry properties mentioned earlier. This is because the complex conjugate of the left-hand side of eqn (8.43) in the limit $s \to 0$ is the original one, whereas in the same limit, the complex conjugate of the right-hand side differs in the three terms proportional to i (since $\chi(\mathbf{r},\mathbf{r})=0$). Again, using the definition of the velocity field given by eqn (8.41b), the last term of eqn (8.43) becomes

$$\frac{1}{m}\lim_{s\to 0}\sigma\frac{\partial\chi}{\partial s_k}\frac{\partial\chi}{\partial s_l} = \rho(\mathbf{r})v_k(\mathbf{r})v_l(\mathbf{r}). \qquad (8.44)$$

The first term on the right-hand side of eqn (8.43) can be represented as the $kl$th element of a tensor $\mathbf{T}^0(\mathbf{r})$ defined as

$$T^0_{kl}(\mathbf{r}) = -\frac{\hbar^2}{2m}\lim_{s\to 0}\frac{\partial^2\sigma}{\partial s_k\partial s_l}. \qquad (8.45)$$

Thus, the derivative of eqn (8.39) with respect to the coordinate **s** in the limit $s \to 0$ leads to the equation

$$\frac{\partial}{\partial t}[\rho(\mathbf{r})\mathbf{v}(\mathbf{r})] + \nabla\cdot[2\mathbf{T}^{(0)}(\mathbf{r}) + \rho(\mathbf{r})\mathbf{v}(\mathbf{r})\mathbf{v}(\mathbf{r})] = \mathbf{F}_v(\mathbf{r}) - \rho(\mathbf{r})\nabla V_{\text{ext}}(\mathbf{r}), \qquad (8.46)$$

where $\mathbf{F}_v(\mathbf{r})$ stands for the first term on the right-hand side of eqn (8.42) and represents the force due to the interparticle Coulomb repulsion. Although written in terms of the two-particle density matrix in eqn (8.42), in density functional prescription it can be expressed, in principle, in terms of the density itself. The kinetic–energy stress tensor $\mathbf{T}^0(\mathbf{r})$ is of quantum origin and is expressed here in terms of the single-particle density matrix $(\sigma(\mathbf{r}',\mathbf{r}'')= |\rho(\mathbf{r},\mathbf{r}'')|)$. Equation (8.46) is the Euler equation of motion for the electron fluid in 3D space. For actual calculations in terms of the density, a self-consistent procedure at a finite temperature for mixed states would be required. More detailed discussion on the quantum stress tensor has been presented by Wong and McDonald.[84]

The derivation given above was originally due to Fröhlich.[14] An alternative, although similar, method is to employ the Taylor expansion of the density matrix

$$\rho(\mathbf{r}+\tfrac{1}{2}\mathbf{s},\ \mathbf{r}-\tfrac{1}{2}\mathbf{s})=\sum_n \frac{1}{n!}(\mathbf{s}\cdot\nabla_s)^n\rho(\mathbf{r}+\tfrac{1}{2}\mathbf{s},\ \mathbf{r}-\tfrac{1}{2}\mathbf{s})|_{s\to 0}$$

into eqn (8.39) and then recover the hydrodynamical equations by collecting terms of different orders of **s**. This has been discussed in detail by Kreuzer.[81] For systems with non-local interactions, QFD equations have been obtained by Kreuzer and Nakamura.[90] A higher-order equation representing the time development of the kinetic energy tensor, etc., has also been obtained. The continuity equation, the Euler equation and this equation represent respectively the mass balance, the momentum balance and the energy balance.

Recasting the Euler equation into the Navier–Stokes form together with further developments have been discussed by Fröhlich,[80] who has also considered their application to superconductivity. Further discussions on the quantum stress tensor and a "par-thermal" stress tensor have been presented by Wong and McDonald[84] with emphasis on their application to nuclear physics. For one- and many-electron systems, the significance of a stress tensor has been highlighted by Deb and Bamzai[22,91] as well as Deb and Ghosh.[23]

The study of such equations had been done exhaustively by Born and Green,[89] in a series of papers, for macroscopic classical and quantum fluids (e.g. liquid helium). The advantage of the density matrix method is that one can systematically obtain the balance equations for kinetic energy, total energy density, pressure, etc. Born and Green[89] further introduced the concept of a generalized temperature in terms of the kinetic energy of a quantum fluid, which has received[25,26] renewed attention recently in connection with the concept of a local temperature for the electron fluid. This aspect will be discussed in Section 8.6.

In the original work of Fröhlich,[14] the Navier–Stokes equation was obtained by specifying the force density, assuming the system to be near equilibrium. A generalization of this work has subsequently been achieved[92] in which no such assumption is made. Also, there has been proposed a series expansion of the second-order reduced density matrix which enters into the expression for force density in the hydrodynamical equations.

### 8.3.2 Quantum hydrodynamics through distribution functions in phase space

An alternative approach to the QFD equations utilizes the kinetic equations in phase space. The additional advantage is that the formalism is more "classical" owing to its employing a function of both position and momentum coordinates. Mathematically of course, the phase-space distribution function in the quantum context, first introduced by Wigner,[15] is a particular representation of the density matrix itself. This is clear from the definition of the Wigner distribution function as

$$f(\mathbf{r},\mathbf{p})=\left(\frac{1}{2\pi\hbar}\right)^3\int \mathrm{d}\mathbf{s}\exp\left(-\frac{\mathrm{i}\mathbf{p}\cdot\mathbf{s}}{\hbar}\right)\rho(\mathbf{r}+\tfrac{1}{2}\mathbf{s},\ \mathbf{r}-\tfrac{1}{2}\mathbf{s}), \tag{8.47}$$

which is the result of a partial Fourier transform of the density matrix with respect to the non-local coordinate $\mathbf{s}=\mathbf{r}-\mathbf{r}'$. Clearly, $f(\mathbf{r},\mathbf{p})$ satisfies the conditions

$$\int f(\mathbf{r},\mathbf{p})\,\mathrm{d}\mathbf{p}=\rho(\mathbf{r}), \tag{8.48a}$$

$$\int f(\mathbf{r},\mathbf{p})\,\mathrm{d}\mathbf{r}=\chi(\mathbf{p}), \tag{8.48b}$$

where $\chi(\mathbf{p})$ is the momentum density. Although the Wigner function cannot be really interpreted as the simultaneous probability of coordinate and momentum, since it can take negative values as well, the classical-like picture still follows from the fact that the expectation values can be evaluated by simple averaging over the variables and no corresponding operators are to be invoked. For example, the kinetic energy can be obtained simply as

$$E_{\mathrm{kin}}=\int\frac{p^2}{2m}f(\mathbf{r},\mathbf{p})\,\mathrm{d}\mathbf{r}\,\mathrm{d}\mathbf{p}. \tag{8.49}$$

In other words, it permits one to replace a quantum mechanical ensemble average by a classical phase-space integration. A more general definition of the phase-space distribution function has, however, been provided by Cohen and Zaparovanny;[93] see also the interesting studies of O'Connell *et al.*[94,95] and Hillery *et al.*[18] The phase-space approach has found[96] several semi-classical applications and has recently led to an interesting route for the calculation of certain quantities within a density functional framework.[27,28] Quantum collision theory has also been formulated through the phase-space distribution function.[79] In what follows, the Wigner phase-space function is employed to derive the equations of quantum hydrodynamics. This approach is originally due to Born and Green[89] as well as Irving and Zwanzig;[97,98] subsequent developments have been suggested by many others.[79,84,99–103]

As in Section 8.3.1, the first step is to obtain an equation of motion for the phase-space function. For this purpose, we start with the equation of motion in the density matrix formalism, eqn (8.39), we then multiply it by $(2\pi)^{-3}\exp[i\mathbf{p}\cdot\mathbf{s}/\hbar]$ and integrate over d**s**. This leads to the equation

$$-i\hbar\frac{\partial f}{\partial t}=-\frac{\hbar^2}{m}\frac{1}{(2\pi)^3}\int d\mathbf{s}\nabla_r\cdot\nabla_s\rho(\mathbf{r}+\tfrac{1}{2}\mathbf{s},\mathbf{r}-\tfrac{1}{2}\mathbf{s})\exp\left(\frac{i\mathbf{p}\cdot\mathbf{s}}{\hbar}\right)$$

$$\frac{1}{(2\pi)^3}\int d\mathbf{s}\exp\left(\frac{i\mathbf{p}\cdot\mathbf{s}}{\hbar}\right)W(\mathbf{r}+\tfrac{1}{2}\mathbf{s},\mathbf{r}-\tfrac{1}{2}\mathbf{s}). \tag{8.50}$$

The kinetic energy term (the first term on the right-hand side) can be simplified on integration by parts. Thus, one obtains

$$-\frac{\hbar^2}{m}\frac{1}{(2\pi)^3}\nabla_r\cdot\int d\mathbf{s}\nabla_s\rho(\mathbf{r}+\tfrac{1}{2}\mathbf{s},\mathbf{r}-\tfrac{1}{2}\mathbf{s})\exp\left(\frac{i\mathbf{p}\cdot\mathbf{s}}{\hbar}\right)$$

$$=\frac{\hbar^2}{m}\frac{1}{(2\pi)^3}\nabla_r\cdot\int d\mathbf{s}\frac{i\mathbf{p}}{\hbar}\exp\left(\frac{i\mathbf{p}\cdot\mathbf{s}}{\hbar}\right)\rho(\mathbf{r}+\tfrac{1}{2}\mathbf{s},\mathbf{r}-\tfrac{1}{2}\mathbf{s})$$

$$=i\frac{\hbar}{m}\mathbf{p}\cdot\nabla_r f(\mathbf{r},\mathbf{p}). \tag{8.51}$$

The simplified result of the kinetic energy has a form identical with the classical result. Therefore, eqn (8.50) simplifies to

$$\frac{\partial f(\mathbf{r},\mathbf{p})}{\partial t}+\frac{\mathbf{p}}{m}\cdot\nabla_r f(\mathbf{r},\mathbf{p})=\frac{i}{\hbar}\frac{1}{(2\pi)^3}\int d\mathbf{s}\exp\left(\frac{i\mathbf{p}\cdot\mathbf{s}}{\hbar}\right)W(\mathbf{r}+\tfrac{1}{2}\mathbf{s},\mathbf{r}-\tfrac{1}{2}\mathbf{s}). \tag{8.52}$$

The potential term on the right-hand side of eqn (8.52) can be further simplified in a number of ways.[97] For the single-particle density matrices and correspondingly the single-particle phase-space functions considered here with $W$ given by eqn (8.40), assuming only single-particle and two-particle interactions to be present, the right-hand side of eqn (8.52) can be written in a simple form (see refs 84 and 97 for derivations) so that eqn

(8.52) becomes

$$\frac{\partial f^{(1)}}{\partial t}+\frac{\mathbf{p}}{m}\cdot\nabla_r f^{(1)}(\mathbf{r},\mathbf{p})=\theta_1 f^{(1)}+\theta_2 f^{(2)}, \tag{8.53}$$

where $\theta_1 f^{(1)}$ and $\theta_2 f^{(2)}$ denote the contributions due to the one-particle and two-particle interaction potential terms, respectively, which are given by

$$\theta_1 f^{(1)}=\frac{2}{\hbar}\sin[\tfrac{1}{2}\hbar\nabla_r^V\cdot\nabla_p^f]V(\mathbf{r})f^{(1)}(\mathbf{r},\mathbf{p}) \tag{8.54a}$$

and

$$\theta_2 f^{(2)}=\frac{2}{\hbar}\iint d\mathbf{r}_2\, d\mathbf{p}_2\,\{\sin[\tfrac{1}{2}\hbar\nabla_r^V\cdot\nabla_p^f]\}V(\mathbf{r},\mathbf{r}_2)f^{(2)}(\mathbf{r},\mathbf{r}_2;\mathbf{p},\mathbf{p}_2), \tag{8.54b}$$

where the differential operator $\nabla_r^V$ acts only on $V(\mathbf{r})$ on the right-hand side and not on $f(\mathbf{r},\mathbf{p})$, whereas $\nabla_p^f$ operates on $f$ only. Here $f^{(1)}$ and $f^{(2)}$ denote the single-particle and two-particle phase-space distribution functions, respectively; $f^{(1)}$ is identical with $f$ defined by eqn (8.47), whereas $f^{(2)}$ is to be defined as the partial Fourier transform of the two-particle matrix, i.e.

$$f^{(2)}(\mathbf{r}_1,\mathbf{r}_2;\mathbf{p}_1,\mathbf{p}_2)=\left(\frac{1}{2\pi\hbar}\right)^6\iint d\mathbf{s}_1\, d\mathbf{s}_2\,\exp\left[-\frac{i}{\hbar}(\mathbf{p}_1\cdot\mathbf{s}_1+\mathbf{p}_2\cdot\mathbf{s}_2)\right]$$
$$\times\rho_2(\mathbf{r}_1+\tfrac{1}{2}\mathbf{s}_1,\mathbf{r}_2+\tfrac{1}{2}\mathbf{s}_2;\mathbf{r}_1-\tfrac{1}{2}\mathbf{s}_1,\mathbf{r}_2-\tfrac{1}{2}\mathbf{s}_2). \tag{8.55}$$

There are various alternative forms (both differential and integral) for the potential energy term of eqn (8.52); see ref. 97 for details.

Equation (8.53) represents the quantum kinetic equation in phase space for the single-particle function and corresponds to the classical Liouville equation. The analogy between the two and the explicit nature of the quantum correction can best be illustrated by considering the kinetic equation for the full Wigner function, i.e. $f^{(N)}(\mathbf{r}_1,\ldots,\mathbf{r}_N;\mathbf{p}_1,\ldots,\mathbf{p}_N)$, which we denote by $f^{(N)}(\mathbf{R},\mathbf{P})$, with $\mathbf{R}$ and $\mathbf{P}$ denoting $\mathbf{r}_1,\ldots,\mathbf{r}_N$ and $\mathbf{p}_1,\ldots,\mathbf{p}_N$ respectively. The complete-space Wigner function is defined as

$$f^{(N)}(\mathbf{R},\mathbf{P})=\left(\frac{1}{2\pi\hbar}\right)^{3N}\iint\cdots\int d\mathbf{S}\exp\left(-\frac{i\mathbf{P}\cdot\mathbf{S}}{\hbar}\right)\rho_N(\mathbf{R}+\tfrac{1}{2}\mathbf{S},\mathbf{R}-\tfrac{1}{2}\mathbf{S}), \tag{8.56}$$

where $\mathbf{S}=(\mathbf{s}_1,\ldots,\mathbf{s}_N)$, $\mathbf{P}\cdot\mathbf{S}=\Sigma_{i=1}^N\,\mathbf{p}_i\cdot\mathbf{s}_i$ and $\rho_N$ is the $N$-body density matrix. The equation of motion for this distribution function is given by[97,104–106]

$$\frac{\partial f^{(N)}}{\partial t}+\frac{\mathrm{P}}{\mathrm{m}}\cdot\nabla_R f^{(N)}=\theta f^{(N)}, \tag{8.57}$$

where

$$\theta f^{(N)} = \frac{2}{\hbar} \sin\left[\tfrac{1}{2}\hbar \nabla_R \cdot \nabla_P\right] U(\mathbf{R}) f^{(N)}(\mathbf{R}, \mathbf{P}), \tag{8.58}$$

where $\nabla_R$ operates only on $U(\mathbf{R})$ on the right-hand side and not on $f^{(N)}(\mathbf{R},\mathbf{P})$. Here $U(\mathbf{R}) \equiv U(\mathbf{r}_1, \ldots, \mathbf{r}_N)$ denotes the total potential energy of the system. By series expansion of eqn (8.58), it is clear that the leading term in $\theta f^{(N)}$ is

$$\theta f^{(N)} = \sum_{i=1}^{N} \nabla_{r_i} U \cdot \nabla_{p_i} f^{(N)} + \mathrm{O}(\hbar^2). \tag{8.59}$$

In the classical limit $\hbar \to 0$ and eqn (8.57) becomes the classical Liouville equation:[104–106]

$$\frac{\partial f^{(N)}}{\partial t} + \sum_{i=1}^{N} \frac{\mathbf{p}_i}{m} \cdot \nabla_{r_i} f^{(N)} = \sum_{i=1}^{N} \nabla_{r_i} U \cdot \nabla_{p_i} f^{(N)}. \tag{8.60}$$

In other words, $\theta = \nabla_R U \cdot \nabla_P$, for the classical case. The same conclusion is also valid for the reduced distribution functions.

Consider now the derivation of the hydrodynamical equations. For the classical case, this has been discussed in detail by Irving and Kirkwood[107] while the quantum hydrodynamics have been obtained by Irving and Zwanzig.[97] The hydrodynamical quantities, namely the density $\rho(\mathbf{r})$ and the current density $\mathbf{j}(\mathbf{r})$, are defined in terms of the phase-space function as

$$\rho(\mathbf{r}) = \int f(\mathbf{r}, \mathbf{p})\, \mathrm{d}\mathbf{p} \tag{8.61a}$$

and

$$\mathbf{j}(\mathbf{r}) = \frac{1}{m} \int \mathbf{p} f(\mathbf{r}, \mathbf{p})\, \mathrm{d}\mathbf{p}. \tag{8.61b}$$

In general, any property density in coordinate space, $\alpha(\mathbf{r})$, is obtainable by taking the average, with respect to $f(\mathbf{r},\mathbf{p})$ of the corresponding dynamical variable $\alpha(\mathbf{r},\mathbf{p})$ over the phase space, i.e.

$$\alpha(\mathbf{r}) = \int \alpha(\mathbf{r}, \mathbf{p}) f(\mathbf{r}, \mathbf{p})\, \mathrm{d}\mathbf{p}. \tag{8.62}$$

For density and current density, $\alpha(\mathbf{r},\mathbf{p})$ is unity and $\mathbf{p}/m$, respectively, as is obvious from eqns (8.61). Similarly, the kinetic energy density $t_k(\mathbf{r})$ is given by

$$t_k(\mathbf{r}) = \int \frac{p^2}{2m} f(\mathbf{r}, \mathbf{p})\, \mathrm{d}\mathbf{p}. \tag{8.63}$$

The hydrodynamical equations can now be obtained by multiplying the kinetic equation (8.53) by the required dynamical variable $\alpha(\mathbf{r}, \mathbf{p})$ and then integrating over $\mathbf{p}$. Following Irving and Zwanzig,[97] one can obtain a general form of the hydrodynamical equation corresponding to any dynamical variable $\alpha(\mathbf{r}, \mathbf{p})$ of the form

$$\alpha(\mathbf{r}, \mathbf{p}) = \alpha_1(\mathbf{r}) + \alpha_2(\mathbf{r})p + \alpha_3(\mathbf{r})\mathbf{p}\cdot\mathbf{p} \tag{8.64}$$

by noting that

$$\frac{\partial\alpha(\mathbf{r}, t)}{\partial t} = \int \alpha(\mathbf{r}, \mathbf{p}) \frac{\partial f(\mathbf{r}, \mathbf{p})}{\partial t}\, \mathrm{d}\mathbf{p}. \tag{8.65}$$

and using the kinetic equation (8.53).

Clearly, one obtains

$$\frac{\partial\alpha(\mathbf{r})}{\partial t} = -\int \alpha(\mathbf{r}, \mathbf{p}) \frac{\mathbf{p}}{m}\cdot\nabla_r f^{(1)}\, \mathrm{d}\mathbf{p} + \int \alpha(\mathbf{r}, \mathbf{p})\theta_1 f^{(1)}\, \mathrm{d}\mathbf{p} + \int \alpha(\mathbf{r}, \mathbf{p})\theta_2 f^{(2)}\, \mathrm{d}\mathbf{p}. \tag{8.66}$$

Before proceeding further, the last two terms in eqn (8.66) may be simplified. Using the differential operator form of Irving and Zwanzig,[97] one can write

$$\int \alpha(\mathbf{r}, \mathbf{p})\theta_1 f^{(1)}\, \mathrm{d}\mathbf{p} = \int A_1(\mathbf{r}, \mathbf{p})\, \mathrm{d}\mathbf{p} \tag{8.67a}$$

and

$$\int \alpha(\mathbf{r}, \mathbf{p})\theta_2 f^{(2)}\, \mathrm{d}\mathbf{p} = \int A_2(\mathbf{r}, \mathbf{p})\, \mathrm{d}\mathbf{p}, \tag{8.67b}$$

where $A_1$ and $A_2$ are defined by

$$A_1(\mathbf{r}, \mathbf{p}) = \frac{\mathrm{i}}{\hbar}\left\{\alpha\left(\mathbf{r}, \mathbf{p} + \frac{\hbar}{2\mathrm{i}}\nabla_r\right) - \alpha\left(\mathbf{r}, \mathbf{p} - \frac{\hbar}{2\mathrm{i}}\nabla_r\right)\right\} V(\mathbf{r})\, f^{(1)}(\mathbf{r}, \mathbf{p}) \tag{8.68a}$$

and

$$A_2(\mathbf{r}, \mathbf{p}) = \frac{\mathrm{i}}{\hbar}\iint \mathrm{d}\mathbf{r}_2\, \mathrm{d}\mathbf{p}_2 \left\{\alpha\left(\mathbf{r}, \mathbf{p} + \frac{\hbar}{2\mathrm{i}}\nabla_r\right) - \alpha\left(\mathbf{r}, \mathbf{p} - \frac{\hbar}{2\mathrm{i}}\nabla_r\right)\right\} \times V(\mathbf{r}, \mathbf{r}_2)\, f^{(2)}(\mathbf{r}, \mathbf{r}_2; \mathbf{p}, \mathbf{p}_2), \tag{8.68b}$$

respectively. Here $V(\mathbf{r})$ and $V(\mathbf{r}, \mathbf{r}_2)$ denote the single-particle and two-particle potentials as has been already mentioned; $\nabla_r$ operates only on $V(\mathbf{r})$ and not on $f(\mathbf{r}, \mathbf{p})$.

Equation (8.66) can now be further simplified for the $\alpha(\mathbf{r}, \mathbf{p})$ given by the form in eqn (8.64). However, in deriving the hydrodynamics, we are only interested in $\alpha=1$, $\mathbf{p}/m$ and $(\mathbf{p}\cdot\mathbf{p})/2m$. Thus, the hydrodynamical equations can be written directly. For $\alpha=1$, both $A_1$ and $A_2$ vanish; using the definitions of the density and the current density, eqns (8.61), eqn (8.66) reduces to

$$\frac{\partial \rho(\mathbf{r})}{\partial t}+\nabla_r\cdot\mathbf{j}(\mathbf{r})=0. \tag{8.69}$$

which is the continuity equation. Similarly, for $\alpha=\mathbf{p}/m$, one obtains

$$m\frac{\partial \mathbf{j}(\mathbf{r})}{\partial t}=-\nabla\cdot\mathbf{T}-\rho(\mathbf{r})\nabla U(\mathbf{r}), \tag{8.70}$$

where the kinetic-energy tensor $\mathbf{T}$ is defined as

$$\mathbf{T}=2\int\frac{\mathbf{p}\cdot\mathbf{p}}{2m}f^{(1)}(\mathbf{r},\mathbf{p})\,\mathrm{d}\mathbf{p} \tag{8.71}$$

and the classical force term is given by

$$\rho(r)\nabla U(\mathbf{r})=\int \mathrm{d}\mathbf{p}[\nabla_r V(\mathbf{r})\;f^{(1)}(\mathbf{r},\mathbf{p})+\int \mathrm{d}\mathbf{r}_2\int \mathrm{d}\mathbf{p}_2\,\nabla_r V(\mathbf{r},\mathbf{r}_2)$$
$$\times f^{(2)}(\mathbf{r},\mathbf{r}_2;\mathbf{p},\mathbf{p}_2)], \tag{8.72}$$

which comprises forces due to the single-particle as well as the pair potential.

In a similar way, for $\alpha=p^2/2m$, one obtains the equation of motion for the energy density (see ref. 97 for details). Born and Green[89] as well as Mazo and Kirkwood[99] have introduced the concept of quantum temperature and thus obtained an equation for the energy flow in terms of the rate of change of quantum temperature. This concept can be further broadened to encompass the electron fluid following the work of Ghosh and Berkowitz[26] (see Section 8.6.3).

### 8.3.3 Kinetic equations and quantum hydrodynamical equations: approximations

It is now clear that the quantum fluid dynamical equations can be obtained through the kinetic equations in terms of either density matrices in coordinate space or the distribution functions in phase space. The equations in both cases are for the 3D charge and current densities. However, the classical force term contains the two-particle density matrix or the phase-space function. Approximate schemes, however, are available to reduce it further. Thus, for classical fluids, the so-called BBGKY hierarchy[108] is closed by approximating the $(n+1)$-particle distribution function in terms of the $n$-particle ones.* In the present context as well, the two-particle reduced density matrix can be expressed in terms of the single-particle ones, namely

$$\rho_2(\mathbf{r}_1,\mathbf{r}_2;\mathbf{r}_1',\mathbf{r}_2')=\rho_1(\mathbf{r}_1,\mathbf{r}_1')\rho_1(\mathbf{r}_2,\mathbf{r}_2')-\rho_1(\mathbf{r}_1,\mathbf{r}_2')\rho_1(\mathbf{r}_2,\mathbf{r}_1'). \tag{8.73}$$

Thus, the kinetic equation can be reduced[110] to a form in terms of single-particle functions alone. The approximation involved in eqn (8.73) is the well-known Hartree–Fock approximation, whereas if only the first term is retained, the Hartree theory results. Similarly, approximations can be introduced for the phase-space function as well. Thus, if one makes the replacement

$$f^{(2)}(\mathbf{r}_1,\mathbf{p}_1;\mathbf{r}_2,\mathbf{p}_2)\approx f^{(1)}(\mathbf{r}_1,\mathbf{p}_1)f^{(1)}(\mathbf{r}_2,\mathbf{p}_2) \tag{8.74}$$

on the right-hand sides of eqns (8.53) and (8.54), one obtains

$$\frac{\partial f^{(1)}}{\partial t}+\frac{\mathbf{p}}{\mathrm{m}}\cdot\nabla_r f^{(1)}=\frac{\mathrm{i}}{\hbar}\left(\frac{1}{2\pi\hbar}\right)^3\int \mathrm{d}\mathbf{s}\exp\left(\frac{\mathrm{i}\mathbf{p}\cdot\mathbf{s}}{\hbar}\right)[U(\mathbf{r}+\tfrac{1}{2}\mathbf{s})-U(\mathbf{r}-\tfrac{1}{2}\mathbf{s})]$$
$$\times\rho_1(\mathbf{r}+\tfrac{1}{2}\mathbf{s},\mathbf{r}-\tfrac{1}{2}\mathbf{s}), \tag{8.75}$$

where $U(\mathbf{r})$ is the Hartree potential, i.e.

$$U(\mathbf{r})=V(\mathbf{r})+\int\frac{\rho|\mathbf{r}'|}{|\mathbf{r}-\mathbf{r}'|}\,\mathrm{d}\mathbf{r}'. \tag{8.76}$$

An alternative form of eqn (8.75), in terms of the sine function analogous to eqn (8.54a), is also possible. Expanding $[U(\mathbf{r}+\tfrac{1}{2}\mathbf{s})-U(\mathbf{r}-\tfrac{1}{2}\mathbf{s})]$ in eqn (8.75) for small $\mathbf{s}$ and neglecting the higher derivatives, one obtains

$$\frac{\partial f}{\partial t}+\frac{\mathbf{p}}{m}\cdot\nabla_r f=\nabla_r U\cdot\nabla_p f, \tag{8.77}$$

which is the quantum Vlasov equation.

In eqn (8.77), the distribution function $f(\mathbf{r},\mathbf{p})$ can be found self-consistently. There are two approximations involved in the derivation. First, $f^{(2)}$ has been written as a product (eqn (8.74)) that corresponds to the Hartree approximation. Secondly, the higher term in the expansion of $U(\mathbf{r}+\tfrac{1}{2}\mathbf{s})$ has been neglected. In eqn (8.75), however, only the first approximation is involved; it may be called the Hartree–Vlasov equation

*For the hierarchy of equations for the quantum density matrix, see ref. 109.

and solved self-consistently since $\rho_1$ and $f$ are related by the partial Fourier transform. One thus has a practical scheme for the calculation of the phase-space functions. The Vlasov equation involves a mean-field-type approximation and is widely used in plasma physics.

The decoupling scheme of eqn (8.74) is rather crude, and other approximations can be employed (see e.g. Singwi *et al.*[111]) for improving it. The derivation of the Boltzmann equation also follows by using a suitable decoupling, as has been discussed by Carruthers and Zachariasen,[79] where one obtains the quantum analogue of the Boltzmann collision integral.

Improvements are also possible by retaining the single-particle independent-type scheme as given by the Hartree approximation, but replacing the Hartree potential of eqn (8.76) by a more accurate effective potential including both exchange and correlation contributions in a DFT framework (see Chapter 6).

The fluid dynamical equations corresponding to the quantum Vlasov equation are straightforward and obtainable as special cases of the general one (eqns (8.69) and (8.70)). These have been discussed in detail by Bertsch[102] and have found various applications in nuclear physics,[112,113] as well as the study of atomic photoabsorption[100] and collision theory.[79]

To summarize this Section, the hydrodynamic equations follow directly from the kinetic equations in terms of the density matrix or the phase-space distribution function. Although mixed states and finite temperature formalisms can be incorporated in both of them, the phase-space formalism brings out the analogy to the classical case in a more transparent manner. This is clear from the discussion presented above. Thus the classical Liouville equation is analogous to the phase-space kinetic equation for quantum distribution. Also, just as the Liouville equation can be reduced to the Boltzmann equation which in turn gives rise to hydrodynamic equations, so does the quantum phase-space equation reduce to the quantum Vlasov equation, from which the hydrodynamics follows.

## 8.4 Phenomenological approach to QFD

The approaches leading to the hydrodynamical equations corresponding to a many-electron system discussed so far are based on derivations involving the Schrödinger equation or its variants, namely the quantum kinetic equations in terms of either reduced density matrices or distribution functions in phase space. There is yet another, phenomenological, route to QFD which was first proposed by Bloch.[114] He had proposed that the oscillations of the electron cloud in a many-electron system may be described by considering the electron gas as a fluid of density $\rho$ and a velocity field $\mathbf{v}$ ($= -\nabla\chi$, where $\chi$ is the velocity potential). Therefore, he found it sufficient to employ the well-known laws of hydrodynamics, namely the continuity equation

$$\frac{\partial \rho}{\partial t}+\nabla\cdot(\rho\mathbf{v})=0 \tag{8.78}$$

and the Euler equation of motion (analogous to Newton's force law)

$$\frac{\partial \mathbf{v}}{\partial t}+(\mathbf{v}\cdot\nabla)\mathbf{v}=\frac{1}{m\rho}[-\rho\nabla V_{\mathrm{Coul}}-\nabla\tilde{p}]. \tag{8.79}$$

In this equation, $\nabla V_{\mathrm{Coul}}$ represents the classical electrostatic field and is obtained from the solution of the Poisson equation

$$\nabla^2 V_{\mathrm{Coul}}=-4\pi\rho \tag{8.80}$$

and, for the pressure term $\tilde{p}$, he proposed the characteristic pressure–density relation for the electron gas within the statistical theory, namely

$$\tilde{p}=\frac{2}{5}\left(\frac{\hbar^2}{2m}\right)(3\pi^2)^{2/3}\rho^{5/3}. \tag{8.81}$$

It may be noted that this follows by equating the virial of the pressure to the Thomas–Fermi[115] expression for the kinetic energy, i.e.

$$T_0=C_k\int\rho^{5/3}\,\mathrm{d}r;\quad C_k=\tfrac{3}{10}(3\pi^2)^{2/3}\ \text{a.u.} \tag{8.82}$$

The derivation, however, follows from considering $\rho$ and $\chi$ as canonically conjugate variables, forming the Lagrangian

$$L=\int\rho\frac{\partial\chi}{\partial t}\,\mathrm{d}\mathbf{r}-H \tag{8.83}$$

and using the variational principle

$$\partial\int_{t_1}^{t_2}L\,\mathrm{d}t=0 \tag{8.84}$$

or Hamilton's equations

$$\frac{\partial H}{\partial\rho}=\frac{\partial\chi}{\partial t};\quad \frac{\partial H}{\partial\chi}=-\frac{\partial\rho}{\partial t}. \tag{8.85}$$

Bloch's equations result[51] from the following expression for the energy functional $H$:

$$H=T_0+\int v_{\mathrm{ext}}(\mathbf{r})\rho(\mathbf{r})\,\mathrm{d}\mathbf{r}+\frac{1}{2}\int\int\frac{\rho(\mathbf{r})\rho(\mathbf{r}')}{|\mathbf{r}-\mathbf{r}'|}\,\mathrm{d}\mathbf{r}\,\mathrm{d}\mathbf{r}'+\frac{1}{2}\int\rho(\nabla\chi)^2\,\mathrm{d}\mathbf{r}, \tag{8.86}$$

where $T_0$ represents the intrinsic kinetic energy of a Thomas–Fermi system (eqn (8.82)) and the last term represents the macroscopic kinetic energy.[32] The functional derivative of the sum of the other two terms on the right-hand side of eqn (8.86) is clearly the electrostatic potential

$$V_{\mathrm{Coul}} = v_{\mathrm{ext}} + \int \frac{\rho(\mathbf{r}')}{|\mathbf{r}-\mathbf{r}'|}\,d\mathbf{r}', \tag{8.87}$$

which is the solution of the Poisson equation (8.80). Clearly, eqn (8.86) is the Thomas–Fermi energy functional except for the last term which comes from dynamical extension.[51]

The hydrodynamical model of Bloch, defined by eqns (8.78) and (8.79),* was originally proposed for studying the density oscillations and the stopping power of electrons. Subsequently, it has found many applications, e.g. in the study of photoabsorption,[116] collective excitations,[117] plasmons of metal clusters,[118] surface plasmons[119] and many other problems of solidstate physics.

An interesting extension of Bloch's theory has been proposed by Ying[120] who replaced the Thomas–Fermi kinetic energy term $T_0$ in the energy expression of eqn (8.86) by a general functional $G[\rho]$. The analogy rests in the dynamical extension of the Hohenberg–Kohn[121] energy functional $E[\rho]$ which gives an "exact" description of the ground state. The energy functional $E[\rho]$ is written as

$$E[\rho] = \int v_{\mathrm{ext}}(\mathbf{r})\rho(\mathbf{r})\,d\mathbf{r} + F[\rho]. \tag{8.88}$$

The universal functional $F[\rho]$ can be separated from the classical electrostatic part:

$$F[\rho] = \frac{1}{2}\iint \frac{\rho(\mathbf{r})\rho(\mathbf{r}')}{|\mathbf{r}-\mathbf{r}'|}\,d\mathbf{r}\,d\mathbf{r}' + G[\rho], \tag{8.89}$$

where the second universal functional $G[\rho]$ consists of the kinetic and the exchange-correlation contributions. The Thomas–Fermi kinetic energy $T_0$ is only an approximation to this exact functional $G[\rho]$. The crux of Ying's work[120] lies in assuming that a universal functional $G[\rho(\mathbf{r}, t)]$ can be written for time-dependent problems as well. Although initially it was an *ad hoc* prescription, it has now been proved first for oscillating time-dependent cases[32] and for a more general time dependence.[33] It has been explicitly demonstated[33] that an equation of the form

$$\frac{\partial \mathbf{j}(\mathbf{r}, t)}{\partial t} = \mathscr{P}[\rho(\mathbf{r}, t)] \tag{8.90}$$

*Note that eqn (8.79) follows from the gradient of the first equation in (8.85).

can be written so that the right-hand side is a functional of density alone. After separating out the classical Coulomb contribution from $\mathscr{P}[\rho]$, one recovers eqn (8.79) with the pressure term defined by

$$\nabla \tilde{p} = \rho \nabla \frac{\delta G}{\delta \rho}. \tag{8.91}$$

This is more general that eqn (8.81) and is exact in principle. Approximating $G$ by $T_0$ leads to eqn (8.81).

Ying's work has been further extended to magnetic vector potentials[122] and the study of surface plasmons.[122,123]

Although the initial thrust of Ying's work was to the study of response, it has also been applied to investigate collective phenomena and photoabsorption in many-electron systems.[19,124–126] The hydrodynamical equations have also been combined with Maxwell's equations in order to study electrodynamics at metal boundaries.[127] Dreizler and coworkers[128–132] have dealt with high-energy proton–atom collisions within the QFD transcription of Thomas–Fermi and Thomas–Fermi–Dirac–Weizsäcker (TFDW) theories. This amounted to numerically solving the corresponding two basic hydrodynamical equations, namely the continuity equation and the equation of motion, to obtain the TD density and velocity field. Such calculations have revealed certain *non-linear* features and other features not obtained from standard calculations relying on the adiabatic picture. In case of proton–argon collisions,[129,130] after the strong disturbance of the electorn cloud by the impinging and receding proton, the atom is left to execute damped oscillations, in which the inner parts of the density have much shorter relaxation times than the outer parts. These oscillations are more complicated than a superposition of simple eigenoscillations of the whole density. There is a discernible effect of rotational coupling on the electron density because the latter does not have axial symmetry with respect to the internuclear axis.[130] The TDTF theory raises the possibility of calculating effective potentials, which depend not only on internuclear separations (or reaction coordinates) but also on impact parameters and velocities. Further, the question whether an ion–atom scattering system behaves quasi-adiabatically at a given collision velocity depends on the impact parameter as well.[132] The above main features, namely density oscillations and axial asymmetry, are also observed with the proton–hydrogen-atom system.[128] The TDTFDW–hydrodynamical theory has also been employed to calculate photoabsorption cross-sections of noble-gas atoms.[131] A more accurate study (see Section 8.2.3) within the QFDFT has been undertaken by Deb and Chattaraj.[34]

In summary, the hydrodynamical approach discussed in this Section consists of eqns (8.78) and (8.79) with the potential given by eqn (8.87) and the pressure defined by eqn (8.91). At this point it may be worthwhile to compare the hydrodynamical equations obtained through different approaches and discuss their equivalence, similarity or difference.

The QFD for a pure quantum system consists of eqns (8.3) and (8.4), or eqns (8.21a) and (8.21b), for single- or many-particle systems, respectively, as obtained from the Schrödinger equation. The generalized QFD obtained through reduced density matrices involves eqns (8.36) and (8.46), while that arising from phase-space considerations consists of eqns (8.69) and (8.70). It is quite clear that except for the pressure term, whose exact form is not known, all the approaches lead to identical results.

Thus, in eqn (8.91) if one assumes the Weizsäcker term,

$$T_{\mathrm{w}} = \frac{\hbar^2}{8m} \int \frac{\nabla\rho \cdot \nabla\rho}{\rho} \, \mathrm{d}\mathbf{r}, \tag{8.92}$$

for the functional $G$, which is exact for a one-electron system, one recovers eqn (8.4). The Bohm potential or the equivalent quantum stress tensor is an important quantity in the hydrodynamical model and is a characteristic of the kinetic energy represented by the Weizsäcker-type term, either in terms of the total density or the orbitals. The natural-orbital, the Kohn–Sham or Hartree–Fock-type schemes for the kinetic energy introduce several single-particle equations of the form of eqn (8.21b). The exchange-correlations are, however, to be incorporated through approximate expressions (see Chapter 6) although the single-particle kinetic energy is treated exactly in the Kohn–Sham scheme.[52]

## 8.5 Miscellaneous consequences and generalizations of QFD

The previous Sections have discussed at length the hydrodynamics corresponding to a quantum system (pure state and mixed states, zero temperature, finite temperature, microscopic and macroscopic). This Section considers several other formalisms which may be regarded as generalizations or consequences of the QFD description.

### 8.5.1 Quantum hydrodynamical equations of change

The first generalization corresponds to the off-diagonal extension of the QFD equations. In other words, while the QFD equations describe the time evolution of the charge density and the current density, the present generalizations involves the transition density and the transition current density. Again, the generalized equations can be obtained through either the wave function or the transition density matrix or the mixed phase-space representation. Since these derivations follow from straightforward extensions of those discussed in the previous Sections, we present here only one of them.

Quantum fluid dynamical equations of change through the wave function have been derived by Hirschfelder.[133] Using the time-dependent Schrödinger equation for two Hamiltonians $H_1$ and $H_2$ with wave functions $\psi_1$ and $\psi_2$, respectively, one can derive[39] the equation of change for any arbitrary operator $\hat{A}$ (ref. 134) as

$$\frac{\mathrm{d}}{\mathrm{d}t}(\psi_1^* \hat{A}\psi_2) = \psi_1^* \left[\frac{\partial \hat{A}}{\partial t} - \frac{\mathrm{i}}{\hbar}(\hat{A}H_2 - H_1\hat{A})\right]\psi_2 - \nabla \cdot \mathbf{F}, \tag{8.93a}$$

where the flux is given by

$$\mathbf{F} = \frac{1}{2m}[\psi_1^* \hat{p}\hat{A}\psi_2 + (\hat{p}\psi_1)^* \hat{A}\psi_2], \tag{8.93b}$$

in terms of the momentum operator $\hat{p}$.

Equation (8.93a), for the case $H_1 = H_2$, is the time-dependent local hypervirial theorem (off-diagonal):

$$\frac{\mathrm{d}}{\mathrm{d}t}(\psi_1^* \hat{A}\psi_2) = \psi_1^* \left\{\frac{\partial A}{\partial t} + \frac{\mathrm{i}}{\hbar}[H, A]\right\}\psi_2 - \nabla \cdot \mathbf{F}. \tag{8.94}$$

The choice of the operator $\hat{A}$ as unity yields the continuity equation for the transition density:

$$\frac{\mathrm{d}}{\mathrm{d}t}(\psi_1^* \psi_2) + \nabla \cdot \mathbf{j}_{12} = 0, \tag{8.95a}$$

where $\mathbf{j}_{12}$ is the transition current density defined by eqn (8.93b) with $\hat{A} = 1$.

On the other hand, for the choice $\hat{A} = \hat{p}$, the real part of eqn (8.94) gives the Euler equation of motion for the transition density:

$$m\frac{\mathrm{d}\mathbf{j}_{12}}{\mathrm{d}t} = -(\psi_1^* \nabla V \psi_2) - \nabla \cdot \sigma_{12}, \tag{8.95b}$$

where $V$ is the potential energy operator in the Hamiltonian and the stress tensor $\boldsymbol{\sigma}$ is given by

$$\sigma_{12} = \frac{1}{4m}\{\psi_1^* \hat{p}\hat{p}\psi_2 + \psi_1 pp\psi_2^* + (\hat{p}\psi_1)^*(\hat{p}\psi_2) + (\hat{p}\psi_1)(\hat{p}\psi_2)^*\}. \tag{8.95c}$$

The diagonal elements of these equations give the now familiar QFD equations of motion for the density and the current density. All these equations can be reduced to 3D space by integrating over $N-1$ coordinates.

Two other forms of the operator $\hat{A}$ which are of importance are the angular momentum operator $\hat{A}=\hat{r}\times\hat{p}$ and the kinetic energy operator $\hat{p}^2/2m$. These lead to the equations[133,135] for the angular momentum transport[136] and the kinetic energy change.

The QFD equations of motion are thus special cases (diagonal elements) of these generalized equations of change. The flux term $\nabla\cdot\mathbf{F}$ vanishes on integration for normal boundary conditions. Thus, $\mathbf{F}$ is the hidden flux and plays a very important role in the QFD equations.

The role and nature of the hidden flux in the local hypervirial relation eqn (8.94), is clear since $\langle\psi|[H,\hat{A}]|\psi\rangle$ vanishes according to the global hypervirial theorem (time-independent)[137] although the commutator $[H,\hat{A}]$ does not vanish. A similar situation arises in the virial theorem[138] as well, where although kinetic and potential energies are globally related by a simple relation $2E_{\mathrm{kin}}+E_{\mathrm{pot}}=0$, the corresponding energy densities are not so related. Also, in the Hellmann–Feynman theorem[139,140] the derivative of the energy is contributed only by the classical electrostatic contributions although the total energy itself consists of the kinetic part as well. It would be interesting to look at the hidden flux or potential terms (vanishing on integration) for the local relations corresponding to the virial and Hellmann–Feynman theorems.

Considering the operator $\hat{A}$ to be the virial operator, i.e. $\hat{r}\cdot\hat{p}$ it can be shown[39] from eqn (8.94) that, for time-independent systems, the local virial theorem[141] is obeyed in the following form

$$2t_{\mathrm{kin}}(\mathbf{r})+V_{\mathrm{pot}}(\mathbf{r})+\rho(\mathbf{r})\sum_{\alpha}R_{\alpha}\cdot\frac{\partial H}{\partial R_{\alpha}}+\nabla\cdot\hat{r}\cdot\sigma=0, \tag{8.96}$$

where $\boldsymbol{\sigma}$ is the quantum stress tensor (the diagonal element from eqn (8.95c)), $V_{\mathrm{po}}{}^{\mathrm{t}}(\mathbf{r})$ is the potential energy density and $t_{\mathrm{kin}}(\mathbf{r})$ is the kinetic energy density[25] defined as

$$t_{\mathrm{kin}}(\mathbf{r})=\tfrac{1}{2}\nabla\psi\cdot\nabla\psi-\tfrac{1}{8}\nabla^2\rho. \tag{8.97}$$

Similarly, if $H_2$ and $H_1$ of eqn (8.93a) differ through a parameter $\lambda$, then on expanding $H_2-H_1$ and $\psi_2-\psi_1$ in powers of $\lambda$, and retaining the first-order term, one obtains for a time-independent case

$$\frac{\partial E}{\partial\lambda}(\psi^*\psi)=\left(\psi^*\frac{\partial V_{\lambda}}{\partial\lambda}\psi\right)+\left(\psi^*\frac{\partial V_{\mathrm{q}}}{\partial\lambda}\psi\right), \tag{8.98}$$

where $V_{\lambda}$ is the external potential characterized by the parameter $\lambda$; $V_{\mathrm{q}}$ is the quantum potential of Bohm defined as $V_{\mathrm{q}}=-(\hbar^2/2m)\nabla^2|\psi|/|\psi|$ and depends on $\lambda$ through the dependence of $\psi$ on $\lambda$. The last term clearly vanishes on integration, yielding the integrated form of the Hellmann–Feynman theorem. This hidden potential in the Hellmann–Feynman theorem has been studied by Nakatsuji.[140]

The generalized QFD equations of change presented here thus offer a tool for studying the nature of hidden potentials and provide a theoretical framework for looking at various transition phenomena.

### 8.5.2 Other generalizations

We have seen that the quantum potential of Bohm and its variants, e.g. the quantum stresss tensor, play a crucial role in the QFD equations. It has also been shown that the resulting quantum force vanishes on integration and therefore does not appear in the global version of the force laws. Thus, the definitions of quantum stress tensor and quantum potential are not unique. Recently, Floyd[143,144] has defined a new modified potential which is also non-unique and describes various possible continuous motions for any particular bound-state energy eigenvalue of the Schrödinger equation. In one dimension $x$, Floyd's modified potential $U(x, E)$ has an explicit dependence on the energy eigenvalue $E$ and is determined by the *non-linear* differential equation

$$U+\frac{\hbar^2}{8m}\frac{\partial^2 U/\partial x^2}{E-U}+\frac{5\hbar^2}{32m}\left[\frac{\partial U/\partial x}{E-U}\right]^2=V, \tag{8.99}$$

where $V(x)$ is the potential. Floyd has established the relationship between his modified potential and the Bohm potential, both characterizing the same Schrödinger equation.

Generalizations of the QFD description have also been proposed corresponding to the generalization of the Schrödinger equation. Thus Schönberg[145] who first presented a detailed development of QFD corresponding to spin-polarized systems also formulated[146,147] the QFD for a nonlinear generalization of the Schrödinger equation (see also Section 8.2.3). In order to obtain the general motion of the quantum fluid he had generalized the velocity field of eqn (8.14c) by introducing Clebsch parameters;

$$\mathbf{v}(\mathbf{r})=\frac{1}{m}\left(\nabla S-\frac{e}{c}\mathbf{A}+\lambda\nabla\mu\right), \tag{8.100}$$

which correspond to a Hamiltonian characterized by additional scalar and vector potentials given respectively by $\lambda\,\partial\mu/\partial t$ and $-\lambda\nabla\mu$. The Euler equation of motion can be obtained by taking the gradient of the corresponding quantum Hamilton–Jacobi equation

$$\frac{\partial S}{\partial t}+\lambda\frac{\partial\mu}{\partial t}+\frac{1}{2m}\left(\nabla S-\frac{e}{c}\mathbf{A}+\lambda\nabla\mu\right)^2+e\phi-\frac{\hbar^2}{2m}\frac{\nabla^2|\psi|}{|\psi|}=0, \tag{8.101}$$

where $\mathbf{A}$ and $\phi$ are the usual vector and scalar potentials. Implications of this generalization have been discussed by Schönberg.[146,147]

Another interesting generalization of QFD has been recently achieved by Joachim,[148] who developed the QFD for an inhomogeneous population of electrons and nuclei (essentially, a two-component fluid). Separate QED equations were obtained to describe the evolution of the nuclear and electronic distributions for non-stationary states. By approximating the equation for the nuclei through a quasi-classical dynamical equation, a non-adiabatic quasi-classical local model of a molecule has been constructed. Through a QFD description, Joachim[148] thus developed a quasi-classical way of studying the intramolecular evolution with electron–nuclear coupling and applied it to the $H_2$ molecule. A density functional theory by treating the electrons and nuclei as two components of a fluid mixture was also developed by Capitani *et al.*[149]

The general QFD description of a non-stationary state reduces to a simple form for stationary states. However, of much interest is the intermediate case, i.e. the quasi-stationary state which is stable but not stationary. Recently, a new concept of stability has been introduced and the theory of "stable" states has been developed within the framework of the hydrodynamic model by Tachibana.[150,151] "Stability" is imposed through the condition $\partial\rho/\partial t = 0$ in the continuity equation and the solution chosen to be a non-stationary function. The hydrodynamic elementary force is shown to vanish and the existence of an external stabilization mechanism is found. The characteristic decaying process of the stable state under the action of the surrounding medium has also been studied. The stable state has an analogy with the soliton-like wave stability of the time-dependent solution of a new type of non-linear Schröedinger equation. A solitary-wave-like[152] property may also be encountered with the solutions of the non-linear Schrödinger equation arising in the QFDFT treatment of atomic collisions[34] (see Section 8.2.3).

Recently, the QFD approach has been extended by Nassar[153,154] to a non-linear Schrödinger equation for the quantum mechanical treatment of dissipative processes, e.g. in a chaotic thermal environment. The corresponding Schrödinger equation characterized by an explicitly time-dependent Hamiltonian is given by (in one-dimension, $x$)

$$i\hbar\frac{\partial\psi}{\partial t} = -\frac{\hbar^2}{2m}\exp(-vt)\,\nabla^2\psi + \exp(vt)\,[V(x) - xA(t)]\psi, \qquad (8.102)$$

where $A(t)$ is a stochastic force (see Section 8.7).

Substitution of the polar form (eqn (8.2)) leads to the fundamental equations, i.e. the continuity equation and the Euler-type equation, given by

$$m\rho\left(\frac{\partial\mathbf{v}}{\partial t} + (\mathbf{v}\cdot\nabla)\mathbf{v}\right) + vm\rho\mathbf{v} = -\rho\nabla[(V(x) - xA(t)) + V_q], \qquad (8.103a)$$

where $\rho = \psi^* \psi$. The velocity field $\mathbf{v}$ and the quantum potential $V_q$ are

$$\mathbf{v} = \exp(-\nu t) \frac{\nabla S}{m} \tag{8.103b}$$

and

$$V_q = -\frac{\hbar^2}{2m} \exp(-2\nu t) \frac{\nabla^2 \rho^{1/2}}{\rho^{1/2}}, \tag{8.103c}$$

respectively. The fact that the quantum force vanishes on integration has been utilized[153,154] to split eqn (8.103a) into the pair of equations

$$\frac{\partial \mathbf{v}}{\partial t} + (\mathbf{v} \cdot \nabla)\mathbf{v} + \nu \mathbf{v} = \frac{\mathbf{x} - \mathbf{X}}{\omega(t)}, \tag{8.104}$$

$$\nabla \left[ \frac{\hbar^2}{2m^2} \exp(-2\nu t) \frac{\nabla^2 \rho^{1/2}}{\rho^{1/2}} \right] = \frac{\mathbf{x} - \mathbf{X}}{\omega(t)}, \tag{8.105}$$

and the solutions have been obtained after certain simplifications. The results lead to the correct classical and quantum limits. Here, $\omega(t)$ is related to $\rho(\mathbf{x}, t)$ and $\mathbf{X} = [1 - \exp(-\nu t)] \nabla S / m\nu$.

In a subsequent work, Nassar[155] has developed the QFD solution of a non-linear Schrödinger–Langevin equation with a multiterm potential of the type (see also refs 156 and 157)

$$i\hbar \frac{\partial \psi}{\partial t} = -\frac{\hbar^2}{2m} \nabla^2 \psi + \left[ \frac{\hbar \nu}{2i} \ln\left(\frac{\psi}{\psi^*}\right) + V(\mathbf{r}, t) \right] \psi, \tag{8.106}$$

describing interactions in non-conservative systems. Again, the corresponding fluid dynamical equations suggest a splitting of the Euler-type equation and hence a relatively easy method to obtain the solution in terms of an auxilliary differential equation has been possible.[155]

### 8.5.3 Streamlines, vorticities, magnetic and electric properties

In analogy with a classical fluid, we have seen that the basic QFD equations are the continuity and Euler equations determining the evolution of the change density $\rho$ and the current density $\mathbf{j}$. One may exploit this analogy further and consider the fluid *streamlines* analogous to classical trajectories. This is possible through integration of the velocity field

$$\frac{d\mathbf{r}}{dt} = \mathbf{v}(\mathbf{r}, t) \tag{8.107}$$

starting with the initial condition $\mathbf{r}=\mathbf{r}_0$ at $t=0$. Analogously to these real streamlines obtained from the real part of the velocity $\mathbf{v}$, one can evaluate imaginary streamlines from the imaginary velocity field of Hirschfelder *et al.*[37] (eq (8.8)).

The streamlines are, however, different from the trajectories of individual particles in that two or more trajectories might pass through a point while the streamline is directed along the weighted average of the directions of these trajectories. Thus the streamlines cannot cross while the trajectories do.[1]

The most significant feature of streamlines is the presence of vortices ("quantum whirlpools") and their studies are important in a number of contexts including magnetic properties and collisions. The vorticity $\mathbf{w}$ defined as

$$\mathbf{w}=\text{curl}\,\mathbf{v}, \tag{8.108}$$

is normally zero at all points where the density is non-zero since $\mathbf{v}$ is defined as the gradient of a scalar. However, curl $\mathbf{v}$ does not vanish at the nodes of the wave function since at those points the phase has no meaning and neither $S$ nor $\nabla S$ is well defined.

The *circulation* $\Gamma$ is defined as the line integral of $\mathbf{v}$ along the contour of a closed loop, i.e.

$$\Gamma=\int_L \mathbf{v}\cdot d\mathbf{r}=\int_A (\nabla\times\mathbf{v})\cdot d\mathbf{S}, \tag{8.109}$$

where the equality to the surface integral follows from Stokes' theorem. Circulation may be non-vanishing if the loop encloses a nodal region where the vorticity is non-zero. Since the phase $S/\hbar$ is defined only up to a multiple of $2\pi$ (at points with no nodes), the circulation becomes

$$\Gamma=\oint \mathbf{v}\cdot d\mathbf{r}=\frac{1}{m}\oint dS=2\pi\frac{\hbar}{m}n, \quad n=0, \pm 1,\ldots, \tag{8.110}$$

when the path of integration does not pass through nodal regions. Thus, the vorticity appearing around the nodes of $\psi$ is quantized and each nodal region is associated with an integer $n$, known as the circulation number.

Different types of vorticity, e.g. line vortices, axial vortices and toroidal vortices have already been reviewed by the present authors[1] and are therefore omitted here. Heller and Hirschfelder[158] have shown that the conditions for the existence of quantized vortices are the same as those necessary for the existence of orbital paramagnetic moments. The physical explanation lies in the fact[158] that the vortex is the charge flow around the nodal region while the orbital paramagnetism arises from the magnetic moment generated by that circulation (see ref. 159 for a detailed discussion of wave function nodes and their physical significance).

Theories for the calculation of various magnetic properties like magnetic susceptibility, chemical shift, etc., in terms of the current density have an advantage that both interpretation and calculation are simultaneously possible, thus making the QFD formulation very attractive. In the presence of magnetic fields, the net current density consists of a paramagnetic part ($\rho \nabla S/m$) and a diamagnetic part proportional to the vector potential. The latter is responsible for diamagnetic shielding in molecular systems subjected to magnetic fields. The permanent magnetic moment of molecules is contributed by the paramagnetic part only.

The magnetic properties of conjugated cyclic molecules have been traditionally explained in terms of the so-called ring currents.[160] However, this approach was criticized by Musher.[161,162] Consequently, even the recent calculations of Lazzeretti *et al.*[163] on benzene were interpreted without endorsing ring current ideas. However, the recent work of Gomes[164] has partially rehabilitated the ring current concept through studies on delocalized magnetic currents in benzene based on a QFD formulation. Gomes has introduced the concept of a delocalized current, which can be compared to the ring current, and has denoted that part of the probability current density as $\mathbf{j}(\mathbf{r})$, which represents a flow of electronic charge around the ring. A rigorous definition has been achieved by considering the current contribution from each natural orbital, which is conserved (divergenceless, in a stationary case) upon inclusion of exchange currents.

The direct coupling of the applied magnetic field with the current density is responsible for making the hydrodynamical formulation attractive in dealing with magnetic problems. However, the QFD approach is versatile enough to be applicable to the study of perturbation by electric fields. Thus, linear response for atoms and molecules has been calculated within the framework of the hydrodynamical model; this has been already discussed in Section 8.2.3. For solids, using the jellium surface model linear response has been studied[165,166] through the phenomenological hydrodynamic model of Ying and its extensions (see Section 8.4). Molecular vibrations have also been treated in terms of the dynamic electric current induced by vibration.[167]

## 8.6 The development of certain chemical concepts from QFD

By now it is clear that QFD requires two key quantities (real), namely the change density $\rho$ and the current density $\mathbf{j}$, because the wave function is complex-valued. For a stationary ground state (or for a static stationary state[1]), the current density vanishes and then the electron density alone suffices to describe the many-electron system; this is equivalent[23] to time-independent ground-state DFT. In this Section, we discuss how interesting insights (see e.g. ref. 168) into the physical foundations of several chemical concepts, related to structure, reactivity and dynamics of many-electron systems, are obtained from the electron density alone. Further, we will see that QFD generates several "thermodynamic" concepts for the electron fluid, such as the chemical potential and the compressibility.

### 8.6.1 "Thermodynamic" quantities for the electron fluid and the physical foundations of chemical concepts

The minimization of the Hohenberg–Kohn energy functional $E[\rho]$ of eqn (8.88) with respect to variation of density subject to the normalization constraint (imposed through a Lagrange multiplier) leads to the definition of the chemical potential (see also Chapter 1):

$$\mu = \frac{\delta E}{\delta \rho}. \tag{8.111}$$

Parr *et al.*[29] have shown that the electronegativity of chemistry is identical with this chemical potential (with a negative sign). This equivalence, i.e.

$$\chi = -\frac{\partial E}{\partial N} = -\int \frac{\delta E}{\delta \rho} \frac{\partial \rho}{\partial N} \mathbf{dr} = -\mu \int \frac{\partial \rho}{\partial N} \mathbf{dr} = -\mu, \tag{8.112}$$

is important in that it defines electronegativity in a manner enabling direct calculations. One can either evaluate $(1/N)\int \mathrm{d}\mathbf{r}\rho(\mathbf{r})\delta E/\delta\rho$ or $\delta E/\delta\rho|_{\mathbf{r}\geqslant\infty}$ for evaluating $\mu$ (see refs 169 and 170). Recall that the electronegativity introduced by Mulliken[171] as $\frac{1}{2}(I+A)$ and defined[172,173] as $-\partial E/\partial N$ could be alternatively obtained only through the finite difference formula of Mulliken. Another importance of this identification is that the conjecture of Sanderson[174] that the electronegativities of atoms are equalized on molecule formation is now rationalized and put on a sound theoretical basis.[175] The equalization of electronegativities of atoms on molecule formation has led to new and interesting methods[176–178] for the calculation of atomic charges[179] in molecules, which help to rationalize various chemical effects.[180]

Since $\mu$ is a constant and independent of position, the chemical potential is the same everywhere in a molecule. However, a paradox arisis if one follows the process of dissociation of a heteronuclear diatomic molecule into its constituent atoms. The question is: if there were only one $\mu$ throughout the molecule, at what point would the molecule dissociate into the atoms[181] having unequal chemical potentials? This is the so-called size-consistency problem. However, this is not a drawback of density functional theory but a consequence of quantum mechanics which does not allow simultaneous description of the combined system as well as the isolated subsystems and is analogous to the famous Einstein–Podolsky–Rosen paradox.[3] A practical solution has, however, been provided by Perdew *et al.*[182] who have also studied the nature of the $E$ versus $N$ curve, and discussed whether the three points corresponding to $N-1$, $N$ and $N+1$ electrons are to be joined by piecewise line segments or by some other curve.

Next, we consider the compressibility of the electron fluid which had received considerable attention in the chemical literature through the concept of hard and soft acids and bases.[183] Although this concept was very successful in "explaining" and classifying a large number of chemical phenomena, its quantitative definition has been provided[184] only recently. Chemical hardness has been defined as the second derivative of energy with respect to the number of electrons,

$$\eta = \frac{1}{2}\frac{\partial^2 E}{\partial N^2}, \tag{8.113}$$

and the conjectures "hard likes hard" and "soft likes soft" were rationalized.[184, 185] However, like electronegativity, eqn (8.113) can be evaluated only through a finite difference formula for the derivative which gives $\eta = \frac{1}{2}(I-A)$. A definition in terms of density functional concepts can be written as

$$\tilde{\eta}(\mathbf{r}) = \frac{1}{2N}\int \frac{\delta^2 F[\rho]}{\delta\rho(\mathbf{r})\delta\rho(\mathbf{r}')}\,\rho(\mathbf{r}')\,\mathrm{d}\mathbf{r}'. \tag{8.114}$$

This is the local hardness[30] and can be integrated to give the global hardness

$$\eta = \int \tilde{\eta}(\mathbf{r})\,\frac{\partial\rho(\mathbf{r})}{\partial N}\,\mathrm{d}\mathbf{r}. \tag{8.115}$$

The significance of the local hardness and the possibility of predicting the site of attack through it have been pointed out. Subsequently, the concept of local softness has been introduced by Yang and Parr[186] as $\partial\rho/\partial\mu$, and its usefulness in catalysis discussed.

The compressibility of the electron fluid expressed in terms of hardness has the following meaning: the more compressible or less hard electron cloud has to absorb more charge to alter the chemical potential by the same amount, and vice versa.

An alternative approach has been introduced by Bader *et al.*,[187] who suggest that the hard–soft behaviour can be rationalized from the properties of $\nabla^2\rho$ which is a measure of the "lumpiness" of the electron density and also the compressibility. According to them a hard site in a molecule corresponds to the appearance of a large extremum of $\nabla^2\rho$.

The electronegativity and hardness quantities discussed here permit one to calculate the partial charges in a heteronuclear molecule. Thus, from the expansion of energy as

$$E_N = E_{N_0} + \mu_0(N - N_0) + \eta_0(N - N_0)^2, \tag{8.116}$$

one obtains the effective chemical potential as

$$\mu_{\text{eff}} = \mu_0 + 2\eta_0(N - N_0). \tag{8.117}$$

Equating the effective chemical potentials of two atoms A and B on molecule formation after a transfer of electrons $\Delta N = \Delta N_A = -\Delta N_B$, one obtains[184]

$$\mu_0^A + 2\eta_0^A \Delta N = \mu_0^B - 2\eta_0^B \Delta N \tag{8.118a}$$

or

$$\Delta N = \frac{\mu_0^B - \mu_0^A}{2(\eta_0^A - \eta_0^B)}, \tag{8.118b}$$

which gives the partial charges $q_A = -\Delta N$ and $q_B = \Delta N$. However, this charge is independent of the internuclear separation $R$; this is not correct. This problem occurs because the expansion (8.116) or (8.117) is not complete, and should be supplemented by the contribution from the change in the surrounding potential.[185] In the spherical density approximation of the atoms in the molecule, one can write

$$\mu_{\text{eff}}^A = \mu_0^A + 2\eta_0 \Delta N_A - \frac{q_B}{R}, \tag{8.119}$$

instead of eqn (8.117), which leads to the *wrong* expression:

$$\Delta N = \frac{\mu_0^B - \mu_0^A}{2(\eta_0^A + \eta_0^B - \frac{1}{2}R)}. \tag{8.120}$$

More accurate expressions have been obtained[176] by taking other factors into account. The recent work of Pearson[188] on electronegativity and hardness is also of interest.

The above approach, however, is not applicable to homonuclear or covalent molecules. In order to calculate the energy by using the electronegativity approach one should take into account the build-up of charge in the binding region during molecule formation.[189]

The chemical potential $\mu$, although constant in an equilibrium situation, can be assigned to any point. Very recently, an approach has been suggested[190] where the bond centre is assigned a bond electronegativity and bond hardness. Thus, the diatomic molecule consists of three point charges in this picture, at the two atomic sites and the "bond site." The first-order energy expansion,

$$\Delta E = -\mu_A q_A - \mu_B q_B - \mu_{AB} q_{AB}, \tag{8.121}$$

has a close analogy with the Hückel theory, which gives

$$E = \sum_r \alpha_r q_r + \sum_{r \neq s} \beta_{rs} p_{rs} \tag{8.122}$$

The equivalence follows from the identifications $\mu_r = -\alpha_r$ and $\mu_{AB} = \beta_{rs}/S_{rs}$ and $q_{AB} = p_{rs} S_{rs}$. What is more interesting is that all modifications of the Hückel method,[191] namely the $\omega$-technique, the modified $\beta$-method, the Pariser–Parr–Pople method, etc. can be generalized[190] by systematically including the higher-order terms in the energy expansion (eqn (8.121)). This gives a new semi-empirical approach to density functional theory. Further, the various reactivity indices based on Hückel theory, introduced by Coulson and Longuet–Higgins,[192] can now be given a density functional interpretation.

The frontier electron theory[193] of chemical reactivity also follows from a density functional approach.[194, 195] The ground-state electronic energy $E[N, v]$ and the chemical potential $\mu[N, v]$ may be considered to be characterized by $N$ and the external potential $v(\mathbf{r})$ from which the fundamental equations for the changes $\mathrm{d}E$ and $\mathrm{d}\mu$ follow,[194] namely

$$\mathrm{d}E = \mu\, \mathrm{d}N + \int \rho(\mathbf{r})\, \mathrm{d}v(\mathbf{r})\, \mathrm{d}\mathbf{r} \tag{8.123}$$

and

$$\mathrm{d}\mu = 2\eta\, \mathrm{d}N + \int f(\mathbf{r})\, \mathrm{d}v(\mathbf{r})\, \mathrm{d}\mathbf{r}. \tag{8.124}$$

The function $f(\mathbf{r})$ has been defined by

$$f(\mathbf{r}) \equiv \left[\frac{\delta\mu}{\delta v(\mathbf{r})}\right]_N = \frac{\partial\rho(\mathbf{r})}{\partial N}, \tag{8.125}$$

where the equality is a Maxwell relation[196] for eqn (8.123). Parr and Yang[194] have argued that large values of $f$ at a site favour reactivity at that site and accordingly have called $f(\mathbf{r})$ the "frontier (or Fukui) function" for a molecule.

In order to predict the preferred direction of attack of a reagent R on the molecule S, they assumed that the preferred direction is the one for which the initial $|d\mu|$ for the species S is a maximum. Neglecting the less sensitive direction dependence of the first term of eqn (8.124), especially at the initial stages of a reaction (i.e. large distance between R and S), one can assume that the preferred direction is the one having largest $f(\mathbf{r})$ at the reaction site.

Now, just as the $E$ versus $N$ curve has a discontinuity of slope at each integral $N$ (see ref.[182]), i.e. there is $\mu^- = (\partial E/\partial N)^-$ (positive-ion side), $\mu^+ = (\partial E/\partial N)^+$ (negative-ion side) and $\mu^0 = \frac{1}{2}(\mu^+ + \mu^-)$, one can observe similar slope discontinuities in the Fukui function $f(\mathbf{r})$. Thus, the electrophilic attack (when $\mu_S > \mu_R$) is governed by

$$f^-(\mathbf{r}) = (\partial \rho(\mathbf{r})/\partial N)^-, \tag{8.126a}$$

the nucleophilic attack (when $\mu_S < \mu_R$) by

$$f^+(\mathbf{r}) = (\partial \rho(\mathbf{r})/\partial N)^+ \tag{8.126b}$$

and the neutral or radical attack (when $\mu_S \simeq \mu_R$) by

$$f^0(\mathbf{r}) = (\partial \rho(\mathbf{r})/\partial N)^0. \tag{8.126c}$$

The frozen-core approximation, i.e. $d\rho = d\rho_{\text{valence}}$ can be introduced leading to the results:[194]

$$\text{electrophilic attack} \quad f(\mathbf{r}) \simeq \rho_{\text{HOMO}}(\mathbf{r}), \tag{8.127a}$$

$$\text{nucleophilic attack} \quad f^+(\mathbf{r}) \simeq \rho_{\text{LUMO}}(\mathbf{r}), \tag{8.127b}$$

$$\text{radical attack} \quad f^0(\mathbf{r}) \simeq \tfrac{1}{2}[\rho_{\text{HOMO}} + \rho_{\text{LUMO}}], \tag{8.127c}$$

which are the rules of frontier electron theory.[193] The simple assumption of Parr and Yang[121] that "$|d\mu|$ big is good" thus generates the whole of frontier electron theory.

In this connection it might be mentioned that the approach of Klopman[197] who developed a general theory of chemical reactivity and distinguished between "charge-controlled" and "frontier-controlled" reactions, can perhaps be reinterpreted in terms of density functional quantities, namely the hardness and the Fukui function, especially in the light of semi-empirical density functional theory.[190]

### 8.6.2 Quantum chemistry through "local" quantities: topology of charge density and current density

It is well known that more detailed insight into a phenomenon can be obtained through the study of "local" quantities,[1] apart from the global integrated numbers. One defines in general a property density $p(\mathbf{r})$ so that on integration it gives the global value of the property concerned, i.e.

$$P = \int p(\mathbf{r})\,d\mathbf{r}. \tag{8.128}$$

Various property densities such as the electron density $\rho$, the current density $\mathbf{j}$, the force density and energy densities, have been discussed by many workers. The role played by the contour plots of the difference density in providing an improved understanding of chemical binding[189] is well known. It has also been possible[189,198] to partition a molecule into various subspaces corresponding to different atoms, through the behaviour of the $\nabla\rho$ field; the virial fragments are separated by dividing surfaces having a zero flux as defined by $\nabla\rho(\mathbf{r})\cdot\mathbf{n}(\mathbf{r})=0$. Consideration of the topological aspects of molecular charge distribution leads to precise definitions and mechanics of the atoms in molecules, as has been established by the pioneering works of Bader[199] (see also ref. 200). The next important quantity which has also been highlighted by Bader[201] is the Laplacian of the charge density, $\nabla^2\rho$. A theory of chemical reactivity has been developed[187,201] based on this quantity.

One of the most widely employed quantities derived from the charge density is the electrostatic potential (ref. 202) $\phi(\mathbf{r})$ defined as

$$\phi(\mathbf{r}) = -\sum_{\alpha} \frac{Z_\alpha}{|\mathbf{r}-\mathbf{R}_\alpha|} + \int \frac{\rho(\mathbf{r}')}{|\mathbf{r}-\mathbf{r}'|}\,d\mathbf{r}'. \tag{8.129}$$

The Hohenberg–Kohn theorem[121] regarding the uniqueness of the potential obtained from density or, alternatively, the cusp condition, i.e.

$$\left.\frac{d\rho_{av}}{dr}\right|_{\mathbf{r}\to\mathbf{R}_\alpha} = -2Z_\alpha \rho_{av}(\mathbf{R}_\alpha), \tag{8.130}$$

gives $\mathbf{R}_\alpha$ from the position of the cusp and $Z_\alpha$ from eqn (8.130). Therefore, $\phi(\mathbf{r})$ is entirely defined through the density. Various inter-relationships studied by Politzer[203] on $\phi(\mathbf{r})$ for atoms are interesting (see also Chapter 3).

The first term of eqn (8.129), the bare-nuclear potential, can be thought to reflect density only. This reflection occurs from the close parallelism observed between the contours of density and that of the bare-nuclear potential as studied by Parr and Berk[204] as well as Tal *et al.*[205]

A generalization of the electrostatic potential leads to the electrostatic stress tensor first defined by Deb and Bamzai.[22,91,206] The importance of this stress tensor as well as of the comprehensive stress tensor[23] including the kinetic and exchange-correlation contributions has been reviewed in refs 2 and 207. The stress concept is a generalization of the force concept[140] and leads directly to the QFD viewpoint.

Among the various other property densities obtainable from the density are the kinetic energy density,[189] nuclear magnetic shielding density,[208] polarizability density[209] and pair polarizabilities.[210,211]

So far, the discussion in this Section has been only in terms of the charge density. A detailed topological analysis of the current density in molecules induced by magnetic fields, has been presented by Gomes.[164,212–215] Apart from a clear view of the distribution of this vector field in the molecule, his analysis is helpful for understanding the contribution of different regions of a molecule to its overall magnetic properties. Gomes has discussed in detail the singularities, vortices and other topological characteristics of the current density. An interconnection between the topological analysis of $\rho$ and $\mathbf{j}$ is also worth looking for.

### 8.6.3 A "local thermodynamics" of the electron fluid

In order to acquire a deeper understanding of the analogy between the electron cloud and a classical fluid it is necessary to look for other equations of classical fluid theory corresponding to the electron fluid. This might also provide alternative practical schemes for electronic structure calculations.[26]

The difficulty in formulating this extension lies in the fact that quite a few equations for a classical fluid involve temperature, whereas one normally ascribes a zero temperature to a pure microscopic quantum system. However, even at this zero external temperature, the electron is moving and has a kinetic energy. This kinetic energy, which is of purely quantum origin, has led to the definition of an electronic temperature,[25] which is local (varying in space) and internal.

An analogous effective temperature was defined first by Born and Green[89] and subsequently by Mazo and Kirkwood[99] for a macroscopic quantum fluid. It was shown that a quantum fluid at a temperature $T$ behaves like a classical fluid of the same density but at a different effective temperature $\tau$ (dependent on $\rho$ and $T$), which in general may be position-dependent. In the same spirit, Ghosh *et al.*[25,26] have shown that a pure microscopic quantum system at zero external temperature is analogous to a classical fluid at a local temperature $T(\mathbf{r})$. The concept of this electronic temperature is consistent with the terminology "electron gas" used in the literature, which was coined by Fermi[216] for atomic electrons in his first paper on the statistical theory for atoms. This temperature shows a close connection with the quantum potential of Bohm[12] and the imaginary velocity field of Hirschfelder *et al.*[37]

The concept of a local temperature has made it possible to formulate a local "thermodynamics" for the electron gas. The motivation for this "thermodynamic" viewpoint can be further illustrated in terms of the inherent similarities between a classical (large-$N$) system and a quantum few-particle system. Although for classical particles, the exact equation of motion is known and is given by Newton's laws, the fact that the number of particles involved is very large makes it prohibitive to solve the equations of motion for the individual particles. Also, so much detailed information about the coordinates and momenta of each particle may be unnecessary. On the other hand, the classical Liouville equation which governs the probability distribution function in phase space might suffice. In equilibrium, the system is characterized by a Maxwellian velocity distribution function corresponding to a temperature $T$. The latter together with other quantities like pressure and chemical potential govern the "thermodynamic" description of the system.

It appears that such a "thermodynamic" picture of a quantum system based on an analogy with the classical situation might be important. Therefore, in this Section, we discuss a quantum "thermodynamics" of many-electron systems using the concept of a local space-dependent temperature. Note, however, that in order to describe the *dynamics* of an evolving system, one should invoke local space–time-dependent "thermodynamic" observables. Although this approach will not be discussed here, an attempt has been made by Deb and Chattaraj[34] to explain high-energy ion–atom collisions through a parallel local "thermodynamics" over a pulsating (time-dependent) potential energy surface. For other developments, see de Broglie[217] and Beretta *et al.*[218, 219]

The concept of temperature plays a crucial role in any thermodynamic description. Different definitions of temperature stem essentially from employing different energy components for the definition, the common feature being that the temperature is a parameter defined so as to characterize a distribution function. The usual thermodynamic temperature determines the Maxwellian distribution for the translational motion of atoms and molecules. When this temperature approaches zero, all classical motions cease. However, even at this zero external temperature, a quantum particle is not at rest and is associated with a kinetic energy. In the present context, the local electronic temperature within an atom or molecule is defined[25] through the local kinetic-energy density associated with the electron distribution. This local temperature characterizes a phase-space distribution which is very close to the exact one and provides an effective means for extracting the momentum density[27] from the electron density in position space, for obtaining the density matrix and hence the exchange energy.[28]

The concept of this electron temperature has opened up the possibility of introducing other equations of classical fluid theory for the electron fluid.

The equations of interest are Ornstein–Zernike (OZ), Yvon, Percus-Yevick (PY), hypernetted-chain (HNC), Wertheim–Lovett–Mou–Buff (WLMB) equations, equation of state, compressibility equation, etc. These equations are valid only in equilibrium situations. Developments for non-equilibrium situations, although of much interest, are still awaited. In view of the fact that the continuity and Euler equations govern the time evolution of density and the current density, a transcription into non-equilibrium thermodynamics would certainly be useful.

Below, the quantum system is treated as a classical fluid subject to a local temperature gradient. The quantum mechanical nature of the problem is thus transferred to the quantum temperature and a purely classical analogy prevails. Using this "thermodynamic" treatment, one can calculate Compton profiles and exchange energies as well as derive certain integral equations for the electron fluid.

Consider the ground state of an $N$-electron system characterized by an external potential $v(\mathbf{r})$ for which the energy functional is given by

$$E[\rho]=T_s[\rho]+E_{\text{pot}}[\rho]=\int t_s(\mathbf{r})\,d\mathbf{r}+\int \varepsilon_{\text{pot}}(\mathbf{r})\,d\mathbf{r}, \tag{8.131}$$

where $T_s[\rho]$ is the kinetic energy functional for an $N$-particle system of non-interacting particles of the same density $\rho(\mathbf{r})$.

Let us take the various density functionals to be the results of ensemble averages over the dynamical variables. The underlying phase-space distribution function $f(\mathbf{r}, \mathbf{p})$ must then satisfy the following properties*:

$$\left.\begin{aligned} \rho(\mathbf{r})=\int f(\mathbf{r},\mathbf{p})\,d\mathbf{p}; \quad \int \rho(\mathbf{r})\,d\mathbf{r}=N; \\ t_s(\mathbf{r})=\int \tfrac{1}{2}p^2 f(\mathbf{r},\mathbf{p})\,d\mathbf{p}. \end{aligned}\right\} \tag{8.132}$$

It is well known[93] that the phase-space function is non-unique. However, the least-biased $f(\mathbf{r}, \mathbf{p})$ can be obtained by maximizing the entropy (e.g. in an information theoretic[220] sense) defined as

$$S(\mathbf{r})=-k\int d\mathbf{p}\, f[\ln f-1]; \quad S=\int S(\mathbf{r})\,d\mathbf{r}, \tag{8.133}$$

subject to the constraints of correct density as well as kinetic energy density

*In a general situation, the kinetic energy density $t_s(\mathbf{r})$ will also contain a term[34] involving the current density. This term vanishes for the ground state and for those excited states which are static stationary states.

(eqn (8.132)) and implemented through position-dependent Lagrange multipliers. Ghosh *et al.*[25] have obtained

$$f(\mathbf{r}, \mathbf{p}) = \frac{\rho(\mathbf{r})}{[2\pi k T(\mathbf{r})]^{3/2}} \exp\left[\frac{-p^2}{2kT(\mathbf{r})}\right], \tag{8.134}$$

where $T(\mathbf{r})$ is a local electronic temperature defined in terms of the kinetic energy density in analogy with the ideal gas law, namely

$$t_s(\mathbf{r}) = \tfrac{3}{2}\rho(\mathbf{r})kT(\mathbf{r}), \tag{8.135}$$

with $k$ the Boltzmann constant.

The entropy density $S(\mathbf{r})$ can be written in terms of a Sackur–Tetrode equation:

$$S(\mathbf{r}) = -k\rho \ln \rho + \tfrac{3}{2}k\rho \ln T(\mathbf{r}) + \tfrac{1}{2}k\rho[5 + 3 \ln (2\pi k)], \tag{8.136}$$

which can alternatively be expressed as indicating a measure of the departure of $t_s(\mathbf{r})$ from the Thomas–Fermi kinetic energy density, i.e.

$$S(\mathbf{r}) = \tfrac{3}{2}k\rho[\text{constant} + \ln (t_s/t_{TF})] \tag{8.137}$$

It has also been shown[25] that the local pressure is given by the ideal-gas equation:

$$\tilde{p}(\mathbf{r}) = \rho k T(\mathbf{r}), \tag{8.138}$$

and a free-energy density has also been defined.

The study of these "thermodynamic" quantities is likely to be useful for the study of chemical binding. The quantity which has already been employed[27,28] is the phase-space distribution function of eqn (8.134). The Maxwellian distribution, characterized by the local temperature $T(\mathbf{r})$, strengthens the classical analogy.

An interesting observation about the local temperature is that it vanishes at infinity as is clear from the expression for hydrogenic atoms ($Z \equiv$ nuclear charge),

$$T(\mathbf{r}) = \frac{Z}{3kr}, \tag{8.139}$$

obtained by employing the following definition for the local kinetic-energy density:

$$t_s(\mathbf{r}) = \frac{1}{8}\sum_i \frac{\nabla\rho_i \cdot \nabla\rho_i}{\rho_i} - \frac{1}{8}\nabla^2\rho. \tag{8.140}$$

An interesting study will be the extension to a finite temperature formalism and to look at the asymptotic behaviour of $T(\mathbf{r})$. The expression (8.140) has been confirmed through the positivity of the kinetic-energy density,[25] the results of Compton profiles,[27] exchange energies[28] and through an expansion of the density matrix.[221]

The $f(\mathbf{r}, \mathbf{p})$ of eqn (8.134) gives a prescription for obtaining the momentum density by the relation

$$\chi(\mathbf{p}) = \int f(\mathbf{r}, \mathbf{p})\, d\mathbf{r}, \tag{8.141}$$

and hence the Compton profile (spherically averaged) from

$$J(q) = \frac{1}{2}\int \frac{\chi(\mathbf{p})}{p}\, d\mathbf{p} = (2\pi)^{-1/2}\int d\mathbf{r}\, \beta(\mathbf{r})^{1/2}\rho(\mathbf{r})\exp[-\tfrac{1}{2}\beta(\mathbf{r})q^2], \tag{8.142}$$

$$\beta(\mathbf{r}) = \frac{1}{kT(\mathbf{r})},$$

within the impulse approximation.[222] The result for the hydrogen atom is

$$J(q) = Z^{-1}\left(\frac{675}{1024}\right)^{1/2}\left[1 + \frac{3}{4}\left(\frac{q}{Z}\right)^2\right]^{-7/2}, \tag{8.143a}$$

which compares favourably (although not identically) with the exact result:

$$J(q) = Z^{-1}\left(\frac{8}{3\pi}\right)\left[1 + \left(\frac{q}{Z}\right)^2\right]^{-3}. \tag{8.143b}$$

Parr *et al.*[27] have also suggested an extension of the definition of the local temperature to the individual components:

$$\beta_x(\mathbf{r}) = [kT_x(\mathbf{r})]^{-1}; \quad \tfrac{1}{2}\rho(\mathbf{r})kT_x(\mathbf{r}) = t_x(\mathbf{r}), \quad \text{etc.} \tag{8.144}$$

By imposing the constraints of yielding the correct components of kinetic energy densities, entropy maximization yields

$$f(\mathbf{r}, \mathbf{p}) = \frac{[\beta_x(\mathbf{r})\beta_y(\mathbf{r})\beta_z(\mathbf{r})]^{1/2}}{(2\pi)^{3/2}}\rho(\mathbf{r})\exp[-\tfrac{1}{2}\beta_x p_x^2 - \tfrac{1}{2}\beta_y p_y^2 - \tfrac{1}{2}\beta_z p_z^2]. \tag{8.145}$$

The directional Compton profile is defined as

$$J_{\mathbf{k}}(q) = \int d\mathbf{p}\,\chi(\mathbf{p})\,\delta(q - \mathbf{p}\cdot\mathbf{k}), \tag{8.146}$$

where $\mathbf{k}$ is a unit vector along the direction in which the Compton profile is to be evaluated. For diatomic molecules, the parallel ($z$-direction) and the perpendicular components of Compton profiles are given by

$$\left.\begin{aligned} J_{\|}(q) &= (2\pi)^{-1/2} \int \mathrm{d}\mathbf{r}\, \beta_{\|}(\mathbf{r})^{1/2} \rho(\mathbf{r}) \exp\left[-\tfrac{1}{2}\beta_{\|}(\mathbf{r})q^2\right], \\ J_{\perp}(q) &= (2\pi)^{-1/2} \int \mathrm{d}r\, \beta_{\perp}(r)^{1/2} \rho(r) \exp\left[-\tfrac{1}{2}\beta_{\perp}(r)q^2\right]. \end{aligned}\right\} \tag{8.147}$$

Parr *et al.*[27] have presented numerical results on the noble-gas atoms as well as the molecules $H_2$ and $N_2$, which show excellent agreement with other standard results. Physical intuition would, of course, suggest that $\rho(\mathbf{r})$ and $t_s(\mathbf{r})$ should contain considerable information about the momentum space. Nevertheless, this method provides an effective prescription for density functional calculation of Compton profiles.

Another direct application of the phase-space function of eqn (8.144) is the calculation of the first-order density matrix and hence the exchange energy.[28] Using the Fourier transform relation

$$\rho_1(\mathbf{r}+\tfrac{1}{2}\mathbf{s}, \mathbf{r}-\tfrac{1}{2}\mathbf{s}) = \int f(\mathbf{r}, \mathbf{p}) \exp(\mathrm{i}\mathbf{p}\cdot\mathbf{s})\, \mathrm{d}\mathbf{p} \tag{8.148a}$$

and $f(\mathbf{r}, \mathbf{p})$ of eqn (8.134), one obtains the density matrix, $\rho_1$, given by

$$\rho_1(\mathbf{r}+\tfrac{1}{2}\mathbf{s}, r'-\tfrac{1}{2}\mathbf{s}) = \rho(\mathbf{r}) \exp\left[-s^2/2\beta(\mathbf{r})\right] \tag{8.148b}$$

and the exchange energy $K[\rho]$ as

$$K[\rho] = \frac{\pi}{2} \int \rho^2(\mathbf{r})\beta(\mathbf{r})\, \mathrm{d}\mathbf{r}. \tag{8.149}$$

The exchange hole corresponding to these relations has been shown[28] to be vastly improved over the local density approximation result. The numerical results of Ghosh and Parr[28] on the $K[\rho]$ values are also promising.

An interesting observation is that if $\beta(\mathbf{r})$ is calculated from the kinetic-energy density relation ot Thomas–Fermi theory, eqn (8.149) yields the well-known local density relation multiplied by a numerical factor ($\frac{10}{9}$) which for the exchange potential gives a value of the parameter $\alpha = 0.74$. This is within the range of $\alpha = 0.77$ for He and $\alpha = 0.70$ for Xe in $X\alpha$ theory. Further, by employing gradient corrections to the kinetic-energy density for evaluating $\beta(\mathbf{r})$ one obtains new relations for gradient expansion of the exchange energy.

Equation (8.149) is in essence a new non-local exchange functional, but with the whole effect of non-locality packed into the local temperature. It may be employed in self-consistent calculations without serious difficulty, in contrast with the gradient expansion, which suffers from divergence difficulties (for recent exchange energy functionals, see ref 223).

The concept of the local temperature described above is reminiscent of an effective kinetic temperature defined for a quantum fluid by Born and Green[89] and Mazo and Kirkwood.[99] Ghosh *et al.*[25, 26] have shown that the density of a many-electron system can be written in terms of a "classical" formula

$$\rho(\mathbf{r}) = [\beta(\mathbf{r})/2\pi]^{-3/2} \exp[\beta(\mathbf{r})\{\mu - v_{\text{eff}}(\mathbf{r})\}], \tag{8.150}$$

where $v_{\text{eff}}$ is an effective potential[26] related to the Kohn–Sham potential and the density functional version of the quantum potential $V_q = \delta T_s/\delta\rho$.

Defining an intrinsic potential

$$u(\mathbf{r}) \equiv \frac{\delta F}{\delta \rho} = \mu - v_{\text{ext}}(\mathbf{r}), \tag{8.151}$$

and employing eqn (8.150), Ghosh and Berkowitz[26] have defined the correlation function $h(\mathbf{r}, \mathbf{r}')$ by

$$\frac{\delta\rho(\mathbf{r})}{\delta u(\mathbf{r}')} = \beta(\mathbf{r}')\rho(\mathbf{r})[\delta(\mathbf{r}-\mathbf{r}') + \rho(\mathbf{r}')h(\mathbf{r}, \mathbf{r}')] \tag{8.152}$$

and also the direct correlation function $C(\mathbf{r}, \mathbf{r}')$ by

$$\frac{\delta u(\mathbf{r})}{\delta\rho(\mathbf{r}')} = \frac{1}{\beta(\mathbf{r})}\left[\frac{\delta(\mathbf{r}-\mathbf{r}')}{\rho(\mathbf{r})} - C(\mathbf{r}, \mathbf{r}')\right]. \tag{8.153}$$

Making use of the identity

$$\int \frac{\delta\rho(\mathbf{r})}{\delta u(\mathbf{r}'')} \frac{\delta u(\mathbf{r}'')}{\delta\rho(\mathbf{r}')} \,\mathrm{d}\mathbf{r}'' = \delta(\mathbf{r}-\mathbf{r}'), \tag{8.154}$$

the Ornstein–Zernike equation follows

$$h(\mathbf{r}, \mathbf{r}') = C(\mathbf{r}, \mathbf{r}') + \int \mathrm{d}\mathbf{r}''\, h(\mathbf{r}, \mathbf{r}'')\rho(\mathbf{r}'')\, C(\mathbf{r}'', \mathbf{r}'). \tag{8.155}$$

The Yvon equation has also been derived,[26]

$$\Delta\rho(\mathbf{r}) = \beta(\mathbf{r})\rho(\mathbf{r})\Delta u(\mathbf{r}) + \rho(\mathbf{r})\int \beta(\mathbf{r}')h(\mathbf{r}, \mathbf{r}')\Delta u(\mathbf{r}')\,\mathrm{d}\mathbf{r}', \tag{8.156}$$

which gives a measure of the density response corresponding to a potential variation $\Delta u(\mathbf{r}')$.

The WLMB integrodifferential equations for the electron fluid, which correspond to the first member of the BBGKY hierarchy for classical fluids, can be written as

$$\beta(\mathbf{r})\nabla u(\mathbf{r}) = \nabla \ln \rho(\mathbf{r}) - \int C(\mathbf{r}, \mathbf{r}')\nabla'\rho(\mathbf{r}')\,\mathrm{d}\mathbf{r}'$$

$$= \nabla \ln \rho(\mathbf{r}) + \int \nabla' C(\mathbf{r}, \mathbf{r}')\rho(\mathbf{r}')\, d\mathbf{r}' \qquad (8.157a)$$

and

$$\nabla \ln \rho(\mathbf{r}) = \beta(\mathbf{r})\nabla u(\mathbf{r}) + \int h(\mathbf{r}, \mathbf{r}')\rho(\mathbf{r}')\beta(\mathbf{r}')\nabla' u(\mathbf{r}')\, d\mathbf{r}'. \qquad (8.157b)$$

These provide a method for the calculation of the density distribution, an alternative to the Euler equation for density[67] in DFT.

A compressibility equation (called the "local inverse compressibility equation") has also been obtained:

$$N\left(\frac{\delta\mu}{\delta\rho}\right)_{v(\mathbf{r})} = \frac{1}{\beta(\mathbf{r})}\left[1 - \int \rho(\mathbf{r}')\, C(\mathbf{r}, \mathbf{r}')\, d\mathbf{r}'\right], \qquad (8.158)$$

where $(\delta\mu/\delta\rho)$ is the local hardness[30] defined in Section 8.6.1 and can be considered as the inverse of "local compressibility".

The local density functional equation of state is given by

$$\tilde{p}(\mathbf{r}) = \rho k T(\mathbf{r}) - \tfrac{1}{3}\mathrm{Tr}\, \sigma_{xc}, \qquad (8.159a)$$

where $\sigma_{xc}$ is defined by

$$\nabla \cdot \sigma_{xc} = -\rho(\mathbf{r})\nabla\left(\frac{\delta E_{xc}}{\delta\rho}\right) \qquad (8.159b)$$

through the exchange-correlation energy $E_{xc}$. The pressure $\tilde{p}(\mathbf{r})$ of eqn (8.159a) is defined[24] in terms of the universal functional $G[\rho]$:

$$\tilde{p}(\mathbf{r}) = \tfrac{1}{3}\mathrm{Tr}\, \sigma; \quad \nabla \cdot \sigma = -\rho\nabla\frac{\delta G}{\delta\rho}. \qquad (8.159c)$$

The virial equation of state for only the exchange-correlation forces is

$$3\int \tilde{p}_{xc}(\mathbf{r})\, d\mathbf{r} = -\int \mathbf{r}\cdot \sigma_{xc}\, d\mathbf{r} + 2T_{xc} + U_{xc}, \qquad (8.159d)$$

where $E_{xc} = T_{xc} + U_{xc}$, with $T_{xc}$ and $U_{xc}$ being the kinetic and the potential contributions respectively to exchange-correlation energy. Equation (8.159d) has been obtained by using the recent result of Ghosh and Parr,[224] namely

$$\int \rho\mathbf{r}\cdot\nabla\frac{\delta E}{\delta\rho}\, d\mathbf{r} = -lE, \qquad (8.160)$$

where $l = 2$ for the kinetic and 1 for the potential energy contribution.

Through this correlation function formalism in terms of local temperature various integral equations for density can be derived. Although OZ and WLMB equations give density, one has to introduce, for practical calculations, some approximations' owing to the lack of exact knowledge of $C$, $h$ and $\beta$ as functionals of density. Using functional integration techniques,[225-229] the following equations which correspond to the HNC and PY approximations respectively, as applied to classical inhomogeneous fluids, can be obtained. Thus, one has

$$\rho(\mathbf{r})=\rho_0(\mathbf{r})\exp\left[\beta(\mathbf{r})\{u(\mathbf{r})-u_0(\mathbf{r})\}\right]$$
$$\times\exp\left[\beta(\mathbf{r})\int d\mathbf{r}'\{\bar{C}_h(\mathbf{r}-\mathbf{r}';\bar{\rho})-\bar{K}_h(\mathbf{r},\mathbf{r}';\bar{\rho})\}\{\rho(\mathbf{r}')-\rho_0(\mathbf{r}')\}\right], \quad (8.161a)$$

which on linearization yields

$$\rho(\mathbf{r})=\rho_0(\mathbf{r})\exp\left[\beta(\mathbf{r})\{u(\mathbf{r})-u_0(\mathbf{r})\}\right]$$
$$\times\left[1+\beta(\mathbf{r})\int d\mathbf{r}'\{\bar{C}_h(\mathbf{r}-\mathbf{r}';\bar{\rho})-\bar{K}_h(\mathbf{r},\mathbf{r}';\bar{\rho})\}\{\rho(\mathbf{r}')-\rho_0(\mathbf{r}')\}\right]. \quad (8.161b)$$

Here $\bar{C}_h$ and $\bar{K}_h$ are the homogeneous gas results for $C(\mathbf{r},\mathbf{r}')$ and $K(\mathbf{r},\mathbf{r}')$, with

$$C(\mathbf{r},\mathbf{r}')=\int_0^1 d\alpha\tilde{C}(\mathbf{r},\mathbf{r}';\alpha)=\int_0^1 d\alpha\frac{C(\mathbf{r},\mathbf{r}';\alpha)}{\beta(\mathbf{r};\alpha)} \quad (8.162a)$$

and

$$K(\mathbf{r},\mathbf{r}')=\int_0^1 d\alpha\frac{\ln[\rho_\alpha(\mathbf{r})/\rho_0(\mathbf{r})]}{\beta(\mathbf{r};\rho_\alpha)^2}\frac{\delta\beta(\mathbf{r};\rho_\alpha)}{\delta\rho_\alpha(\mathbf{r}')}, \quad (8.162b)$$

where $\alpha$ is the parameter characterizing the path of functional integration, i.e. $\rho_\alpha\equiv\rho(\mathbf{r};\alpha)$, which is $\rho_0(\mathbf{r})$ at $\alpha=0$ and $\rho(\mathbf{r})$ at $\alpha=1$; $\bar{\rho}$ is an average density which can be $\rho(\frac{1}{2}(\mathbf{r}+\mathbf{r}'))$ or $\frac{1}{2}(\rho(\mathbf{r})+\rho(\mathbf{r}'))$. Ghosh and Berkowitz[26] have reformulated these approximations for $C$, $\beta$ and $K$. They have also introduced the spin-polarized version of these integrodifferential equations by employing the spin-polarized version of DFT and considering the electron fluid to be a fluid mixture of two-component fluids corresponding to the two spins, analogous in spirit to the work of Schönberg.[145]

The works discussed in this Section advance the fluid dynamical analogy, suggest a self-consistent scheme for density calculation, introduce an equation of state and also relate hydrodynamical variables to other quantities of chemical significance. The basic underlying philosophy has been to formulate a classical-fluid-like and "thermodynamic" description of a microscopic quantum system. Interpretive problems associated with quantum systems have been debated from time to time. There have also been attempts to unify microphysics and macrophysics.[80] Bohm and Hiley[230] have employed the quantum potential interpretation of QFD to show a continuity in the perception of reality between the quantum and classical worlds. Other approaches relating the quantum description of the micro-world to the classical description of matter in bulk have been discussed by George *et al.*[231]

## 8.7 The stochastic route to QFD

The apparent "radical" departure of the concepts involved in quantum mechanics from those of classical physics prompted many scientists to search for a classical interpretation of quantum phenomena. The QFD formalism is one outcome of such attempts; here, quantum mechanics has been shown to be equivalent to two fluid-dynamical equations provided the effect of an additional quantum force arising from a quantum potential[12] is incorporated. Originally, the quantum potential was thought of as arising from an internal stress in the fluid; subsequently, Bohm and Vigier[8] attempted a causal interpretation by assuming the electron fluid to undergo random fluctuations about the Madelung motion as the mean, where the fluctuations were assumed to arise from interaction with a sub-quantum medium. Several authors[3] attempted to propose that instead of assuming a quantum mechanical potential, the motion could be understood in terms of a Markov process. This finally led[9,10] to a classical "derivation" and interpretation of the Schrödinger equation through the theory of stochastic processes by assuming the particle to undergo a Brownian motion under the action of an "external" force and a stochastic force.

Such a stochastic mechanical equivalent of quantum mechanics was first proposed by Schrödinger himself (see ref. 232) and thereafter has been formulated in various forms by different authors; there is a vast literature on this subject.[3,9,10,20,21,233–241] There have also been criticisms of this approach.[239,240]

The formulation of the stochastic approach of Nelson[9,10] provides an essentially straightforward particle interpretation of the Madelung fluid. The question as to whether or not this underlying particle picture[8] has a physical existence is apparently not settled; however, what has been clarified is that the underlying stochastic process shows a direct equivalence to the Schrödinger equation. Since there are excellent reviews[3,11,20,21,237] on this subject, we shall present only a brief outline of the stochastic approach as formulated by Nelson.

It is assumed that the quantum mechanical motion of a particle in a potential $v(\mathbf{x}, t)$ can be described by a continuous Markov process with constant isotropic diffusion $Q$. Considering $\mathbf{x}(t)$ to be a random variable, one can write

$$\mathrm{d}\mathbf{x}(t)=\boldsymbol{\mu}(\mathbf{x},t)\,\mathrm{d}t+\mathrm{d}\mathbf{w}(\mathbf{x},t) \tag{8.163a}$$

and

$$\mathrm{d}\mathbf{x}(t)=\boldsymbol{\mu}_*(\mathbf{x},t)\,\mathrm{d}t+\mathrm{d}\mathbf{w}_*(\mathbf{x},t) \tag{8.163b}$$

for the forward and backward descriptions respectively, where $\mathrm{d}\mathbf{w}$ is a random variable normally distributed with mean zero and variance $\sigma^2\,\mathrm{d}t(\mathrm{Q}=\sigma^2)$ and $\boldsymbol{\mu}(\mathbf{x},t)$ is the drift.

If $\rho(\mathbf{x},t)$ is the probability density associated with $\mathbf{x}(t)$, the forward and backward Fokker–Planck equations give, respectively,

$$\frac{\delta\rho(\mathbf{x},t)}{\partial t}=-\mathrm{div}\,[\rho(\mathbf{x},t)\,\boldsymbol{\mu}(\mathbf{x},t)]+\frac{1}{2}Q\nabla^2\rho(\mathbf{x},t) \tag{8.164a}$$

and

$$\frac{\partial\rho(\mathbf{x},t)}{\partial t}=-\mathrm{div}\,[\rho(\mathbf{x},t)\boldsymbol{\mu}_*(\mathbf{x},t)]-\frac{1}{2}Q\nabla^2\rho(\mathbf{x}\,t). \tag{8.164b}$$

Addition of the two equations (8.164a) and (8.164b) leads to the continuity equation

$$\frac{\partial\rho}{\partial t}=-\mathrm{div}\,[\rho\mathbf{v}], \tag{8.165}$$

where the velocity field $\boldsymbol{v}$ of the probability current is defined as

$$\boldsymbol{v}=\tfrac{1}{2}(\boldsymbol{\mu}+\boldsymbol{\mu}_*). \tag{8.166}$$

Now the forward and backward derivatives of a function $f(\mathbf{x},t)$ are given by

$$\mathrm{D}f(\mathbf{x},t)=\left[\frac{\partial}{\partial t}+\boldsymbol{\mu}(\mathbf{x},t)\cdot\boldsymbol{\nabla}+\frac{1}{2}Q\nabla^2\right]f(\mathbf{x},t), \tag{8.167a}$$

$$\mathrm{D}_*f(\mathbf{x},t)=\left[\frac{\partial}{\partial t}+\boldsymbol{\mu}_*(\mathbf{x},t)\cdot\boldsymbol{\nabla}-\frac{1}{2}\mathrm{Q}\nabla^2\right]f(\mathbf{x},t). \tag{8.167b}$$

Using the relation between $\boldsymbol{\mu}$ and $\boldsymbol{\mu}_*$,

$$\boldsymbol{\mu}_*(\mathbf{x},t)=\boldsymbol{\mu}(\mathbf{x},t)-\mathrm{Q}\frac{\nabla\rho(\mathbf{x},t)}{\rho(\mathbf{x},t)}, \tag{8.168a}$$

the stochastic velocity defined as

$$\mathbf{u}(\mathbf{x},t)=\tfrac{1}{2}[\boldsymbol{\mu}(\mathbf{x},t)-\boldsymbol{\mu}_*(\mathbf{x},t)] \tag{8.168b}$$

can be rewritten as

$$\mathbf{u}(\mathbf{x}, t) = \frac{Q}{2}\frac{\nabla\rho}{\rho} \equiv \frac{Q}{2}\nabla \ln \rho(\mathbf{x}, t). \tag{8.168c}$$

This velocity $\mathbf{u}$ is the velocity acquired by the Brownian particle in equilibrium with respect to an external force; it balances the osmotic force and hence has been called the *osmotic velocity* by Nelson.[9]

Defining the mean acceleration as

$$\mathbf{a} = \tfrac{1}{2}(\mathrm{DD}_* + \mathrm{D}_*\mathrm{D})\,\mathbf{x}(t) \tag{8.169a}$$

and using the relations

$$\mathrm{D}\mathbf{x}(t) = \mu(\mathbf{x}, t); \quad \mathrm{D}_*\mathbf{x}(t) = \mu_*(\mathbf{x}, t), \tag{8.169b}$$

together with eqn (8.167a) and (8.167b) for $f = \mu_*$ and $\mu$, respectively, one obtains the mean acceleration as

$$\mathbf{a} = \frac{\partial \mathbf{v}}{\partial t} + (\mathbf{v}\cdot\nabla)\mathbf{v} + (\mathbf{u}\cdot\nabla)\mathbf{u} + \frac{Q}{2}\nabla^2\mathbf{u}. \tag{8.170}$$

Now, imposing the dynamical condition that Newton's law holds in the form

$$\mathbf{F} = m\mathbf{a},$$

where $\mathbf{F}$ is the external force, one can write from eqn (8.170)

$$m\left[\frac{\partial \mathbf{v}}{\partial t} + (\mathbf{v}\cdot\nabla)\mathbf{v}\right] = \mathbf{F} + m\left(\frac{Q}{2}\nabla^2\mathbf{u} + (\mathbf{u}\cdot\nabla)\mathbf{u}\right). \tag{8.171}$$

Now, choosing the diffusion constant to be given by $Q = \hbar/m$, and hence $\mathbf{u} = (\hbar/2m)(\nabla\rho/\rho)$, eqn (8.171) becomes

$$\begin{aligned} m\left[\frac{\partial \mathbf{v}}{\partial t} + (\mathbf{v}\cdot\nabla)\mathbf{v}\right] &= \mathbf{F} + m\nabla\left(\frac{Q}{2}\nabla\cdot\mathbf{u} + \frac{1}{2}\mathbf{u}^2\right) \\ &\mathbf{F} + \frac{\hbar^2}{4m}\nabla\left[\nabla\cdot\left(\frac{\nabla\rho}{\rho}\right) + \frac{1}{2}\left(\frac{\nabla\rho}{\rho}\right)^2\right] \\ &= \mathbf{F} - \nabla\left[\frac{\hbar^2}{8m}\frac{\nabla\rho\cdot\nabla\rho}{\rho^2} - \frac{\hbar^2}{4m}\frac{\nabla^2\rho}{\rho}\right]. \end{aligned} \tag{8.172}$$

If we identify $\rho(\mathbf{x}, t)$ with the correct quantum mechanical probability density $(=|\psi|^2)$ and $\mathbf{v}(\mathbf{x}, t)$ with the correct velocity field $\mathbf{v}$, then the continuity equation (8.165) and eqn (8.172) are identical with the QFD equations (8.3) and (8.4). The external force **F** arises from the gradient of the Coulomb potential and the last bracketted quantity in eqn (8.172) represents the quantum potential. The osmotic velocity of eqn (8.168b) is the same as the imaginary velocity field of eqn (8.9). It is easy to show that the Schrödinger equation follows from eqn (8.165) and (8.172) in configuration space. In the same spirit, Heisenberg's uncertainty relations are implied[242] by certain inequalities relating the mean square deviations of position and osmotic velocity.

Thus, we see that in Nelson's derivation the quantum potential term appears naturally, and the QFD equations follow. Since the QFD equations are equivalent to the Schrödinger equation, this gives a classical "derivation" of the Schrödinger equation. Various other considerations like stochastic interpretation and quantum indeterminism as well as quantum separability have been reviewed by Ghirardi *et al.*[20] A variational formulation of Nelson's probabilistic representation of quantum dynamics within the realm of stochastic calculus of variations has been suggested by Yasue;[238,243] he and Weaver[244] have applied the stochastic interpretation to the problem of tunnelling. The stochastic approach has been developed in phase space by Schroeck,[236] which claims to avoid the difficulties encountered in Wigner's approach.[15] A relativistic extension of the analogy between quantum mechanics and Brownian motion has been obtained by Gaveau *et al.*[245] (see also ref. 246). The generalized Schrödinger–Langevin-type equation in the hydrodynamical formulation has been investigated via this stochastic mechanics by Nassar.[247–249] Among other developments, mention may be made of the recent work[250] on introducing momentum variables into the stochastic scheme. The work on stochastic electrodynamics and quantum theory is also of interest.[251]

Nelson's approach and the claim that "the radical departure from classical physics produced by the introduction of quantum mechanics was unnecessary" has, however, been criticized.[239,240] The main objection is that the backward process introduced is not truley Markovian, although it has some properties of a Markovian process; this has been discussed by Grabert *et al.*[239] Nevertheless, Ghiradi *et al.*[20] feel that the backward process is indeed Markovian. We also refer the reader to Schrödinger's[252] original work on the stochastic approach and its interpretation by Claverie and Diner[253] in the light of Nelson's work.

The stochastic approach is comparable to the Feynman path-integral approach[254] in that both attempt a particle interpretation. The former is based on random trajectories of stochastic processes while the latter rests on deterministic trajectories of classical mechanics, the quantum features being manifested through the summation over all possible paths.

## 8.8 Quantum fluid dynamics within a relativistic density functional framework

The QFD formulation has also been extended to include relativistic effects. Based on the hydrodynamical interpretation of the single-particle Dirac equation, given by Takabayashi,[255] analogous hydrodynamics have been developed[256] for many-particle systems within the framework of relativistic DFT.[257] We start with the one-particle Dirac-like equations,

$$\{-i\hbar c\boldsymbol{\alpha}\cdot\nabla+\beta mc^2+V^{\text{eff}}[\rho(\mathbf{r},t)]+e\boldsymbol{\alpha}\cdot\mathbf{A}^{\text{eff}}[\mathbf{J}(\mathbf{r},t)]\}\psi_j(\mathbf{r},t)=i\hbar\frac{\partial\psi_j}{\partial t},\tag{8.173}$$

where

$$V^{\text{eff}}[\rho(\mathbf{r},t)]=V_{\text{ext}}(\mathbf{r},t)+e^2\int\frac{\rho(\mathbf{r}',t)}{|\mathbf{r}-\mathbf{r}'|}\,\mathrm{d}\mathbf{r}'+\frac{\delta E_{\text{xc}}[\rho,\mathbf{J}]}{\delta\rho},\tag{8.174a}$$

$$\mathbf{A}^{\text{eff}}[\mathbf{J}(\mathbf{r},t)]=\mathbf{A}^{\text{ext}}(\mathbf{r},t)+e\int\frac{J(\mathbf{r}',t)}{|\mathbf{r}-\mathbf{r}'|}\,\mathrm{d}\mathbf{r}'+\frac{1}{c}\frac{\delta E_{\text{xc}}[\rho,\mathbf{J}]}{\delta\mathbf{J}}.\tag{8.174b}$$

The particle density $\rho$ and the current density $j_k$ are defined through the four-component spinors $\psi_j(\mathbf{r},t)$ as

$$\rho(\mathbf{r},t)=\sum_j\psi_j^\dagger\psi_j;\quad j_k(\mathbf{r},t)=\sum_j\psi_j^\dagger\alpha_k\psi_j.\tag{8.175}$$

In terms of the Dirac $\gamma$ matrices, i.e. $\gamma_k=-i\beta\alpha_k$ and $\gamma_4=\beta$, eqn (8.173) can be compactly written as

$$\sum_\mu\gamma_\mu\left[\frac{\partial}{\partial x_\mu}+\frac{ie}{\hbar c}\phi_\mu^{\text{eff}}\right]\psi_j+\frac{mc}{\hbar}\psi_j=0,\tag{8.176}$$

and (8.175) becomes

$$J_0=\rho=\sum_j\bar{\psi}_j\gamma_4\psi_j;\quad J_k=i\sum_j\bar{\psi}_j\gamma_k\psi_j.\tag{8.177}$$

Here $\phi_\mu^{\text{eff}}\equiv(\phi_0^{\text{eff}},\ A_k^{\text{eff}})$ denotes the effective four-potential, where $e\phi_0^{\text{eff}}=V^{\text{eff}}$

By premultiplying eqn (8.176) by $\bar{\psi}_j\gamma^A$ and postmultiplying the conjugate of eqn (8.176) by $\gamma^A\psi_j$, followed by algebraic manipulation, one obtains the following set of independent hydrodynamical quantities ($\gamma^A$ stands for a suitable combination of $\gamma_\mu$). The hydrodynamical equations are[175] (summation over $\mu$ is implied and subscript $j$ denotes the $j$th orbital contribution):

(a) *scalar equation*

$$\partial_\mu(J_\mu)_j=0;$$

(b) *pseudoscalar equation*

$$\partial_\mu (J'_\mu)_j + (2mc/\hbar)(Q')_j = 0;$$

(c) *scalar equation*

$$(T_{\mu\mu})_j + (Q)_j = 0;$$

(d) *pseudoscalar equation*

$$(T'_{\mu\mu j} = 0;$$

(e) *vector equation*

$$\begin{aligned}
&(J_\gamma)_j \partial_\nu (J_\mu)_j - (J'_\nu)_j \partial_\nu (J'_\mu)_j \\
&= -(P)_j \partial_\mu (P)_j - (P)_j^{-2} \{(J_\mu)_j (J'_\nu)_j - (J_\nu)_j (J'_\mu)_j\} (Z_\nu)_j \\
&+ \mathrm{i}(P)_j^{-2}\, \varepsilon_{\mu\nu\kappa\lambda} \times (J_\nu)_j (J'_k)_j (K_\lambda)_j;
\end{aligned}$$

(f) *pseudovector equation*

$$\begin{aligned}
&(J_\nu)_j \partial_\nu (J'_\mu)_j - (J'_\nu)_j \partial_\nu (J_\mu)_j \\
&= -(Z_\mu)_j - (P)_j^{-1} \partial_\nu (P)_j \{(J_\mu)_j (J'_\nu)_j - (J_\nu)_j (J'_\mu)_j\} \\
&+ \mathrm{i}(P)^{-2j} \varepsilon_{\mu\nu\kappa\lambda} (J_\nu)_j (J'_k)_j (R_\lambda)_j.
\end{aligned} \tag{8.178}$$

Here each of the last two equations represents only two linearly independent equations because of the restrictions from the relationships among various $J_\mu$:

$$\begin{aligned}
(J_\mu)_j (J_\mu)_j &= -(P)_j^2, \\
(J'_\mu)_j (J'_\mu)_j &= (P)_j^2, \\
(J_\mu)_j (J'_\mu)_j &= 0,
\end{aligned}$$

where

$$(P)_j^2 = (Q)_j^2 + (Q')_j^2. \tag{8.179}$$

The various quantities appearing in eqn (8.178) are defined in ref. 256, and hence are not repeated here. However, the physical significance of the QFD quantities is mentioned below.

The quantity $(J_\mu)_j$ is the four-current vector corresponding to the $j$th spinor; this on multiplication by $c$ yields the actual current density. The electron density ($j$th spinor) is $\rho_0 = (J_0)_j = -\mathrm{i}(J_4)_j$. The net current and electron densities come from summation over $j$. The quantity $(\overline{Q}_\mu)_j$ can be interpreted as the convection current density for the $j$th spinor and $T_{\mu\nu}$ represents the energy–momentum tensor.

The quantum stress tensor and the Navier–Stokes-type equation, etc., have been discussed by Ghosh and Deb.[256] Such relativistic hydrodynamics corresponds to a "spinning" fluid. The net many-electron fluid consists of components each of which is characterized by fluid dynamical quantities corresponding to each spinor and each fluid component obeys the set of hydrodynamical equations presented. The 3D QFD picture of non-relativistic quantum mechanics thus extends its validity to the relativistic domain as well. Although the relativistic QFD equations have a complicated appearance, they are nevertheless "classical", and valid in 3D space or 4D time–space.

## 8.9 Conclusion

This Chapter has reviewed the main developments on the quantum fluid dynamical significance of the single-particle density (see also Chapters 9 and 10). These include both formal developments and applications to many-electron systems. Ever since the formulation of Schrödinger's equation, QFD has been playing an increasingly important role in the continuing search for a classical interpretation of quantum mechanics. In the case of non-relativistic quantum mechanics of many-electron systems, a two-way transcription from the former to fluid dynamics and vice versa is now possible, through a Madelung-type approach and a stochastic approach (see also ref. 258) respectively. For a pure state of the system, the polar form of the wave function is a crucial element in this transcription. However, till today, a similar two-way transcription does not seem to have been worked out for relativistic quantum mechanics (e.g. the Dirac equation). Another problem is how to incorporate Fermi–Dirac statistics into the QFD viewpoint. Further developments,* especially in the stochastic viewpoint, are awaited because there are certain unanswered questions of both philosophical and practical nature.

One of the most significant aspects of the QFD viewpoint is that it enlarges the scope and the range of applicability of DFT which, in almost all practical applications so far, has been unable to rise above the ground state or deal with general time-dependent phenomena. The interesting way of examining structures, properties and dynamics of many-electron systems through a joint QFDFT approach is beginning to take shape. These formalisms would work in 3D space, rather than a multi-dimensional space and thus retain the computational as well as interpretive advantages of density-based formalisms over the wave-function-based ones. However, a

*Recently, both Madelung's and Nelson's approaches have been interpreted through a new variational principle[259] in terms of which the wave–particle duality has been re-examined.[260]

major complication emerges in that the QFDFT equations turn out to be highly non-linear, demanding new analytical or numerical methods for their solution. However, this could also be a blessing in disguise because such non-linear equations may be ideally suited to investigate non-linear microscopic (quantum) phenomena which would otherwise be very difficult to handle.

## Acknowledgments

B.M.D. would like to thank Pratim K. Chattaraj and Dr Harjinder Singh for helpful discussions. A research grant to B.M.D. from the Council of Scientific and Industrial Research, New Delhi, is gratefully acknowledged.

## References

1. Ghosh, S.K. and Deb, B.M. (1982). *Phys. Rep.*, **92**, 1.
2. Bamzai, A.S. and Deb, B.M. (1981). *Rev. Mod. Phys.*, **53**, 96. (1981). *Ibid.*, **53**, 593.
3. Jammer, M. (1974). *The Philosophy of Quantum Mechanics*, John Wiley, New York.
4. Madelung, E. (1926). *Z. Phys.*, **40**, 332.
5. de Broglie, L. (1960). *Nonlinear Wave Mechanics*, Elsevier, Amsterdam.
6. Lochak, G. (1982). *Found. Phys.* **12**, 931.
7. Bohm, D. and Hiley, B.J. (1982). *Found. Phys.*, **12**, 1001.
8. Bohm, D. and Vigier, J.P. (1954). *Phys. Rev.*, **96**, 208.
9. Nelson, E. (1967). *Dynamical Theory of Brownian Motion*, Princeton University Press.
10. Nelson, E. (1985). *Quantum Fluctuations*, Princeton University Press.
11. Lavenda, B.H. and Santamato, E. (1984). *Int. J. Theor. Phys.*, **23**, 585.
12. Bohm, D. (1952). *Phys. Rev.*, **85**, 166, 180.
13. Takabayashi, T. (1952). *Prog. Theor. Phys.*, **8**, 143.
14. Frölich, H. (1967). *Physica*, **37**, 215.
15. Wigner, E.P. (1932). *Phys. Rev.*, **40**, 749.
16. Moyal, E. (1949). *Proc. Camb. Phil. Soc.*, **45**, 99.
17. Takabayashi, T. (1954). *Prog. Theor. Phys.* **11**, 341.
18. Hillery, M., O'Connell, R.F., Scully, M.O. and Wigner, E.P. (1984). *Phys. Rep.*, **106**, 121.
19. Lundqvist, S. (1983). In *Theory of Inhomogeneous Electron Gas*, eds Lundqvist, S. and March, N.H., p. 149, Plenum, New York.
20. Ghirardi, G.C., Omero, C., Rimini, A. and Weber, T. (1978). *Riv. Nuovo Cim.*, **1**, 1.
21. Guerra, F. (1981). *Phys. Rep.*, **77**, 263.
22. Deb, B.M. and Bamzai, A.S. (1979). *Molec. Phys.*, **38**, 2069.
23. Deb, B.M. and Ghosh, S.K. (1979). *J. Phys.*, **B12**, 3857.
24. Bartolotti, L.J. and Parr, R.G. (1980). *J. Chem. Phys.*, **72**, 1593.

25. Ghosh, S.K., Berkowitz, M. and Parr, R.G. (1984). *Proc. Natl Acad. Sci. U.S.A.*, **81**, 8028.
26. Ghosh, S.K. and Berkowitz, M. (1985). *J. Chem. Phys.*, **83**, 2976.
27. Parr, R.G., Rupnik, K. and Ghosh, S.K. (1986). *Phys. Rev. Lett.*, **56**, 1555.
28. Ghosh, S.K. and Parr, R.G. (1986). *Phys. Rev.* **A34**, 785.
29. Parr, R.G., Donnelly, R.A., Levy, M. and Palke, W.E. (1978). *J. Chem. Phys.*, **68**, 3801.
30. Berkowitz, M., Ghosh, S.K. and Parr, R.G. (1985). *J. Am. Chem. Soc.*, **107**, 6811.
31. Ghosh, S.K. and Deb, B.M. (1982). *Int. J. Quant. Chem.*, **22**, 871.
32. Deb, B.M. and Ghosh, S.K. (1982). *J. Chem. Phys.*, **77**, 342.
33. Runge, E. and Gross, E.K.U. (1984). *Phys. Rev. Lett.*, **52**, 897.
34. Deb, B.M. and Chattaraj, P.K. (to be published); see also refs. 261, 262.
35. Ghosh, S.K. and Deb, B.M. (1983). *J. Molec. Struc. Theochem.*, **103**, 163.
36. Buckingham, A.D. (1978). In *Intermolecular Interactions: From Diatomics to Polymers.* ed. B. Pullman, John Wiley, New York, p. 1.
37. Hirschfelder, J.O., Christoph, A.C. and Palke, W.E. (1974). *J. Chem. Phys.*, **61**, 5435.
38. Kan, K.K. and Griffin, J.J. (1977). *Phys. Rev.*, **C15**, 1126.
39. Ghosh, S.K. (1987). *J. Chem. Phys.* (in press).
40. Bader, R.F.W. (1985). *Acc. Chem. Res.*, **18**, 9.
41. Fernandex, F.M. and Castro, E.A. (1981). *Phys. Rev.*, **A24**, 2344.
42. Levy, M. and Perdew, J.P. (1985). *Phys. Rev.*, **A32**, 2010.
43. Wong, C.Y. (1976). *J. Math. Phys.*, **16**, 1008.
44. Bartolotti, L.J. (1981). *Phys. Rev.*, **A24**, 1661.
45. Bartolotti, L.J. (1982). *Phys. Rev.*, **A26**, 2243.
46. Li, T. and Tong, P. (1985). *Phys. Rev.*, **A31**, 1950.
47. Li, T. and Li, Y. (1985). *Phys. Rev.*, **A31**, 3970.
48. Xu, B.X. and Rajagopal, A.K. (1985). *Phys. Rev.*, **A31**, 2682.
49. Kohl, H. and Dreizler, R.M. (1986). *Phys. Rev. Lett.*, **56**, 1991.
50. Dhara, A.K. and Ghosh, S.K. (1987). *Phys. Rev*; **A35**, 442.
51. Lundqvist, S. (1975). *Int. J. Quant. Chem.*, **S9**, 23.
52. Kohn, W. and Sham, L.J. (1965). *Phys. Rev.*, **140**, A1133.
53. Deb, B.M. (1984). In *Local Density Approximations in Quantum Chemistry and Solid State Physics*, eds Dahl, J.P. and Avery, J., p. 75, Plenum, New York.
54. Levy, M., Perdew, J.P. and Sahni, V. (1984). *Phys. Rev.*, **A30**, 2745.
55. Kreuzer, H.J. (1978). *Nuovo Cim.*, **45B**, 169.
56. Starace, A.F. (1971). *Phys. Rev.*, **A3**, 1242.
57. Deb, B.M. (1981). *Current Sci.*, **50**, 973.
58. Chattaraj, P.K. and Deb, B.M. (1984). *J. Sci. Ind. Res.*, **43**, 238; see ref. 263.
59. March, N.H. (1975). *Self-Consistent Fields in Atoms*, Pergamon, New York.
60. Chattaraj, P.K., Mukherjee, A., Das, M.P. and Deb, B.M. (1986). *Proc. Ind. Acad. Sci.* (*Chem. Sci.*), **96**, 231.
61. Haq, S., Chattaraj, P.K. and Deb, B.M. (1984). *Chem. Phys. Lett.*, **111**, 79.
62. Deb, B.M. and Chattaraj, P.K. (to be published).
63. Bader, R.F.W. and Essen, H. (1984). *J. Chem. Phys.*, **80**, 1943.
64. Ghosh, S.K. and Balbas, L.C. (1985). *J. Chem. Phys.*, **83**, 5778.
65. Chattaraj, P.K. and Deb, B.M. (1985). *Chem. Phys. Lett.*, **121**, 143.
66. (a) Deb, B.M. and Chattaraj, P.K. (1986). *Theoret. Chim. Acta* (*Berl.*), **69**, 259 (1986). (b) Lee, C. and Ghosh, S.K. (1986). *Phys. Rev.*, **A33**, 3506.
67. Deb, B.M. and Ghosh, S.K. (1983). *Int. J. Quant. Chem.*, **23**, 1.
68. Chattaraj, P.K., Rao Koneru, S. and Deb, B.M. (1987). *J. Comp. Phys.* (in press).

69. Singh, H. and Deb, B.M. (1986). *Pramana J. Phys.*, **27**, 337.
70. Ghosh, S.K. and Deb, B.M. (1982). *Chem. Phys.*, **71**, 295; (1984). *Ibid.*, **91**, 478.
71. Ghosh, S.K. and Deb, B.M. (1983). *Theoret. Chim. Acta* (*Berl.*), **62**, 209; (1985). *Ibid.*, **68**, 336.
72. Soven, P. (1985). *J. Chem. Phys.*, **82**, 3289.
73. Zangwill, A. (1983). *J. Chem. Phys.*, **78**, 5926.
74. Ghosh, S.K. (1983). *Current Sci.*, **52**, 769.
75. Peuckert, V. (1978). *J. Phys.*, **C11**, 4945.
76. Levy, M. (1979). *Proc. Natl Acad. Sci* U.S.A., **76**, 6062.
77. Davidson, E.R. (1976). *Reduced Density Matrices in Quantum Chemistry*, Academic Press, New York.
78. Liboff, R.L. (1969). *Introduction to the Theory of Kinetic Equations*, John Wiley, New York.
79. Carruthers, P. and Zachariasen, F. (1983). *Rev. Mod. Phys.*, **55**, 245.
80. Fröhlich, H. (1973). *Riv. Nuovo Cim.*, **3**, 490.
81. Kreuzer, H.J. (1975). *Physica*, **80A**, 585.
82. Kreuzer, H.J. (1981). *Nonequilibrium Thermodynamics and its Statistical Foundations*, Clarendon, Oxford, Chap. 8.
83. Wong, C.Y., Maruhn, J.A. and Welton, T.A. (1975). *Nucl. Phys.*, **A253**, 469.
84. Wong, C.Y. and McDonald, J.A. (1977). *Phys. Rev.*, **C16**, 1196.
85. Singwi, K.S., Sjolander, A., Tosi, M.P. and Land, R.H. (1970). *Phys. Rev.*, **B1**, 1044.
86. March, N.H. and Tosi, M.P. (1972). *Proc. Roy. Soc.*, **A330**, 373.
87. Hang, H. and Weiss, K. (1972). *Physica*, **59**, 29.
88. Thouless, D.J. (1972). *The Quantum Mechanics of Many-Body Systems*, p. 14, Academic Press, New York.
89. Born, M. and Green, H.S. (1949). *A General Kinetic Theory of Liquids*, Cambridge University Press.
90. Kreuzer, H.J. and Nakamura, K. (1974). *Physics*, **73**, 600.
91. Deb, B.M. and Bamzai, A.S. (1978). *Molec. Phys.*, **35**, 1349.
92. Terreaux, C. and Lal, P. (1972). In *Cooperative Phenomena*, eds Haken, H. and Wagner, M., p. 192, Springer, Berlin.
93. Cohen, L. and Zaparovanny, Y.I. (1980). *J. Math. Phys.*, **21**, 794.
94. Nacrowich, F.J. and O'Connell, R.F. (1986). *Phys. Rev.*, **A34**, 1.
95. O'Connell, R.F. and Walls, D.F. (1984). *Nature*, **312**, 257.
96. Heller, E.J. (1976). *J. Chem. Phys.*, **65**, 1289.
97. Irving, J.H. and Zwanzig, R.W. (1951). *J. Chem. Phys.*, **19**, 1173.
98. Zwanzig, R.W. (ed.) (1967). *Selected Topics in Statistical Mechanics*, Gordon & Breach, New York.
99. Mazo, R.M. and Kirkwood, J.G. (1958). *J. Chem. Phys.*, **28**, 644.
100. Hoodbhoy, P.A. (1981). *Phys. Rev.*, **A24**, 3136.
101. McLafferty, F. (1985). *J. Chem. Phys.*, **83**, 5043.
102. Bertsch, G.F. (1978). In *Nuclear Physics with Heavy Ions and Mesons*, Vol. 1, eds Balian, R., Rho, M. and Ripka, G., p. 176, North-Holland, Amsterdam.
103. Kerman, A.K. and Koonin, S.E. (1976). *Ann. Phys.* (*N.Y.*), **100**, 332.
104. Green, H.S. (1952). *The Molecular Theory of Fluids*, North-Holland, Amsterdam.
105. Green, H.S. (1951). *J. Chem. Phys.*, **19**, 955.
106. Mazo, R.M. (1967). *Statistical Mechanical Theories of Transport Processes*, Chap. 3, Pergamon, Oxford.

107. Irving, J.H. and Kirkwood, J.G. (1950). *J. Chem. Phys.*, **18**, 817.
108. Hansen, J.P. and McDonald, I.R. (1976). *Theory of Simple Liquids*, Academic Press, New York.
109. Nakatsuji, H. (1976). *Phys. Rev.*, **A14**, 41.
110. Cohen, L. and Frishberg, C. (1976). *Phys. Rev.*, **A13**, 927.
111. Singwi, K.S., Tosi, M.P., Land, R.H. and Sjolander, A. (1968). *Phys. Rev.*, **176**, 589.
112. Nix, J.R. and Sierk, A.J. (1980). *Phys. Rev.*, **C21**, 396.
113. Di Nardo, M., Di Toro, M., Giansiracusa, G., Lombardo, U. and Russo, G. (1983). *Phys. Rev.*, **C28**, 929.
114. Bloch, F. (1933). *Z. Phys.*, **81**, 363.
115. March, N.H. (1957). *Adv. Phys.*, **6**, 1.
116. Ball, J.A., Wheeler, J.A. and Firemen, E.L. (1973). *Rev. Mod. Phys.*, **45**, 333.
117. Walecka, J.D. (1976). *Phys. Lett.*, **58A**, 83.
118. Ruppin, R. (1978). *J. Phys. Chem. Solids*, **39**, 233.
119. Bennett, A.J. (1970). *Phys. Rev.*, **B1**, 203.
120. Ying, S.C. (1974). *Nuovo Cim.*, **23B**, 270.
121. Hohenberg, P. and Kohn, W. (1964). *Phys. Rev.*, **136**, B864.
122. Eguiluz, A. and Quinn, J.J. (1976). *Phys. Rev.*, **B14**, 1347.
123. Equiluz, A., Ying, S.C. and Quinn, J.J. (1975). *Phys. Rev.*, **B11**, 2118.
124. Mukhopadhyay, G. and Lundqvist, S. (1975). *Nuovo Cim.*, **27B**, 1.
125. Lundqvist, S. and Mukhopadhyay, G. (1980). *Physica Scripta*, **21**, 503.
126. Lundqvist, S. (1983). In *Electron Correlations in Solids, Molecules and Atoms*, eds Devreese, J.T. and Brosens, F., p. 361, Plenum, New York.
127. Forstmann, F. and Stenschke, H. (1977). *Phys. Rev. Lett.*, **38**, 1365.
128. Lüdde, H.J., Horbatsch, M., Gross, E.K.U. and Dreizler, R.M. (1979). *Proceedings of XVII International Winter Meeting on Nuclear Physics*, Bormio, p. 120.
129. Horbatsch, M. and Dreizler, R.M. (1981). *Z. Phys.*, **A300**, 119.
130. Horbatsch, M. and Dreizler, R.M. (1982). *Z. Phys.*, **A308**, 329.
131. Malzacher, P. and Dreizler, R.M. (1981). *Z. Phys.*, **A307**, 211.
132. Horbatsch, M. (1983). *J. Phys.*, **B16**, 4643.
133. Hirschfelder, J.O. (1978). *J. Chem. Phys.*, **68**, 5151.
134. Hayes, E.F. and Parr, R.G. (1965). *J. Chem. Phys.*, **43**, 1831.
135. Folland, N.O. (1981). *Int. J. Quant. Chem.*, **S15**, 369.
136. Dahler, J.S. (1963). *Phys. Rev.*, **129**, 1464.
137. Hirschfelder, J.O. (1960). *J. Chem. Phys.*, **33**, 1462.
138. Marc, G. and McMillan, W.G. (1985). *Adv. Chem. Phys.*, **58**, 209.
139. Deb, B.M. (1973). *Rev. Mod. Phys.*, **45**, 22.
140. Deb, B.M. (ed.) (1981). *The Force Concept in Chemistry*, Van Nostrand-Reinhold, New York.
141. Bader, R.F.W. and Nguyen-Dang, T.T. (1981). *Adv. Quant. Chem.*, **14**, 63.
142. Nakatsuji, H. (1977). *J. Chem. Phys.*, **67**, 1312.
143. Floyd, E.R. (1982). *Phys. Rev.*, **D26**, 1339.
144. Floyd, E.R. (1984). *Phys. Rev.*, **D29**, 1842.
145. Schönberg, M. (1959). *Nuovo Cim.*, **12**, 103.
146. Schönberg, M. (1954). *Nuovo Cim*, **11**, 674.
147. Schönberg, M. (1954). *Nuovo Cim.*, **12**, 649.
148. Joachim, C. (1984). *Molec. Phys.*, **52**, 1191.
149. Capitani, J.F., Nalewajski, R.F. and Parr, R.G. (1982). *J. Chem. Phys.*, **76**, 568.

150. Tachibana, A. (1982). *Int J. Quant. Chem.*, **22**, 191.
151. Tachibana, A. (1983). *Int. J. Quant. Chem.*, **23**, 195.
152. Dodd, R.K., Eilbeck, J.C., Gibbon, J.D. and Morris, H.C. (1982). *Solitons and Nonlinear Wave Equations*, Academic Press, New York.
153. Nassar, A.B. (1984). *Phys. Lett.*, **106A**, 43.
154. Nassar, A.B. (1984). *Lett. Nuovo Cim.*, **41**, 476.
155. Nassar, A.B. (1985). *J. Phys.*, **A18**, L423; (1985). *Ibid.*, **A18**, 2861.
156. Kan, K.K. and Griffin, J.J. (1974). *Phys. Lett.*, **50B**, 241.
157. Hasse, R.W. (1978). *Rep. Prog. Phys.*, **41**, 1027.
158. Heller, D.F. and Hirschfelder, J.O. (1977). *J. Chem. Phys.*, **66**, 1929.
159. Riess, J. (1970). *Ann. Phys.* (*N.Y.*), **57**, 301; (1971). *Ibid.*, **67**, 347.
160. Gomes, J.A.N.F. (1980). *Molec. Phys.*, **40**, 765.
161. Musher, J.I. (1965). *J. Chem. Phys.*, **43**, 4081.
162. Musher, J.I. (1968). *Adv. Magn. Reson.*, **2**, 177.
163. Lazzeretti, P., Rossi, E. and Zanasi, R. (1982). *J. Chem. Phys.*, **77**, 3129.
164. Gomes, J.A.N.F. (1983). *J. Chem. Phys.*, **78**, 3133.
165. Das Sarma, S. and Quinn, J.J. (1979). *Phys. Rev.*, **B20**, 4872.
166. Das Sarma, S. (1982). *Phys. Rev.*, **B26**, 6559.
167. Tachibana, A., Hori, K., Asai, Y., Yamabe, T. and Fukui, K. (1985). *J. Molec. Struc. Theochem.* **123**, 267.
168. Parr, R.G. (1985). In *Density Functional Methods in Physics*, eds Dreizler, R.M. and da Providencia, J., Plenum, New York.
169. March, N.H. (1979). *J. Chem. Phys.*, **71**, 1004.
170. March, N.H. and Bader, R.F.W. (1980). *Phys. Lett.*, **78A**, 242.
171. Mulliken, R.S. (1935). *J. Chem. Phys.*, **3**, 573.
172. Pritchard, H.O. and Sumner, F.H. (1956). *Proc. Roy. Soc.*, **A235**, 136.
173. Iczkowski, R.P. and Margrave, J.L. (1961). *J. Am. Chem. Soc.*, **83**, 3547.
174. Sanderson, R.T. (1955). *Science*, **121**, 207.
175. Mortier, W.J. In *Electronegativity*, ed. Sen, K.D., Springer, Berlin (to be published).
176. Mortier, W.J., Ghosh, S.K. and Shankar, S. (1986). *J. Am. Chem. Soc.*, **108**, 4315.
177. Mortier, W.J., Van Genechten, K. and Gasteiger, J. (1985). *J. Am. Chem. Soc.*, **107**, 829.
178. Mullay, J. (1986). *J. Am. Chem. Soc.*, **108**, 1770.
179. Hall, G.G. (1985). *Adv. Atom. Molec. Phys.*, **20**, 41.
180. Fliszar, S. (1983). *Charge Distributions and Chemical Effects*, Springer, Berlin.
181. Perdew, J.P. (1985). In *Density Functional Methods in Physics*, eds Dreizler, R.M. and da Providencia, J., Plenum, New York.
182. Perdew, J.P., Parr, R.G., Levy, M. and Balduz, Jr., J.L. (1982). *Phys. Rev. Lett.*, **49**, 1691.
183. Pearson, R.G. (1973). *Hard and Soft Acids and Bases*, Dowden, Hutchinson and Ross, Stroudenberg, PA.
184. Parr, R.G. and Pearson, R.G. (1983). *J. Am. Chem. Soc.*, **105**, 7512.
185. Nalewajski, R.F. (1984). *J. Am. Chem. Soc.*, **106**, 944.
186. Yang, W. and Parr, R.G. (1985). *Proc. Natl Acad. Sci.* U.S.A., **82**, 6723.
187, Bader, R.F.W., MacDougall, P.J. and Lau, C.D.H. (1984). *J. Am. Chem. Soc.*, **106**, 1594.
188. Pearson, R.G. (1985). *J. Am. Chem. Soc.*, **107**, 6801.
189. Bader, R.F.W. (1981). In *The Force Concept in Chemistry*, ed. Deb, B.M., Chap. 2, Van Nostrand-Reinhold, New York.

190. Ghosh, S.K. and Parr, R.G. (1987). *Theoret. Chim Acta* (*Berl*) (in press).
191. Heilbronner, E. and Brock, H. (1976). *The HMO Model and its Application*, Vols 1–3, John Wiley, New York.
192. Coulson, C.A. and Longuet-Higgins, H.C. (1948). *Proc. Roy. Soc.*, **A192**, 16.
193. Fukui, K. (1982). *Science*, **218**, 747.
194. Parr, R.G. and Yang, W. (1984). *J. Am. Chem. Soc.*, **106**, 4049.
195. Yang, W., Parr, R.G. and Pucci, R. (1984). *J. Chem. Phys.*, **81**, 2862.
196. Nalewajski, R.F. and Parr, R.G. (1982). *J. Chem. Phys.*, **77**, 399.
197. Klopman, G. (1983). *J. Molec. Struc. Theochem.*, **12**, 121.
198. Bader, R.F.W. (1975). *Acc. Chem. Res.*, **8**, 34.
199. Bader, R.F.W. (1985). *Acc. Chem. Res.*, **18**, 9.
200. Collard, K. and Hall, G.G. (1977). *Int. J. Quant. Chem.*, **12**, 623.
201. Bader, R.F.W. and MacDougall, P.J. (1985). *J. Am. Chem. Soc.*, **107**, 6788.
202. Politzer, P. and Truhlar, D.G. (eds) (1981). *Chemical Applications of Atomic and Molecular Electrostatic Potentials*, Plenum, New York.
203. Politzer, P. (1980). *J. Chem. Phys.*, **73**, 3264.
204. Parr, R.G. and Berk, A. in ref. 202.
205. Tal, Y., Bader, R.F.W. and Erkku, J. (1980). *Phys. Rev.*, **A21**, 1.
206. Bamzai, A.S. and Deb. B.M. (1981). *Int. J. Quant. Chem.*, **20**, 1315.
207. Deb, B.M. (1984). *Proc. Ind. Acad. Sci.* (*Chem. Sci.*), **93**, 965.
208. Jameson, C.J. and Buckingham, A.D. (1980). *J. Chem. Phys.*, **73**, 5684.
209. Orttung, W.H. and Vosooghl, D. (1983). *J. Phys. Chem.*, **87**, 1432, 1438.
210. Orttung, W.H. (1985). *J. Phys. Chem.*, **89**, 3011.
211. Oxtoby, D.W. (1978). *J. Chem. Phys.*, **69**, 1184.
212. Gomes, J.A.N.F. (1983). *J. Chem. Phys.*, **78**, 4585.
213. Gomes, J.A.N.F. (1983). *Phys. Rev.*, **A28**, 559.
214. Gomes, J.A.N.F. (1983). *J. Molec. Struc. Theochem.*, **93**, 111.
215. Gomes, J.A.N.F. (1984). In *Local Density Approximations in Quantum Chemistry and Solid State Physics*, eds Dahl, J.P. and Avery, J., Plenum, New York.
216. Fermi, E. (1928). *Z. Phys.*, **48**, 73 (reprinted in ref. 59).
217. de Broglie, L. (1967). *Compt. Rend.*, **264**, 1041.
218. Beretta, G.P., Gyftopoulos, E.P., Park, J.L. and Hatsopoulos, G.N. (1984). *Nuovo Cim.*, **82B**, 169.
219. Beretta, G.P., Gyftopoulos, E.P. and Park, J.L. (1985). *Nuovo Cim.*, **87B**, 77.
220. Levine, R.D. and Tribus, M. (eds) (1979). *The Maximum Entropy Formalism*, MIT, Massachusetts.
221. Berkowitz, M. (to be published).
222. Williams, B.G. (eds), (1977). *Compton Scattering*, McGraw-Hill, New York.
223. Perdew, J.P. and Yue, W. (1986). *Phys. Rev.*, **B33**, 8800.
224. Ghosh, S.K. and Parr, R.G. (1985). *J. Chem. Phys.*, **82**, 3307.
225. Lebowitz, J.L. and Percus, J.K. (1963). *J. Math. Phys.*, **4**, 116.
226. Saam, W.F. and Eoner, C. (1977). *Phys. Rev.*, **A15**, 2566.
227. Evans, R. (1979). *Adv. Phys.*, **28**, 143.
228. Rowlinson, J.S. and Widom, B. (1982). *Molecular Theory of Capillarity*, Clarendon, Oxford.
229. Percus, J.K. (1982). In *The Liquid State of Matter: Fluids, Simple and Complex*, eds Montroll, E.W. and Lebowitz, J.L., p. 31. North-Holland, Amsterdam.
230. Bohm, D. and Hiley, B.J. (1985). *Phys. Rev. Lett.*, **55**, 2511.
231. George, C., Prigogine, I. and Rosenfeld, L. (1972). *Nature*, **240**, 25.
232. Zambrini, J.C. (1986). *Phys. Rev.*, **A33**, 1532.

233. Davidson, M. (1979). *Lett. Math. Phys.*, **3**, 367.
234. de la Pena, L. and Cetto, A.M. (1975). *Found. Phys.*, **5**, 355.
235. de la Pena, L. and Cetto, A.M. (1982). *Found. Phys.*, **12**, 1017.
236. Schroeck, Jr., F.E. (1982). *Found. Phys.*, **12**, 825.
237. de la Pena, L. and Cetto, A.M. (1985). *Rev. Mex. Fis.*, **31**, 551.
238. Yasue, K. (1982). In *Dynamical Systems and Microphysics*, eds Avez, A., Blaquiere, A. and Marzollo, A., Academic Press, New York, p. 461.
239. Grabert, H., Hauggi, P. and Talkner, P. (1979). *Phys. Rev.*, **A19**, 2440.
240. Mielnik, B. and Tengstrand, G. (1980). *Int. J. Theor. Phys.*, **19**, 239.
241. Nelson, E. (1979). *Lecture Notes in Physics*, Vol. 100 (Einstein Symposium, Berlin), p. 168, Springer, Berlin.
242. de Falco, D., De Martino, S. and de Siena, S. (1982). *Phys. Rev. Lett.*, **49**, 181; (1983). *Ibid.*, **50**, 704.
243. Yasue, K. (1978). *Phys. Rev. Lett.*, **40**, 665.
244. Weaver, D.L. (1978). *Phys. Rev. Lett.*, **40**, 1473.
245. Gaveau, B., Jacobson, T., Kac, M. and Schulman, L.S. (1984). *Phys. Rev. Lett.*, **53**, 419.
246. Maddox, J. (1984). *Nature*, **311**, 101.
247. Nassar, A.B. (1985). *Phys. Rev.*, **A32**, 1862.
248. Nassar, A.B. (1986). *Phys. Rev.*, **A33**, 2134.
249. Nassar, A.B. (1986). *J. Math. Phys.*, **27**, 755.
250. Golin, S. (1986). *J. Math. Phys.*, **27**, 1549.
251. de la Pena–Auerbach, L. and Cetto, A.M. (1977). *Int. J. Quant. Chem.*, **12** (Suppl. 1), 23.
252. Schrödinger, E. (1929). *Collected Papers on Wave Mechanics*, Blackie, Glasgow.
253. Claverie, P. and Diner, S. (1976). In *Localization and Delocalization in Quantum Chemistry*, Vol. 2, eds Chalvet, O., Daudel, R., Diner, S. and Malrieu, J.P., p. 395, Reidel, Dordrecht.
254. Feynman, R.P. and Hibbs, A.R. (1975). *Quantum Mechanics and Path Integrals*, McGraw-Hill, New York.
255. Takabayashi, T. (1957). *Prog. Theor. Phys. Suppl.*, **4**, 1.
256. Ghosh, S.K. and Deb, B.M. (1984). *J. Phys.* **A17**, 2463.
257. Ramana, M.V. and Rajagopal, A.K. (1984). *Adv. Chem. Phys.*, **54**, 231.
258. Gardiner, C.W. (1983). *Handbook of Stochastic Methods*, Springer, Berlin.
259. Rosenbrock, H.H. (1985). *Phys. Lett.*, **110A**, 343.
260. Rosenbrock, H.H. (1986). *Phys. Lett.*, **114A**, 163.
261. Deb, B.M. and Chattaraj, P.K. (1987). In *Solitons: Introduction and Applications*, eds Lakshmanan, M. and Kaw, P.K., Springer, Berlin.
262. Deb, B.M. and Chattaraj, P.K. (1987). *Proc. Ind. Acad. Sci.* (*Chem. Sci.*) (in press).
263. Deb, B.M. and Chattaraj, P.K. (1987). *Phys. Rep.* (submitted).

# 9 The Hydrodynamical Formulation of Time-Dependent Kohn–Sham Orbital Density Functional Theory

L.J. BARTOLOTTI

*Department of Chemistry, University of Miami, Coral Gables, Florida 33124, U.S.A.*



## 9.1 Introduction

The ability to calculate the properties associated with atoms and molecules interacting with time-varying electric and/or magnetic fields is very important in our understanding of the electronic structure of matter.[1] The *a priori* prediction of these properties is possible using time-dependent wave function functional theory (Schrödinger theory).[2] Unfortunately, the many-particle nature of wave-function functional theory prevents the calculation of accurate results, except for atoms and molecules having just a few electrons. Time-dependent Hartree–Fock theory[2] does provide a computational scheme which is applicable to many-particle systems. However, Hartree–Fock theory is itself an approximation, and it is necessary to go beyond this model to obtain accurate results.

Time-dependent density functional theory provides a convenient alternative to the conventional wave-function functional approach. Although time-dependent density functionals have long been used to study the dynamical properties of matter,[3] they have only recently been formally justified. In 1978 Peuckert[4] developed a time-dependent density functional theory based on the principle of the stationary action functional. Since a stationary principle is much weaker than a minimization principle, severe limitations had to be imposed and questions about its validity persist.[3,4] A time-dependent density functional theory based upon wave-function functional minimization principles has recently been presented by Bartolotti[5–7] and by Ghosh and Deb.[8,9] As with the corresponding wave-function

Single-Particle Density in Physics and Chemistry ISBN 0-12-470518-9

functional theory, certain restrictions apply. Only potentials with a periodic time dependence can be considered and the minimization principle is strictly valid only for frequencies of the incident radiation below that required to produce transitions to excited states. Many dynamic electronic properties can be calculated within this theory, since time-varying electric and magnetic fields belong to the above class of potentials. Note that under these conditions the action principle is a minimization principle and the procedure of Peuckert becomes rigorous. Runge and Gross[10] have recently attempted to generalize time-dependent density functional theory to include arbitrary time-dependent potentials. Their generalization has been criticized by Xu and Rajagobal.[11]

The most useful construction of time-dependent functional theory for doing practical calculations is the orbital-density description, similar to the Kohn–Sham[12] orbital-density construction in time-independent density functional theory. This form of the theory has been successfully employed in calculations of properties of atoms and molecules, which are manifested by the polarization induced by a time-varying oscillating electric field.[7–9,13–22] As in time-dependent wave-function functional theory, there are various (equivalent) ways to express the time-dependent Kohn–Sham orbital-density functional. The method for separating the time dependence into positive and negative Fourier components has recently been reviewed by Zangwill.[19] We describe here the hydrodynamic formulation of time-dependent Kohn–Sham theory[6–9] and show how it can be used to calculate dynamic optical susceptibilities. Treating the time-varying electric field as a perturbing potential, one can generate a set of functionals whose stationary points are the order-by-order corrections to the time-dependent Kohn–Sham amplitudes and phases.[7,8] These functionals have some interesting properties. They determine the order-by-order corrections to the energy of the system arising from interaction with an applied field. If evaluated with approximate solutions, the functionals provide lower bounds to the dynamic multipole susceptibilities when the frequency of the incident radiation is less than that necessary to cause a singularity in the susceptibilities. Since the functionals depend only on even powers of the applied frequency, we can directly obtain values of the susceptibilities evaluated at imaginary frequencies. In addition, the correct time-independent Kohn–Sham equations are recovered in the static limit (see also Chapter 8).

## 9.2 Hydrodynamical formulation of time-dependent Kohn–Sham theory

The derivation of the hydrodynamical formulation of time-dependent Kohn–Sham theory[6–9] parallels the Levy[23–25] construction for the stationary case. We begin with the definition of the kinetic-energy functional for a system of $N$ non-interacting electrons:

$$T_s[\{\chi_i(\mathbf{r},t), S_i(\mathbf{r},t)\}] = \sum_{i=1}^{N} \langle \chi_i(\mathbf{r},t)| -\tfrac{1}{2}\nabla^2 + \tfrac{1}{2}(\nabla S_i(\mathbf{r},t)\cdot \nabla S_i(\mathbf{r},t))|\chi_i(\mathbf{r},t)\rangle, \tag{9.1}$$

where $\chi_i$ and $S_i$ are the time-dependent Kohn–Sham amplitudes and phases, respectively. They are real functions of space and time,* and they represent the polar decomposition of the complex time-dependent Kohn–Sham orbitals, $\psi_k(\mathbf{r},t) = \chi_k(\mathbf{r},t)\exp(iS_k(\mathbf{r},t))$. The functional $T_s[\{\chi_i, S_i\}]$ is minimized with respect to independent variations of the $\chi_i$ and the $S_i$. During the minimization, the $\chi_i$ are constrained to satisfy

$$\langle \chi_i|\chi_i\rangle = 1, \tag{9.2}$$

and

$$\rho(\mathbf{r},t) = \sum_{i=1}^{N} \chi_i(\mathbf{r},t)^2, \tag{9.3}$$

while the $S_i$ must satisfy

$$\frac{\partial \chi_i(\mathbf{r},t)^2}{\partial t} + \nabla\cdot[\chi_i(\mathbf{r},t)^2 \nabla S_i(\mathbf{r},t)] = 0, \tag{9.4}$$

or

$$\frac{\partial \rho(\mathbf{r},t)}{\partial t} + \nabla\cdot \sum_{i=1}^{N} \chi_i(\mathbf{r},t)^2 \nabla S_i(\mathbf{r},t) = 0. \tag{9.5}$$

This minimization process is equivalent to the variational principle

$$\delta J[\{\chi_i, S_i\}] - \sum_{i=1}^{N} \frac{\partial \langle \chi_i|\delta S_i|\chi_i\rangle}{\partial t} = 0, \tag{9.6}$$

where

$$\begin{aligned} J[\{\chi_i(\mathbf{r},t), S_i(\mathbf{r},t)\}] = \sum_{i=1}^{N} \langle \chi_i(\mathbf{r},t)| &-\tfrac{1}{2}\nabla^2 + \tfrac{1}{2}(\nabla S_i(\mathbf{r},t)\cdot \nabla S_i(\mathbf{r},t)) \\ &+ \frac{\partial S_i(\mathbf{r},t)}{\partial t}|\chi_i(\mathbf{r},t)\rangle + J[\rho(\mathbf{r},t)] + E_{xc}[\rho(\mathbf{r},t)] \\ &+ \int \rho(\mathbf{r},t)\, v_{ne}(\mathbf{r})\, d\mathbf{r} + \int \rho(\mathbf{r},t)\, H^{(1)}(\mathbf{r},t)\, d\mathbf{r}. \end{aligned} \tag{9.7}$$

*This does not prevent one from using complex spherical harmonics to represent the angular portion of the functions. This is a choice of convenience and not one of necessity. The radial portion of the functions will always be real.

Here $v_{\text{ne}}(\mathbf{r})$ is the nuclear–electron attraction potential,

$$J[\rho(\mathbf{r},t)]=\frac{1}{2}\iint\frac{\rho(\mathbf{r}_1,t)\rho(\mathbf{r}_2,t)}{r_{12}}\,d\mathbf{r}_2\,d\mathbf{r}_1 \tag{9.8}$$

is the Coulomb energy density functional, $E_{\text{xc}}[\rho(\mathbf{r},t)]$ is the unknown exchange-correlation energy density functional and $H^{(1)}(\mathbf{r},t)$ is the as yet undefined external periodic time-dependent single-particle potential. Performing the indicated variations in eqn (9.6), we obtain the following set of coupled non-linear differential equations:

$$\left\{-\frac{1}{2}\nabla^2+\frac{1}{2}(\nabla S_i\cdot\nabla S_i)+\frac{\delta\{J[\rho]+E_{\text{xc}}[\rho]\})}{\delta\rho}+v_{\text{ne}}+H^{(1)}\right\}\chi_i$$

$$=-\frac{\partial S_i}{\partial t}\chi_i=\varepsilon_i\chi_i \tag{9.9}$$

and

$$\frac{\partial\chi_i^2}{\partial t}+\nabla\cdot(\chi_i^2\nabla S_i)=0. \tag{9.10}$$

The quantities $\partial S_i(\mathbf{r},t)/\partial t=-\varepsilon_i(\mathbf{r},t)$ act like Lagrange multipliers that ensure all the constraints, eqns (9.2)–(9.4), are satisfied. Unlike time-independent Kohn–Sham theory[12] they are functions of space and time. This set of coupled non-linear differential equations is consistent with the time-independent Kohn–Sham equation. In the limit for which the time dependence is "turned off", $\nabla S_i$ vanishes and eqns (9.6), (9.7), (9.9) and (9.10) reduce to their (correct) time-independent counterparts.

Equation (9.6) is not in the form of the variation of a functional which vanishes. Consequently, it is not obvious that it can lead to a minimization principle. For the special case of periodic time-dependent potential, a time average causes the second term in eqn (9.7) to vanish and we obtain a variational principle of the desired form. We shall perform such time-average below.

The above time-dependent Kohn–Sham theory can be used to calculate physical properties associated with the interaction between an atom or molecule and a periodic time-dependent electric field

$$\mathbf{E}=\mathbf{F}\cos\omega t, \tag{9.11}$$

where $\mathbf{F}$ is the field strength and $\omega$ is the frequency of incident radiation. This interaction can be described by the electric-dipole polarization vector:[27–30]

$$\mathbf{P}(t)=\boldsymbol{\mu}+\boldsymbol{\alpha}\cdot\mathbf{E}+\frac{1}{2!}\,\boldsymbol{\beta}:\mathbf{EE}+\frac{1}{3!}\,\boldsymbol{\gamma}:\mathbf{EEE}+\cdots, \tag{9.12}$$

The optical susceptibility tensors are defined as follows: $\boldsymbol{\mu}$ is the dipole moment, $\boldsymbol{\alpha}$ is the dipole polarizability (a second-rank tensor which is first order in the applied field), $\boldsymbol{\beta}$ is a third-rank tensor and defines the non-linear electric susceptibility of order two, and $\gamma$ is a fourth-rank tensor which defines the third-order non-linear electric susceptibility. For an atom in an electric field oriented in the $z$-direction, the polarization parallel to the applied field can be written as a scalar quantity, namely,

$$P(t) = \alpha(\omega) F \cos \omega t + \tfrac{1}{24}\{\gamma(3\omega; \omega, \omega, \omega) \cos 3\omega t + 3\gamma(\omega; \omega, \omega, -\omega) \cos \omega t\} F^3 + \cdots. \tag{9.13}$$

The even terms in eqn (9.12) identically vanish since the atom is spherically symmetric. The first term is the frequency-dependent polarizability $\alpha(\omega)$, $\gamma(3\omega; \omega, \omega, \omega)$ is the third-harmonic generation coefficient and $\gamma(\omega; \omega, \omega, -\omega)$ is the intensity-dependent refractive index.

The electric-dipole polarization for an atom as defined in eqn (9.13) can be obtained from the time-dependent electron density and

$$P(t) = -\int \rho(\mathbf{r}, t) h^{(1)}(\mathbf{r}) \, d\mathbf{r}, \tag{9.14}$$

where $h^{(1)}(\mathbf{r})$ is the dipole-moment operator. Comparing eqn (9.14) with eqn (9.13) and noting that for small field strengths, $\rho(\mathbf{r}, t)$ can be expanded as a power series in the applied field,

$$\rho(\mathbf{r}, t) = \sum_{j=0}^{\infty} F^j \rho^{(j)}(\mathbf{r}, t), \tag{9.15}$$

we obtain expressions for the optical susceptibilities in terms of the order-by-order corrections to $\rho$ (and the order-by-order corrections to the $\chi_i$).

Treating $H^{(1)}(\mathbf{r}, t)$ as an adiabatically switched-on harmonic perturbation arising from monochromatic radiation of a uniform electric field oriented in the $z$-direction,

$$H^{(1)}(\mathbf{r}, t) = h^{(1)}(\mathbf{r}) F \cos \omega t, \tag{9.16}$$

perturbation theory can be used to calculate the order-by-order corrections to the $\chi_i$ and the $S_i$. To proceed, we expand the $\chi_i(\mathbf{r}, t)$ and the $S_i(\mathbf{r}, t)$ as power series in $F$:

$$\chi_i(\mathbf{r}, t) = \sum_{j=0}^{\infty} F^j \chi_i^{(j)}(\mathbf{r}, t) \tag{9.17}$$

and

$$S_i(\mathbf{r}, t) = \sum_{j=0}^{\infty} F^j S_i^{(j)}(\mathbf{r}, t). \tag{9.18}$$

Substituting eqns (9.15)–(9.18) into eqns (9.6) and (9.7), we obtain the order-by-order corrections to $J[\{\chi_i, S_i\}]$. The zeroth-order correction $J_0[\{\chi_i^{(0)}(\mathbf{r})\}]$ is just the familiar time-independent Kohn–Sham functional[12] and leads to the unperturbed ground-state electronic density $\rho^{(0)}(\mathbf{r})$ and the ground-state electronic energy. In general, the $j$th-order correction to $J[\{\chi_i, S_i\}]$ represents the $j$th-order correction to the electron energy arising from the perturbation of the applied field. Since the $2j+1$ rule is obeyed, knowledge of the $j$th-order corrections to the $\chi_i$ and the $S_i$ determine the energy up to the $(2j+1)$th order. To calculate the dynamic polarizability given in eqn (9.13), which depends on the first-order correction to the density, we only need to consider the second-order correction to $J[\{\chi_i, S_i\}]$, namely,

$$\begin{aligned} J_2[\{\chi_i^{(1)}, S_i^{(1)}\}] = \sum_{i=1}^{N} \{ & \langle \chi_i^{(1)} | h^{(0)} - \varepsilon_i^{(0)} | \chi_i^{(1)} \rangle \\ & + 2\langle \chi_i^{(1)} | h^{(1)} \cos \omega t + \tfrac{1}{2} v_i^{(1)} + \frac{\partial S_i^{(1)}}{\partial t} | \chi_i^{(0)} \rangle \\ & + \tfrac{1}{2} \langle \chi_i^{(0)} | \nabla S_i^{(1)} \cdot \nabla S_i^{(1)} | \chi_i^{(0)} \rangle \}, \end{aligned} \tag{9.19}$$

where

$$v_i^{(1)} = 2 \sum_{j=1}^{N} \int \chi_j^{(1)}(2) \frac{\delta^2 \{J[\rho] + E_{xc}[\rho]\}}{\delta\rho(1)\delta\rho(2)} \bigg|_{\substack{\rho=\rho^{(0)} \\ N}} \chi_j^{(0)}(2)\, d\tau_2, \tag{9.20}$$

and $\chi_i^{(0)}$ are the solutions to the time-independent Kohn–Sham equations:[12]

$$h^{(0)} \chi_i^{(0)} = \varepsilon_i^{(0)} \chi_i^{(0)}, \tag{9.21}$$

with

$$h^{(0)} = -\frac{1}{2}\nabla^2 + \frac{\delta\{J[\rho] + E_{xc}[\rho]\}}{\delta\rho} \bigg|_{\substack{\rho=\rho^{(0)} \\ N}} + v_{ne}. \tag{9.22}$$

The functional derivatives in eqns (9.20) and (9.22) are performed with the number of electrons kept fixed and they are evaluated with the zeroth-order ground-state density

$$\rho^{(0)} = \sum_{i=1}^{N} |\chi_i^{(0)}|^2. \tag{9.23}$$

The Euler equations associated with independent variations of $J_2[\{\chi_i^{(1)}, S_i^{(1)}\}]$ with respect to the $\chi_i^{(1)}$ and $S_i^{(1)}$ are given by

$$(h^{(0)} - \varepsilon_i^{(0)})\chi_i^{(1)} + \left( h^{(1)} \cos \omega t + v_i^{(1)} + \frac{\partial S_i^{(1)}}{\partial t} \right) \chi_i^{(0)} = 0, \tag{9.24}$$

and

$$2\chi_i^{(0)}\frac{\partial\chi_i^{(1)}}{\partial t}+\nabla\cdot(\chi_i^{(0)}\chi_i^{(0)}\nabla S_i^{(1)})=0. \tag{9.25}$$

From eqn (9.24) we see that the time dependence of the $\chi_i^{(1)}$ can be written

$$\chi_i^{(1)}(\mathbf{r},t)=\phi_i^{(1)}(\mathbf{r},\omega)\cos\omega t, \tag{9.26}$$

and from eqns (9.25) and (9.26) we see that we can write

$$S_i^{(1)}(\mathbf{r},t)=\frac{1}{\omega}s_i^{(1)}(\mathbf{r},\omega)\sin\omega t. \tag{9.27}$$

Putting eqns (9.26) and (9.27) into the functional $J_2[\{\chi_i^{(1)},S_i^{(1)}\}]$ and then time-averaging over one period $(0\leqslant t\leqslant 2\pi/\omega)$, we obtain the functional

$$\begin{aligned}J_2[\{\phi_i^{(1)},s_i^{(1)}\}]=\sum_{i=1}^{N}\{&\langle\phi_i^{(1)}|h^{(0)}-\varepsilon_i^{(0)}|\phi_i^{(1)}\rangle\\&+2\langle\phi_i^{(1)}|h^{(1)}+\tfrac{1}{2}v^{(1)}+s_i^{(1)}|\chi_i^{(0)}\rangle\\&+\frac{1}{2\omega^2}\langle\chi_i^{(0)}|\nabla s_i^{(1)}\cdot\nabla s_i^{(1)}|\chi_i^{(0)}\rangle\},\end{aligned} \tag{9.28}$$

with

$$v^{(1)}=2\sum_{j=1}^{N}\int\phi_j^{(1)}(2)\left.\frac{\delta^2\{J[\rho]+E_{xc}[\rho]\}}{\delta\rho(1)\delta\rho(2)}\right|_{\rho=\rho_N^{(0)}}\chi_j^{(0)}(2)\,d\tau_2. \tag{9.29}$$

The functional $J_2[\{\phi_i^{(1)},s_i^{(1)}\}]$ now depends upon $\omega$ instead of $t$. Equation (9.28) has also been derived by Ghosh and Deb;[8] however, their expression involving $v^{(1)}$ neglects the factor $\frac{1}{2}$. The Euler equations associated with the minimization of this functional, $\delta J_2[\{\phi_i,s_i\}]=0$, are

$$(h^{(0)}-\varepsilon_i^{(0)})\phi_i^{(1)}+(h^{(1)}+v^{(1)}+s_i^{(1)})\chi_i^{(0)}=0 \tag{9.30}$$

and

$$\phi_i^{(1)}\chi_i^{(0)}-\frac{1}{2\omega^2}\nabla\cdot(\chi_i^{(0)}\chi_i^{(0)}\nabla s_i^{(1)})=0. \tag{9.31}$$

Since we are also interested in calculating quadrupole and octupole dynamic polarizabilities, we replace the dipole-moment operator in eqn (9.28) by the more general $2^l$-pole operator[30]

$$h^{(1)}=-r^lP_l(\cos\theta), \tag{9.32}$$

where $P_l(\cos\theta)$ is a Legendre polynomial of order $l$. Letting $l=1$, 2 or 3, we can obtain the dipole, quadrupole or octupole dynamic polarizabilities, respectively (see also Section 8.2.3 for a calculation of these quantities).

Evaluating $J_2[\{\phi_i^{(1)}, s_i^{(1)}\}]$ at the solution points gives the second-order correction to the energy:

$$J_2[\{\phi_i^{(1)}, s_i^{(1)}\}] = \sum_{i=1}^{N} \langle \phi_i^{(1)} | h^{(1)} | \chi_i^{(0)} \rangle = -\frac{1}{2\omega^2} \sum_{i=1}^{N} \langle \chi_i^{(0)} | \nabla s_i^{(1)} \cdot \nabla \imath^{(1)} | \chi_i^{(0)} \rangle$$

$$= -\frac{1}{\omega^2} \sum_{i=1}^{N} \langle \chi_i^{(0)} s_i^{(1)} | (h^{(0)} - \varepsilon_i^{(0)}) h^{(1)} | \chi_i^{(0)} \rangle, \tag{9.33}$$

or

$$J_2[\{\phi_i^{(1)}, s_i^{(1)}\}] = \frac{1}{2} \int \rho^{(1)} h^{(1)} \, d\tau = -\frac{1}{2} \alpha_{2^l}(\omega), \tag{9.34}$$

where $\rho^{(1)}$ is the first-order correction to the density and $\alpha_{2^l}(\omega)$ is the $2^l$-pole polarizability.[30] Imaginary-frequency multipole polarizabilities $\alpha_{2^l}(i\omega)$ can be calculated by replacing $\omega^2$ with $-\omega^2$ in eqns (9.28) and (9.31). In the limit as $\omega$ goes to zero (static limit) the quantities $\nabla s_i$ vanish and we recover the time-independent Kohn–Sham equation[12] for an atom in a static electric field. Finally, we note that for approximate solutions $\tilde{\phi}_i^{(1)}$ and $\tilde{s}_i^{(1)}$, $J_2[\{\tilde{\phi}_i^{(1)}, \tilde{s}_i^{(1)}\}]$ is an upper bound to the exact second-order energy when $\omega$ is less than the first excitation frequency. Alternatively, we have a lower bound to the dynamic $2^l$-pole polarizability $\alpha_{2^l}(\omega)$. The proof for the bounding properties is straightforward; consequently, we present here only the arguments involved. The above equations are similar to those in the hydrodynamical formulation of time-dependent wave-function functional theory.[31] In wave-function functional theory, the hydrodynamical formulation is equivalent to the standard Hylleraas-type variational method.[32] The bounding properties of the Hylleraas-type functional are well known[2] and apply to $J_2[\{\phi_i^{(1)}, s_i^{(1)}\}]$ as well. Similar bounding properties will hold for the higher-order corrections fo $J[\{\phi_i, s_i\}]$, except that the interval of validity decreases with each $2j$-order. For instance, the fourth-order functional is strictly valid for frequencies that are one-half the normal dispersion interval.[2]

An alternative to the direct variational approach is obtained by expanding the first-order Kohn–Sham amplitudes and phases as power series in the applied frequency:

$$\phi_i^{(1)} = \sum_{k=0}^{\infty} \psi_{k,i} \omega^{2k} \tag{9.35}$$

and

$$\frac{1}{\omega} s_i^{(1)} = \sum_{k=1}^{\infty} \eta_{k-1,i} \omega^{2k-1}. \tag{9.36}$$

The multipole polarizability is now given by the Cauchy series

$$\alpha_{2^l}(\omega) = \sum_{k=0}^{\infty} \alpha_{2^l,k}\,\omega^{2k}, \tag{9.37}$$

where the Cauchy moments are given by

$$\alpha_{2^l,k} = -2\sum_{i=1}^{N} \langle \psi_{k,i}|h^{(1)}|\chi_i^{(0)}\rangle = \sum_{i=1}^{N} \langle \chi_i^{(0)}|\nabla\eta_{k,i}\cdot\nabla h^{(1)}|\chi_i^{(0)}\rangle$$
$$= 2\sum_{i=1}^{N} \langle \chi_i^{(0)}\eta_{k,i}|(h^{(0)}-\varepsilon_i^{(0)})h^{(1)}|\chi_i^{(0)}\rangle. \tag{9.38}$$

The functions $\psi_{k,i}$ and $\eta_{k,i}$ are the sequential solutions to the following set of equations:

$$L_2[\{\psi_{0,i}\}] = \sum_{i=1}^{N} \{\langle \psi_{0,i}|h^{(0)}-\varepsilon^{(0)}|\psi_{0,i}\rangle + 2\langle \psi_{0,i}|h^{(1)}+\tfrac{1}{2}v_0^{(1)}|\chi_i^{(0)}\rangle\}, \tag{9.39}$$

with

$$(h^{(0)}-\varepsilon_i^{(0)})\psi_{0,i} + (h^{(1)}+v_0^{(1)})\chi_i^{(0)} = 0 \tag{9.40}$$

and

$$L_2[\{\eta_{k-1,i}\}] = \sum_{i=1}^{N} \{\langle \chi_i^{(0)}\eta_{k-1,i}|h^{(0)}-\varepsilon_i^{(0)}|\eta_{k-1,i}\chi_i^{(0)}\rangle$$
$$+\, 2\langle \psi_{k-1,i}|\eta_{k-1,i}|\chi_i^{(0)}\rangle\}, \tag{9.41}$$

with

$$(h^{(0)}-\varepsilon_i^{(0)})\eta_{k-1,i}\chi_i^{(0)} + \psi_{k-1,i} = 0 \tag{9.42}$$

and

$$L_2[\{\psi_{k,i}\}] = \sum_{i=1}^{N} \{\langle \psi_{k,i}|h^{(0)}-\varepsilon_i^{(0)}|\psi_{k,i}\rangle + 2\langle \psi_{k,i}|\tfrac{1}{2}v_{k-1}^{(1)}+\eta_{k-1,i}|\chi_i^{(0)}\rangle\}, \tag{9.43}$$

with

$$(h^{(0)}-\varepsilon_i^{(0)})\psi_{k,i} + (v_k^{(1)}+\eta_{k-1,i})\chi_i^{(0)} = 0. \tag{9.44}$$

In the above equations, the quantity $v_k^{(1)}$ is defined

$$v_k^{(1)} = 2\sum_{j=1}^{N} \int \psi_{k,j}(2)\left.\frac{\delta^2\{J[\rho]+E_{xc}[\rho]\}}{\delta\rho(1)\delta\rho(2)}\right|_{\substack{\rho=\rho^{(0)}\\ N}} \chi_j^{(0)}(2)\,d\tau_2. \tag{9.45}$$

The convergence of the Cauchy series, eqn (9.37), can be improved by using Padè summation techniques.[33–37] The Padè approximant sums the series within its radius of convergence and furnishes an analytic continuation outside the radius of convergence. The $[n_l, n_l-1]_\alpha$ and $[n_l, n_l-1]_\beta$ Padè approximants, as defined by Langhoff,[34–37] provide convenient methods for obtaining bounds to the multipole polarizabilities and dispersion coefficients. A Padè approximant defines a pseudo(effective)-spectra from which spectral sums $S(l, -k)$ can be obtained. In terms of an $n$th-order $[n_l, n_l-1]_\alpha$ or $[n_l, n_l-1]_\beta$ Padè approximant, the $S(l, -k)$ are given by

$$S(l,-k)=\sum_{i=1}^{n_l} f_{l,i}\,\omega_{l,i}^{-k}, \tag{9.46}$$

where the $f_{l,i}$ and the $\omega_{l,i}$ are the pseudo(or fictitious)-oscillator strengths and transition frequencies, respectively. The multipole Cauchy moments defined by eqn (9.37) are the spectral sums

$$S(l,-2k-2)=\alpha_{2^l,k}. \tag{9.47}$$

The pseudo-spectra also determine the dynamic multipole polarizability

$$\alpha_{2^l}(\omega)=\sum_{i=1}^{n_l} \frac{f_{l,i}}{(\omega_{l,i}^2-\omega^2)}. \tag{9.48}$$

Likewise, the two-body dispersion coefficient $\varepsilon_2(l,L)$, which represents the interaction between the induced $2^l$-pole of atom I and the induced $2^L w$-pole of atom II, can be obtained from $f_{l,i}$ and $\omega_{l,i}$, i.e.

$$\varepsilon_2(l,L)=\frac{(2l+2L)}{4(2l)!\ (2L)!}\sum_{i=1}^{n_l}\sum_{j=1}^{n_L}\frac{f_{l,i}f_{L,j}}{\omega_{l,i}\omega_{L,j}(\omega_{l,i}+\omega_{L,j})}, \tag{9.49}$$

Alternatively, $\varepsilon_2(l, L)$ can be obtained from the integration over $\omega$ of the product of multipole polarizabilities of the interacting atoms evaluated at imaginary frequencies.[38–41]

The direct variational approach leads to a set of pseudo-oscillator strengths and transition frequencies,[42,43] which can also be used in eqns (9.46), (9.48) and (9.49). However, the size of this set of pseudo-oscillator strengths and transition frequencies increases with the number of Kohn–Sham amplitudes. For the properties discussed above, the small set (usually $n_l=6$) of Padè approximant's $\{f_{l,i}\}$ and $\{\omega_{l,i}\}$ provide a convenient and compact way to tabulate all the information needed to calculate any of the above quantities. Alternatively, one can calculate $\alpha_{2^l}(\omega)$ from eqn (9.34), $S(l, -2k-2)$ from eqn (9.47) and $\varepsilon_2(l, L)$ from an integration over $\alpha_{2^l}(\mathrm{i}\omega)\alpha_2^L(\mathrm{i}\omega)$. The pseudo-spectrum should not be confused with the exact oscillator strengths and transition frequencies. Only the first pseudo-transition frequency has any physical significance. It is an upper bound to the exact first-transition frequency and it is exact when the exact solutions are known.

We see from the Euler equations, eqns (9.30), (9.40) and (9.44), that the first-order amplitudes of one orbital are coupled to the first-order amplitudes of all the other orbitals through the potentials $v^{(1)}$ and $v_k^{(1)}$. This suggests that uncoupling schemes such as found in Hartree–Fock theory[44] can be defined. The present theory allows two uncoupling schemes, whereas there are three in Hartree–Fock theory. This occurs because the functional derivative of $E_{xc}[\rho]$ in $h^{(0)}$ is local as opposed to the non-local exchange operator in Hartree–Fock theory. The first uncoupling procedure neglects all terms in $v^{(1)}$ and $v_k^{(1)}$ when $j \neq i$. Neglecting all terms in $v^{(1)}$ and $v_k^{(1)}$ leads to the second uncoupling scheme. These uncoupling procedures are similar to the methods b and c, respectively, discussed by Langhoff, Karplus and Hurst.[44]

## 9.3 A variational solution

The determination of the first-order corrections to the dynamic Kohn–Sham amplitudes and phases requires either solving the appropriate differential equations or applying variational techniques. The latter approach provides a method of solution that is equally applicable to atoms and molecules. With a suitable choice of trial functions, the results can be as accurate as desired. The trial functions have the general form of a radial function multiplied by the appropriate spherical harmonic. The angular parts are rigidly fixed by the selection rules associated with the matrix elements of the multipole polarizability, $\langle \phi_i^{(1)} | h^{(1)} | \chi_i^{(0)} \rangle \neq 0$.[31,46] It is the radial functions which are varied, and they should be as flexible as possible. We have recently employed a simple set of trial functions in which the radial portion of the phases was approximated by a polynomial in $r$ and the product *ansatz* was used to approximate the amplitudes.[7] Within this approximation each of the trial amplitudes is written as the product of the zeroth-order functions and a polynomial; although this approximation is somewhat restrictive and not generally applicable,[44,46] it has the advantage of simplifying the calculations:

$$\phi_i^{(1)} = \chi_i^{(0)} \sum_{j=1}^{M} \{C_j^{(i)} g_{j,0}^{(l,i)} + \Delta_2^{(l,i)} E_j^{(i)} g_{j,2}^{(l,i)} + \Delta_4^{(l,i)} F_j^{(i)} g_{j,4}^{(l,i)}\}, \tag{9.50}$$

$$s_i^{(1)} = \sum_{j=1}^{M'} \{C_{j+M}^{(i)} g_{j,0}^{(l,i)} + \Delta_2^{(l,i)} E_{j+M}^{(i)} g_{j,2}^{(l,i)} + \Delta_4^{(l,i)} F_{j+M}^{(i)} g_{j,4}^{(l,i)}\}, \tag{9.51}$$

$$\psi_{k,i} = \chi_i^{(0)} \sum_{j=1}^{M_k} \{C_j^{(i,k)} g_{j,0}^{(l,i)} + \Delta_2^{(l,i)} E_j^{(i,k)} g_{j,2}^{(l,i)} + \Delta_4^{(l,i)} F_j^{(i,k)} g_{j,4}^{(l,\hat{\imath})}\} \tag{9.52}$$

and

$$\eta_{k,i}=\sum_{j=1}^{M'_k}\{C^{(i,k)}_{j+M_k}g^{(l,i)}_{j,0}+\Delta^{(l,i)}_2 E^{(i,k)}_{j+M_k}g^{(l,i)}_{j,2}+\Delta^{(l,i)}_4 F^{(i,k)}_{j+M_k}g^{(l,i)}_{j,4}\}, \tag{9.53}$$

where the constants $\{C\}$, $\{E\}$ and $\{F\}$ are the linear variational parameters to be determined from the minimization of the appropriate functional. The $g^{(l,i)}_{j,\lambda}$ are defined by

$$g^{(l,i)}_{j,\lambda}=r^{j+l-\lambda-1}\frac{Y^{(m_i)}_{l_i+l-\lambda}}{Y^{(m_i)}_{l_i}}, \tag{9.54}$$

where the $Y^{(m_i)}_{l_i}$ are spherical harmonics and the $l_i$ and $m_i$ are the angular momentum and magnetic quantum numbers, respectively, associated with the $\chi^{(0)}_i$. The quantities $\Delta^{(j,i)}_\lambda$ are either zero or one:

$$\Delta^{(l,i)}_\lambda=\begin{cases}1 & \text{if } l+l_i-|m_i|-\lambda\geqslant 0 \text{ and } l_i>\frac{1}{2}(\lambda-2),\\ 0 & \text{otherwise.}\end{cases} \tag{9.55}$$

The trial functions defined by the above equations are sufficient to determine atomic dipole, quadrupole and octupole polarizabilities provided the zero-order solutions $\chi^{(0)}_i$ have angular momentum values in the range $0\leqslant l_i\leqslant 2$.

The minimization of $J_2[\{\phi^{(1)}_i, s^{(1)}_i\}]$ with respect to the linear variational parameters yields a set of simultaneous equations. The minimization of $L_2[\{\psi_{0,i}\}]$, $L_2[\{\eta_{k-1,i}\}]$ and $L_2[\{\psi_{k,i}\}]$ leads to a sequential set of simultaneous equations. Both minimization procedures involve the same matrix elements

$$A^{(i,\lambda)}_{\zeta,\xi}=\langle\chi^{(0)}_i|h^{(0)}-\varepsilon^{(0)}|\chi^{(0)}_i\rangle=\tfrac{1}{2}\langle\chi^{(0)}_i|\nabla g^{(l,i)}_{\zeta,\lambda}\cdot\nabla g^{(l,i)}_{\xi,\lambda}|\chi^{(0)}_i\rangle, \tag{9.56}$$

$$B^{(i,\lambda)}_{\zeta,\xi}=\langle\chi^{(0)}_i|g^{(l,i)}_{\zeta,\lambda}\,g^{(l,i)}_{\xi,\lambda}|\chi^{(0)}_i\rangle, \tag{9.57}$$

$$D^{(i,\lambda)}_{\zeta,\xi}=\tfrac{1}{2}\langle\chi^{(0)}_i|\nabla g^{(l,i)}_{\zeta,\lambda}\cdot\nabla g^{(l,i)}_{\xi,\lambda}|\chi^{(0)}_i\rangle, \tag{9.58}$$

$$H^{(i,\lambda)}_{\zeta}=\langle\chi^{(0)}_i|g^{(l,i)}_{\zeta,\lambda}\,r^l P_l(\cos\theta)|\chi^{(0)}_i\rangle \tag{9.59}$$

and

$$T^{(i,\lambda,j,\gamma)}_{\zeta,\xi}=2\iint\chi^{(0)}_i(1)^2 g^{(l,i)}_{\zeta,\lambda}(1)\frac{\delta^2(J[\rho]+E_{\mathrm{xc}}[\rho])}{\delta\rho(1)\delta\rho(2)}\bigg|_{\substack{\rho=\rho^{(0)}\\ N}}\chi^{(0)}_j(2)^2 g^{(l,j)}_{\xi,\gamma}(2)\,\mathrm{d}\tau_2\,\mathrm{d}\tau_1. \tag{9.60}$$

However, the size of the matrices involved in the solution of the simultaneous equations is much smaller for the Cauchy moment approach.

Since eqns (9.50) and (9.52) are approximate variational solutions, they do not necessarily satisfy the orthogonalization constraints (as the exact solutions would)[46,47]

$$\langle \phi_i^{(1)} | \chi_j^{(0)} \rangle + \langle \chi_i^{(0)} | \phi_j^{(1)} \rangle = 0, \tag{9.61}$$

$$\langle \psi_{i,k} | \chi_j^{(0)} \rangle + \langle \chi_i^{(0)} | \psi_{j,k} \rangle = 0. \tag{9.62}$$

When necessary, these constraints are introduced with Lagrange multipliers (for a numerical solution of the perturbative equations see ref. 60).

## 9.4 Numerical results

Up to now, we have been assuming that the exact $E_{xc}[\rho]$ is known. Unfortunately it is not known, and approximations to it must be used in the above equations. Consequently, the solutions obtained are only as good as the chosen approximation to $E_{xc}[\rho]$. The bounding properties discussed above no longer refer to the exact functional, but rather to a functional with a given approximation to $E_{xc}[\rho]$. Fortunately, reasonable approximations to the exchange-correlation functional are known.[48–50]

In our previous work,[7] we studied four approximations to $E_{xc}[\rho]$:
(i) *the Dirac functional*[51]

$$E_{xc}[\rho] = C_D \int \rho^{4/3}\, d\tau, \tag{9.63}$$

where

$$C_D = -\tfrac{3}{4}(3/\pi)^{1/3}; \tag{9.64}$$

(ii) *the* $X\alpha$ *functional*[52]

$$E_{xc}[\rho] = \frac{3}{2}\alpha C_D \int \rho^{4/3}\, d\mathbf{r}, \tag{9.65}$$

where $\alpha$ is a parameter chosen to yield the Hartree–Fock energy;
(iii) *the Gunnarsson–Lundqvist functional*[53]

$$E_{xc}[\rho] = C_D \int \rho^{4/3}\, d\mathbf{r} - C_P \int \rho \left\{ \left(1 + \frac{x^3}{\rho}\right) \ln\left(1 + \frac{\rho^{1/3}}{x}\right) + \frac{1}{2}\rho^{-1/3} x - \rho^{-2/3} x^2 - \frac{1}{3} \right\} d\mathbf{r}, \tag{9.66}$$

with

$$C_P = 0.0333, \tag{9.67}$$

$$R_P = 11.4 \tag{9.68}$$

and

$$x = \frac{1}{R_P}\left(\frac{3}{4\pi}\right)^{1/3}; \tag{9.69}$$

(iv) *the gradient expansion for atoms*[50]

$$E_{xc}[\rho] = C_D \int \rho^{4/3}\left\{C(N) + D(N)\left(r\frac{\nabla\rho}{\rho}\right)^2\right\} d\mathbf{r}, \tag{9.70}$$

with

$$C(N) = 1 + C_2 N^{-2/3}, \tag{9.71}$$

$$C_2 = -\{1 - (\tfrac{1}{3}\pi^2)^{1/3}\}, \tag{9.72}$$

$$D(N) = D_2 N^{-2/3} \tag{9.73}$$

and

$$D_2 = -\frac{4}{2187}\left(\frac{\pi^2}{3}\right)^{1/3}. \tag{9.74}$$

The trial functions employed in the calculations are given by eqns (9.50)–(9.53). First, it was necessary to determine for which atoms the product *ansatz* had sufficient flexibility. Static polarizabilities $\alpha_{2^l}(0)$ ($l=1,2$ and 3) were calculated (using the $E_{xc}[\rho]$ defined by eqns (66)–(69)) for He, Be, Ne, Mg, Ar, Ca and Kr. These variationally determined static multipole polarizabilities were compared with the $\alpha_2(0)$ calculated by Stott and Zaremba[54] and the $\alpha_{2^l}(0)$ calculated by Mahan.[55] They also used eqns (9.66)–(9.69) to approximate $E_{xc}[\rho]$ in their time-independent Kohn–Sham calculations, but solved the differential equations numerically. Their results are therefore "exact" for this approximation to $E_{xc}[\rho]$. Comparing the variationally determined $\alpha_{2^l}$ with the "exact" $\alpha_{2^l}(0)$ shows that the product *ansatz* was satisfactory for He, Ne and Ar. A more flexible trial function is needed for other atoms. This can be achieved by replacing $\chi_i^{(0)}$ in eqns (9.50) and (9.52) with a set of non-linear exponential variational parameters: preferably, one set of non-linear variational parameters for each set of linear variational parameters.

Next, calculations were performed to study the applicability of the above approximate $E_{xc}[\rho]$ to time-dependent problems. For the atoms He, Ne and Ar, the major source of error in the calculations comes from the approximation to $E_{xc}[\rho]$, and not from using the product *ansatz* in the trial functions. Values of $\alpha_{2^l}(0)$ calculated using the four approximations to $E_{xc}[\rho]$ suggests that the gradient expansion for atoms, eqns (9.70)–(9.74), provides the best overall results. Most striking are the results for He. For

**Table 9.1** Dipole sum rules (in a.u.) for the helium atom

| | $S(1,0)$ | $S(1,-1)$ | $S(1,-2)$ | $S(1,-3)$ | $S(1,-4)$ |
|---|---|---|---|---|---|
| Present* | 2 | 1.484 | 1.363 | 1.414 | 1.582 |
| Previous† | 2 | 1.504 | 1.383 | 1.415 | 1.542 |

*Calculated from the pseudo-spectrum generated by the $[6,5]_\beta$ Padè approximant. The $E_{xc}[\rho]$ defined by eqns (9.70–(9.74) was used in the calculations.
†From ref. 56. This was a correlated wave function functional calculation.

Ne and Ar, there was not that much difference between the results obtained using the $X\alpha'$ Gunnarsson–Lundqvist, and atomic gradient expansion $E_{xc}[\rho]$ functionals.

Table 9.1 gives some dipole sum rules calculated for the He atom. The atomic gradient expansion to $E_{xc}[\rho]$ was used in the calculations. These results are compared to the very accurate wave-function functional results obtained by Thakkar,[56] who used a compact but correlated wave function in solving the Schrödinger equation by variational techniques. Some dipole–dipole dispersion (van der Waals) coefficients $\varepsilon_2(1,1)$ are given in Table 9.2. Again, the results are compared with accurate values of $\varepsilon_2(1,1)$. The agreement between the density functional results and the accurate values is excellent. A more complete listing of $\alpha_{2^l}(0)$, $S(l,-k)$ and $\varepsilon_2(l,L)$ can be found in ref. 7.

The calculated values of $\alpha_4(0)$ and $\alpha_8(0)$ (and therefore the calculated $\varepsilon_2(l,L)$) for Ne and Ar are in very poor agreement with values calculated by Doran.[58] This poor agreement is obtained independent of the approximation to $E_{xc}[\rho]$. We observed that if we neglect the $\{E\}$ and $\{F\}$ linear variational parameters (for instance, the $\overrightarrow{p \to p}$ transition is neglected in $\alpha_4(0)$), we obtain results that are in better agreement with the many-body calculations of Doran. Reinsch and Meyer[59] also question Doran's value of $\alpha_4(0)$. We believe that Doran may have neglected some important transitions in his calculations.

**Table 9.2** Dipole–dipole dispersion coefficients (in a.u.)

| | He–He | $Li^+ - Li^+$ | Ne–Ne | Ar–Ar |
|---|---|---|---|---|
| Present* | 1.410 | 0.07652 | 6.775 | 68.53 |
| Previous | 1.461† | 0.07821† | $6.87 \pm 0.40$‡ | $67.2 \pm 3.6$‡ |

*Calculated from the pseudo-spectrum generated by the $[6,5]_\beta$ Padē approximant. The $E_{xc}[\rho]$ defined by eqns (9.70)–(9.74) was used in the calculations.
†From ref. 56. This was a correlated wave-function functional calculation.
‡From ref. 57.

The hydrodynamical formulation is not the only form of time-dependent Kohn–Sham theory, nor is it the most widely used. A generalization of the linear response method in which the time dependence has been factored into its positive and negative Fourier components has been successfully applied to a number of problems.[13–22] The formal justification of this approach has recently been given by Ghosh and Deb.[8]

Ahlberg and Goscinski[13] used this form of time-dependent Kohn–Sham theory, with the X$\alpha$ exchange-correlation energy density functional, to calculate $\alpha_2(0)$, $S(1, -3)$ and $\alpha_2(i\omega)$ for various atoms. Most recently, Soven and co-workers[14–17, 19–22] have used this form of the theory to study optical absorption and photoionization of atoms and molecules. They used the $E_{xc}[\rho]$ given in eqns (9.66)–(9.69) in their calculations and call the method the time-dependent local density approximation (TDLDA). Zangwill[19] has recently given an excellent review of the application of TDLDA to the calculation of atomic photoionization cross-sections.

The TDLDA method has been used by Mahan[18] to calculate $\alpha_2(i\omega)$ for various atoms and $\varepsilon_2(1,1)$ for various pairs of atoms. Zangwill[20] has used TDLDA to calculate the third-harmonic generation coefficient $\gamma(3\omega; \omega, \omega, \omega)$ for the noble gases. His results for $\gamma(3\omega; \omega, \omega)$ evaluated with $\omega = 6943$ Å, the ruby laser frequency, are in general agreement with experiment. The only exception was for the neon atom.

## 9.5 Conclusion

For chemical-systems interaction with external potentials which have a harmonic time dependence, time-dependent Kohn–Sham theory is in principle exact. However, the necessity of approximating the exchange-correlation energy density functional makes the application of the theory inexact. What effect does using approximate $E_{xc}[\rho]$ have on the usefulness of the theory? The calculations reported to date, using various approximations to $E_{xc}[\rho]$ are very encouraging and they do show that time-dependent Kohn–Sham functionals provide a viable alternative to the conventional wave-function functional approach. As more and more calculations are performed, especially on molecules, the full potential of this approach will become apparent. The future prospects of time-dependent Kohn–Sham theory are bright.

## References

1. Heitler, W. *The Quantum Theory of Radiation*, Oxford University Press, London.

2. Langhoff, P.W., Epstein, S.T. and Karplus, M. (1972). *Rev. Mod. Phys.*, **44**, 602.
3. Lundqvist, S. (1983). In *Theory of Inhomogeneous Electron Gas*, eds Lundqvist, S. and March, N.H.,, p. 149, Plenum, New York.
4. Peuckert, V. (1978). *J. Phys.* **3C11**, 4945.
5. Bartolotti, L.J. (1981). *Phys. Rev.*, **A24**, 1661.
6. Bartolotti, L.J. (1983). *Phys. Rev.*, **A26**, 2243; (1983). *Ibid.*, **27**, 2248.
7. Bartolotti, L.J. (1984). *J. Chem. Phys.*, **80**, 5687.
8. Ghosh, S.K. and Deb, B.M. (1982). *Chem. Phys.*, **71**, 295; (1984). *Ibid.*, **91**, 478.
9. Ghosh, S.K. and Deb, B.M. (1983). *Theoret. Chim. Acta* (*Berl.*), **62**, 209; (1985). *Ibid.*, **68**, 336.
10. Runge, E. and Gross, E.K.U. (1984). *Phys. Rev. Lett.*, **52**, 997.
11. Xu, Bu-Xing, and Rajagopal, A.K. (1985). *Phys. Rev.*, **A31**, 2682.
12. Kohn, W. and Sham, L.J. (1965). *Phys. Rev.*, **140**, A1133.
13. Ahlberg, R. and Goscinski, O. (1975). *J. Phys.*, **B8**, 2149.
14. Zangwill, A. and Soven, P. (1980). *Phys. Rev.*, **A21**, 1561.
15. Zangwill, A. and Soven, P. (1981). *Phys. Rev. Lett.*, **45**, 204.
16. Zangwill, A. and Soven, P. (1980). *J. Vac. Sci. Technol.*, **17**, 159.
17. Zangwill, A. and Soven, P. (1981). *Phys. Rev.*, **B24**, 4121.
18. Mahan, G.D. (1982). *J. Chem. Phys.*, **76**, 493.
19. Zangwill, A. (1983). In *Atomic Physics 8*, eds. Lindgren, I. Rosen, A. and Svanberg S., p. 339, Plenum, New York.
20. Zangwill, A. (1983). *J. Chem. Phys.*, **78**, 5926.
21. Levine, Z.H. and Soven, P. (1983). *Phys. Lett.*, **50A**, 2074.
22. Levine, Z.H. and Soven, P. (1984). *Phys. Rev.*, **A29**, 625.
23. Levy, M. (1979). *Proc. Natl Acad. Sci.* U.S.A., **76**, 6062.
24. Levy, M. (1979). *Bull. Am. Phys. Soc.*, **24**, 626.
25. Levy, M. and Perdew, J.P. (1983). In *Density Functional Methods in Physics*, eds Dreizler, R.M. and da Providencia, J., Plenum, New York.
26. Flytzanis, C. (1975). In *Quantum Electronics*, Part A, eds Rabin H. and Tang, T.L., p. 9, Academic Press, New York.
27. Kielich, S. (1976). In *Molecular Electro-Optics*, Part I. *Theory and Methods*, ed. O'Konski, C.T., p. 391, Marcel Dekker, Inc., New York.
28. Reintjes, J.F., (1984). *Nonlinear Optical Parametric Processes in Liquids and Gases*, Academic Press, New York.
29. Shen, Y.R. (1984). *The Principles of Nonlinear Optics*, John Wiley, New York.
30. Dalgarno, A. (1967). *Adv. Chem. Phys.*, **12**, 143.
31. See for instance, Bartolotti, L.J. and Mollmann, J.C. (1979). *Molec. Phys.*, **38**, 1359.
32. Bartolotti, L.J. and Epstein, S.T. (1979). *Molec. Phys.*, **38**, 1311.
33. Langhoff, P.W. and Karplus, M. (1969). *J. Opt. Soc. Am.*, **59**, 863.
34. Langhoff, P.W. and Karplus, M. (1970). *J. Chem. Phys.*, **52**, 1435.
35. Langhoff, P.W. and Karplus, M. (1970). *J. Chem. Phys.*, **53**, 233.
36. Langhoff, P.W. and Karplus, M. (1970). In *The Padè Approximants in Theoretical Physics*, eds Baker, Jr., G.A. and Gammel, J.L., p. 41, Academic Press, New York.
37. Langhoff, P.W., Gordon, R.G. and Karplus, M. (1971). *J. Chem. Phys.*, **55**, 2126.
38. Bartolotti, L.J. and Tyrrell, J. (1976). *Chem. Phys. Lett.*, **39**, 19.
39. Bartolotti, L.J. and Tyrrell, J. (1978). *Molec. Phys.*, **36**, 97.
40. Dalgarno, A. and Davison, W.D. (1966). *Adv. Atomic Molec. Phys.*, **2**, 1.
41. Dalgarno, A. (1965). *Perturbation Theory and its Applications in Quantum Mechanics*, ed. Wilcox, C.H., p. 145, John Wiley & Sons, Inc., New York.

42. Bartolotti, L.J. (1979). *Chem. Phys. Lett.*, **60**, 507.
43. Bartolotti, L.J. (1980). *J. Chem. Phys.*, **73**, 3666.
44. Langhoff, P.W., Karplus, M. and Hurst, R.P. (1966). *J. Chem. Phys.*, **44**, 505.
45. Foley, H.M., Sternheimer, R.M. and Tycko, D. (1954). *Phys. Rev.*, **93**, 734.
46. Langhoff, P.W. and Hurst, R.P. (1965). *Phys. Rev.*, **139**, A1415.
47. Epstein, S.T. (1974). *The Variational Method in Quantum Chemistry*, p. 175, Academic Press, New York.
48. Rajagopal, A.K. (1980). *Adv. Chem. Phys.*, **41**, 59.
49. Perdew, J.P. and Zunger, A. (1981). *Phys. Rev.*, **B23**, 5048.
50. Bartolotti, L.J. (1982). *J. Chem. Phys.*, **76**, 6057.
51. Dirac, P.A.M. (1930). *Proc. Camb. Phil. Soc.*, **26**, 376.
52. Slater, J.C. (1974). *The Self-Consistent Field for Molecules and Solids*, McGraw-Hill, New York.
53. Gunnarsson, O. and Lundqvist, B.I. (1976). *Phys. Rev.*, **B13**, 4274.
54. Stott, M.J. and Zaremba, E. (1980). *Phys. Rev.*, **A21**, 12; (1980). *Ibid.*, **A22**, 2293.
55. Mahan, G.D., (1980). *Phys. Rev.*, **A22**, 1780.
56. Thakkar, A.J. (1981). *J. Chem. Phys.*, **75**, 4496.
57. Tang, K.T., Norbeck, J.M. and Certain, P.R. (1976). *J. Chem. Phys.*, **64**, 3063.
58. Doran, M.B. (1974). *J. Phys.*, **B7**, 558.
59. Reinsch, E.A. and Meyer, W. (1978). *Phys. Rev.*, **A18**, 1793.
60. Ghosh, S.K. and Deb, B.M. (1983). *J. Molec Struc. Theochem.*, **103**, 163.

# 10 Inhomogeneous Liquids: Mainly Liquid Surfaces

N.H. MARCH

*Theoretical Chemistry Department, University of Oxford, 1 South Parks Road, Oxford OX13TG, England*



Single Particle Density in Physics
and Chemistry ISBN 0-12-470518-9

## 10.1 Introduction

The purpose of this the final Chapter is to provide an introduction to the statistical mechanical theory of inhomogeneous systems. While the main focus will be classical statistical mechanics, in the final Section an outline will be given of the currently available theory of inhomogeneous quantum systems. Inevitably, some of the theory to date remains formal and therefore it will be useful to introduce the discussion by two elementary examples to illustrate the way the single-particle density can be employed to characterize the properties of inhomogeneous liquids.

The first of these is, in fact, quantal: the description of the inhomogeneous electron gas at the surface of a simple liquid metal. However, the work covered in the previous Chapters is entirely adequate for this purpose; as already mentioned the quantal liquid case will then be taken up again in the final Section 10.14, including a closely related metal-surface example. The second example treated in this context of illustrating the use of the single-particle density is classical: the Gouy–Chapman theory, relating to electrical double layers and coagulation. Beyond these examples, the dominant applications considered in this Chapter are the various statistical mechanical methods for calculating liquid surface tension and some results from the theory of freezing treated as an application of the statistical mechanics of inhomogeneous systems.

## 10.2 Two examples of use of single-particle density profiles

Let us turn then immediately to the first of the two examples cited above; the electron density profile through the surface of a simple liquid metal.

### 10.2.1 Inhomogeneous electron gas at a metal surface

The simplest theories of the surface tension of liquid metals focus all attention on the behaviour of the conduction electrons. At the most elementary level, one constructs, following Brown and March,[1] the total energy $E$ of the inhomogeneous system as

$$E=\int \mathrm{d}\mathbf{r}\left[\varepsilon(\rho)+\frac{\lambda\hbar^2}{8m}\frac{(\nabla\rho)^2}{\rho}\right], \tag{10.1}$$

where $\lambda$ is a numerical factor. The slowly varying electron gas theory, in fact gives $\lambda=\frac{1}{9}$, whereas von Weizsäcker's original work would correspond to the choice $\lambda=1$ (see Appendix A 1.1). In this eqn (10.1) the local energy term $\varepsilon(\rho)$ is the kinetic, exchange and correlation energy of a homogeneous electron gas at density $\rho$, while the inhomogeneity term in $(\nabla\rho)^2$ is associated with kinetic energy. The variational principle used extensively in Chapter 1, namely

$$\frac{\delta(E-\mu N)}{\delta\rho}=0, \tag{10.2}$$

yields from eqn (10.1) the Euler equation

$$-\frac{\lambda\hbar^2}{4m}\left[\frac{\rho''}{\rho}-\frac{1}{2}\left(\frac{\rho'}{\rho}\right)^2\right]+\frac{\mathrm{d}\varepsilon}{\mathrm{d}\rho}=\mu, \tag{10.3}$$

from which the conduction electron density profile $\rho(a)$ through the liquid metal surface can be determined. Setting $\rho=\psi^2$, this is formally equivalent to a Schrödinger equation of the form

$$-\frac{\lambda\hbar^2}{2m}\frac{\mathrm{d}^2\psi}{\mathrm{d}z^2}+\left(\frac{\mathrm{d}\varepsilon}{\mathrm{d}\rho}-\mu\right)\psi=0. \tag{10.4}$$

The surface tension, defined as the energy difference between the inhomogeneous electron gas and a homogeneous electron gas, per unit surface area (cf. the analogy with eqn (10.60) below) is given by

$$\sigma=\frac{\lambda\hbar^2}{m}\int_{-\infty}^{\infty}\mathrm{d}z\,[\psi'(z)]^2. \tag{10.5}$$

This elementary approach has the merit that it can be completely solved analytically for certain simple forms of $\varepsilon(\rho)$ and in particular for the case

$$\varepsilon(\rho) = \rho\varepsilon_0[(\rho/\rho_0)^{2/3} - 2(\rho/\rho_0)^{1/3}], \tag{10.6}$$

which corresponds to the sum of a kinetic term $\sim\rho^{5/3}$, discussed in Chapter 1, and an exchange term $\sim\rho^{4/3}$. Clearly $\rho_0$ is the equilibrium density at zero pressure and the corresponding electron gas compressibility is given by $K^{-1} = \frac{2}{9}\rho_0\varepsilon_0$. The solution of eqn (10.3) is then

$$\rho(z)/\rho_0 = (1 + B\exp(z/l))^{-3}, \tag{10.7}$$

where $B$ is a constant, while the length $l$, which clearly measures the surface thickness, is given by

$$l = (9\lambda h^2/8m\varepsilon_0)^{1/2}. \tag{10.8}$$

From eqn (10.5) one finally obtains

$$\sigma K = \tfrac{3}{4}l. \tag{10.9}$$

An important advance in the more refined theories of the electronic surface density profile $\rho(z)$ is to avoid a density-gradient treatment for the electronic kinetic energy. The equilibrium condition for the profile can still be formulated as a Schrödinger equation in which, however, the Laplacian term accounts for the full single-particle kinetic energy while the potential term includes exchange and correlation contributions. A Hartree-like term for the electron–ion interaction also appears when the ionic and electronic density profiles are allowed to differ somewhat, as will indeed be the case in liquid metals. For specific work on liquid metals, including some account of the transition from metallic to localized electron states in the transition region from liquid to vapour, the work of Rice *et al.*[2] should be consulted. As already mentioned, the final Section includes an account of the metal surface problem via the full many-body theory of an inhomogeneous quantal liquid. In concluding this example, it is of interest to note that Singwi and Tosi[3] have adapted the above method to calculate the surface tension of the electron–hole liquid which exists in a semiconductor like Ge. For the background to their work, the account in the book by March and Parrinello[4] may be consulted.

### 10.2.2 Solvation forces, electrical double layer and density profiles

In this Section, the role of liquid structure in determining the forces between solids immersed in liquids will be treated, following, for example, the account of Rickayzen.[5] As he stresses, the problem is of particular importance for understanding the stability of colloids, especially sols. Sols comprise solid particles dispersed through a liquid medium. Although the particles are large compared with the liquid molecules, they are still microscopic and their surfaces play an important part in their behaviour.

Seen from a distance, these particles are usually neutral and the dominant long-range force between them is the van der Waals interaction. This force arises from the polarizability of the particles and of the medium through which they interact. Because of such polarizability, and owing to the long range of the interaction of the dipoles, the polarization of the whole has normal modes whose frequencies depend on the separation of the particles. The zero-point energies of these oscillations contribute to the total energy of the system and since the zero-point energies are the lower the closer are the particles, this results in an attraction between the particles. For separations $r$ that are large compared with their size, but not so large that retardation effects become important, the potential energy of interaction has the well-known form

$$V = -C/r^6, \tag{10.10}$$

$C$ depending on the dynamic polarizabilities of the particles and of the liquid. On the other hand, for separations small compared with their size, the particles can be treated as thick flat plates. In this case the interaction energy is given by

$$V = -A/12\pi r^2, \tag{10.11}$$

where $A$ depends on the polarizabilities. The constant $A$ is known as the Hamaker constant, for which there is an extensive theory, which however will not be treated here.

Since this force is attractive, its tendency will be to bring the suspended particles together into a coagulate. However many sols have very long lives and give the appearance of being stable. Thus it must be the case that this attractive force is opposed by a shorter range repulsive force which prevents coagulation. In electrolytes this force is widely regarded as an electrostatic interaction due to electric double layers surrounding the colloidal particles. The solid particles acquire a net charge, which the surrounding ions in the liquid will tend to neutralize. However, because of their finite size and kinetic energy there will be some separation of the net charge on the particle from the net charge in the fluid. This constitutes the electrical double layer. On the approach of two such layers, like charges in the liquid will come closest together, and this is the origin of the repulsion that opposes coagulation. This Section is concerned with the classical approach to calculating this repulsion, and to modern modifications of this treatment.

Throughout the Section, and for simplicity, only the forces between two thick parallel plates separated by a distance $h$ will be considered. Derjaguin has demonstrated that the force between two large spherical particles of radius $R$ and separation $h$ ($\ll R$) can be found from knowledge of that between two plates. In fact, if $V(h)$ is the potential between the plates, the force between the spheres is

$$F(h)=\pi R V(h), \tag{10.12}$$

and it therefore follows that the problem of flat plates is already of practical relevance. One important result of the classical theory, to be outlined below, is the existence of a critical concentration of electrolyte, $\rho_{0c}$ say, below which the colloid will be stable. In the theory, the critical concentration satisfies the so-called Schulze–Hardy rule:[5]

$$\rho_{0c} \propto T^5/Z^6, \tag{10.13}$$

$T$ being, as usual, the temperature and $Z$ the valency of the ions. This rule does seem to accord with experiment although it has probably not been very stringently tested to date.

### 10.2.3 Gouy–Chapman theory and coagulation

For simplicity, the case of two equally charged surfaces will be treated below. In the theory of Gouy and Chapman,[6] one views the solution as a medium of diectric constant $\varepsilon_r$ in which the ions are embedded. If there are different kinds of ions of charge $e_i$ and number density $\rho_i(x)$ at position $x$, there is an electrostatic potential $V(x)$ which satisfies Poisson's equation

$$\nabla^2 V(x) = -\sum_i e_i \rho_i(x)/\varepsilon_0 \varepsilon_r. \tag{10.14}$$

At low density the ions can be treated as independent and in a potential $V(x)$ each ion has energy $e_i V(x)$. The Boltzmann distribution then gives the ionic density profile $\rho_i$ in terms of the number density $\rho_{0i}$ in the bulk, corresponding to $V(x)=0$, as

$$\rho_i(x) = \rho_{0i} \exp(-\beta e_i V(x)). \tag{10.15}$$

Given that the surface charge density on each plate is $Q$, the potential is fixed by overall charge neutrality through

$$\sum_i e_i \int \rho_i(x)\,\mathrm{d}x = -2Q. \tag{10.16}$$

Once the densities $\rho_{0i}$ are specified, eqns (10.14)–(10.16) are sufficient to determine $V(x)$ completely. Since the bulk is neutral these densities must satisfy

$$\sum_i e_i \rho_{0i} = 0 \tag{10.17}$$

*Limit of Debye–Hückel theory:* $\beta e_i V(x) \ll 1$ for all $i$

Employing eqn (10.15) in eqn (10.14) yields

$$\frac{\mathrm{d}^2 V(x)}{\mathrm{d}x^2} = -\sum_i \frac{e_i \rho_{0i} \exp(-\beta e_i V(x))}{\varepsilon_0 \varepsilon_r}. \tag{10.18}$$

Given the assumption that the potential is weak, the exponentials can be expanded to first order in $V(x)$ to yield after using condition (10.17) to eliminate the zeroth-order terms:

$$\frac{\mathrm{d}^2 V(x)}{\mathrm{d}x^2} = \sum_i \frac{e_i^2 \rho_{0i} \beta V(x)}{\varepsilon_0 \varepsilon_r} = +\frac{V(x)}{\lambda^2}, \tag{10.19}$$

the Debye length $\lambda$ being defined by (cf. screening of test charges in solids in Chapter 5)

$$\lambda^{-2} \equiv \kappa^2 = \beta \sum_i e_i^2 \frac{\rho_{0i}}{\varepsilon_0 \varepsilon_r}. \tag{10.20}$$

With plates at separation $h$, placed at $x = \pm\frac{1}{2}h$, the solution of eqn (10.19), symmetric about the origin, is readily verified to be

$$V(x) = A \cosh(x/\lambda). \tag{10.21}$$

The electric field $\mathscr{E}(x)$ at any point in the fluid is readily obtained:

$$\mathscr{E}(x) = -\frac{\mathrm{d}V}{\mathrm{d}x} = -\frac{A}{\lambda} \sinh\left(\frac{x}{\lambda}\right). \tag{10.22}$$

However, just outside the left-hand plate the field is given by

$$\mathscr{E}(-\tfrac{1}{2}h) = \frac{Q}{\varepsilon_0 \varepsilon_r} = \frac{A}{\lambda} \sinh\left(\frac{h}{2\lambda}\right), \tag{10.23}$$

and hence it follows that

$$A = \frac{\lambda Q}{\varepsilon_0 \varepsilon_r \sinh(h/2\lambda)} \tag{10.24}$$

and

$$V(x) = \frac{\lambda Q \cosh(x/\lambda)}{\varepsilon_0 \varepsilon_r \sinh(h/2\lambda)}. \tag{10.25}$$

This solution automatically ensures charge conservation.

It will be clear from the above that the Debye length characterizes the range over which the ionic density profiles, and the potential, vary.

### 10.2.4 Energy of system

The energy in the system, $\Omega$ say, is the energy required for the plates to acquire charge $Q$ from zero and is given by

$$\Omega(Q)=2\int_0^Q V(Q')\,\mathrm{d}Q', \tag{10.26}$$

where $V(Q')$ is the potential of a plate when the charge is $Q'$. Since $V(Q)$ is proportional to $Q$, this yields

$$\Omega(Q)=V(Q)Q=(\lambda Q^2/\varepsilon_0\varepsilon_r)\coth(h/2\lambda), \tag{10.27}$$

which is the potential energy from which the force can be obtained at constant charge on the plates.

For the more usual case of constant potential on the plates, one makes use of the function $\mathscr{F}(V_0)$ related to $\Omega(Q)$ by a Legendre transformation:

$$\mathscr{F}(V_0)=\Omega(Q)-2V_0Q=-(\varepsilon_0\varepsilon_r V_0^2/\lambda)\tanh(h/2\lambda), \tag{10.28}$$

where $V_0$ is the potential on the plate such that

$$V_0=(\lambda Q/\varepsilon_0\varepsilon_r)\coth(h/2\lambda). \tag{10.29}$$

Measuring relative to the potential energy when the separation is infinite, one finds explicitly

$$\mathscr{F}(V_0)=\frac{\varepsilon_0\varepsilon_r V_0^2}{\lambda}\left[1-\tanh\left(\frac{h}{2\lambda}\right)\right]. \tag{10.30}$$

Since this is positive, the plates tend to repel each other as expected from the fact that like charges are being brought close together. The force of repulsion when $V_0$ is held fixed is given by

$$f_v(h)=-\frac{\partial\mathscr{F}}{\partial h}=\frac{\varepsilon_0\varepsilon_r V_0^2}{2\lambda^2}\operatorname{sech}^2\left(\frac{h}{2\lambda}\right). \tag{10.31}$$

As the separation $h$ between the plates tends to infinity this tends to $(2\varepsilon_0\varepsilon_r V_0^2/\lambda^2)\exp(-h/\lambda)$ and hence decays exponentially with decay distance determined by the Debye length $\lambda$.

With regard to the magnitude of the Debye length, following, for example, Rickayzen,[5] when $T=300\,\mathrm{K}$, $\varepsilon_r=80$ and with a $1-1$ electrolyte of 0.5 M, $\lambda=4\,\text{Å}$, and it should also be noted that $\lambda\propto\rho^{-1/2}$. For $\lambda\sim4\,\text{Å}$, however, one ought to account for the solvent structure, a point to which we shall return later.

### 10.2.5 Gouy–Chapman results

Consider the case of an electrolyte with two ions with $e_1 = -e_2 = Ze$, where $e$ is the elementary charge, and $\rho_{01} = \rho_{02} = \frac{1}{2}\rho_0$. Equations (10.14) and (10.15) then lead to the explicit results

$$\frac{d^2V}{dx^2} = -\frac{Ze\rho_0}{2\varepsilon_0\varepsilon_r}[\exp(-\beta ZeV) - \exp(\beta ZeV)]$$

$$= \frac{Ze\rho_0}{\varepsilon_0\varepsilon_r}\sinh(\beta ZeV). \tag{10.32}$$

Hence

$$\left(\frac{dV}{dx}\right)^2 = \frac{2\rho_0}{\beta\varepsilon_0\varepsilon_r}[\cosh(\beta ZeV) - \cosh(\beta ZeV_m)], \tag{10.33}$$

where $V = V_m$ when $dV/dx = 0$, at the midpoint between the plates. This eqn (10.33) can be integrated further using elliptic integrals, but the details will not be developed here.

For infinite separation $h, V_m$ is zero and eqn (10.33) can be solved to yield

$$\frac{dV}{dx} = \pm\left(\frac{4\rho_0}{\beta\varepsilon_0\varepsilon_r}\right)^{1/2}\sinh(\tfrac{1}{2}\beta ZeV). \tag{10.34}$$

If one chooses the origin so that the plate is at $x = 0$, this equation can be integrated to yield

$$\tanh(\tfrac{1}{4}\beta ZeV) = \tanh(\tfrac{1}{4}\beta ZeV_0)\exp(-x/\lambda), \tag{10.35}$$

where $V_0$ denotes the potential on the plate. When $V_0$ is so small that

$$\beta ZeV_0 \ll 1, \tag{10.36}$$

then eqn (10.35) reduces to the Debye–Hückel result

$$V = V_0\exp(-x/\lambda). \tag{10.37}$$

### 10.2.6 Critical concentration: the Schulze–Hardy rule

The interaction between two plates will now be obtained, the argument below following closely the account of Rickayzen.[5] First, there is a repulsive contribution, arising from the Coulomb interaction of the charges, which decays exponentially at large separations and which tends to a finite limit as the plates are brought together. The second part of the interaction is the van der Waals attraction, which has a power-law dependence on the separation. The combination of the two terms results in the sum, denoted by total potential energy $W$, having a maximum, say $W_m$. If this maximum value is negative, there is no obstacle to the plates coming together and coagulation will take place rapidly. But if $W_m > 0$, there is a macroscopic barrier to be surmounted and coagulation will be slowed down. One can conclude, therefore, that a rough criterion for stability is $W_m > 0$. The implication then is that there is a critical ion concentration $\rho_{0c}$ determined by

$$\frac{dW(h)}{dh} = 0, \quad W(h) = 0 \tag{10.38}$$

at which the colloid becomes unstable. For $\rho_0 < \rho_c$ the colloid is stable while for $\rho > \rho_c$ it is unstable.

The total potential energy between the walls, including the van der Waals contribution, is $(Z\beta eV_0 \ll 1)$

$$W(h) = \varepsilon_0 \varepsilon_r \frac{V_0^2}{\lambda} \left[1 - \tan\left(\frac{h}{2\lambda}\right)\right] - \frac{A}{12\pi h^2}. \tag{10.39}$$

For $\exp(h/\lambda) \gg 1$ this becomes

$$W(h) = \frac{2\varepsilon_0 \varepsilon_r V_0^2}{\lambda} \exp\left(\frac{-h}{\lambda}\right) - \frac{A}{12\pi h^2}. \tag{10.40}$$

Using the above equation for the critical concentration, one obtains

$$h/\lambda = 2 \tag{10.41}$$

and

$$2\varepsilon_0 \varepsilon_r V_0^2 e^{-2}/\lambda = A/12\pi h^2. \tag{10.42}$$

Since $\lambda$ depends on $\rho_0$ this is an equation for the critical concentration. This needs refinement, because it has been assumed that $\exp(h/\lambda) \gg 1$ and $Z\beta eV_0/2 \ll 1$ (i.e. $V_0 \ll 100/Z$ mV). Neither of these approximations is entirely satisfactory. Solving the Gouy–Chapman equation yields the new condition for a $Z$–$Z$ electrolyte

$$\rho_c = \frac{(192)^2 (4\pi\varepsilon_0 \varepsilon_r)^3 (k_B T)^5 \exp(-4)}{\pi Z^6 e^6 A^2}. \tag{10.43}$$

For water at 25°C,

$$\rho_c = 17.3 \times 10^{-39}/Z^6 A^6,$$

which is the Schulze–Hardy rule. Data are given in Table 10.1, showing that agreement with experiment is good, though the value of $A$ is overestimated.

Having given these two examples of the value of the description of simple

**Table 10.1** Data relevant to Schulze–Hardy rule (adapted from ref. 5). Quantity actually tabulated is critical coagulation concentration in millimoles per litre.

| *Charge Z for counter-ions* | *Sol of $As_2S_2$ (negatively charged)* | | *Sol of Au (negatively charged)* | | *Sol of $Fe_2O_3$ (positively charged)* | | *Sol of $Al_2O_3$ (positively charged)* | |
|---|---|---|---|---|---|---|---|---|
| 1 | LiCl | 58 | | | | | | |
| | NaCl | 51 | NaCl | 24 | NaCl | 9.25 | NaCl | 43.5 |
| | KCl | 49.5 | | | KCl | 9.0 | KCl | 46 |
| | | | $KNO_3$ | 25 | $KNO_3$ | 12 | $KNO_3$ | 60 |
| 2 | $MgCl_2$ | 0.72 | | | | | | |
| | $MgSO_4$ | 0.81 | | | $MgSO_4$ | 0.22 | | |
| | $BaCl_2$ | 0.69 | $BaCl_2$ | 0.35 | | | | |
| | $CaCl_2$ | 0.65 | $CaCl_2$ | 0.41 | | | | |
| | | | | | $K_2SO_4$ | 0.205 | $K_2SO_4$ | 0.30 |
| | | | | | $K_2Cr_2O_7$ | 0.195 | $K_2Cr_2O_7$ | 0.63 |
| 3 | $AlCl_3$ | 0.093 | | | | | | |
| | $\frac{1}{2}Al_2(SO_4)_3$ | 0.096 | $\frac{1}{2}Al_2(SO_4)_3$ | 0.009 | | | | |
| | | | | | | | $K_2Fe(CN)_6$ | 0.080 |

inhomogeneous systems by the single-particle density, in much of the remainder of this Chapter emphasis will be placed on the classical statistical mechanics of the liquid–vapour surface, in which the average atomic density varies rapidly as one passes from bulk liquid through to the vapour phase. To approach this problem, the classical statistical mechanics of inhomogeneous systems will first be set up within the density functional framework.[7] In the case of the surface[8,9] of a simple liquid like argon, it will emerge that the average atomic density $\rho(z)$ through the surface is related directly to the so-called direct Ornstein–Zernike correlation in the presence, however, of the interface. From the standpoint of the single-particle density which is the main theme here, one can say that the Ornstein–Zernike correlation function of the inhomogeneous system is, itself, a function of the average atomic density. Unfortunately, this functional relation is not presently known. Therefore, a lot of attention will be given, below, to the development of a method appropriate for a slowly varying particle density; namely the method of low-order gradient expansion, which is the statistical mechanical analogue of the electron density gradient approach considered in some detail in the earlier Chapters. This electron density gradient method, in fact, has been used for the surfaces of liquid metals in the previous Section, where it was remarked that a complete solution of the problem will require knowledge of both electron and ionic density profiles.

After a discussion of the one-component classical liquid density profile, and the way in which this determines the surface tension, the same basic theory of the statistical mechanics of inhomogeneous systems will be applied to a number of related problems. These embrace, briefly, the theory of freezing, spinodal decomposition, the metal–electrolyte interface, and finally the problem of liquid crystals.

## 10.3 Thermodynamics of liquid surfaces

The atomic density profile $\rho(z)$ must vary continuously across the interface from the value $\rho_l$ of the bulk liquid to the value $\rho_v$ of the bulk vapour. This variation can be expected to take place over a few interatomic distances, at least if one is far from the critical point.

The anisotropy of the profile implies a net attraction to the liquid phase for an atom in the transition region: one must do work to bring an atom from the bulk of the liquid to the surface; i.e. an excess of free energy is associated with the creation of the interface, namely the surface free energy. It also implies that the tangential pressure, defined as the force per unit area transmitted perpendicularly across an area element in the $y$–$z$ or $x$–$z$ plane is a function $p_t(z)$ of position in the transition region. The difference between the components $p_t(z)$ and $p_n = p$ of the stress tensor in the transition region is negative, i.e. it has the nature of a tension, namely surface tension.

The surface free energy, as discussed, for example in the reviews by Brown and March,[10] or that by Tosi,[11] is defined by comparing the free energy $F$ of the actual system with the sum of the free energies of suitably chosen amounts of homogeneous liquid and vapour phases. If $f_l$ and $f_v$ are the free energies per unit volume of the homogeneous phases, one can write the surface free energy $\sigma$ per unit area as

$$\sigma = \frac{1}{A}(F - f_l V_l - f_v V_v), \tag{10.44}$$

where $A$ is the surface area while $V_l$ and $V_v$ are the volumes of the homogeneous phases. These are fixed by

$$V_l + V_v = V, \quad \rho_l V_l + \rho_v V_v = N, \tag{10.45}$$

which therefore attributes no excess matter to the interface. These conditions can be re-expressed as

$$\int_{-\infty}^{z_G} dz(\rho_l - \rho(z)) + \int_{z_G}^{\infty} (\rho_v - \rho(z)) = 0, \tag{10.46}$$

which fixes the location $z_G$ of the Gibbs surface dividing the two hypothetical homogeneous fluids.

The thermodynamic definition (10.44) has some consequences:

(i) $\sigma\, dA$ is the work required to increase the surface area by an amount $dA$ in any isothermal reversible process; and
(ii) the excess surface entropy is given by $s = -d\sigma/dT$ and hence the excess surface energy $u = \sigma - T\, d\sigma/dT$.

With respect to (i), it should be stressed that the work done against the surface tension in expanding the surface area by stretching is equal to the surface free energy of the same area of the new surface; i.e. the surface tension and the surface free energy, which are usually expressed in $\text{dyn cm}^{-1}$ and in $\text{erg cm}^{-2}$, respectively, are numerically the same for an interface between two fluids. This line of argument can be employed to express $\sigma$ in terms of the integrated deficit of tangential pressure as

$$\sigma = \int dz\,(p - p_t(z)), \tag{10.47}$$

a derivation of this result being given, for instance, in the book by March and Tosi. The argument assumes that alternative processes lead to the same equilibrium surface structure, as will be the case when diffusion is allowed.

It is worth summarizing here a few facts and some phenomenology, before turning to the theory in terms of the density gradient approach. First, as treated by Faber,[12] the values of $\sigma$ for liquid metals correlate with the latent heat of vaporization. This can be reconciled intuitively by the notion that the work done in bringing an atom from the bulk liquid up to the surface involves breaking a fraction of its bonds. The corresponding surface entropy, obtained as already discussed above from the measured temperature dependence of $\sigma$, is about $k_B$ per surface atom. Such an amount can be estimated by considering the effect of replacing an appropriate number of bulk sound waves by capillary waves, as discussed, for example, in Faber's book.[12] Some notable exceptions should be noted though; e.g. Zn where $d\sigma/dT$ is positive over a limited range of temperatures above the triple point. One anticipates that $\sigma(T)$ should decrease with increasing temperature and vanish at the critical point, where the distinction between liquid and vapour no longer obtains.

It is also of interest to record that a "law of corresponding states" is observed, as summarized by Buff and Lovett,[13] for the surface tension of simple atomic and molecular liquids. In particular, in terms of critical volume $V_c$ and temperature $T_c$, the quantity $\sigma(T)V_c^{2/3}/T_c$ is practically a universal function of $T/T_c$. The data turn out to be fitted quite usefully by

$$\sigma(T) \propto (1 - T/T_c)^{1.27 \pm 0.02}. \tag{10.48}$$

The final useful fact we note here is that molten salts, the values of $\sigma(T)$ being given by Janz,[14] have a large cohesive energy relative to a free-ion state but comparatively modest values of $\sigma$, which can be taken to indicate that ionic bonding is largely preserved in the surface.

Following this introduction, we return now to the central theme as to the way the liquid–vapour interface can be characterized, along of course with the surface tension, by the properties of the density profile.

## 10.4 Phenomenology and density gradient theory: Cahn–Hilliard theory of width of liquid–vapour interfaces

It will be useful to approach the description of surface tension in terms of the density profile by considering the phenomenology associated with the names of Cahn and Hilliard.[15] As will emerge below, these workers, essentially, characterize the atomic density profile by a single parameter, its "width" $l$ (cf. Section 10.2.1). The precise meaning of this will only become completely clear when we turn from phenomenology to a microscopic theory based on the classical statistical mechanics of inhomogeneous fluids. The correlation established by Cahn and Hilliard between surface tension $\sigma$

**Table 10.2** Relation between surface tension and bulk compressibility of metals near melting point (after Alonso and March[17]).

| *Metal* | *Surface tension* $\sigma$ *(dyn cm$^{-1}$)* | *Isothermal compressibility* $K_{T_m}$ $(10^{-12}\,dyn^{-1}\,cm^2)$ | *Product* $\sigma K_{T_m}$ (Å) *(see eqn (10.9))* |
|---|---|---|---|
| Li | 410 | (11) | 0.45 |
| Na | 200 | 18.6 | 0.37 |
| K | 110 | 38.2 | 0.42 |
| Rb | 85 | 49.3 | 0.42 |
| Cs | 70 | 68.8 | 0.48 |
| Be | (1350) | (1.94) | 0.26 |
| Mg | 570 | 5.06 | 0.29 |
| Ca | 350 | 11.0 | 0.38 |
| Sr | 295 | 13.1 | 0.39 |
| Ba | 255 | 17.8 | 0.45 |
| Cu | 1310 | 1.49 | 0.19 |
| Ag | 910 | 2.11 | 0.19 |
| Zn | 770 | 2.50 | 0.19 |
| Cd | 590 | 3.24 | 0.19 |
| Hg | 485 | 3.75 | 0.18 |
| Al | 865 | 2.42 | 0.21 |
| Ga | 715 | 2.19 | 0.16 |
| In | 560 | 2.96 | 0.17 |
| Tl | 465 | 3.83 | 0.18 |
| Sn | 570 | 2.71 | 0.15 |
| Pb | 460 | 3.49 | 0.16 |
| Sb | 390 | 4.90 | 0.19 |
| Bi | 380 | 4.21 | 0.16 |
| Fe | 1830 | 1.43 | 0.26 |

and isothermal compressibility $K_T$ was shown by Egelstaff and Widom[16] to be strikingly verified experimentally for a whole class of liquids at or near the triple point. A compilation made subsequently by Alonso and March[17] is reproduced in Table 10.2.

Because of this empirical correlation, it is of obvious interest to understand the problem more deeply from first-principles theory. Since, however the motivation for the argument below, due to Bhatia and March,[18] lies in the Cahn–Hilliard approach, let us recall briefly that $\sigma$ is written as the sum of two terms:

$$\begin{aligned}\sigma &= \sigma_1 + \sigma_2 \\ &= [l(\Delta\rho)^2/2\rho^2 K_T] + Bl[\Delta\rho/l]^2. \qquad (10.49)\end{aligned}$$

In this expression, the first term arises from the treatment of the surface inhomogeneity as an "accidental fluctuation" while the second evidently comes from the density gradient. In eqn (10.49), $l$ is the effective thickness of the interface, $\Delta\rho$ is the fluctuation in the number density $\rho$, while $B$ is a constant.

Minimizing $\sigma$ with respect to $l$ yields the alternative forms

$$\sigma_{\min} = 2Bl(\Delta\rho/l)^2 \tag{10.50}$$

or

$$\sigma_{\min} = l(\Delta\rho)^2/\rho^2 K_T \quad \text{or} \quad \sigma_{\min} K_T \sim l. \tag{10.51}$$

The objective below is to expose the origin of this correlation (10.51) from first-principles theory. To do this, let us seek a parallel with the above phenomenology, in that

(a) one wishes to minimize the free energy of the inhomogeneous system with respect to the density profile $\rho(x)$; and
(b) one again wants to separate density gradient terms clearly from bulk or local density contributions.

Such an approach has been developed by Yang *et al.*[19] who write the free energy of a non-uniform system in terms of a local free-energy density $\psi(\mathbf{r})$ as

$$F = \int \mathrm{d}\mathbf{r}\, \psi(\mathbf{r}). \tag{10.52}$$

Gradient expansion of $\psi(\mathbf{r})$ then yields, for a flat interface of area $\tilde{a}$ in zero external potential:

$$F = \tilde{a} \int_a^b \mathrm{d}x\, \psi(x), \tag{10.53}$$

where

$$\psi(x) = \psi[\rho(x)] + \tfrac{1}{2}A[\rho(x)]\rho'(x)^2, \tag{10.54}$$

with $A(\rho)$ given by

$$A(\rho) = \tfrac{1}{6}k_\mathrm{B} T \int \mathrm{d}\mathbf{r}\, r^2 c(\mathbf{r}, \rho). \tag{10.55}$$

Here $a$ and $b$ define the boundaries of the system of volume $V$. In eqn (10.54), $\psi(\rho)$ is the free-energy density of a uniform system of density $\rho$, while $c(\mathbf{r},\rho)$ in eqn (10.55) is the direct correlation function of a uniform system of density $\rho$. Minimizing $F$, with the chemical potential $\mu$ introduced as the Lagrange multiplier taking care of the normalization of $\rho(x)$, Yang *et al.*[19] obtain the Euler equation of the variational problem as

$$\mu = \mu[\rho(x)] - A[\rho(x)]\rho''(x) - \tfrac{1}{2}A'[\rho(x)]\rho'(x)^2. \tag{10.56}$$

In this equation $\mu(\rho)$ is the chemical potential of a uniform system of density $\rho$, while

$$A'(\rho)=\frac{\partial A(\rho)}{\partial \rho};\quad \mu(\rho)=\frac{\partial \psi(\rho)}{\partial \rho}. \tag{10.57}$$

Equation (10.56) is equivalent to the constancy of the pressure $p$ across the inhomogeneity, $p$ being given by an integration of eqn (10.56) as

$$p=\mu\rho(x)-\psi[\rho(x)]+\tfrac{1}{2}A[\rho(x)]\rho'(x)^2. \tag{10.58}$$

Integrating eqn (10.58) over the total volume $V$ they obtain the free energy as[19]

$$F=\mu N-pV+\sigma\tilde{a}, \tag{10.59}$$

where

$$\sigma=\int_a^b \mathrm{d}x\, A[\rho(x)]\rho'(x)^2, \tag{10.60}$$

a result which goes back, in essence, to van der Waals.

It is to be stressed at this point that the form (10.60) is a more sophisticated version of the result (10.50), whereas the result analogous to eqn (10.51) is what one seeks here. By analogy with the decomposition (10.49), one now divides the constant interfacial pressure $p$ into two parts:

$$p=\mu\rho(x)-\psi[\rho(x)]+p_{\text{density gradient}}(x). \tag{10.61}$$

Using eqns (10.58), (10.60) and (10.61) it follows first that

$$\sigma=2\int_a^b \mathrm{d}x\, p_{\text{density gradient}}(x), \tag{10.62}$$

and hence one obtains the alternative form

$$\sigma=2\int_a^b \mathrm{d}x\,(p-\mu\rho(x)+\psi[\rho(x)]). \tag{10.63}$$

Equation (10.63) is the generalization of the form (10.51) one is seeking, as demonstrated below. First, since $p_{\text{density gradient}}(x)$ is non-zero only over the effective thickness of the interface, it follows from eqn (10.61) that $l$ is the extent of the range of integration in eqn (10.63). Secondly, the pressure $p$ is given by

$$p=\mu\rho_{\mathrm{l}}-\psi(\rho_{\mathrm{l}}=\mu\rho_{\mathrm{v}}-\psi(\rho_{\mathrm{v}}), \tag{10.64}$$

$\rho_{\mathrm{l}}$ being the bulk liquid density and $\rho_{\mathrm{v}}$ the vapour density. One now Taylor expands $\psi[\rho(x)]$ around the bulk liquid density, and correctly to

$O([\rho(x)-\rho_l]^2)$ one finds

$$\sigma \sim \frac{1}{K_T}\int_{\text{interface}} dx[\rho(x)-\rho_l]^2/\rho_l^2$$
$$+ \text{ higher-order terms.} \tag{10.65}$$

Here $K_T^{-1}$ is $\rho_l^2(\partial^2\psi/\partial\rho^2)|_{\rho_l}$. Thus to lowest order, one has established that eqn (10.63) is the counterpart of the Cahn–Hilliard relation (10.51). One notes that eqn (10.63) can be evaluated from a knowledge of the density profile $\rho(x)$, plus thermodynamic information as a function of density over the range from $\rho_v$ to $\rho_l$. In contrast, eqn (10.60) requires knowledge of $A(\rho)$ or the direct correlation function $c(r,\rho)$ over the same density range.

We note finally that Stott and Young[20] have made use of a similar approach in subsequent work on liquids with van der Waals interactions, quite explicitly.

## 10.5 Classical statistical mechanics of non-uniform systems

One can trace the development of the classical statistical mechanics of non-uniform fluids at least back to Morita and Hiroike[21] and to independent work by De Dominicis.[22] These studies used techniques of functional differentiation and cluster expansions. Later work by Stillinger and Buff[23] and by Lebowitz and Percus[24] led to explicit results for thermodynamic potentials of an inhomogeneous fluid in an external potential in terms of the density-dependent Ornstein–Zernike direct correlation function. While the general formalism thereby developed proved of value in the theory of the structure of bulk uniform liquids, it was almost ten years later that results were forthcoming on the liquid–vapour interface.[25–27]

In this account of the density functional theory of the problem of inhomogeneous fluids, we shall first refer to a variational principle for the grand potential. The argument below is, in essence, that found in the work of Yang *et al.*[28]

### 10.5.1 Variational principle for the grand potential

The argument for setting up the variational principle in ref 28 hinges on the introduction of two quantities; essentially an "intrinsic" Helmholtz free energy $F$ and the grand potential $\Omega$ itself. Both these quantities are functionals of the single-particle density.

It turns out, in fact, that $F$ and $\Omega$ act as generating functionals for hierarchies of correlation functions. The direct correlation functions, a special case already having been referred to, turn out to arise from the functional derivatives of $F$ with respect to the density $\rho(\mathbf{r})$, while, on the other hand, functional derivatives of $\Omega$ with respect to the external potential yeild the $n$-particle distribution functions.

But the most directly useful results which emerge are formally exact integro-differential equations for the equilibrium density; the original derivations are sometimes attributed to Yvon but are found, it seems, explicitly in the work of Triezenberg and Zwanzig,[25] Lovett *et al.*[26] and Wertheim.[27] In fact, because of applications later in the Chapter, we shall present the derivation of these equations directly for a multicomponent system, following Bhatia *et al.*[29] (see Section 10.8).

To make use of this formalism, we then turn almost immediately to the case of slowly varying density. This means we employ gradient expansions of the general formulae. While, for much of the Chapter, we shall be content with low-order terms in such expansions, in fact it is possible to perform partial summation of the gradient expansion. Ebner *et al.*[30] have used this to calculate free energies of non-uniform systems.

## 10.6 Surface tension in terms of pair potentials and the direct correlation function

### 10.6.1 Pair potentials

The burden of this Section is to present the formally exact theory of Kirkwood and Buff[31] within a framework of pairwise interatomic forces. This approach unavoidably rests on knowledge of the pair function $g(\mathbf{r},\mathbf{r}')$ of atoms in the now inhomogeneous system. This is the generalization of, say, the bulk liquid pair function, which is the much simpler limit where $g$ defined above becomes $g(|\mathbf{r}-\mathbf{r}'|)$. In the case of a planar interface, which is the geometry we deal with predominantly below, we can conveniently regard the full pair function as dependent on this vector difference $\mathbf{R}=\mathbf{r}-\mathbf{r}'$ and of the coordinate $z$ of the first atom measured perpendicular to the planar surface. An interatomic potential of the form $\phi(R)$, depending therefore only on the relative distance $R$ of a pair of atoms, is assumed to underlie the treatment below.

The tangential pressure $p_t(z)$ already introduced can then be expressed as

$$p_t(z)=k_B T\rho(z)-\frac{1}{2}\int d\mathbf{R}\ \frac{X^2}{R}\phi'(R)g_2(\mathbf{R},z), \tag{10.66}$$

where the components of $\mathbf{R}$ have been written as $(X,Y,Z)$ and $\phi'(R)=d\phi(R)/dR$. When one is far from the interface, it is plain that the $z$

dependence must disappear and in this limit eqn (10.66) reduces to the well-known formula for the bulk pressure, following almost immediately from the virial theorem:

$$p=\rho k_B T-\tfrac{1}{6}\int \mathrm{d}\mathbf{R}\, R\phi'(R)n_2(R); \quad n_2=\rho^2 g(r), \tag{10.67}$$

with $g(r)$ the usual pair function normalized to tend to unity at large $r$.

An expression similar to eqn (10.66) holds for the normal pressure $p_n$. By imposing the condition for hydrostatic equilibrium $p_n(z)=p$, one derives an expression for the density profile $\rho(z)$:

$$k_B T\frac{\mathrm{d}\rho(z)}{\mathrm{d}z}=\int \mathrm{d}\mathbf{R}\frac{Z}{R}\phi'(R)n_2(R,z). \tag{10.68}$$

Use of eqn (10.47) discussed earlier in this Chapter leads to the expression for surface tension $\sigma$ given by

$$\sigma=\frac{1}{2}\int_{-\infty}^{\infty}\mathrm{d}z\int \mathrm{d}\mathbf{R}\frac{X^2-Z^2}{R}\phi'(R)n_2(R,z). \tag{10.69}$$

It remains a major source of difficulty to calculate the pair function $n_2$ in the presence of the surface, and therefore, to date, the analytical progress made in evaluating this formula, due to Kirkwood and Buff,[31] has related $n_2$ to the bulk pair function $g(r)$. The most drastic simplification is to take the liquid as homogeneous up to the Gibbs surface and to assume the vapour has negligible density. One thereby recovers from eqn (10.69) the result of Fowler:[32]

$$\gamma=\tfrac{1}{8}\pi\rho_l^2\int_0^{\infty}\mathrm{d}R\,\phi'(R)R^4 g(R). \tag{10.70}$$

In evaluating an integral such as in eqn (10.70), it has emerged from a variety of different directions that it is important to input a $g(r)$ wholly consistent with the assumed $\phi(R)$. This can be done by using computer-simulation data, an example being that in the study of McDonald and Freeman.[33] These workers find reasonable results from eqn (10.70), and they exhibit an improvement when they allow a smooth density profile of finite thickness. For a detailed discussion of this approach, reference should be made to the work of Berry *et al.*[34]

### 10.6.2 Direct correlation function

Consider a monatomic fluid such as argon. In this case, a formally exact equation for the density profile $\rho(\mathbf{r})$ in an inhomogeneous fluid state can be written (see Section 10.8 for the derivation of the multi-component generalization)

$$\nabla\rho(\mathbf{r})=\rho(\mathbf{r})\int d\mathbf{r}'\,c(\mathbf{r},\mathbf{r}')\nabla\rho(\mathbf{r}'), \tag{10.71}$$

where however $c(\mathbf{r},\mathbf{r}')$ is the Ornstein–Zernike direct correlation function in the presence of the inhomogeneity.

For the case of a planar liquid–vapour interface (perpendicular to the $z$ axis say) the matrix $c(\mathbf{r},\mathbf{r}')$ depends only on the three quantities $z, z'$ and $s=[(x-x')^2+(y-y')^2+(z-z')^2]^{1/2}$. Taking the two-dimensional Fourier transform

$$\hat{c}(k;z,z')=\int d^2s\,\exp(ik\cdot s)c(s;z,z'), \tag{10.72}$$

one then finds

$$\frac{d\rho(z)}{dz}=\rho(z)\int_{-\infty}^{\infty}dz'\,c_0(z,z')\,\frac{d\rho(z')}{dz'}, \tag{10.73}$$

where $c_0(z,z')\equiv\hat{c}(k=0;z,z')=\int d^2s\,c(\mathbf{r},\mathbf{r}')$. This eqn (10.73) is the equilibrium condition in the fluctuation approach to surface tension of Triezenberg and Zwanzig.[25] Their derivation hinges on examining the effect of a long-wavelength fluctuation of the planar surface with wave-vector along the surface (see also Section 10.8 for the derivation of the multi-component results generalizing what follows in this Section). To lowest order in the wavenumber of the fluctuation, this represents a rigid translation of the system and eqn (10.73) follows from the fact that neither a free-energy change nor a surface area change accompany such a translation.

These changes arise instead at the next order in the wavenumber and their evaluation results in the desired expression for the surface tension. In particular the free-energy change $\Delta F$ for a small fluctuation $\Delta\rho(\mathbf{r})$ is given by

$$\Delta F=\tfrac{1}{2}k_B T\iint d\mathbf{r}\,d\mathbf{r}'\,\Delta\rho(\mathbf{r})K(\mathbf{r},\mathbf{r}')\Delta\rho(\mathbf{r}'), \tag{10.74}$$

where

$$K(\mathbf{r},\mathbf{r}')=\frac{\delta(\mathbf{r}-\mathbf{r}')}{\rho(r')}-c(\mathbf{r},\mathbf{r}') \tag{10.75}$$

for a classical fluid. A detailed calculation for the planar surface (cf. Section 10.8 and ref. 35) yields for the surface tension

$$\sigma=k_B T\int_{-\infty}^{\infty}\int dz\,dz'\,\frac{d\rho(z)}{dz}\,c_2(z,z')\frac{d\rho(z')}{dz'}, \tag{10.76}$$

where

$$c_2(z,z') = \frac{1}{4}\int d^2 s\, s^2 c(r,r')$$

$$= -\frac{1}{4}\left[\frac{d^2\,\hat{c}(k;z,z')}{dk^2}\right]_{k=0}. \tag{10.77}$$

Equations (10.73) and (10.77) are formally exact and independent of the detailed nature of the interionic forces. Of course, $c(\mathbf{r},\mathbf{r}')$ is not presently known in the inhomogeneous fluid, which means in turn that $c_0$ and $c_2$ are not currently available. The theory only becomes practicable when these are related, albeit approximately, to properties of the homogeneous fluid.

In concluding this Section, it is of interest to record that the equivalence of the Kirkwood–Buff and Triezenberg–Zwanzig formulae for surface tension presented here has been demonstrated directly by Schofield.[35]

## 10.7 Surface tension from variational principle for free energy

A different approach to the problem of surface tension calculation has been given by Bhatia and March.[36] Define a function $F'$, following Yang *et al.*,[28] related to the Helmholtz free energy $F$ by

$$F' = \beta F - \beta \int d\mathbf{r}\; U(\mathbf{r})\rho(\mathbf{r}), \tag{10.78}$$

where $U(\mathbf{r})$ is the external potential. $F'$ and $F$ are regarded as functionals of the density profile $\rho(\mathbf{r})$ and one then has

$$\frac{\delta F'}{\delta\rho(\mathbf{r})} = \beta\mu - \beta U(\mathbf{r}) \tag{10.79}$$

or

$$\frac{\delta F'}{\delta\rho(\mathbf{r})} = \ln\rho(\mathbf{r}) + (\ln\Lambda^3) - C(\mathbf{r}); \quad \Lambda = \left(\frac{\beta h^2}{2\pi M}\right)^{1/2}, \tag{10.80}$$

where $\mu$ is the chemical potential while eqn (10.80) serves to define $C(\mathbf{r})$. Its functional derivatives generate the various-order direct correlation functions; thus

$$c^{(2)}(\mathbf{r}_1,\mathbf{r}_2) = \frac{\delta C(\mathbf{r}_1)}{\delta\rho(\mathbf{r}_2)} = \frac{\delta C(\mathbf{r}_2)}{\delta\rho(\mathbf{r}_1)}, \tag{10.81}$$

$$c^{(3)}(\mathbf{r}_1,\mathbf{r}_2,\mathbf{r}_3) = \frac{\delta C^{(2)}(\mathbf{r}_1,\mathbf{r}_2)}{\delta\rho(\mathbf{r}_3)}, \quad \text{etc.} \tag{10.82}$$

If one writes $F'$ as

$$F' = F'_0 - \Phi[\rho(\mathbf{r})], \tag{10.83}$$

where

$$F'_0 = \int \mathrm{d}\mathbf{r}\,\rho(\mathbf{r})\{\ln(\rho(\mathbf{r})\Lambda^3) - 1\}, \tag{10.84}$$

then $\Phi$ is a functional of $\rho(\mathbf{r})$ and

$$\frac{\partial \Phi}{\delta \rho(\mathbf{r})} = C(\mathbf{r}), \quad \frac{\delta^2 \Phi}{\delta \rho(\mathbf{r}_1)\delta \rho(\mathbf{r}_2)} = c^{(2)}(\mathbf{r}_1, \mathbf{r}_2), \quad \text{etc.} \tag{10.85}$$

It will be convenient in what follows also to write $F'$ in the absence of the external potential as

$$F' = \beta F = \beta \int \psi(\rho(\mathbf{r}))[\mathrm{d}\mathbf{r} - \phi[\rho(\mathbf{r})], \tag{10.86}$$

where $\psi(\rho(\mathbf{r}))$ is a function of the local density $\rho(\mathbf{r})$ and $\phi[\rho(\mathbf{r})]$ is a functional of $\rho(\mathbf{r})$ such that if $\mathrm{d}\rho/\mathrm{d}x$, $\mathrm{d}\rho/\mathrm{d}y$ and $\mathrm{d}\rho/\mathrm{d}z$ are all identically zero then $\phi = 0$.

### 10.7.1 Density profile

Taking the gradient of eqn (10.79) one finds

$$-\beta \nabla U(\mathbf{r}) = \frac{\nabla \delta F'}{\delta \rho(\mathbf{r})} = \frac{1}{\rho(\mathbf{r})} \nabla \rho(\mathbf{r}) - \nabla C(\mathbf{r}). \tag{10.87}$$

From the property that $C(\mathbf{r})$ is a functional of $\rho(\mathbf{r})$ it follows that

$$\nabla C(\mathbf{r}) = \int c^{(2)}(\mathbf{r}, \mathbf{r}') \nabla \rho(\mathbf{r}') \mathrm{d}\mathbf{r}' \tag{10.88}$$

and hence

$$-\beta \nabla U(\mathbf{r}) = \frac{1}{\rho(\mathbf{r})} \nabla \rho(\mathbf{r}) - \int c^{(2)}(\mathbf{r}, \mathbf{r}') \nabla \rho(\mathbf{r}')\, \mathrm{d}\mathbf{r}', \tag{10.89}$$

which is the equation for the density profile given by Lovett *et al.*[26] In the absence of the external potential ($\nabla U(\mathbf{r}) = 0$), eqn (10.89) reduces to equation (10.71) of Triezenberg and Zwanzig.[25]

It should be noted that when $U = 0$, $F' = \beta F$, the equation for equilibrium is from eqn (10.79),

$$\frac{\delta F}{\delta \rho(\mathbf{r})} = \mu, \tag{10.90}$$

which is the Euler equation for the problem.

### 10.7.2 Formula for surface tension

One knows from thermodynamics (cf. Section 10.3) that in the presence of a planar surface, which is again taken perpendicular to the $z$-axis here, $F$ has the form

$$F = \mu N - pV + \sigma \tilde{a}, \tag{10.91}$$

where $\tilde{a}$ is the area of the planar interface. This expression leads to a formula for the surface tension $\sigma$ in terms of the quantity $\phi$ introduced in eqn (10.86).

Setting $U(r) = 0$, eqns (10.79) and (10.86) give

$$\beta\mu = \beta \frac{\mathrm{d}\psi(\rho(\mathbf{r}))}{\mathrm{d}\rho} - \frac{\delta\phi}{\delta\rho(\mathbf{r})}. \tag{10.92}$$

Multiplying this equation by $\mathrm{d}\rho/\mathrm{d}z$, one has

$$\beta \frac{\mathrm{d}}{\mathrm{d}z} [\psi(\rho) - \mu\rho(z)\} = \frac{\mathrm{d}\rho}{\mathrm{d}z} \frac{\delta\phi}{\delta\rho(\mathbf{r})}, \tag{10.93}$$

and introducing

$$\frac{\mathrm{d}\rho}{\mathrm{d}z} \frac{\delta\phi}{\delta\rho(\mathbf{r})} \equiv \frac{\mathrm{d}G}{\mathrm{d}z}, \tag{10.94}$$

one finds

$$\frac{\mathrm{d}}{\mathrm{d}z} \left\{ \mu\rho(z) - \psi(\rho) + \frac{G}{\beta} \right\} = 0. \tag{10.95}$$

Now when $\phi = 0$, $G = 0$, the term in curly brackets in eqn (10.95) is just the pressure for a homogeneous system. Hence one can identify the pressure $p$ as

$$p = \mu\rho(z) - \psi(\rho) + \frac{G}{\beta}. \tag{10.96}$$

Integrating over all the volume, using eqn (10.86) to eliminate $\psi(\rho)$ and recalling that $F' = \beta F$, one finds

$$pV = \mu N - F + \frac{1}{\beta}\left\{\int G\,\mathrm{d}\mathbf{r} - \phi\right\}. \tag{10.97}$$

Comparing with eqn (10.91) one obtains

$$\sigma = \frac{k_{\mathrm{B}}T}{\tilde{a}}\left\{\int G\,\mathrm{d}\mathbf{r} - \phi\right\}, \tag{10.98}$$

which is the desired expression. If $\phi$ is a known functional of the single-particle density $\rho(\mathbf{r})$, then the surface tension may be calculated from this expression.

First, one observes that if $\phi$ were of the local form

$$\phi = \int \mathscr{F}(\rho(\mathbf{r}))\,\mathrm{d}\mathbf{r}, \tag{10.99}$$

where $\mathscr{F}$ is an arbitrary function, i.e. not a functional, of $\rho(\mathbf{r})$, then

$$\frac{\delta\phi}{\delta\rho(\mathbf{r})} = \mathscr{F}'(\rho), \quad \mathscr{F}' = \frac{\mathrm{d}\mathscr{F}}{\mathrm{d}\rho}, \tag{10.100}$$

$$\frac{\mathrm{d}G}{\mathrm{d}z} = \frac{\mathrm{d}\rho}{\mathrm{d}z}\mathscr{F}'(\rho) = \frac{\mathrm{d}}{\mathrm{d}z}\mathscr{F}(\rho), \tag{10.101}$$

or $G = \mathscr{F}$, since when $\mathscr{F} = 0$, $G = 0$ by definition. Hence the surface tension $\sigma$ is zero. Therefore one can add a local contribution to $\phi$ without affecting the value of $\sigma$.

In particular, one can replace $\phi$ by $\Phi$ of eqn (10.83) and write $\sigma$ in the form

$$\sigma = \frac{k_{\mathrm{B}}T}{\tilde{a}}\left\{\int \tilde{G}\,\mathrm{d}\mathbf{r} - \Phi\right\}, \tag{10.102}$$

where $\tilde{G}$ is such that

$$\frac{\mathrm{d}\tilde{G}}{\mathrm{d}z} = \frac{\mathrm{d}\rho}{\mathrm{d}z}\frac{\delta\Phi}{\delta\rho(\mathbf{r})} = \frac{\mathrm{d}\rho}{\mathrm{d}z}C(\mathbf{r}). \tag{10.103}$$

One can view eqn (10.102) as an alternative to the Triezenberg–Zwanzig formula. It is re-iterated here that the Treizenberg–Zwanzig derivation starts with changes in $F$ due to fluctuations in $\rho(z)$ from the equilibrium profile. If one makes a Taylor (functional derivative) expansion of $\Phi$, $\tilde{G}$ in eqn (10.102), one notices that $\sigma$ can be expressed as a series involving increasingly high order of direct correlation functions of a homogeneous system. It is clear that the sum of this series must yield again the pair direct correlation function in the inhomogeneous system, in terms of which the Triezenberg–Zwanzig result of Section 10.6.2 is expressed.

## 10.8 Generalization to multicomponent systems

It is of interest to generalize the treatment of a monatomic fluid in Section 10.6.2 to multicomponent systems. The treatment below follows that of Bhatia *et al.*[29] These workers set out a first-principles theory of the density profiles in a multicomponent system in terms of the partial direct correlation functions $c_{ij}$ in the presence of a planar surface. From these same quantities, the surface tension of liquid mixtures is calculated.

### 10.8.1 Density profile equations

The argument below generalizes that of Lovett *et al.* to multicomponent mixtures.

Let $u_i(r)$ denote the dimensionless one-body potential per particle for species $i$:

$$u_i(r) = \beta(\mu_i - U_i(r)); \quad \beta = (k_B T)^{-1}. \tag{10.104}$$

Here $U_i(r)$ is the external potential for species $i$, while $\mu_i$ is its chemical potential.

The system is considered to be open and at constant $V$ and $T$. The single-particle densities are then

$$\rho_i(\mathrm{r}) = \langle \hat{\rho}_i(r) \rangle, \tag{10.105}$$

where $\langle \cdots \rangle$ denotes the ensemble average.

Given all the $u_i(r)$, the various $\rho_j(r)$ are uniquely determined and vice versa, at given volume $V$ and temperature $T$. Hence the quantities $u_i(r)$ can be regarded as functionals of the various $\rho_j(r)$ and vice versa. One has then

$$\frac{\delta \rho_i(r)}{\delta u_j(r')} = \langle \hat{\rho}_i(r) \hat{\rho}_j(r') \rangle - \langle \hat{\rho}_i(r) \rangle \langle \hat{\rho}_j(r') \rangle \tag{10.106}$$

and

$$\begin{aligned} \frac{\delta u_i(r)}{\delta \rho_j(r')} &= \delta_{ij} \frac{\delta(r - r')}{\rho_i(r)} - c_{ij}(r, r') \\ &\equiv K_{ij}(r, r'), \end{aligned} \tag{10.107}$$

$c_{ij}(r,r')$ being the direct correlation functions already referred to. If $u_i(r)$ is written as

$$u_i(r)=\ln(\rho_i(r)\Lambda_i^3)-C_i(r), \tag{10.108}$$

where $\Lambda_i=h(2\pi m_i k_{\rm B}T)^{-1/2}$, then

$$c_{ij}(r,r')=\frac{\delta C_i(r)}{\delta\rho_j(r')}. \tag{10.109}$$

Denoting the functional dependence of the quantities $u_i$ by, say,

$$u_i(r_1,[\rho_1(r),\ldots,\rho_\nu(r)])\equiv u_i(r_1),$$

then translational invariance implies

$$u_i(r_1,[\rho_1(r+\delta),\ \ldots,\ \rho_\nu(r+\delta)])\equiv u_i(r_1+\delta). \tag{10.110}$$

Hence one finds

$$u_i(r_1+\delta)-u_i(r_1)=\sum_{j=1}^{\nu}\,\mathrm{d}\mathbf{r}\frac{\delta u_i(r_1)}{\delta\rho_j(r)}(\rho_j(r+\delta)-\rho_j(r))+\cdots. \tag{10.111}$$

Passing to the limit $\delta\to 0$ yields

$$\begin{aligned}\nabla u_i(r_1)&=\sum_{j=1}^{\nu}\int\mathrm{d}^3r\,\frac{\delta u_i(r_1)}{\delta\rho_j(r)}\nabla\rho_j(r)\\ &=\sum_{j=1}^{\nu}\int\mathrm{d}^3r\,K_{ij}(r_1,r)\nabla\rho_j(r),\end{aligned} \tag{10.112}$$

where in eqn (10.112), use has been made of the definition (10.107) of $K_{ij}$.

If the external potential is reduced to zero at this stage, i.e. $\nabla u_i(r)=0$, then eqn (10.112) becomes

$$\sum_{j=1}^{\nu}\int\mathrm{d}^3r\,K_{ij}(r_1,r)\nabla\rho_j(r)=0 \tag{10.113}$$

or using eqn (10.107)

$$\frac{\nabla\rho_i(\mathrm{r}_1)}{\rho_i(\mathrm{r}_1)}=\sum_{j=1}^{\nu}\int\mathrm{d}^3r\,c_{ij}(r_1,r)\nabla\rho_j(r). \tag{10.114}$$

These then are the basic equations determining the density profiles; the partial direct correlation functions (cf. Appendix A 10.1) being those in the presence of the surface. These equations are the desired generalization to mixtures of the one-component equation (10.71).

### 10.8.2 Surface tension of mixtures

In Section 10.6.2, the way the surface tension could be calculated in a one-component liquid was sketched. The generalization of the Triezenberg–Zwanzig argument to liquid mixtures is given below.

To calculate the surface energy and surface tension, consider that the equilibrium profiles $\rho_i$ have gradients along the $x$-axis. Then the total energy depends on the cross-sectional area of the fluid perpendicular to the $x$-axis. Attention is now focused on that surface which satisfies the Gibbs equimolar criteria with respect to the species $i$; the origin of the Gibbs surface is located at $(0, y, z)$. Then, taking the system in the form of a cylinder of basal area $a_0$, one can write, for extension $a$ from the Gibbs surface in phase I and extension $b$ similarly in phase II:

$$a_0 \int_{-a}^{b} \rho_i(x)\,\mathrm{d}x = a_0[a\rho_i^{\mathrm{I}} + b\rho_i^{\mathrm{II}}] = N_i, \tag{10.115}$$

where $\rho_i^{\mathrm{I}}$ and $\rho_i^{\mathrm{II}}$ are the densities of species $i$ in the bulk phases I and II.

Let us write $r$ for the two-dimensional vector $\mathbf{r}\ (y, z)$. Now if there is a small fluctuation $\Delta\rho_j(x)$ in the various $\rho_j(x)$, $j = 1, 2, \ldots, \nu$, so that

$$\int \Delta\rho_j(r, x)\,\mathrm{d}^2 r\,\mathrm{d}x = 0 \quad \text{for all } j, \tag{10.116}$$

then the change in the Helmholtz free energy $F$ is

$$\Delta F = \tfrac{1}{2} k_{\mathrm{B}} T \sum_{ij} \int \Delta\rho_i(\mathrm{r}_1, x_1)\Delta\rho_j(r_2, x_2) \times K_{ij}(r_1, x_1, r_2, x_2)\,\mathrm{d}^2 r_1\,\mathrm{d}^2 r_2\,\mathrm{d}x_1\,\mathrm{d}x_2. \tag{10.117}$$

One next Fourier analyses $\Delta\rho_j$ as

$$\Delta\rho_j(r_2, x_2) = \sum_q \rho_j(q, x_2)\exp(\mathrm{i}\,q \cdot r_2) \tag{10.118}$$

and recalls that $K_{ij}(r_1, x_1, r_2, x_2) = K_{ij}(0, x_1, r_2 - r_1, x_2)$. One also defines then

$$K_{ij}(q, x_1, x_2) = \int K_{ij}(0, x_1, r_2 - r_1, x_2)\exp(\mathrm{i}q \cdot (r_2 - r_1))\,d^2(r_2 - r_1). \tag{10.119}$$

From a direct generalization of the Triezenberg–Zwanzig argument one has, using eqn (10.107)

$$K_{ij}(q, x_1, x_2) = \delta_{ij}\frac{\delta(x_1 - x_2)}{\rho_i(x_1)} - \hat{c}_{ij}(q, x_1, x_2), \tag{10.120}$$

with

$$\hat{c}_{ij}(q, x_1, x_2) = \int c_{ij}(0, x_1, r_1, x_2) \exp(iq \cdot r) \, d^2 r. \tag{10.121}$$

Substituting eqn (10.118) and eqn (10.121) in eqn (10.117) yields

$$\Delta F = \tfrac{1}{2} a_0 k_B T \sum_{ij} \sum_q \int \rho_i^*(q, x_1) \, \rho_j(q, x_2) K_{ij}(q, x_1, x_2) \, dx_1 \, dx_2. \tag{10.122}$$

In this quadratic approximation for $\Delta F$ there is no mixing of different $q$ values and one can consider the fluctuation of each $q$ separately.

For small $q$,

$$K_{ij}(q, x_1, x_2) = K_{ij}^0(x_1, x_2) + q^2 K_{ij}^{(2)}(x_1, x_2) + \ldots, \tag{10.123}$$

$$K_{ij}^0 = \frac{\delta_{ij}\delta(x_1 - x_2)}{\rho_i(x_1)} - \int c_{ij}(r_1 = 0, x_1, r, x_2) \, d^2 r, \tag{10.124}$$

$$K_{ij}^{(2)}(x_1, x_2) = \tfrac{1}{4} \int c_{ij}(r = 0, x_1, r, x_2) \, r^2 d^2 r, \tag{10.125}$$

and eqn (10.122) for $\Delta F$ becomes, omitting the sum over $q$:

$$\Delta F = \tfrac{1}{2} a_0 k_B T \sum_{ij} \rho_i^*(q, x_1) \rho_j(q, x_2) K_{ij}^0(x_1, x_2) \, dx_1 \, dx_2$$

$$+ \tfrac{1}{2} a_0 k_B T q^2 \sum_{ij} \int \rho_i^*(q, x_1) \rho_j(q, x_2) K_{ij}^{(2)}(x_1, x_2) \, dx_1 \, dx_2 + \cdots. \tag{10.26}$$

Now because of the fluctuation $\Delta \rho_i(r, x)$ in the density of species $i$, the location of the Gibbs surface also fluctuates. If $x_0(\mathbf{r})$ is the shift in the Gibbs surface at $\mathbf{r}$, one has

$$\int_{-a}^{b} [\rho_i(x) + \Delta \rho_i(r, x)] dx = [a + x_0(r)] \rho_i^{\mathrm{I}} + [b - x_0(r)] \rho_i^{\mathrm{II}} \tag{10.127}$$

or

$$\int_{-a}^{b} \Delta \rho_i(r, x) \, dx = x_0(r)[\rho_i^{\mathrm{I}} - \rho_i^{\mathrm{II}}] \equiv \Delta \rho_i(x_0(r)). \tag{10.128}$$

Thus

$$x_0(r) = \frac{1}{\Delta \rho_i} \int \Delta \rho_i(r, x) \, dx \tag{10.129}$$

where $\Delta \rho_i = \rho_i^{\mathrm{I}} - \rho_i^{\mathrm{II}}$.

The area of the new surface is

$$a = a_0 + \tfrac{1}{2}\int d^2 r\, |\nabla_r x_0(r)|^2, \tag{10.130}$$

where the integration is over the original area $a_0$.

With the Fourier expansion

$$x_0(r) = \sum_q x_0(q) \exp(iq\cdot r), \tag{10.131}$$

one can write

$$x_0(q) = \frac{1}{\Delta\rho_i}\int \rho_i(q, x)\, dx. \tag{10.132}$$

The change in area due to the fluctuation is thus

$$a - a_0 = \tfrac{1}{2} a_0 \sum_q q^2 |x_0(q)|^2 \tag{10.133}$$

or for fluctuations of $\Delta\rho_i(r, x)$ of a given $q$

$$a - a_0 = \tfrac{1}{2} a_0 q^2 |x_0(q)|^2. \tag{10.134}$$

One can, in principle, determine $\rho_j(q, x)$ by minimizing the equation for $\Delta F$ with respect to them, subject to a given change in the area of the surface (see eqn (10.134)). According to Triezenberg and Zwanzig, this gives the same result as their heuristic procedure for a one-component system.

Below, the heuristic procedure is again followed for the multi-component case. Consider that a small fluctuation $x_0(q)$ in the surface has occurred. For small $q$, this amounts to virtually a vertical shift. Then it is not unreasonable to suppose that the density profiles are all bodily shifted, i.e.

$$\rho_j(x) \to \rho_j(x - x_0(q)) \quad \text{for all } j. \tag{10.135}$$

Substituting eqn (10.135) in the equation for $\Delta F$ yields

$$\Delta F = \tfrac{1}{2} a_0 k_B T |x_0(q)|^2 \left\{ \sum_{ij} \int \frac{d\rho_i}{dx_1}\frac{d\rho_j}{dx_2} K_{ij}^{(0)}(x_1, x_2)\, dx_1\, dx_2 \right.$$
$$\left. + \sum_{ij} q^2 \int \frac{d\rho_i}{dx_1}\frac{d\rho_j}{dx_2} K_{ij}^{(2)}(x_1, x_2)\, dx_1\, dx_2 + \cdots \right\}. \tag{10.136}$$

In the limit $q \to 0$, $\Delta a \to 0$ and $\Delta F = 0$. In other words one must have

$$\sum_{ij} \int \frac{d\rho_i}{dx_1}\frac{d\rho_j}{dx_2} K_{ij}^{(0)}(x_1, x_2)\, dx_1 dx_2 = 0 \tag{10.137}$$

or

$$\sum_i \frac{d\rho_i}{dx_1}\left(\int \sum_j \frac{d\rho_j}{dx_2} K_{ij}^{(0)}(x_1, x_2)\, dx_2\right) dx_1 = 0. \tag{10.138}$$

Sufficient conditions for eqn (10.138) to be true are

$$\sum_j \frac{d\rho_j}{dx_2} K_{ij}^{(0)}(x_1, x_2)\, dx_2 = 0 \quad \text{for each } i\ (=1, 2, \ldots, \nu). \tag{10.139}$$

These $\nu$ equations are just eqns (10.114) determining the density profiles. Substituting the result (10.138) in eqn (10.136), using $a - a_0 = \frac{1}{2} a_0 q^2 x_0^2$, and noting that $\sigma$ is given by

$$\sigma = \Delta F/(a - a_0) \tag{10.140}$$

one finds

$$\sigma = k_B T \sum_{ij} \int \frac{d\rho_i}{dx_1} \frac{d\rho_j}{dx_2} K_{ij}^{(2)}(x_1, x_2)\, dx_1\, dx_2), \tag{10.141}$$

which is the desired generalization given by Bhatia *et al.* of the Treizenberg–Zwanzig formula to multi-component mixtures. The same result can also be obtained by generalizing the one-component treatment based on the pressure difference across a curved surface, due to Lovett *et al.*[26]

The relation of the above treatment to the phenomenology of Bhatia and March, treated in Section 10.9, is discussed in ref. 29. Also there, the model of conformal solutions, due to Longuet-Higgins,[37] is employed to consider the total density and surface segregation profiles. Such a model, though limited in its range of validity, should apply, for example, to the Na–K alloy system.[38]

## 10.9 Surface segregation in liquid binary alloys

Progress will be reported in this Section that has proved possible by rather simple analytical theory in the problem of surface segregation. Following Bhatia and March,[39] the phenomenological treatment of Cahn and Hilliard[75] (see also, for example, the review by Brown and March[10]), which relates the product $K_T\sigma$ of the isothermal compressibility $K_T$ and the surface tension $\sigma$ to the thickness $l$ of the liquid surface will first be generalized to liquid alloys. The result thereby obtained is

$$\sigma \sim \frac{l}{K_T}\left[1 + \frac{\delta^2 S_{cc}(0)}{\rho k_B T K_T}\right]^{-1}, \tag{10.142}$$

where $\delta = (1/V)(\partial V/\partial c)$ is the size difference factor, $S_{cc}(0)$ represents the concentration fluctuations $\rho V\langle(\Delta c)^2\rangle$, while $\rho$ denotes the number density in the alloy. As we shall indicate below, this formula already affords a ready explanation of the remarkable variation of surface tension in the amalgans of the alkali metals.

It should be noted here that Gibbs long ago pointed out that from thermodynamics the excess surface concentration is related to the dependence of surface tension $\sigma$ on concentration $c$, so that the problem of surface segregation is that of finding $d\sigma/dc$.

Following the Cahn–Hilliard approach, one considers a fluctuation $\Delta\rho$ in the number density $\rho$, occurring in a volume $V$ say. Then from fluctuation theory one associates with it a free-energy contribution (cf. eqn (10.49))

$$F_1 = \frac{AV}{K_T}\left(\frac{\Delta\rho}{\rho}\right)^2, \tag{10.143}$$

where $A$ is equal to $\frac{1}{2}$ for "accidental" fluctuations. For the generalization to binary systems effected below, it is useful at this point to write eqn (10.143) in an alternative, but equivalent, form:

$$F_1 = A\left(\frac{\delta^2 F}{\delta N^2}\right)_{T,V}(\Delta N)^2, \tag{10.144}$$

where $F$ is the Helmholtz free energy of volume $V$, containing a mean number of particles $N$, and $\Delta N$ is the fluctuation in $N$. Since $V$ is held fixed, $\Delta\rho = \Delta N/V$.

If the effective thickness of the liquid interface is $l$, and if one assumes that eqn (10.143) is applicable to the real inhomogeneity at a liquid surface, then the corresponding contribution, $\sigma_1$ say, to the surface energy is (see also eqn (10.49))

$$\sigma_1 = Al(\Delta\rho)^2/\rho^2 K_T. \tag{10.145}$$

Clearly, minimization of eqn (10.145) with respect to $l$ would yield the manifestly erroneous result that $l = 0$.

Therefore, in the Cahn–Hilliard approach one adds a term to the free energy which is proportional to the square of the density gradient, as elaborated by, say, Yang *et al.*[28] and Bhatia and March:[36].

$$F_2 = BV(\Delta\rho/l)^2, \tag{10.146}$$

with a corresponding contribution, $\sigma_2$ say, to the surface energy given by

$$\sigma_2 = Bl(\Delta\rho/l)^2. \tag{10.147}$$

The next step is to minimize the sum $\sigma_1 + \sigma_2$ with respect to $l$, to obtain (see eqn (10.50))

$$\sigma=\sigma_1+\sigma_2=\frac{2A(\Delta\rho)^2 l}{\rho^2 K_T}. \tag{10.148}$$

To complete the argument, $\Delta\rho$ is estimated as $\rho-\rho_v$, the difference between the bulk liquid and vapour densities. For dense liquids $\rho \gg \rho_v$ and hence

$$\sigma=2Al/K_T, \tag{10.149}$$

or taking $A\sim\frac{1}{2}$ as mentioned above we have the earlier result (10.51).

The length $l$ turns out for a whole variety of liquids near the triple point to be $\sim 1$ Å, as is illustrated in Table 10.2. Equation (10.51) is therefore interesting in that it clearly correlates a bulk property, $K_T$, with a surface property, $\sigma$.

### 10.9.1 Generalization to two components

For two components, eqn (10.144) can be generalized to read

$$F_1=A\left\{\left(\frac{\partial^2 F}{\partial N_1^2}\right)_{T,V,N_2}(\Delta N_1)^2+\left(\frac{\partial^2 F}{\partial N_2^2}\right)(\Delta N_2)^2 + 2\left(\frac{\partial^2 F}{\partial N_1 \partial N_2}\right)\Delta N_1 \Delta N_2\right\}; \tag{10.150}$$

$N_1$ and $N_2$ being the number of particles of types 1 and 2, respectively, while $\Delta N_1$ and $\Delta N_2$ measure fluctuations in them. Next, eqn (10.146) is generalized to read

$$F_2=l^{-2}\{b_{11}(\Delta N_1)^2+2b_{12}\Delta N_1 \Delta N_2+b_{22}(\Delta N_2)^2\}. \tag{10.151}$$

In eqn (10.150), by analogy with the one-component case, for fluctuation theory $A=\frac{1}{2}$. Hence

$$\sigma=V^{-1}[AXl+l^{-1}Y], \tag{10.152}$$

where $X$ and $Y$ denote the contents of the curly brackets in the above equations for $F_1$ and $F_2$, respectively. As before the minimum $d\sigma/dl=0$ yields

$$l^2=Y/XA \tag{10.153}$$

or

$$\sigma=2Al(X/V). \tag{10.154}$$

In the manner of Bhatia and Thornton,[38] we can now obtain

$$\frac{X}{V}=\frac{1}{K_T N^2}[\Delta N+N(\Delta c)\delta]^2+\frac{1}{V}\left(\frac{\partial^2 G}{\partial c^2}\right)_{T,p,N}(\Delta c)^2, \qquad (10.155)$$

where $G$ is the Gibbs free energy for volume $V$, $K_T$ is the alloy isothermal compressibility, i.e.

$$K_T=-\frac{1}{V}\left(\frac{\partial V}{\partial p}\right)_{T,N,c}, \qquad (10.156)$$

while $\delta$ is the size factor

$$\delta=\frac{1}{V}\left(\frac{\partial V}{\partial c}\right)_{T,p,N}=\frac{v_1-v_2}{cv_1+(1-c)v_2}, \qquad (10.157)$$

where $v_1$ and $v_2$ are the partial molar volumes of the two species. Furthermore in eqn (10.155)

$$\Delta N=\Delta N_1+\Delta N_2, \qquad (10.158)$$

while

$$\Delta c=N^{-1}[(1-c)\Delta N_1-c\Delta N_2], \qquad (10.159)$$

with $N=N_1+N_2$ and $c$ denoting the concentration of species 1.

Hence, from eqns (10.154) and (10.155) one finds

$$\sigma\sim\frac{2Al}{K_T N^2}\left(\{\Delta N+N(\Delta c)\delta\}^2+\frac{K_T N^2}{V}\left(\frac{\partial^2 G}{\partial c^2}\right)_{T,p,N}(\Delta c)^2\right). \qquad (10.160)$$

As a result of treating a liquid alloy, it can be seen from eqn (10.160) that a specifically new feature arises, namely the determination of $\Delta c$. Clearly, however, in accord with the previous minimization of $\sigma$, $\Delta c$ should be allowed to adjust in order again to minimize the surface free-energy. Assuming $l$ constant, one determines $\Delta c$ from

$$\frac{\partial\sigma}{\partial\Delta c}=0. \qquad (10.161)$$

Using eqn (10.160) in (10.161), it is found that

$$\Delta c=\left(\frac{-\delta/K_T}{(1/V)(\partial^2 G/\partial c^2)+(\delta^2/K_T)}\right)\frac{\Delta N}{N}. \qquad (10.162)$$

This result for $\Delta c$ is not to be confused with the excess surface concentration which is differently defined. The corresponding minimum value of $\sigma$ is then found by substituting this value for $\Delta c$ into eqn (10.160). With $A$ taken again as $\sim\frac{1}{2}$ and $\Delta N\sim N$ following precisely the Cahn–Hilliard work, plus $\Delta N_i\sim N_i$, $i=1, 2$, $|\Delta c|\ll 1$, the desired result already quoted in eqn (10.142) is obtained. In this formula, we have introduced $S_{cc}(0)$ in place of $\partial^2 G/\partial c^2$ through

$$S_{cc}(0)=\rho k_B T\left(\frac{1}{V}\frac{\partial^2 G}{\partial c^2}\right)^{-1}, \tag{10.163}$$

which represents the zero-wave number limit of the liquid structure factor describing concentration correlations or fluctuations.

Via eqn (10.142), one is provided with an almost quantitative route to $\sigma$ as a function of concentration from a knowledge of

(a) the isothermal compressibility $K_T$ as a function of concentration $c$;
(b) the concentration fluctuations $S_{cc}(0)$; and
(c) the size factor $\delta$.

Choosing as an illustrative example the non-conducting system acetone–chloroform, shown in Fig. II.8 from Adamson,[39] there is no difficulty in explaining these results by means of eqn (10.142) since here $\delta^2/\theta \sim 0.25$ and $(\delta^2/\theta)S_{cc}(0)$, with $\theta=\rho k_B T K_T$, can be expected to be negligible at all concentrations.

In view of this, it will be useful to emphasize below some of the salient qualitative features predicted by eqn (10.142),* which it should be possible to bring into contact with experiment.

### 10.9.2 Qualitative predictions of surface tension formula

Let us construct from eqn (10.142) the slope $\mathrm{d}\sigma/\mathrm{d}c$:

$$\frac{1}{\sigma}\frac{\mathrm{d}\sigma}{\mathrm{d}c}=\frac{1}{l}\frac{\mathrm{d}l}{\mathrm{d}c}-\frac{1}{K_T}\frac{\mathrm{d}K_T}{\mathrm{d}c}-\frac{S_{cc}(0)\dfrac{\mathrm{d}}{\mathrm{d}c}\left(\dfrac{\delta^2}{\theta}\right)+\dfrac{\delta^2}{\theta}\dfrac{\mathrm{d}S_{cc}(0)}{\mathrm{d}c}}{1+(\delta^2/\theta)S_{cc}(0)}. \tag{10.164}$$

It is useful to first examine the limiting cases $c\to 0$ and $c\to 1$ from this formula (10.164). Since the concentration fluctuations $S_{cc}(0)\to c$ as $c\to 0$ and to $1-c$ as $c\to 1$, one finds

$$\frac{1}{\sigma}\frac{\mathrm{d}\sigma}{\mathrm{d}c}\sim\frac{1}{l}\frac{\mathrm{d}l}{\mathrm{d}c}-\frac{1}{K_T}\frac{\mathrm{d}K_T}{\mathrm{d}c}-\frac{\delta^2}{\theta}\quad \text{at } c=0 \tag{10.165}$$

and

$$\frac{1}{\sigma}\frac{\mathrm{d}\sigma}{\mathrm{d}c}\sim\frac{1}{l}\frac{\mathrm{d}l}{\mathrm{d}c}-\frac{1}{K_T}\frac{\mathrm{d}K_T}{\mathrm{d}c}+\frac{\delta^2}{\theta}\quad \text{at } c=1. \tag{10.166}$$

*The relation of this phenomenological theory to first-principles statistical mechanics is treated by the present author in ref. 105.

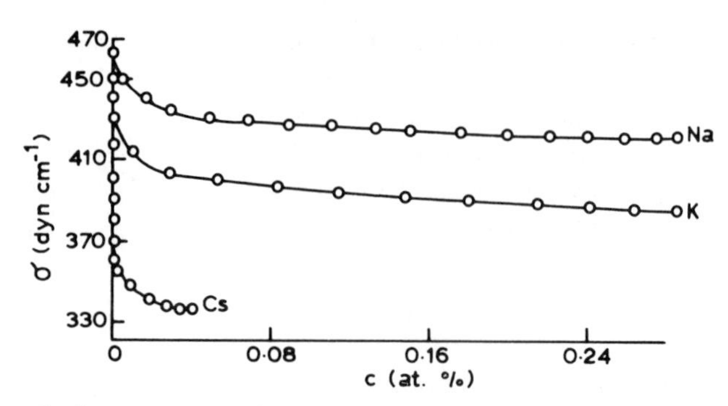


**Fig. 10.1.** Surface tension as function of $c$ for amalgams of alkali metals Na, K and Cs.

### 10.9.3 Dilute alkali solutions in mercury

As an example, consider the rather remarkable variation of surface tension in amalgams of the alkali metals. Thus, in Fig. 10.1 the results of Pugachevich and Timofeevicheva[40] for dilute Na, K and Cs in Hg are reproduced. From the size factor term $\delta^2/\theta$, one estimates that for potassium $-(1/\sigma)(\mathrm{d}\sigma/\mathrm{d}c)_{c=0}$ is of the order of 100, while for caesium a value 3 or 4 times as large results. It seems evident that the behaviour shown in Fig. 10.1. is therefore accounted for in a natural physical manner as being due to the size factor appearing in eqn (10.142).

### 10.9.4 Mg–Sn and Mg–Pb systems

Though somewhat less striking than the above example of alkalis in Hg, the systems Mg–Sn and Mg–Pb will next be considered since there are measurements of surface tension available across the entire concentration range. According to eqns (10.165) and (10.166), even if $\mathrm{d}l/\mathrm{d}c$ and $\mathrm{d}K_T/\mathrm{d}c$ retain the same sign over the whole concentration range, which is to be expected physically in many, though not necessarily all, liquid binary alloys, one sees that the size factor term $\delta^2/\theta$ could lead to a change in sign from negative to positive as one goes across from $c=0$ to $c=1$, and such behaviour is indeed observed in Mg–Sn and Mg–Pb.

### 10.9.5 Data on surface segregation and its interpretation

Though, as already stressed, the problem of surface segregation in dilute alloys is the same as the problem of the sign of $(\mathrm{d}\sigma/\mathrm{d}c)_{c=0}$, where $c$ is the solute concentration, the problem is of sufficient importance in materials science and for aspects of catalysis[41] to warrant dwelling a little further on the above phenomenological theory in this area.

Equation (10.165) is evidently a suitable basis for the discussion of surface segregation in dilute alloys. It shows immediately that the condition favouring surface segregation is

$$\left(\frac{\mathrm{d}K_T}{\mathrm{d}c}\right)_{c=0} > 0. \tag{10.167}$$

It should be noted that, according to the phenomenological theory, this inequality (10.167) is a sufficient condition for surface segregation provided $(\mathrm{d}l/\mathrm{d}c)_{c=0}$ can be taken as zero. For liquid-metal binary alloys with atoms of different valence, Bhatia and March[39] give arguments that Friedel screening[42] requires this condition. The size factor always assists and hence, as shown schematically in Fig. 10.2, makes the slope of $\mathrm{d}\sigma/\mathrm{d}c$ more negative.

Of course, the size factor $-\delta^2/\theta$ can be sufficiently large and negative in eqn (10.165) to allow $\mathrm{d}\sigma/\mathrm{d}c$ to also be negative. Thus the inequality (10.167) is not a necessary condition for surface segregation.

Table 10.3 contains a collection of data[39,43] showing the effect of alloying on the surface tension in dilute solutions of two metals in the first two columns. The final column shows the prediction of the phenomenology, obtained in the simplest possible way by examining if $K_1 > K_2$; $K = K_T$; "2" referring to the solvent metal and "1" to the solute in its own pure metal. Since accurate data on the liquids are still not always available, Table 10.3 was constructed using the data for crystals at 20°C for $K_1$ and $K_2$.

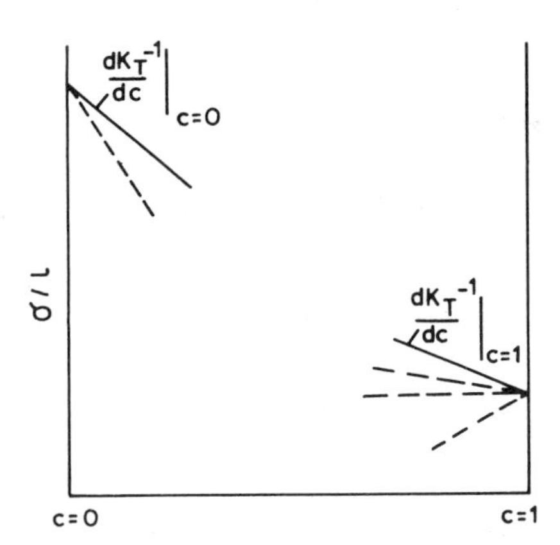


**Fig. 10.2.** Solid lines show possible behaviour of slope of $K_T^{-1}$ near end-points $c=0$ and 1. Dashed lines indicate predictions[39] of possible slopes of surface tension $\sigma$ as function of concentration $c$. (Schematic only.)

**Table 10.3** Influence of alloying on surface tension in dilute solutions of two metals (solvent is 2; solute is 1). $K_T = K$ is bulk compressibility, in units of $\times 10^{-12}\,\mathrm{cm^2\,dyn^{-1}}$ (Taken from ref. 39.)

| *Solvent* (2) | *Type* | *Compressibility properties* |
|---|---|---|
| Hg | Surface-active: Cd, Ag, Sn, Pb, Mg, Tl, Sr, Ba, Na, Li, K, Rb, Cs | $K_{Hg} = K_2 = 3.8$. $K_1 > K_2$ for Sr, Ba and alkalis. For others, $K_1 < K_2$ and size effect must dominate |
| | Surface-inactive: Co, Bi, Zn, Cu | All have $K_1 < K_2$ |
| Sn | Active: Bi, Na, Pb, Sb, Tl | $K_2 = 1.9$. All have $K_1 > K_2$ |
| | Inactive: Cd, Zn, Al, Mn, Cu | Cd has $K_1 > K_2$: rest have $K_1 < K_2$. |
| Bi | Active: Na, K | $K_2 = 3.0$. $K_1 > K_2$ |
| | Inactive: Pb, Zn. | $K_1 < K_2$ |
| Fe | Active: Cu, Sn | $K_2 = 0.6$, $K_1 > K_2$ |
| | Inactive: | |
| Al | Active: Zn, Li, Bi, Pb | $K_2 = 1.4$, $K_1 > K_2$ |
| | Inactive: Mg, Sb, Sn | All have $K_1 > K_2$ |
| Cu | Active: Sb, Sn, Ag, Au | $K_2 = 0.75$, $K_1 > K_2$ except for Au, for which size difference is important |
| | Inactive: | |
| Ag | Active: | |
| | Inactive: Cu | $K_2 = 1.0$, $K_1 < K_2$ |
| Sb | Active: | |
| | Inactive: Cd, Zn, Pb | $K_2 = 2.7$, $K_1 < K_2$ |
| Zn | Active: Sb, Sn, Bi, Pb, Li | $K_2 = 1.7$, $K_1 > K_2$ |
| | Inactive: | |
| Pb | Active: Bi, K, Na, Ca | $K_2 = 2.3$, $K_1 > K_2$. |
| | Inactive: Sn | $K_1 < K_2$. |

NB: Out of the 19 surface inactive cases, 4 alloys disagree, on this basis, with theoretical predictions. These are Cd in Sn and Mg, Sb and Sn in Al.

As can be seen from detailed inspection of Table 10.3, the correlation with the criterion (10.167) is quite encouraging. The most decisive cases are those which are surface-inactive; reference should be made to the notes below Table 10.3 for these.

### 10.9.6 Binary liquid transition-metal alloys containing Mn and Cr

The present writer[44] has drawn attention anew to the fact that the bulk moduli of the 3d, 4d and 5d transition series in the solid state exhibit a pronounced dependence on the filling of the d band, as is clear, for example, in the review by Gschneidner.[45] In contrast with the 4d and 5d series, the 3d series exhibits a pronounced dip occurring at Mn. The formula (10.142), in the pure-metal limit, predicts a close correlation between $\sigma$ and $B$ and it therefore comes as no surprise that $\sigma$ plotted against d-shell occupancy also shows a dip at Mn for the 3d series, as seen, for instance, in Fig. 5.6 of the review by Brown and March.[10]

Rewriting the criterion (10.167) as

$$\left(\frac{\mathrm{d}B}{\mathrm{d}c}\right)_{c=0}<0; \tag{10.168}$$

in the absence of evidence to the contrary the assumptions will now be made that (a) for pure transition metals the bulk modulus variation with the filling of the d shell retains the same character just above the melting point as in the solid phase, and (b) based on known specific examples the sign of $\mathrm{d}B/\mathrm{d}c$ can be predicted from the bulk moduli of the two pure metals involved. Then given assumptions (a) and (b), the criterion (10.168) predicts that in binary liquid transition-metal alloys with Mn as solute in dilute concentration, Mn will segregate to the surface in 21 out of the theoretically possible 26 alloys. The remaining 5 are specifically Sc, Y, La, Tc and Re and because condition (10.168) is sufficient but not necessary, surface segregation of Mn may or may not occur in these cases.

Though less interesting than Mn, it is also worth adding that according to the criterion (10.168), 9 dilute Cr-liquid transition-metal alloys should show Cr segregation to the surface and the remaining 17 possibilities, listed in ref. 44, may or may not.

## 10.10 Theory of freezing: density and direct correlation function

Lovett and Buff[46] revived interest in the question asked originally by Kirkwood and Monroe[47] as to whether classical statistical mechanical equations, such as the first member of the Born–Green–Yvon hierarchy which connects the singlet density $\rho(\mathbf{r})$ and the liquid pair-correlation function $g(r)$, can admit more than one solution for $\rho$ for a given $g$. Actually, these workers focused on eqn (10.71) relating $\rho$ and the Ornstein–Zernike direct-correlation function $c(r)$ of a liquid. This latter equation has been derived by a number of workers[25–27] and has the advantage over the equation relating $g(r)$ and $\rho(\mathbf{r})$ that no assumption of pairwise interactions need be invoked.

This equation relating $\rho(\mathbf{r})$ and $c(r)$, namely

$$\ln \rho(\mathbf{r}_1)=\int \mathrm{d}\tau_2\, c(r_{12})\rho(\mathbf{r}_2)+\text{const}, \tag{10.169}$$

is the central tool employed in the present work. The integrated form (10.169) has been discussed, for example, by Lovett.[46] We shall demonstrate below, by direct solution of this equation, that for a given liquid direct-correlation function, eqn (10.169) admits not only a solution for which the singlet density is uniform, with value $\rho(\mathbf{r})=\rho_l$, but also a co-existing periodic solution $\rho_p(\mathbf{r})$. If we then linearize the equation determining $\rho_p(\mathbf{r})$, we regain the bifurcation condition of Lovett and Buff.[46]

We then go on to set up the free-energy difference corresponding to the two types of singlet density. Somewhat surprisingly, by using the equation for $\rho_p(\mathbf{r})$ in this free energy, we find a result obtained earlier in a theory of freezing by Ramakrishnan and Yussouff,[48] which was derived by making use of the hypernetted chain approximation. The presentation below[49,50] shows that this approximation is inessential in their theory of freezing.

### 10.10.1 Free-energy difference between homogeneous and periodic phases

Evidently there is for a given liquid direct-correlation function $c(r)$ a solution of eqn (10.169) for which $\rho(\mathbf{r})$ is constant with a value $\rho_l$ say. What is more important for our present purposes is to prove that a periodic solution $\rho_p(\mathbf{r})$ for the singlet density, for a given liquid $c(r)$, also exists.

To do this, we apply eqn (10.169) to both the singlet densities $\rho_l$ and $\rho_p(\mathbf{r})$ and then subtract to find

$$\ln\left(\frac{\rho_p(r_1)}{\rho_l}\right)-\int d\tau_2\, c(|\mathbf{r}_1-\mathbf{r}_2|)[\rho_p(\mathbf{r}_2)-\rho_l]=0, \qquad (10.170)$$

where it is assumed that the constants in eqn (10.169) are equal for the two phases in co-existence. Since, by assertion, $\rho_p$ is periodic, we expand in a Fourier series using the reciprocal lattice vectors $\mathbf{G}$ to obtain

$$\rho_p(\mathbf{r})=\rho_0+V^{-1}\sum_{\mathbf{G}\neq\mathbf{0}}\rho_{\mathbf{G}}\exp(i\mathbf{G}\cdot\mathbf{r}), \qquad (10.171)$$

where $V$ is the total volume. Inserting eqn (10.171) into (10.170) and integrating over $\mathbf{r}_2$ yields

$$\ln\left(\frac{\rho_p(\mathbf{r})}{\rho_l}\right)=\frac{\rho_0-\rho_l}{\rho_l}\tilde{c}(0)+(\rho_l V)^{-1}\sum_{\mathbf{G}\neq\mathbf{0}}\rho_{\mathbf{G}}\tilde{c}(\mathbf{G})\exp(i\mathbf{G}\cdot\mathbf{r}), \qquad (10.172)$$

where $\tilde{c}(k)$ is the Fourier transform of $c(r)$. This eqn (10.172) is, for a given set of Fourier components of the liquid direct-correlation function, to be solved for the Fourier components $\rho_G$ of the singlet density.

Without seeking a specific solution of eqn (10.172), a matter to which we shall refer again below, it will be useful at this point to regard eqn (10.172) as the Euler equation of a minimum-free-energy principle. Of course, the thermodynamic requirement for the two phases to be in equilibrium is that this free-energy difference should be zero. Actually, we shall work with the thermodynamic potential $\Omega$, related to the Helmholtz free energy $F$ and chemical potential $\mu$ by

$$\Omega = F - N\mu. \tag{10.173}$$

The Helmholtz free energy can be conveniently divided into two parts, one corresponding to free particles and the other taking account of the interparticle interactions via the direct-correlation function $c(r)$. The first part is well known for uniform density and we merely take the free-energy density over into the local density $\rho(\mathbf{r})$. The second part is also available in essence, for example in ref. 2, and thus we can write

$$\frac{\Delta\Omega}{k_B T} = \int \mathrm{d}\tau \left[ \rho_p(\mathbf{r}) \ln\left(\frac{\rho_p(\mathbf{r})}{\rho_1}\right) - (\rho_p(\mathbf{r}) - \rho_1) \right]$$
$$- \tfrac{1}{2} \iint \mathrm{d}\tau_1\, \mathrm{d}\tau_2\, [\rho_p(\mathbf{r}_1) - \rho_1]\, c(|\mathbf{r}_1 - \mathbf{r}_2|)\, [\rho_p(\mathbf{r}_2) - \rho_1]. \tag{10.174}$$

Performing the variation of $\Delta\Omega$ with respect to $\rho_p(\mathbf{r})$ is readily verified to lead back to eqn (10.170).

At this stage, we insert the Fourier expansion of $\rho$ and $c$ into (10.174) to find, with $N = \rho_1 V$,

$$\frac{\Delta\Omega}{k_B T} = \frac{1}{2N} \sum_{\mathbf{G}} \tilde{c}(\mathbf{G}) |\rho_{\mathbf{G}}|^2 - \frac{N(\rho_0 - \rho_1)}{\rho_1} + \frac{1}{2} N \tilde{c}(0) \frac{(\rho_0^2 - \rho_1^2)}{\rho_1^2}. \tag{10.175}$$

This is the desired expression for the free-energy difference in terms of the Fourier components $\rho_G$ of the periodic density, and the volume change reflected in the difference between $\rho_1$ and $\rho_0$.

The possibility of co-existence of homogeneous liquid and periodic phases is clear from eqn (10.175) because of the balance between positive contributions from the first and third terms on the right-hand side and the negative term from the volume change, provided the periodic phase has the higher density. That these terms are strongly coupled is clear from the highly non-linear nature of the Euler equation (10.172). The actual co-existence point is evidently determined by the properties of $\tilde{c}(\mathbf{G})$, including $\mathbf{G} = \mathbf{0}$, linking the liquid structure intimately with the appearance of the periodic phase.

### 10.10.2 Discussion and summary

The above discussion of eqns (10.172) and (10.175) has, of course, been focused on what is, in principle, possible from the structure of these two equations. It is remarkable that the work of Ramakrishnan and Yussouff[48] using as it does the hypernetted chain (HNC) approximation nevertheless leads to the same Euler eqn (10.172). Work[49,50] based directly on eqn (10.71) shows clearly that this use of HNC is inessential to their theory of freezing. However, a significant difference between their treatment and the one presented here resides in the variational principle from which the Euler equation derives. The basic variation here is of $\Delta\Omega$ in eqn (10.174). Their variation is on an equation resembling, but not identical with, eqn (10.175). This latter equation, in the present treatment, already embodies the result for the periodic density determined by the non-linear Euler equation (10.170). The specific difference between the present eqn (10.175) and their form for $\Delta\Omega$ is that they have an additional term of order $\rho_0-\rho_1$, giving their result as

$$\frac{\Delta\Omega}{k_B T}=\frac{1}{2N}\sum_{\mathbf{G}}\tilde{c}(\mathbf{G})\,|\rho_{\mathrm{G}}|^2-\tfrac{1}{2}N[1-\tilde{c}(0)]\,\frac{(\rho_0^2-\rho_1^2)}{\rho_1^2}. \tag{10.176}$$

This form is appealing because $1-\tilde{c}(0)$ is essentially the inverse compressibility of the liquid. Their work[48] in fact shows how an explicit periodic solution, for a given liquid direct-correlation function $c(r)$, can be derived from eqn (10.172).

The final comment concerns the relation of the non-linear eqn (10.172) to the work of Lovett and Buff[46] on bifurcation. Whereas the above treatment is evidently describing a first-order transition, with a volume change and finite Fourier components $\rho_{\mathbf{G}}$ actually at the freezing point, their work explores the condition under which the $\rho_{\mathbf{G}}$ develop continuously from the homogeneous phase. The procedure they use corresponds to linearizing eqn (10.172), which then has a solution of periodic form provided the condition $1-\rho_1\,\tilde{c}(\mathbf{G})=0$ is satisfied. This corresponds in fact to the structure factor $S(\mathbf{G})=(1-\rho_1\tilde{c}(\mathbf{G}))^{-1}$ becoming infinite. This is an instability of the liquid phase.

It is not our purpose here to enter into the not inconsiderable technical aspects of actually calculating the freezing properties of some liquid systems from this theory. Suffice it to say that successful applications have been made, and that, although the convergence of the reciprocal lattice sum is not as fast as the success of Verlet's rule,[51] depending essentially only on the first such vector, might suggest at first sight, nevertheless, one can have confidence that a first-principles statistical mechanical theory of freezing based on $c(r)$ is now laid down, through the work of Ramakrishnan and Yussouff[48] and later authors[49,50] cited in this Section.

### 10.10.3 Verlet's rule related to Lindemann's law of melting

For Lennard-Jones liquids like argon, Verlet[51] drew attention to the fact that the liquids freeze when the principal peak height of the structure factor reaches a value of about 2.8 as the temperature is lowered. With regard to Na and K, Ferraz and March[52] pointed out that the classical one-component plasma would freeze under roughly the same condition, and that this condition was borne out by comparison with experiment for these two alkali metals. Later work has elaborated the relation of the one-component plasma model to the structure factors of the alkali metals.

Thus, for freezing, one has the above criterion that $S(q)$ at its maximum, occurring at position $q_m$ say, is approximately 2.8, whereas Lindemann's law of melting states that crystals melt when the root mean square vibrational amplitude becomes a fixed fraction of the lattice spacing.

It is of interest to elaborate here on the relation between these criteria for this first-order phase transition. The work of Bhatia and March,[53] though designed more generally to relate principal peak height, position and width of the structure factor $S(q)$ of dense monatomic liquids, supplies such a relation. That is not to say that, in practice, Lindemann's law is as good a criterion for characterizing the phase transition quantitatively as is the rule of Verlet; that is a matter of detailed numerical study, whereas the point emphasized below is that a definite correlation can be established between the two different rules.

*Relation between principal peak height, position and width of $S(q)$*

In dense classical liquids, the pair function $g(r)$ must satisfy the condition $g(0)=0$. Using the Fourier-transform relation between the total correlation function $g(r)-1$ and $S(q)-1$, this condition $g(0)=0$ reads, for $N$ atoms in volume $V$,

$$-2\pi^2\frac{N}{V}=-2\pi^2\rho=\int_0^\infty (S(q)-1)q^2\,\mathrm{d}q. \qquad (10.177)$$

An approximate evaluation of the integral in eqn (10.177) in such dense liquids can be carried out as follows. With $q_m$, as above, denoting the position of the principal peak of $S(q)$, let the peak width, $2\Delta q$ say, be measured by the distance between the two adjacent nodes of $S(q)-1$ which embrace $q_m$. Furthermore, it will be assumed that any asymmetry of the peak about $q_{\rm m}$ may be neglected. If eqn (10.177) is now written as

$$-2\pi^2\rho=\int_0^{q_{\rm m}-q}\{S(q)-1]q^2\,dq+\int_{q_{\rm m}-\Delta q}^{q_{\rm m}+\Delta q}[S(q)-1]q^2\,dq$$
$$+\int_{q_{\rm m}+\Delta q}^{\infty}[S(q)-1]q^2\,dq, \tag{10.178}$$

then for an $S(q)$ appropriate to dense fluids, such as argon near the triple point, or Na and K near the melting point, the following approximations prove useful:

(i) to replace $S(q)-1$ by $-1$ over the range of the first integral in eqn (10.178);
(ii) to neglect the third integral, because of the oscillations about zero of $S(q)-1$; and
(iii) to estimate the second integral by the triangular area

$$[S(q_{\rm m})-1]q_{\rm m}^2\Delta q.$$

Using these simplifications and introducing the mean interatomic separation $R_{\rm A}$ through $\rho=3/4\pi R_{\rm A}^3$, it is readily shown that

$$S(q_{\rm m})\,q_{\rm m}^2\,\Delta q \doteqdot \tfrac{1}{3}q_{\rm m}^3\left[1-\frac{9\pi}{2}\frac{1}{(R_{\rm A}q_{\rm m})^3}\right]. \tag{10.179}$$

Empirically, in dense liquids, $R_{\rm A}q_{\rm m}\cong 4.4$, and the second term in square brackets in eqn (10.179) contributes 0.15 compared with unity. Thus one is left with the result

$$S(q_{\rm m})\sim\frac{0.3q_{\rm m}}{\Delta q}. \tag{10.180}$$

It is instructive to confront the approximate prediction (10.180) with the accurate diffraction data of Yarnell *et al.*[54] on liquid argon at 85 K. One finds from this data that $S(q_{\rm m})=2.70$, $q_{\rm m}=2.00$ and $\Delta q=0.275$, $q$ being in Å$^{-1}$. This yields $S(q_{\rm m})/(q_{\rm m}/\Delta q)=0.37$, which is nearer $\frac{3}{8}$ than the predicted 0.3 in eqn (10.180). It is satisfactory that the data of Greenfield *et al.*[55] on liquid potassium at 65°C yields $S(q_{\rm m})=2.73$, $q_{\rm m}=1.62$ and $\Delta q=0.225$, and hence a constant of 0.38; while for the experimental results obtained at 135°C the constant is 0.37. Similarly for Na at 100°C and 200°C the measured data yield 0.37 and 0.36, respectively. Thus for the five experiments referred to above, it turns out that eqn (10.180) is quantitative when 0.3 is replaced by $\frac{3}{8}$. The fact that this is greater than 0.3 seems to indicate that the third integral in eqn (10.178) actually has a non-zero negative value. Just at the time of writing, Silbert (private communication) has found that for some 50 dense liquids the relation (10.180) is well obeyed, the constant ranging between 0.3 and 0.4 in about 40 of these cases.

Turning to $g(r)$, it is worth noting that an argument in which $S(0)$ is evaluated via $\int_0^\infty [g(r)-1]r^2\,dr$ can be carried through with assumptions paralleling (i)–(iii) above for calculating $g(0)$. The result is

$$g(r_m)r_m^2\Delta r = \tfrac{1}{3}r_m^3 - \tfrac{1}{3}R_A^3(1-S(0)), \tag{10.181}$$

with definitions precisely analogous to those for $S(q)$. Since $g(r)$ is less readily accessible than $S(q)$, no comparison of eqn (10.181) with experimental data will be attempted as it stands. However, for the data of Yarnell *et al.*[54] on liquid argon at 85 K, one finds

$$g(r_m)=3.05, \quad \Delta r = 0.545\,\text{Å}, \quad r_m = 3.68\,\text{Å}, \tag{10.182}$$

and the value of $r_m/\Delta r$ is found to be 6.7, to be compared with $q_m\Delta q = 7.2$. Thus, very approximately,

$$\frac{q_m}{\Delta q} \cong \frac{r_m}{\Delta r}. \tag{10.183}$$

### 10.10.4 Melting and freezing criteria related

To complete this section, one must now return to the Verlet criterion that simple liquids freeze when $S(q_m)=2.8$, this value, as seen, being also true for Na and K. In these cases of Ar, Na and K, where freezing involves only modest changes in local coordination then, at the melting temperature $T_m$, use of $S(q)|_{T_m}=2.8$ yields from eqns (10.180) and (10.183) the estimate $(\Delta r/r_m)_{T_m} \cong 0.11$. But Lindemann's law of melting, according to Faber,[12] gives $(\Delta r/R_A)_{T_m} \cong 0.2$ if one here identifies $\Delta r$ as the root mean square displacement of the atoms. Since $r_m \cong 1.8\,R_A$, these results are seen to be roughly consistent. Thus, there is no conflict between freezing criteria based on $S(q_m)|_{T_m}=2.8$ on the one hand and Lindemann's law on the other.

### 10.10.5 Statistical mechanical theory of freezing of alkali halides

This Section is based on the work of March and Tosi.[56,57] Their starting point is the set of equations relating the singlet densities $\rho_i$ of a two-component system to the three partial direct correlation functions, set out in Section 10.8. For $\nu$ components, these equations take the explicit form (10.114). Following the work of Lovett,[56] one now integrates eqns (10.114) under the assumption that $c_{ij}(\mathbf{r}_1, \mathbf{r})$ depends only on $|\mathbf{r}_1 - \mathbf{r}|$ as in a bulk liquid. The result is then

$$\ln \rho_i(\mathbf{r}_1) = \sum_{j=1}^{\nu} \int \mathrm{d}\mathbf{r}\, c_{ij}(|\mathbf{r}_1 - \mathbf{r}|)\rho_j(\mathbf{r}) + A_i, \tag{10.184}$$

where $A_i$ is a constant of integration.

The question one now poses again is whether, given the liquid partial direct correlation functions $c_{ij}$, this eqn (10.184) can exhibit a periodic solution, $\rho_{ip}(\mathbf{r})$ say, in co-existence with the obvious homogeneous solution $\rho_{il}$, when the constant $A_i$ is the same in the two phases. To answer this question, it is helpful to construct the difference between eqns (10.184) for periodic and for homogeneous one-particle or singlet densities. Regarding the result

$$\ln \frac{\rho_{ip}(\mathbf{r}_1)}{\rho_{il}} = \sum_{j=1}^{\nu} \int \mathrm{d}\mathbf{r}\, c_{ij}(|\mathbf{r}_1 - \mathbf{r}|)[\rho_{jp}(\mathbf{r}) - \rho_{j1}] \tag{10.185}$$

as the Euler equation of a variational problem again, one can construct the difference $\Delta\Omega$ in thermodynamic potential, $\Omega$ being specifically $F - \Sigma_i N_i \mu_i$, with $F$ the Helmholtz free energy, $N_i$ the number of particles and $\mu_i$ the chemical potential of species $i$. Using the separation into free-particle terms and those arising from the interparticle interactions reflected in $c_{ij}$, one obtains (cf. eqn (10.174))

$$\begin{aligned} \frac{\Delta\Omega}{k_B T} = \sum_{i=1}^{\nu} \int \mathrm{d}\tau \left\{ \rho_{ip}(\mathbf{r}) \ln \left| \frac{\rho_{ip}(\mathbf{r})}{\rho_{il}} \right| - [\rho_{ip}(\mathbf{r}) - \rho_{il}] \right\} \\ - \frac{1}{2} \sum_{i,j=1}^{\nu} \iint \mathrm{d}\tau_1\, \mathrm{d}\tau_2 [\rho_{ip}(\mathbf{r}_1) - \rho_{il}] c_{ij}(|\mathbf{r}_1 - \mathbf{r}_2|)[\rho_{jp}(\mathbf{r}_2) - \rho_{jl}]. \end{aligned} \tag{10.186}$$

Varying this expression for $\Delta\Omega$ with respect to the periodic singlet densities, the correct Euler equation is regained.

One now uses eqn (10.185) to remove $\ln[\rho_{ip}(\mathbf{r})/\rho_{il}]$ from eqn (10.186), when one obtains

$$\begin{aligned} \frac{\Delta\Omega_{\min}}{k_B T} = -\sum_{i=1}^{\nu} \int \mathrm{d}\tau\, [\rho_{ip}(\mathbf{r}) - \rho_{il}] + \frac{1}{2} \sum_{i,j=1}^{\nu} \iint \mathrm{d}\tau_1\, \mathrm{d}\tau_2 \\ \times [\rho_{ip}(\mathbf{r}_1) + \rho_{i1}] c_{ij}(|\mathbf{r}_1 - \mathbf{r}_2|)[\rho_{jp}(\mathbf{r}_2) - \rho_{j1}]. \end{aligned} \tag{10.187}$$

This quantity should vanish at the co-existence point at which the periodic singlet densities are to be found from eqn (10.185).

### 10.10.6 Singlet densities and charge-number correlation functions for alkali halides

This is the point at which specialization to the case of an alkali halide will be made; so far the discussion is appropriate to any multicomponent system (e.g. a binary alloy). It is convenient to work with the total density $\rho(\mathbf{r})=\rho_1(\mathbf{r})+\rho_2(\mathbf{r})$ and the charge density $Q(\mathbf{r})=\rho_1(\mathbf{r})-\rho_2(\mathbf{r})$ for reasons discussed elsewhere. In the alkali halide, the condition of charge neutrality ensures that $\int d\tau\, Q(\mathbf{r})=0$. Similarly, one can introduce the number-charge direct-correlation functions, which are related to $c_{ij}(\mathbf{r})$ by

$$c_{NN}(r)=\tfrac{1}{2}[c_{11}(r)+c_{22}(r)+2c_{12}(r)], \tag{10.188}$$

$$c_{NQ}(r)=\tfrac{1}{2}[c_{11}(r)-c_{22}(r)] \tag{10.189}$$

and

$$c_{QQ}(r)=\tfrac{1}{2}[c_{11}(r)+c_{22}(r)-2c_{12}(r)]. \tag{10.190}$$

It is straightforward then to rewrite the above theory in terms of these $c_{NQ}$ functions.

After some manipulation using Fourier analysis, eqn (10.187) can be rewritten as

$$\frac{\Delta\Omega_{\min}}{k_B T}=-\frac{(\rho_0-\rho_1)N}{\rho_1}+\tfrac{1}{2}N\tilde{c}_{NN}(0)\frac{\rho_0^2-\rho_1^2}{\rho_1^2}$$
$$+\frac{2}{N}\sum_{\mathbf{G}}{}'\,[\tilde{c}_{NN}(\mathbf{G})|\rho_{1\mathbf{G}}|^2\cos^2(\tfrac{1}{2}\mathbf{G}\cdot\mathbf{h})+\tilde{c}_{QQ}(\mathbf{G})|\rho_{1\mathbf{G}}|^2\sin^2(\tfrac{1}{2}\mathbf{G}\cdot\mathbf{h})], \tag{10.191}$$

where $\rho_{1\mathbf{G}}$, $\rho_{2\mathbf{G}}$ and $\tilde{c}_{NN}$, etc., are defined by

$$\rho_{1\mathrm{p}}(r)=\rho_{10}+\frac{1}{V}\sum_{\mathbf{G}}{}'\,\rho_{1\mathbf{G}}\exp[i\mathbf{G}\cdot\mathbf{r}], \tag{10.192}$$

$$\rho_{2\mathrm{p}}(r)=\rho_{20}+\frac{1}{V}\sum_{\mathbf{G}}{}'\,\rho_{2\mathbf{G}}\exp[i\mathbf{G}\cdot(\mathbf{r}+\mathbf{h})] \tag{10.193}$$

and

$$c_{NN}(\mathbf{r})=\frac{2}{N}\sum_{\mathbf{q}}\tilde{c}_{NN}(q)\exp[i\mathbf{q}\cdot\mathbf{r}]. \tag{10.194}$$

Here the prime on the summation means that the reciprocal-lattice vector $\mathbf{G}=\mathbf{0}$ is omitted; $N$ is the total number of ions, while $V$ is the total volume. $\mathbf{h}$ is the vector joining the two ions in the unit cell of the crystal structure.

For the alkali halides, $\rho_{1G}=\rho_{2G}$ and furthermore

$$\rho_{10}=\rho_{20}=\tfrac{1}{2}\rho_0, \qquad (10.195)$$

where $\rho_0$ is the average density of the periodic phase which is clearly different from the liquid density $\rho_1$ in a first-order transition.

Equation (10.191) already demonstrates the possibility that $\Delta\Omega_{min}$ can be zero, because of the balance between the first term involving the volume change, which will be negative in the alkali halides, and the positive terms from the Fourier components $\rho_{1G}$, to be determined, along with the volume change, by solution of eqn (10.185), written in terms of Fourier components. The condition that $\Delta\Omega_{min}$ vanishes then suffices to determine the freezing temperature. Numerical consequences of these equations, when experimental data for $c_{ij}$ in molten RbCl are used as input, are summarized in refs 58 and 61.

### 10.10.7 Freezing theory of molten $BaCl_2$ into a superionic phase

March and Tosi[59] have generalized their freezing theory of alkali halides summarized above with the specific case of $BaCl_2$ in mind for which neutron diffraction data are available from the work of Edwards *et al.*[60] The important new point here is that the liquid $BaCl_2$ freezes into a superionic phase. It is apparent from the neutron data on the liquid structure factors $S_{Ba-Ba}$, etc., that the Ba cations exhibit a marked degree of ordering. The structure factors $S_{Cl-Cl}$ and $S_{Ba-Cl}$ have much less pronounced features, occurring at somewhat different wavenumbers. This observed behaviour is to be contrasted with the situation occurring in molten alkali halides such as RbCl discussed elsewhere where the three partial structure factors have main peaks of comparable height at essentially the same wave number. Of course, this is not wholly surprising in view of the different charges and sizes of the ionic constituents. In the work of March and Tosi one of the interesting points is the way in which these structural differences are reflected in the freezing process, with specific reference to $BaCl_2$. Additional relevant data concern the magnitudes of the relative volume change at freezing. For $BaCl_2$, the volume increase on melting is 3.5% to be contrasted with the 14% increase for RbCl, or the 25% increase for NaCl.

These contrasting features between the alkali halides and $BaCl_2$ lead one to think of freezing into the superionic phase in the following manner. The freezing of $BaCl_2$ specifically is regarded as being driven by the marked cation ordering in the liquid. This leads to a cation sublattice, with the volume contraction associated with this first-order transition. In turn, the anions are subjected thereby to (i) the volume change and (ii) modulation of their singlet ionic density, by this cation sublattice order. Below, we, therefore, present (following ref. 49) a generalization of the theory of alkali halide freezing given above to embody the specific features appropriate to $BaCl_2$ and liquids with similar properties.

Let us denote the singlet densities in liquid and in superionic phases by $\rho_{i1}$ and $\rho_{is}(\mathbf{r})$, respectively. Then eqn (10.185) holds, with $\nu=2$ and with $\rho_{ip}$ replaced simply by $\rho_{is}$. The two phases must be taken at the co-existence point; the condition of equilibrium of the two phases being that the difference $\Delta\Omega$ of the thermodynamic potential be zero. This difference, when the above Euler equations are used, turns out to be

$$\frac{\Delta\Omega}{k_{\mathrm{B}}T}=-(\rho_0-\rho_1)V+\frac{1}{2}\sum_{i,j=1}^{2}\iint \mathrm{d}\tau_1\,\mathrm{d}\tau_2\,[\rho_{is}(\mathbf{r}_1)+\rho_{il}]$$
$$\times c_{ij}(|\mathbf{r}_1-\mathbf{r}_2|)[\rho_{js}(\mathbf{r}_2)-\rho_{j1}], \qquad (10.196)$$

where $\rho_0$ and $\rho_1$ are the mean number densities in solid and liquid phases, while $V$ is the volume.

Again, the singlet densities are Fourier-analysed, to yield

$$\rho_{is}(\mathbf{r})=\rho_{i0}+\frac{1}{V}\sum_{\mathbf{G}}{}'\,\rho_{i\mathrm{G}}\exp[\mathrm{i}\mathbf{G}\cdot\mathbf{r}],\quad i=1,2, \qquad (10.197)$$

where $\rho_0=\rho_{10}+\rho_{20}$ and, as usual, the **G**s are reciprocal-lattice vectors. The direct correlation functions are also Fourier-decomposed:

$$c_{ij}(\mathbf{r})=\frac{1}{V(\rho_{i1}\rho_{j1})^{1/2}}\sum_{\mathbf{q}}\tilde{c}_{ij}(\mathbf{q})\exp[\mathrm{i}\mathbf{q}\cdot\mathbf{r}]. \qquad (10.198)$$

Substituting in the Euler equations, one then finds for the $\mathbf{q}=\mathbf{0}$ component

$$\ln\frac{\rho_{i0}}{\rho_{il}}+\frac{1}{V}\int\mathrm{d}\tau\,\ln\left[1+\frac{1}{\rho_{i0}V}\sum_{\mathbf{G}}{}'\rho_{i\mathrm{G}}\exp[\mathrm{i}\mathbf{G}\cdot\mathbf{r}]\right]$$
$$=\sum_{j=1}^{2}\frac{\rho_{j0}-\rho_{jl}}{(\rho_{i1}\rho_{j1})^{1/2}}\tilde{c}_{ij}(0). \qquad (10.199)$$

The divergent Coulomb terms in the $\tilde{c}_{ij}(q)$ for $q$ tending to zero cancel in eqn (10.199) on account of the electrical-neutrality condition

$$\frac{\rho_{10}}{\rho_{11}}=\frac{\rho_{20}}{\rho_{21}}. \qquad (10.200)$$

After the divergent terms in $\tilde{c}_{ij}(q)$ at small $q$ are accounted for as above, there is a finite limit as $q \to 0$, with value $\tilde{c}_{ij}^{0}(0)$ say. Then one finds

$$\ln \frac{\rho_{io}}{\rho_{il}} + \frac{1}{V} \int d\tau \ln \left[ 1 + \frac{1}{\rho_{io} V} \sum_{\mathbf{G}}{}' \rho_{i\mathbf{G}} \exp [i\mathbf{G} \cdot \mathbf{r}] \right]$$

$$= \frac{\rho_0 - \rho_l}{\rho_l} \sum_{j=1}^{2} (\rho_{jl}/\rho_{il})^{1/2} \tilde{c}_{ij}(0). \tag{10.201}$$

The combination of $\tilde{c}_{ij}$ entering eqn (10.201) involves the compressibility and the difference in partial molar volumes for the two components of the liquid.

Turning to the case $G \neq 0$, from the Euler equations one obtains

$$\int d\tau \ln \left[ 1 + \frac{1}{V \rho_{io}} \sum_{\mathbf{G}'}{}' \rho_{i\mathbf{G}'} \exp [i\mathbf{G}' \cdot \mathbf{r}] \right] \exp [-i\mathbf{G} \cdot \mathbf{r}]$$

$$= \sum_{j=1}^{2} (\rho_{il} \rho_{jl})^{-1/2} \rho_{j\mathbf{G}} \tilde{c}_{ij}(\mathbf{G}). \tag{10.202}$$

Finally the expression (10.196) for $\Delta\Omega$ can be rewritten as

$$\frac{\Delta\Omega}{k_B T} = -(\rho_0 - \rho_l)V + \frac{V}{2} \left( \frac{\rho_0^2}{\rho_l^2} - 1 \right) \sum_{i,j=1}^{2} (\rho_{il} \rho_{jl})^{1/2} \tilde{c}_{ij}^{0}(0)$$

$$+ \frac{1}{2V} \sum_{i,j=1}^{2} \sum_{\mathbf{G}}{}' \tilde{c}_{ij}(\mathbf{G}) \frac{\rho_{i\mathbf{G}} \rho_{j\mathbf{G}}}{(\rho_{il} \rho_{jl})^{1/2}}. \tag{10.203}$$

This form of $\Delta\Omega$ demonstrates that the phase transition can occur as a balance between a favourable term from the volume contraction and an increase in $\Omega$ from the Fourier-component modulation of the singlet densities. Again, the particular combination of $\tilde{c}_{ij}^{0}(0)$ in eqn (10.203) is related to the compressibility of the liquid.

### 10.10.8 Linear response theory connecting Fourier components of anion and cation singlet densities

So far, the development has been formally exact. The next step is to find out how the anion density modulation, described by the Fourier components $\rho_{2\mathbf{G}}$, say, can be related to the cation sublattice Fourier components $\rho_{1\mathbf{G}}$. For the special case $\mathbf{G} = \mathbf{0}$, the charge neutrality condition (10.200) determines $\rho_{20}$ exactly, given $\rho_{10}$.

If we recall the experimental results that the Ba cations are highly ordered in the liquid phase, a reasonable starting point is to argue that the freezing into a cation sublattice can be treated as a perturbation on the anion singlet density, which is homogeneous, with density $\rho_{21}$ in the liquid phase.

One, therefore, is led to consider eqn (10.202) for $i=2$ and, linearizing the left-hand side in the $\rho_{2G}$s, the integration can be performed to yield

$$\rho_{2G}=\frac{(\rho_{21}/\rho_{11})^{1/2}\tilde{c}_{12}(G)}{\rho_{21}/\rho_{20}-\tilde{c}_{22}(G)}\rho_{1G}. \tag{10.204}$$

As was to be expected, the coupling of the sublattice 1 to the modulations $\rho_{2G}$ is determined by the cross-correlation function $\tilde{c}_{12}(\mathbf{G})$. The denominator in eqn (10.204) represents in essence the response function of the system 2, i.e. the anions in $BaCl_2$.

The numerical consequences of this theory are summarized in refs 61 and 57. However, using the data of Edwards *et al.* as input, one obtains from eqn (10.204) for the first reciprocal-lattice vector $\mathbf{G}_1$ of the Ba sublattice the estimate $\rho_{2G_1}\cong 0.4\ \rho_{1G_1}$. Experimental tests of the prediction are not available at the time of writing.

## 10.11 Spinodal decomposition

The purpose of this short section is to develop the van der Waals–Cahn–Hilliard approach referred to above to treat the problem of spinodal decomposition (see for instance Figs 10.4 and 10.5 below). In pargicular, a summary will be given of the work of Abraham[62] who has studied the stability of an initially homogeneous fluid to infinitesimal (one-dimensional) density fluctuations. Then, following Cahn and Hilliard,[64] it is assumed that there exists a function $f(z)$ that can be identified as the local Helmholtz free energy per unit volume at position $z$ in the non-uniform fluid.

Then, as discussed above, the van der Waals–Chan–Hilliard, theory has

$$f(z)=f_1(\rho(z))+A\rho'(z)^2, \quad A>0, \tag{10.205}$$

where if $f(\rho)$ denotes the Helmholtz free-energy density as a function of $\rho$ in the one-phase region, the first term $f_{\text{local}}=f_1(\rho)$ is the analytical continuation of the local function of $\rho$ into the two-phase region and is related to the true free-energy density by the double tangent construction.

Cahn's pioneering work on spinodal decomposition is based on the Helmholtz free-energy density of the form (10.205). Following this work, Abraham writes, by expanding $f_1$ about the average density $\rho_0$

$$f_1[\rho]\cong f_1(\rho_0)+(\rho-\rho_0)\left(\frac{\partial f_1}{\partial\rho}\right)_0+\tfrac{1}{2}(\rho-\rho_0)^2\left(\frac{\partial^2 f_1}{\partial\rho^2}\right)_0. \tag{10.206}$$

If one assumes an infinitesimal density fluctuation of the form

$$\rho-\rho_0=\alpha\cos\beta z, \tag{10.207}$$

then one finds the result

$$f(z)-f_1(\rho_0)=\alpha\cos\beta z\left\{\left(\frac{\partial f_1}{\partial\rho}\right)_0+2\pi\rho_0\int_0^\infty[\cos\beta\omega-1]\Omega_0(\omega)\,\mathrm{d}\omega\right.$$

$$\left.+\frac{\alpha^2}{2}\cos^2\beta z\left\{\left(\frac{\partial^2 f_1}{\partial\rho^2}\right)_0+4\pi\int_0^\infty[\cos\beta\omega-1]\Omega_0(\omega)\,\mathrm{d}\omega\right\}\right., \tag{10.208}$$

where

$$\Omega_0(\omega)=\int_\omega^\infty\phi_1(\xi)g_0(\xi,\rho_0)\xi\,\mathrm{d}\xi, \tag{10.209}$$

$\phi_1(\xi)$ being the perturbation from the underlying reference liquid and $g_0$ the radial distribution function of the unperturbed system.*

The difference in the free energy per unit volume between the initially uniform fluid $\rho_0$ and one with varying density is then

$$\frac{\Delta F}{V}=\frac{\alpha^2}{4}\left\{\left(\frac{\partial^2 f_1}{\partial\rho^2}\right)_0+4\pi\int_0^\infty[\cos\beta\omega-1]\Omega_0(\omega)\,\mathrm{d}\omega\right\}. \tag{10.210}$$

With $(\partial^2 f_1/\partial\rho^2)_0>0$, $\Delta F>0$ for all $\beta$ and the fluid is stable with respect to infinitesimal sinusoidal fluctuations of all wavelengths. Within the spinodal region, where $(\partial^2 f_1/\partial\rho^2)_0<0$, the fluid is unstable with respect to infinitesimal sinusoidal fluctuations of wavelength greater than $\lambda_c=2\pi\beta_c$ where

$$-\left(\frac{\partial^2 f_1}{\partial\rho^2}\right)_0=4\pi\int_0^\infty[\cos\beta_c\omega-1]\Omega_0(\omega)\,\mathrm{d}\omega. \tag{10.211}$$

If $\lambda_c$ is sufficiently large, or $\beta_c$ sufficiently small so that

$$\cos(\beta_c\omega)-1\cong-\tfrac{1}{2}(\beta_c\omega)^2 \quad \text{for } \omega\leqslant\omega_f$$

and

$$\Omega_0(\omega)\cong 0 \quad \text{for } \omega\geqslant\omega_f,$$

the integral of eqn (10.211) becomes

$$-2\pi\beta_c^2\int_0^{\omega_f}\omega^2\Omega(\omega,\rho_0)\,\mathrm{d}\omega=2\beta^2 A_0, \tag{10.212}$$

*Details are to be found in ref. 62, the treatment being based on ref. 106, where a generalization of bulk-liquid-state perturbation theory was given.

where evidently

$$A_0 \cong -\pi \int_0^{\omega_f} \omega^2 \Omega(\omega, \rho_0)\, d\omega = \text{const.} > 0. \tag{10.213}$$

From eqns (10.211) and (10.212) the critical wavelength fluctuation is

$$\lambda_c = \frac{2\pi}{\beta_c} = \left[\frac{-8\pi^2 A_0}{(\partial^2 f_1/\partial \rho^2)_0}\right]^{1/2}. \tag{10.214}$$

In Fig. 10.4, taken from the work of Abraham, the spinodes $\partial^2 f_1/\partial \rho^2 = 0$ of each free-energy density curve are marked as solid circles. The spinodes have the greatest separation on the density scale at the lowest temperatures and approach one another as the temperature increases, converging to the critical point at the critical temperature. In Fig. 10.5, the locus of spinodes from Fig. 10.4 is depicted by the dashed curve on the phase diagram—the spinodal. For other references to this area, the reader is referred to the short review by Abraham;[65] and especially to the work of Langer referred to there.

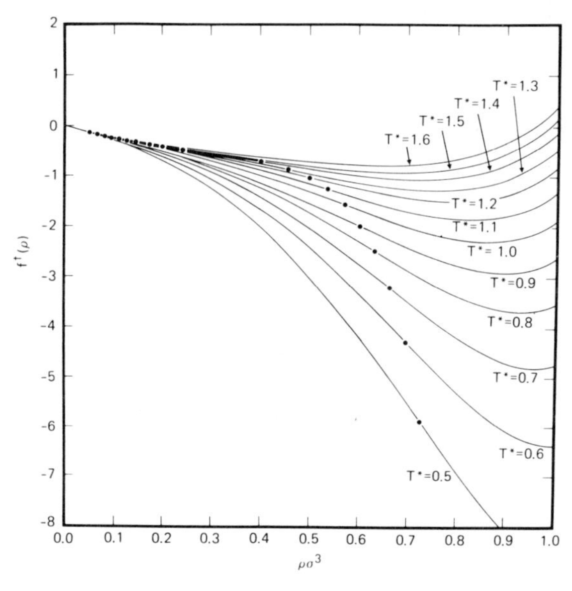


**Fig. 10.4.** Helmholtz free energy $f^\dagger(\rho)$ of a single-phase Lennard-Jones fluid at density $\rho$ and temperature $T^*$ using liquid-state perturbation theory. The ideal-gas contribution to the free-energy density is not included, only the configurational part being shown in the figure. The unit of energy is $k_B T/\sigma^3$. The spinodes, $\partial^2 f^\dagger/\partial \rho^2 = 0$ of each free-energy density curve are denoted by solid circles. (After Abraham[62].)

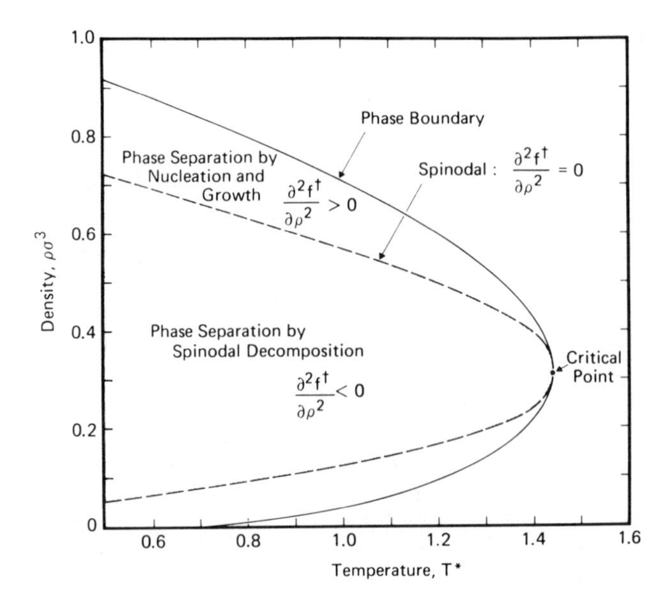


**Fig. 10.5** The locus of spinodes from Fig. 10.4 is depicted by dashed curve on the phase diagram of a Lennard-Jones fluid. (After Abraham[62].)

## 10.12 Metal–electrolyte interface

In this Section, use will be made of the example of Gouy–Chapman theory[11] with which this Chapter began. These workers, long ago, introduced a simple model for an electrode–electrolyte interface. The model considers a fluid of point ions in a uniform dielectric medium, which builds up a diffuse layer of screening charge in front of a charged hard wall.

In the Gouy–Chapman theory, as seen in Section 10.2, the approximation is made that the particle densities $\rho_\alpha(z)$ are simply related to the electrical potential energy $\phi\,(z)$ by the Boltzmann form

$$\rho_\alpha(z) = \rho_\alpha^0 \exp\left(-Z_\alpha \phi(z)/k_B T\right), \tag{10.215}$$

as for a system of non-interacting particles in an external potential. Here the $\rho_\alpha^0$ denote the particle densities in the bulk electrolyte, where one has chosen $\phi(\infty)=0$. Self-consistency is introduced by requiring that $\phi(z)$ is determined both by the charge on the wall and by the charge density $\rho_0 = \Sigma_\alpha Z_\alpha e \rho_\alpha(z)$ induced in the electrolyte, the whole being screened by the dielectric constant $\varepsilon$ of the uniform medium, which represents the solvent.

Thus $\phi(z)$ satisfies Poisson's equation

$$\frac{d^2\phi(z)}{dz^2} = -\frac{4\pi e}{\varepsilon}\rho_0(z)$$

$$= -\frac{4\pi e^2}{\varepsilon}\sum_\alpha Z_\alpha \rho_\alpha^0 \exp\left(-\frac{Z_\alpha \phi(z)}{k_B T}\right) \quad (z>0), \tag{10.216}$$

with the boundary condition following from Gauss' theorem as

$$q = -\frac{\varepsilon}{4\pi e}\left[\frac{d\phi(z)}{dz}\right]_{z=0}. \tag{10.217}$$

For a 1–1 electrolyte, where $|Z_\alpha| = 1$ and $\rho_1^0 = \rho_2^0 = \rho_0$ say, the solution of eqn (10.216) is

$$\phi(z) = 4k_B T \tanh^{-1}(s \exp(-\kappa z)) \tag{10.218}$$

where $s = \tan h[\phi(0)/4k_B T]$ and $\kappa = (8\pi\rho_0 e^2/\varepsilon k_B T)^{1/2}$ is the Debye–Hückel inverse screening length. This result corresponds to an approximately exponential decay of $\phi(z)$, as can be seen in the case where $s$ is small, when $\phi(z) = \phi(0)\exp(-\kappa z)$.

The surface charge density on the wall is related to $\phi(0)$ by eqn (10.217):

$$q = \left(\frac{2k_B T\varepsilon\rho_0}{\pi}\right)^{1/2} \sinh\frac{\phi(0)}{2k_B T}, \tag{10.219}$$

and hence the capacitance $C_d$ of the diffuse layer is given by

$$C_d = e\left(\frac{\partial q}{\partial \phi(0)}\right)_{\text{composition}} = \left(\frac{\varepsilon\rho_0 e^2}{2\pi k_B T}\right)^{1/2} \cosh\frac{\phi(0)}{2k_B T}. \tag{10.220}$$

This yields $C_d = (\varepsilon/4\pi)\kappa$ at the point of zero charge, followed by an initially parabolic rise with voltage. The surface excess of particles of either species can also be calculated:

$$\rho_\alpha = \int_0^\infty dz\,[\rho_\alpha(z) - \rho_\alpha^0]$$

$$= 2\rho_0 Z_\alpha \int_0^\infty dz \exp\left(-\frac{Z_\alpha\phi(z)}{2k_B T}\right)\frac{d\phi(z)}{dz}\left(\frac{32\pi k_B T\rho_0 e^2}{\varepsilon}\right)^{1/2}$$

$$= \left(\frac{\varepsilon k_B T\rho_0}{2\pi e^2}\right)^{1/2}\left\{\exp\left(-\frac{Z_\alpha\phi(0)}{2\,k_B T}\right) - 1\right\}, \tag{10.221}$$

the last step following from a change of integral over $z$ to an integral over $\phi$. Lastly, the interfacial tension, relative to its value at the potential of zero charge, is easily evaluated by integration of eqns (10.219) and (10.221) with the thermodynamic relations as follows:

$$C \equiv \left(\frac{\partial q}{\partial \phi}\right)_{\mu} = -\left(\frac{\partial^2 \sigma}{\partial \phi^2}\right)_{\mu}, \tag{10.222}$$

the value of $\sigma$ as a function of $\phi$ being found to have a maximum, which defines a fundamental reference potential for the electrified interface, the point of zero charge (cf. $q = -(\partial\sigma/\partial\phi)_{\mu}$).

Hencc onc finally obtains

$$\sigma = -2k_B T\left(\frac{2k_B T\varepsilon\rho_0}{\pi e^2}\right)^{1/2}\left[\cosh\left(\frac{\phi(0)}{2k_B T}\right) - 1\right]. \tag{10.223}$$

In applying the predictions of the Gouy–Chapman theory to an analysis of data on electrode–electrolyte interfaces, it has become usual, following the lead of Stern,[66] to recognize that in addition to the diffuse layer contribution $C_d$, a second, so-called inner-layer, contribution $C_i$ also exists. From the simplest point of view, this accounts for finite ionic sizes by allowing for a non-vanishing value of the distance of closest approach of the ions to the wall. The interface is then viewed as a series of two capacitors, according to

$$C = \left(\frac{1}{C_i} + \frac{1}{C_d}\right)^{-1}. \tag{10.224}$$

Much effort has been devoted to modelling the inner layer, the Gouy–Chapman theory frequently being employed to isolate the corresponding capacitance on the assumption that eqn (10.224) is valid. The points below[11] refer to some of the problems that have been discussed; with additional references to relevant work.

(1) Models of the dipolar orientation contribution in the inner layer and effects of dielectric saturation are treated in the article by Reeves in ref. 67 and in the work of Liu.[68]
(2) Preferential adsorption of ions and solvent molecules at the electrode, implying a multiplicity of Stern layers, is discussed in the article by Habib and Bockris in ref. 67.
(3) Non-local effects in dielectric screening and spilling out of the electron distribution from the electrode are treated in Kornyshev and Vorotyntsev;[69] see also Schmickler and Henderson.[70]
(4) Improved theories of the ionic screening cloud are provided by the work of Blum and others,[71] and by Grimson and Rickayzen,[72] the latter especially in connection with the problem of solvation forces and the behaviour of colloidal dispersions referred to in Section 10.2.

To conclude this Section, brief reference will be made to the work of Painter *et al.*[73] and of Roman *et al.*,[74] specifically on the alkali-halide–metal-electrode interface. There a further discussion of the formula (10.224) can be found. The subsequent studies of Ballone *et al.*[75] make substantial progress by showing how the computer-simulation results of Torrie and Valleau[76] can be understood in terms of refinements of the hypernetted chain theory of liquid structure to include the so-called bridge function. For a discussion of such a treatment in bulk isotropic liquids, the reader should consult Appendix A 10.1.

Specifically, Ballone *et al.* studied interfacial properties of an ionic fluid next to a uniformly charged planar wall in the restricted primitive model by both theoretical and Monte Carlo methods; for earlier work in this same general area the other references in their paper should be consulted. Ballone *et al.*[75] treated a 1:1 fluid of charged hard spheres having equal diameters in a state appropriate to a 1 M aqueous electrolyte solution. The interfacial density profiles of counter-ions and co-ions were evaluated by extending the hypernetted chain approximation (cf. Appendix A 10.1) to include the leading bridge diagrams for the wall–ion correlations. The theoretical results compare well with those of the grand canonical Monte Carlo calculations of Torrie and Valleau[76] (see also the work of van Megen and Snook[77]), over the whole range of surface density considered by these authors. In relation to the above discussion, the differential capacitance of the model interface was evaluated.

## 10.13 Application of density functional theory to liquid crystals

As the final classical application of this Chapter, we shall consider in some detail the results of the density functional theory of liquid crystals. These constitute a state of matter occurring on the phase diagram between the ordered solid phase, where the mobility of individual molecules is restricted severely, and the isotropic phase where the molecules are highly mobile and only short-range order exists.

Briefly, liquid crystals may be of the nematic, cholesteric or smectic type. In the nematic phase, the molecular centres of gravity are disordered as in the liquid, but one has a statistically parallel orientation of the long axes of the molecules along an exis (the director $\hat{\mathbf{n}}$). There appears to be complete rotational symmetry around the preferred axis. The nematic phase usually occurs only with materials that do not distinguish between right and left and whose molecules have rod-like shapes. The cholesteric phase is a helically ordered nematic phase.

Smectic liquid crystals have stratified structures, with the long axes of the rod-like or lath-like molecules parallel to each other in the layers. Since a variety of molecular arrangements are possible within each stratification, a number of smectic phases are possible (eight are known). In smectic A, the molecules are upright in each layer with their centres irregularly specified in a liquid-like fashion. Smectic B differs from A in that the molecular centres in each layer are hexagonal close-packed. Smectic C is a tilted form of smectic A. If in addition to the tilt there is an ordered arrangement within each layer, it is labelled $B_C$. Other smectic modifications have more complex structure, as discussed in the work of Luckhurst and Gray.[78]

In introducing the density functional treatment of liquid crystals, it is worth referring to the fact that the theory of liquid crystals was originally developed in two different directions. One of these directions was the phenomenological theory of Landau and de Gennes,[79] which is referred to again later in this Section. The Helmholtz free energy $F$ is, in this treatment, expanded in powers of the order parameter and its gradient. Unfortunately, in the process, one has five or more adjustable parameters to be determined from experiment. While this approach is physically appealing and convenient for mathematical development, one drawback is its lack of quantitative predictive power about the phase diagram, as is clear from the discussion of Senbetu and Woo.[80] In the molecular field theory, given by Maier and Saupe[81] and subsequently by McMillan,[82] one begins with a model, whether in the form of rods or characterized by interparticle potentials, and proceeds to calculate a solvent-mediated anisotropic external potential (effective one-body potential or pseudo-potential) acting on each individual molecule. Such calculations require full knowledge of pair-correlation functions, as discussed, for instance, by Singh and Singh.[83]

With this brief introduction, we shall present the results of the density functional theory (see Appendix A 10.2), following the work of Singh[84] rather closely. In the text, attention will be focused on one aspect primarily to illustrate the use of the theory, namely the Frank elastic constants of nematic liquid crystals.[84]

### 10.13.1 Frank elastic constants of nematic liquid crystals

The Frank elastic constants measure the free energy associated with longwavelength distortions of the nematic state in which the local preferred direction of molecular orientation varies in space. If the local preferred direction at the point $R$ is parallel to the director $\hat{\mathbf{n}}(R)$, the free energy associated with the distortion may be written, from the discussions of de Gennes or of Stephen and Straley:[85]

$$\Delta F = \frac{1}{2}\int \mathrm{d}\mathbf{R}(K_1(\nabla\cdot\hat{\mathbf{n}})^2 + K_2(\hat{\mathbf{n}}\cdot\nabla\times\hat{\mathbf{n}})^2 + K_3(\hat{\mathbf{n}}\times\nabla\times\hat{\mathbf{n}})^2). \quad (10.225)$$

The distortions corresponding to $K_1, K_2$ and $K_3$ are referred to as splay, twist and bend, respectively. The Frank elastic constants $K_i$ in eqn (10.225) characterize the free-energy increase associated with the three normal modes of deformation of the ordered nematic state.

To derive molecular expressions for the $K_i$, one first chooses an arbitrary point $R=0$ in the deformed liquid crystal as the origin of a space-fixed coordinate system. The $z$-axis of this system is taken parallel to the director at the origin, i.e. $\hat{z}=\hat{\mathbf{n}}(R=0)$. For pure splay, twist and bend deformations, the variations in $\hat{\mathbf{n}}(R)$ are always confined to a plane. If the $x$-axis is chosen such that $(x, z)$ is the plane containing $\hat{\mathbf{n}}(R)$ then, following Gelbart and Ben-Shaul,[86]

$$\hat{\mathbf{n}}(R) = \hat{\mathbf{x}} \sin\chi(R) + \hat{\mathbf{z}}\cos\chi(R). \quad (10.226)$$

In this equation, $\chi(R)$ is the angle between the director at $R$ and the director at the origin, i.e. $\cos\chi(R) = \hat{\mathbf{z}}\cdot\hat{\mathbf{n}}(R)$.

With any distortion, one may associate a wavenumber $q$. Since the increase in free energy arising from a long-wavelength distortion is even in $q$, because the symmetry of the system ensures that the distortions corresponding to $q$ and $-q$ are equivalent, the simplest assumption is that $\Delta F \propto q^2$. The long-wavelength distortion corresponds to the change in the director over some characteristic length of the system being small, i.e. $\chi \sim qd \ll 1$, $d$ being taken, say, as the range of the direct correlation function. In this limit one finds, following Singh,[84]

$$\chi(R) = \begin{cases} qx - q^2xz + \mathrm{O}(q^3) & \text{splay,} \\ qy & \text{twist,} \\ qz + q^2xz + \mathrm{O}(q^3) & \text{bend;} \end{cases} \quad (10.227)$$

and for distortion-free energy density around the origin

$$\Delta a(0) = \begin{cases} \tfrac{1}{2}K_1q^2 + \mathrm{O}(q^4) & \text{splay,} \\ \tfrac{1}{2}K_2q^2 + \mathrm{O}(q^4) & \text{twist,} \\ \tfrac{1}{2}K_3q^2 + \mathrm{O}(q^4) & \text{bend.} \end{cases} \quad (10.228)$$

Since in pure splay, bend and twist, the deformed nematic has the same local structure everywhere, the results given above do not depend on the choice of origin.

In a deformed nematic phase, the orientational distribution function depends on position in space. Let it be written as $f(\Omega, R)$ for the distribution at point $R$. Taking the undeformed nematic phase as reference system, one rewrites eqn (10.225) in a form useful for the present application as

$$\beta F=\int \mathrm{d}\mathbf{R}\ \beta a(R)$$
$$=\int \mathrm{d}\mathbf{R}[\beta a_{\mathrm{u}}(R)+\beta\Delta a(R)], \tag{10.229}$$

where[84] (see also Appendix A 10.2)

$$\beta a_{\mathrm{u}}(R)=\rho_0\int \mathrm{d}\Omega_1\ f(\Omega_1,R)[\ln f(\Omega_1,R)+\ln\rho_0+\ln\Lambda-1]$$
$$-\rho_0^2\int \mathrm{d}\mathbf{r}\,\mathrm{d}\Omega_1\ \mathrm{d}\Omega_2 f(\Omega_1,\bar{R})f(\Omega_2,\bar{R})\bar{c}_2(\rho(\bar{r}_1\Omega)), \tag{10.230*}$$

$$\beta\Delta a(R)=-\rho_0^2\int \mathrm{d}\mathbf{r}\,\mathrm{d}\Omega_1\,\mathrm{d}\Omega_2 f(\Omega_1,\bar{R})\,[f(\Omega_2,\bar{R}+\bar{r})-f(\Omega_2,\bar{R})]\bar{c}_2(\rho_0)$$
$$-\rho_0^3\int \mathrm{d}\bar{r}\,\mathrm{d}\Omega_1\,\mathrm{d}\Omega_2 f(\Omega_1,\bar{R})[f(\Omega_2,\bar{R}+\bar{r})-f(\Omega_2,R)]\frac{\delta\bar{c}_2(\rho_0)}{\delta\rho_0}$$
$$-\frac{1}{2}\rho_0^3\int \mathrm{d}\bar{r}\,\mathrm{d}\Omega_1\,\mathrm{d}\Omega_2\,[f(\Omega_2;\bar{R}+\bar{r})-f(\Omega_2,\bar{R})]^2$$
$$\times\left[\frac{\delta\bar{c}_2(\rho_0)}{\delta\rho_0}+\frac{1}{2}\rho_0\frac{\delta^2\bar{c}_2(\rho_0)}{\delta\rho_0^2}\right]. \tag{10.231}$$

Here $\beta a_{\mathrm{u}}(R)$ is the free-energy density at position $R$ of an undeformed nematic liquid crystal and $\beta\Delta a(R)$ is the free-energy density of deformation. In writing eqn (10.231) it has been assumed that the direct correlation function in the undeformed liquid crystal is very similar to the isotropic liquid at the same mean density.

Since $\chi(R)$ is the angle between the director at $\mathbf{R}$ and the director at the origin, one can write $f(\Omega, R)=f(\Omega, \chi(R))$. For an undeformed system,

*For the definition of $\bar{c}_2$ see Appendix A 10.2, eqn (A 10.2.14) as well as ref. 84.

$\chi(r)=0$. Thus the distortion free energy at $R=0$ is[84]

$$\beta \Delta a(0) = -\rho_0^2 \int d\bar{r} \int d\Omega_1 \, d\Omega_2 \, f(\Omega_1, 0) [f(\Omega_2, \chi(\bar{r})) - f(\Omega_2, 0)]$$

$$\times \left( \bar{c}_2(\rho_0) + \rho_0 \frac{\partial \bar{c}_2(\rho_0)}{\partial \rho_0} \right)$$

$$- \frac{1}{2} \rho_0^3 \int d\bar{r} \, d\Omega_1 \, d\Omega_2 \, [f(\Omega_2, \chi(\bar{r})) - f(\Omega_2, 0)]^2$$

$$\times \left( \frac{\partial \bar{c}_2(\rho_0)}{\partial \rho_0} + \frac{1}{2} \rho_0 \frac{\partial^2 \bar{c}_2(\rho_0)}{\partial \rho_0^2} \right). \tag{10.232}$$

If the distortion angle is small, i.e. $\chi(r) \ll 1$, one can expand $f(\Omega, \chi(r))$ as a power series in $\chi$. Keeping terms to $O(q^2)$, one finds

$$f(\Omega_2, \chi(r)) - f(\Omega_2, 0) = f'(\Omega_2 0) \begin{pmatrix} qz - q^2 xz \\ qy \\ qz + q^2 xz \end{pmatrix} + \frac{1}{2} f''(\Omega_2, 0) \begin{pmatrix} q^2 x^2 \\ q^2 y^2 \\ q^2 z^2 \end{pmatrix}, \tag{10.233}$$

where $x$, $y$ and $z$ are the Cartesian coordinates of $\bar{r}$ in the space-fixed system.

From eqns (10.232) and (10.233) one finds

$$\beta \Delta a(0) = -\rho_0^2 \int d\bar{r} \int d\Omega_1 \, d\Omega_2 \, f(\Omega_1, 0)$$

$$\left\{ f'(\Omega_2, 0) \begin{pmatrix} qx + q^2 xz \\ qy \\ qz + q^2 xz \end{pmatrix} + \frac{1}{2} f''(\Omega_2, 0) \begin{pmatrix} q^2 x^2 \\ q^2 y^2 \\ q^2 z^2 \end{pmatrix} \right\}$$

$$\times \left( \bar{c}_2(\rho_0) + \rho_0 \frac{\partial \bar{c}_2(\rho_0)}{\partial \rho_0} \right) - \frac{1}{2} \rho_0^3 \int d\bar{r} \, d\Omega_1 \, d\Omega_2 \, [f'(\Omega_2, 0)]^2 \begin{pmatrix} q^2 x^2 \\ q^2 y^2 \\ q^2 z^2 \end{pmatrix}$$

$$\times \left( \frac{\partial \bar{c}_2(\rho_0)}{\partial \rho_0} + \frac{1}{2} \rho_0 \frac{\partial^2 \bar{c}_2(\rho_0)}{\partial \rho_0^2} \right). \tag{10.234}$$

Thus one has the final results for the Frank elastic constants[84]

$$\begin{pmatrix} K_1 \\ K_2 \\ K_3 \end{pmatrix} = -2\rho_0^2 k_B T \int d\bar{r}\, d\Omega_1\, d\Omega_2\, f(\Omega_1, 0)$$

$$\times \left[ f'(\Omega_2, 0) \begin{pmatrix} -xz \\ 0 \\ xz \end{pmatrix} + \frac{1}{2} f''(\Omega_2, 0) \begin{pmatrix} x^2 \\ y^2 \\ z^2 \end{pmatrix} \right] \left( \bar{c}_2(\rho_0) + \rho_0 \frac{\partial \bar{c}_2(\rho_0)}{\partial \rho_0} \right)$$

$$-\rho_0^3 k_B T \int d\bar{r}\, d\Omega_1\, d\Omega_2\, f'(\Omega_2, 0)^2 \begin{pmatrix} x^2 \\ y^2 \\ z^2 \end{pmatrix} \left( \frac{\partial \bar{c}_2(\rho_0)}{\partial \rho_0} + \frac{1}{2} \rho_0 \frac{\partial^2 \bar{c}_2(\rho_0)}{\partial \rho_0^2} \right). \tag{10.235}$$

In deriving the above result for the Frank elastic constants, it has been assumed, following Singh,[84] that the direct-correlation function of an ordered nematic phase can be approximated by the direct-correlation function of the isotropic liquid at the same mean number density $\rho_0$.

Using the method discussed by Singh for the calculation of the direct-correlation function, one can evaluate $K_i$ from eqn (10.235). The expression (10.235) is a generalization of that reported by Gelbart and Ben-Shaul[86] by Priest[87] and by Straley.[88]

### 10.13.2 Molecular theory of freezing: an illustrative example

The density functional theory outlined above and in Appendix A 10.2 has been utilized by Singh and Singh[83] to discuss the molecular theory of freezing of a system of hard ellipsoids of revolution, conveniently parametrized by the length-to-width ratio $X_0 = a/b$, where $2a$ and $2b$ denote the lengths of the major and minor axes of the ellipsoids. This model includes as limiting cases hard spheres, hard platelets and hard needles. All these systems are of some physical interest as primitive models of liquids, solids, plastics and liquid crystals. It is also highly relevant that constant-pressure Monte Carlo simulation experiments have been employed by Frenkel *et al.*[89] to study some of these systems.

Using a suitable representation for the potential energy of interaction of a pair of hard ellipsoids of revolution, with the distance of closest approach of two molecules with given relative orientation taken from the Gaussian overlap model (see ref. 107), Singh and Singh find that the plastic phase is stable first for $0.57 \lesssim X_0 \lesssim 1.75$ and the nematic phase for $X_0 < 0.57$ and $> 1.75$. These findings are in reasonable accord with computer-simulation results. Also, and again in agreement with computer simulation, these workers find symmetry between systems with inverse length-to-width ratios.

### 10.13.3 Relation to Landau–de Gennes theory

The above treatment, based on density functional ideas, can in fact be brought into close contact with phenomenological theory. In the Landau–de Gennes approach, reviewed fully in the book by de Gennes[79] and in the article by Stephen and Straley,[85] the Gibbs free energy is written as a power-law expansion around the isotropic equilibrium position. To make contact with density functional theory, it is more convenient to use the Helmholtz free energy per particle. Then one has (isotropic-nematic case)

$$\frac{\beta \Delta E}{N} = -a\overline{P}_2^2 + b\overline{P}_2^3 + c\overline{P}_2 + d\overline{P}_2\overline{P}_4 + f\overline{P}_2^2\overline{P}_4. \tag{10.235}$$

If the density functional theory is now used[84] and only terms involving $\overline{P}_2$ and $\overline{P}_4$ are retained, then the result may be written (see Appendix A 10.2)

$$a = \frac{5}{2} - 25\rho_0 \int \mathrm{d}\bar{r}_{12}\,\mathrm{d}\Omega_1\,\mathrm{d}\Omega_2\,P_2(\cos\theta_1)P_2(\cos\theta_2)\bar{c}_2(\bar{r}_{12},\Omega_1,\Omega_2;\rho_0), \tag{10.236}$$

with similar expressions given by Singh[84] for the other phenomenological constants.

This author draws attention to the fact that the power-law expansion for the full one-particle orientational distribution entropic contribution converges only slowly. This may well limit the range of applicability of this type of treatment. However, in tackling pretransitional effects, where terms of higher order than quadratic are usually small, eqn (10.235) is very useful, as is clear from the account of de Gennes.[79]

## 10.14 Inhomogeneous quantum liquids

This discussion of the single-particle density in liquids will conclude with a brief outline of progress that has proved possible in treating inhomogeneous quantum liquids. It is fair to say that, based on the success of variational methods, one can regard the ground state of quantum liquids such as the

He isotopes and the electron liquid as fairly well understood. Doubtless it is now the case that the solution of the many-body Schrödinger equation for these systems by means of the Green-function Monte Carlo (GFMC) method provides the most reliable answer to date for the ground state of these systems. Nevertheless, this in no way denies the interest of many-body methods based on variational wave functions, which, for example for Bose systems, are the "product of pairs" Jastrow-type wave functions, cluster expansions and resummation techniques, etc. Encouraged by the success of such methods in the theory of bulk quantum liquids, various workers have sought extensions to inhomogeneous quantum systems, notable work being that of Krotscheck *et al.*[90] and of Pieper *et al.*[91]

In both these references, this theory of inhomogeneous quantum liquids has been applied to the static properties of Bose liquids. This work will be briefly discussed immediately below, followed by an application of this theory to Fermi liquids; the example chosen being the electron density profile for the jellium model of a metal surface (cf. Section 10.2.1 above).

### 10.14.1 Pair function and energy-density functional for liquid $^4$He drops

In the work of Pieper *et al.*,[91] the pair-distribution function $g(\mathbf{r}_1, \mathbf{r}_2)$ in drops of liquid $^4$He has been studied by the variational Monte Carlo method. These authors demonstrate that in drops having 70 or more atoms, $g(\mathbf{r}_1, \mathbf{r}_2)$ can be approximated as a functional of the single-particle density $\rho(\mathbf{r})$ and the separation $r_{12} = |\mathbf{r}_1 - \mathbf{r}_2|$. Pieper *et al.* use a parametrization of the single-particle density which is, in fact, a generalization of the form (10.7) to read

$$\rho(r) = c\left[1 + \exp\left(\frac{r-R}{a}\right)\right]^{-p}, \qquad (10.237)$$

where $R$, $a$ and $p$ are determined by minimizing the energy. Their main result is to express $g(\mathbf{r}_1, \mathbf{r}_2)$ in terms of the bulk liquid pair function $g_b$, but known as a function of density and interparticle separation $r_{12}$, say $g_b(\rho, r_{12})$ as

$$g(\mathbf{r}_1, \mathbf{r}_2) \doteqdot g_b(\rho_{\text{local}}, r_{12}). \qquad (10.238)$$

Here it is permissible to insert for the local density either the average $\frac{1}{2}\{\rho(r_1) + \rho(r_2)\}$, the geometric mean $\{\rho(r_1)\rho(r_2)\}^{1/2}$ or $\rho(R_{\text{cm}})$, where $R_{\text{cm}}$ is the centre of mass $\frac{1}{2}(\mathbf{r}_1 + \mathbf{r}_2)$. Energy-density functionals based on this approach to the pair function give a fairly accurate description of the ground-state properties of the drops.

Pieper *et al.*[91] also calculated the surface tension and thickness of a plane infinite surface of liquid $^4$He. They obtained quite good agreement with the experimentally determined surface tension, the theoretical estimates being somewhat smaller (within 5% or so) of the measured value. They obtained 5.7–5.8 Å for the surface thickness but there is presently no way of checking this estimate. Earlier calculations of the surface properties have been reviewed by Edwards and Saam;[92] the related study of Krotscheck *et al.*[93] is in some respects more detailed than that of Pieper *et al.*, but because of the approximations made in the calculations the surface tension they calculate is not in quantitative agreement with experiment.

### 10.14.2 Fermi liquids: example of density profile for the jellium model of a metal surface

Krotscheck *et al.*[90] have performed variational calculations for the jellium model of metal surfaces. The ground-state wave function is represented by a product of local one- and two-body functions and a model Slater function. The correlation functions and the single-particle orbitals entering the Slater determinant are calculated by an unconstrained optimization procedure. Results for the surface energy and the work function are somewhat higher than values obtained by previous workers.

While density functional theory, as stressed above, can be viewed as formally exact, in order to use it approximations to $E_{xc}[\rho(\mathbf{r})]$ must be made. Most of the useful approximations, as reviewed in this Volume, are really appropriate for the case of a slowly spatially varying single-particle density $\rho(\mathbf{r})$.

In contrast, Krotscheck *et al.*[90] approach the problem of inhomogeneous systems using the (Fermi) hypernetted chain (FHNC) equations (for the classical case, Appendix A 10.1 may again be consulted for a brief survey of this approach to liquid structure). Whereas density functional theory is, by construction, exact for homogeneous systems, FHNC is not because it approximates correlation effects by a two-particle Jastrow correlation function $u_2(\mathbf{r}_i - \mathbf{r}_j)$ and calculates the energy expectation value in an approximate manner. However, the errors are known to be pretty small for the homogeneous electron gas by comparison with the computer-simulation results of Ceperley and Alder.[94] In contrast with the usual approximate forms of density functional theory, the FHNC treatment does not assume slow spatial changes in the single-particle density $\rho(\mathbf{r})$. One may therefore anticipate that, when applied to strongly non-uniform systems, it will yield results comparable to the high accuracy it yields for homogeneous assemblies.

The achievement of Krotscheck *et al.*[90] is to show that the solution of the FHNC equations, including the optimization of the correlation function $u_2(\mathbf{r}_1, \mathbf{r}_2)$ for inhomogeneous systems, is in fact feasible, although the computational effort is very much larger than in a density functional calculation.

It should be noted that the variational application to metal surfaces has been pioneered by Woo and collaborators.[95] They start from precisely the same form of the many-body wave function. Krotscheck *et al.* differ from the approach of Woo and colleagues in that they use, as stressed above, the FHNC method for the summation of infinite classes of diagrams and this involves the optimization of the single-particle basis and the two-body correlations through Euler–Lagrange equations. One of the most attractive aspects of the HNC theory is that the above optimization scarcely complicates the calculation.

Without giving details of the elaborate calculations, we summarize by

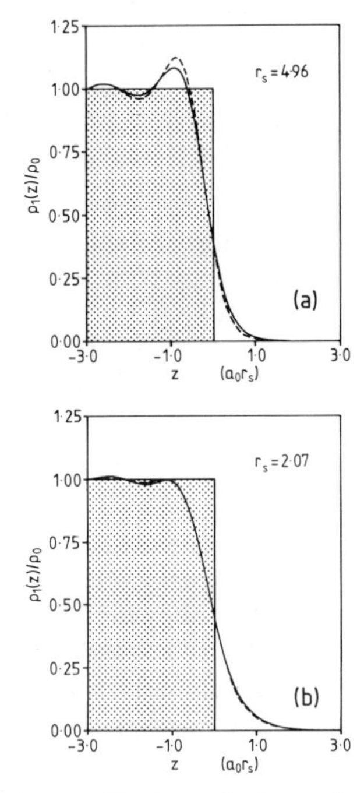


**Fig. 10.6** Electron density profile through the surface of a metal (semi-infinite jellium model: shaded area is jellium background). (a) Mean interelectronic spacing $r_s = 4.96$ atomic units. Solid line: many-body theory of Krotscheck *et al.*[93] Dashed line: Lang–Kohn result[96] for $r_s = 5$ atomic units. (b) $r_s = 2.07$ atomic units. Solid line: Result of Krotscheck *et al.*[93] Dashed line: Lang–Kohn result[96] for $r_s = 2$ atomic units.

showing in Fig. 10.6 the single-particle density $\rho(z)$ through a jellium planar surface for two different values of the mean interelectronic spacing $r_s$. There is rather good agreement with the earlier density functional results of Lang and Kohn,[96] though, as already noted, the values of the surface energy and the work function are higher than those obtained previously.

## Appendix A 10.1

### Hypernetted chain approximation and its generalizations

The purpose of this Appendix is to set out briefly the approximate theory of liquid structure known as the hypernetted chain (HNC) method. As the name implies, the procedure has its origin in diagrammatic theory. However, below, it will be considered, first of all, from the standpoint of uniform liquids, as an approximate procedure resulting from the first non-trivial member of the statistical mechanical hierarchy of equations describing classical liquids. Within a pair-potential framework, this lowest member relates the pair potential $\phi(r)$ say, pair correlation function $g(r)$ and three-particle correlation function $g_3(\mathbf{r}_1, \mathbf{r}_2, \mathbf{r}_3)$. The resulting equation is conveniently written in terms of the potential of mean force, denoted $U_m$ below, which is defined by

$$g(r_{12}) = \exp(-U_m/k_B T). \quad \text{(A 10.1.1)}$$

Then one has

$$-\frac{\partial U_m(r_{12})}{\partial \mathbf{r}_1} = -\frac{\partial \phi(r_{12})}{\partial \mathbf{r}_1} - \int \frac{g_3(\mathbf{r}_1, \mathbf{r}_2, \mathbf{r}_3)}{g(r_{12})} \frac{\partial \phi(r_{13})}{\partial \mathbf{r}_1} \mathrm{d}\mathbf{r}_3. \quad \text{(A 10.1.2)}$$

This exact equation (A 10.1.2) has a straightforward interpretation.[97] The total force on atom 1, written explicitly on the left-hand side is made up of two contributions: (i) from the pair force resulting from atom 2 at distance $r_{12}$, and (ii) from the fact that the probability of finding a third atom at $\mathbf{r}_3$ if there are atoms at $\mathbf{r}_1$ and $\mathbf{r}_2$ is the ratio $g_3/g$ shown in eqn (A 10.1.2).

Of course, it is necessary to "decouple" $g_3$ in terms of $g$ to obtain an explicit relation between $g(r)$ and $\phi(r)$ representing an approximate theory of bulk liquid structure. The most elementary assumption, going back to Kirkwood, is

$$g_3(\mathbf{r}_1, \mathbf{r}_2, \mathbf{r}_3) \doteqdot g(r_{12}) g(r_{23}) g(r_{31}) \quad \text{(A 10.1.3)}$$

and inserting this into eqn (A 10.1.2) one can integrate to find

$$\frac{U_m(r)}{k_B T} = \frac{\phi(r)}{k_B T} - \rho \int E(\mathbf{r} - \mathbf{r}') h(\mathbf{r}') \mathrm{d}\mathbf{r}'. \quad \text{(A 10.1.4)}$$

In this equation, the form of $E(r)$ is explicitly

$$E(r)=\frac{1}{k_{\mathrm{B}}T}\int_{r}^{\infty} g(r)\frac{\partial\phi}{\partial r}\,\mathrm{d}r,$$

and, with this form, eqn (A 10.1.4) constitutes the so-called Born–Green approximation to liquid bulk structure.

The HNC method has the same structure as eqn (A 10.1.4), but with $E$ replaced by the direct correlation function $c(r)$. Then using the Ornstein–Zernike convolution relation

$$h(r)=c(r)+\rho\int c(\mathbf{r}-\mathbf{r}')h(r')\,\mathrm{d}\mathbf{r}', \qquad \text{(A 10.1.5)}$$

one is led to the HNC equation:

$$h(r)-c(r)=\frac{\phi(r)-U_{\mathrm{m}}(r)}{k_{\mathrm{B}}T}. \qquad \text{(A 10.1.6)}$$

The replacement of $E$ by $c$ is appropriate, at least asymptotically, to recover $c(r)=-\phi(r)/\ k_{\mathrm{B}}T$ at sufficiently large $r$, provided one is not near to the critical point, when other considerations than the range of the force law dominate.

*Bridge term*

Still treating a uniform bulk liquid, one corrects the HNC equation (A 10.1.6) by adding a so-called bridge term, discussed, for example, in some detail by Ashcroft and Rosenfeld.[98] This term, written $B(r)$ below, can again be approached diagrammatically; the name again stemming from such a method. What is important in what follows in this Appendix is that a suitable asymptotic form of the bridge term can be obtained in terms of the total correlation function $h(r)=g(r)-1$, namely

$$B(r)\sim\text{const}\iint h(r)h(r')h(|\mathbf{r}-\mathbf{r}'|)h(|\mathbf{r}-\mathbf{r}''|)h(|\mathbf{r}'-\mathbf{r}''|)\,\mathrm{d}\mathbf{r}'\,\mathrm{d}\mathbf{r}''. \qquad \text{(A 10.1.7)}$$

This term has been considered by various workers; as expected in the bulk liquid, it is a functional of the pair function. For a liquid metal, it has been suggested[99] that the bridge term falls off asymptotically at large distances as $r^{-4}$, but more work is required on this point. In contrast though, Ashcroft and Rosenfeld[98] have argued that the bridge term is insensitive to the details of the force law, and can be evaluated for hard spheres therefore. Presumably, the usefulness of such an approach depends on what one is aiming to calculate.

*HNC approach to non-uniform systems*

To complete this account of the HNC approximation and its generalizations, a brief survey will be given of a treatment of non-uniform systems. The summary below follows closely the work of Ballone *et al.*[75] whose work on the hard-wall electrolyte problem was referred to in Section 10.12.

As discussed in the main text, the free energy of an inhomogeneous multi-component system in a set of external potentials $\phi_\alpha(\mathbf{r})$ acting on $\alpha$-type particles is a functional of the particle density profiles $\rho_\alpha(\mathbf{r})$. At equilibrium, with a homogeneous bulk state having particle densities $\rho_\alpha$, the equilibrium conditions for the density profiles can be written by means of a formally exact functional expansion[24,100,101] in the form

$$\ln\left\{\frac{\rho_\alpha(\mathbf{r})}{\rho_\alpha}\exp\left(\frac{\phi_\alpha(\mathbf{r})}{k_B T}\right)\right\}=\sum_\beta \rho_\beta\int d\mathbf{r}'\, c^{(2)}_{\alpha\beta}(|\mathbf{r}-\mathbf{r}'|)h_\beta(\mathbf{r}')$$
$$+\frac{1}{2}\sum_{\beta,\gamma}\rho_\beta\rho_\gamma\iint d\mathbf{r}'\,d\mathbf{r}''\, c^{(3)}_{\alpha\beta\gamma}(\mathbf{r},\mathbf{r}',\mathbf{r}'')h_\beta(\mathbf{r}')h_\gamma(\mathbf{r}'')+\cdots. \qquad (\mathrm{A}\,10.1.8)$$

In eqn (A 10.1.8), the definition of $h_\alpha(\mathbf{r})$ is

$$h_\alpha(\mathbf{r})=(\rho_\alpha(\mathbf{r})-\rho_\alpha)/\rho_\alpha, \qquad (\mathrm{A}\,10.1.9)$$

and the $c$s as usual are the bulk two- and three-body direct-correlation functions. Truncation of the right-hand side of eqn (A 10.1.8) at the two-body terms yields the HNC approximation, while the three-body and higher-order terms are the counterparts of the bridge diagram contribution in the theory of correlations in homogeneous fluids discussed at the beginning of this Appendix.

Experience has built up on the estimation of bridge functions in calculations of bulk liquid structure, as mentioned earlier. In particular, Bacquet and Rossky[102] working on electrolyte solutions and Iyetomi and Ichimaru[103,104] studying the classical one-component plasma have demonstrated that a useful approximation consists in including the three-body direct-correlation function in terms of the bulk total correlation functions $h_{\alpha\beta}(\mathbf{r})$:

$$c^{(3)}_{\alpha\beta\gamma}(\mathbf{r},\mathbf{r}',\mathbf{r}'')=h_{\alpha\beta}(|\mathbf{r}-\mathbf{r}'|)h_{\beta\gamma}(|\mathbf{r}'-\mathbf{r}''|)h_{\gamma\alpha}(|\mathbf{r}''-\mathbf{r}|). \qquad (\mathrm{A}\,10.1.10)$$

In their discussion of electrolyte solutions, Bacquet and Rossky note that their inclusion of leading bridge diagrams by the above approximation successfully corrects for an underestimate of the correlation hole between like ions in the HNC method and for its consequences on the distribution of unlike ions around a given ion.

In their application to the local structure of an electrolyte next to a charged hard wall, referred to in the text, Ballone *et al.* include the bridge corrections to the HNC treatment for the wall–ion correlations, that is for the single-particle density profiles, by the approximation displayed in eqn (A 10.1.10).

## Appendix A10.2

### Density functional theory of liquid crystals

The purpose of this Appendix is to develop the density functional theory of liquid crystals. This will allow the construction of formally exact expansions for thermodynamic functions in terms of the direct-correlation functions and single-particle densities. The direct-correlation function which appears in these expansions is a functional of the single-particle density distribution and the pair interaction. Functional Taylor expansion will then be employed to derive expressions for the single-particle density distribution and the free energy for a non-uniform classical system subjected to an external potential. Finally the density distribution is expressed in terms of suitable order parameters. This can be done in a sufficiently general manner to characterize crystalline solids, different phases of liquid crystals and isotropic liquids.

*Formal statistical mechanical theory*

Following the presentation of Singh[84] closely, let us consider a system of non-spherical molecules of arbitrary symmetry in volume $V$ and at temperature $T$. With $N$ particles in the system, the configurational energy $U$ is approximated by

$$U(\mathbf{x}_1,\ldots,\mathbf{x}_N)=\sum_{i=1}^{N} U^{\mathrm{e}}(\mathbf{x}_i)+\sum_{i>j=1}^{N} U(\mathbf{x}_i,\mathbf{x}_j), \qquad \text{(A 10.2.1)}$$

where the vector $\mathbf{x}_i$ indicates both the location $\mathbf{r}_i$ of the centre of the $i$th molecule and its relative orientation $\Omega_i$ described by the Euler angles $\theta_i$, $\phi_i$ and $\chi_i$. The volume element $\mathrm{d}\mathbf{x}_i$ is equivalent to $\mathrm{d}\mathbf{r}_i\,\mathrm{d}\Omega_i$ where $\mathrm{d}\mathbf{r}=\mathrm{d}x\,\mathrm{d}y\,\mathrm{d}z$ and

$$\mathrm{d}\Omega=(1/8\pi^2)\sin\theta\,\mathrm{d}\theta\,\mathrm{d}\phi\,\mathrm{d}\chi\,.$$

In eqn (A 10.2.1), $U^{\mathrm{e}}(\mathbf{x}_i)$ is the potential energy of a molecule at position $r_i$

with relative orientation $\Omega_i$ arising from external forces and $U(\mathbf{x}_i, \mathbf{x}_j)$ is the intermolecular pair potential for molecules $i$ and $j$.

The grand partition function of the system, written as $\exp W$, is now set up, and with $\mu$ the chemical potential one then obtains

$$\frac{\delta W}{\delta \psi(\mathbf{x}_i)} = \langle \rho(\mathbf{x}_i) \rangle \equiv \rho(\mathbf{x}_i), \tag{A 10.2.2}$$

where

$$\psi(\mathbf{x}_i) \equiv \beta\mu - \beta U^{e}(\mathbf{x}_i); \quad \beta = (k_B T)^{-1}. \tag{A 10.2.3}$$

Also

$$\frac{\delta^2 W}{\delta \psi(\mathbf{x}_i)\delta \psi(\mathbf{x}_j)} = \langle \delta\rho(\mathbf{x}_i)\delta\rho(\mathbf{x}_j) \rangle$$

$$= \rho(\mathbf{x}_i)\delta(\mathbf{x}_i - \mathbf{x}_j) + \rho(\mathbf{x}_i)\rho(\mathbf{x}_j)h(\mathbf{x}_i, \mathbf{x}_j), \tag{A 10.2.4}$$

where $\langle \cdots \rangle$ denotes the ensemble average, $\delta\rho(\mathbf{x}_i)$ represents the fluctuation in the single-particle density distribution at $\mathbf{x}_i$ and $h(\mathbf{x}_i, \mathbf{x}_j)$ is the total pair correlation function.

One next defines a reduced Helmholtz free energy by

$$\beta F = \int \rho(\mathbf{x})\psi(\mathbf{x})\, d\mathbf{x} - W. \tag{A 10.2.5}$$

The functional derivative of $\beta F$ with respect to $\rho(\mathbf{x})$ is

$$\frac{\delta(\beta F)}{\delta \rho(\mathbf{x})} = \psi(\mathbf{x}), \tag{A 10.2.6}$$

and thus $\beta F$ is the natural functional of $\rho(\mathbf{x})$.

In general one writes for the single-particle density

$$\rho(\mathbf{x}) = \exp(\psi(\mathbf{x}) + c_1(\mathbf{x}))/\Lambda, \tag{A 10.2.7}$$

where

$$\Lambda = \int \exp(-\beta E_k)\, dp_1 \ldots .\, dp_s / h^s, \tag{A 10.2.8}$$

where $s$ is the number of degrees of freedom of a molecule and $E_k$ is its kinetic energy. In eqn (A 10.2.7), $-k_B T c_1(\mathbf{x})$ can be thought of as a solvent-mediated effective potential field acting at $\mathbf{x}$. Let $\Delta f = \beta(F - F^0)$, where $\beta F^0$ is the reduced Helmholtz free energy for the system without intermolecular interactions:

$$\beta F^0 = \int d\mathbf{x}\, \rho(\mathbf{x})\, [\ln \rho(\mathbf{x})\Lambda + \beta U^{e}(\mathbf{x}) - 1]. \tag{A 10.2.9}$$

$\Delta f$ is evidently then the excess reduced Helmholtz free energy arising from the intermolecular interactions and is in general a functional of the single-particle density $\rho(\mathbf{x})$ and the pair potential $U$.

In terms of Mayer graphs, $\Delta f$ is the sum of all distinct connected irreducible graphs with no labelled points and with a factor of the single-particle density $\rho(\mathbf{x}_i)$ for every field point $i$. The Mayer function $f(\mathbf{x}_i, \mathbf{x}_j)$ which in fact enters the grand partition function referred to above, connects the points:

$$f(x_i, x_j) = \exp(-\beta U(x_i, x_j)) - 1. \qquad \text{(A 10.2.10)}$$

The function $\Delta f$ can be utilized as a generating functional for the correlation functions: in particular,

$$\frac{\delta \Delta f}{\delta U(\mathbf{x}_1, \mathbf{x}_2)} = -\tfrac{1}{2}\rho_2(\mathbf{x}_1, \mathbf{x}_2;\ \{\rho, U\}) \equiv -\tfrac{1}{2}\rho_2\{\rho\}, \qquad \text{(A 10.2.11)}$$

where the two-particle density distribution function $\rho_2$ gives the probability of finding simultaneously a molecule in a volume element $\mathrm{d}\mathbf{r}_i\,\mathrm{d}\Omega_i$ centred at $(\mathbf{r}_i, \Omega_i)$ and a second molecule in a volume element $\mathrm{d}\mathbf{r}_j\,\mathrm{d}\Omega_j$ centred at $(\mathbf{r}_j, \Omega_j)$.

It is possible to write $\rho_2$ divided by the product of densities as $g(\mathbf{x}_1, \mathbf{x}_2; \{\rho, U\}) = 1 + h(\mathbf{x}_1, \mathbf{x}_2;\ \{\rho, U\})$, $g$ being the pair-correlation function whose functional dependence on $\rho$ and $U$ is explicitly indicated. Similarly the function $c_2$ can be generated from $h$ by means of a straightforward generalization of the usual Ornstein–Zernike relation.

The equilibrium single-particle density is determined by minimizing the free energy $\beta F$ for a given external field subject to the constraint

$$\int \rho(\mathbf{x})\,\mathrm{d}\mathbf{x} = \langle N \rangle. \qquad \text{(A 10.2.12)}$$

*Non-uniform system*

In a non-uniform system, $\rho(\mathbf{x})$ is a function of position and orientation and $c_2$ is a functional of $\rho(\mathbf{x})$. Performing a functional Taylor expansion of $c_2$ about the local density distribution $\rho(\mathbf{x}_1)$, one finds

$$c_2\{\rho\} = c_2(\rho(\mathbf{x}_1)) + \int \mathrm{d}\mathbf{x}_3\,[\rho(\mathbf{x}_3) - \rho(\mathbf{x}_1)]c_3(\rho(\mathbf{x}_1))$$

$$+ \frac{1}{2}\int \mathrm{d}\mathbf{x}_3\,\mathrm{d}\mathbf{x}_4\,[\rho(\mathbf{x}_3) - \rho(\mathbf{x}_1)][\rho(\mathbf{x}_4) - \rho(\mathbf{x}_1)]c_4(\rho(\mathbf{x}_1)) + \cdots,$$

$$\text{(A 10.2.13)}$$

where the $c_n(\rho(\mathbf{x}_1)) \equiv c_n(\mathbf{x}_1, \ldots, \mathbf{x}_n;\ \rho(\mathbf{x}_1))$ are local, i.e. they represent the correlation functions of a uniform fluid with constant density $\rho(\mathbf{x}_1)$. Substituting eqn (A 10.2.13) into the equation for $\Delta f$, namely

$$\Delta f(\rho, U)=\int \mathrm{d}\mathbf{x}_1\,\mathrm{d}\mathbf{x}_2\,\rho(\mathbf{x}_1)\rho(\mathbf{x}_2)\bar{c}_2(\rho)\bar{c}_2=\int_0^1 \mathrm{d}y\int_0^y \mathrm{d}x\, c_2(x,\rho), \tag{A 10.2.14}$$

and assuming that the change in density is small over the range of $c_2$, one is led to the result

$$\beta U^{\mathrm{e}}(\mathbf{x}_1)+\beta[\mu(\rho(\mathbf{x}_1))-\mu_0]-\int \mathrm{d}\mathbf{x}_2[\rho(\mathbf{x}_2)-\rho(\mathbf{x}_1)]c_2(\rho(\mathbf{x}_1))$$

$$-\frac{1}{4}\int \mathrm{d}\mathbf{x}_2\,[\rho(\mathbf{x}_2)-\rho(\mathbf{x}_1)]^2\,\frac{\partial c_2(\rho(\mathbf{x}_1))}{\partial\rho(\mathbf{x}_1)}=0. \tag{A 10.2.15}$$

This form is a generalization of that given by Singh and Abraham[84] for a non-uniform atomic fluid to a system of molecules of arbitrary symmetry. As developed to this point, the above treatment can be employed to describe the properties of crystalline solids, liquid crystals, transitions from one phase to the other as well as the properties of interfaces.

Singh then proceeds to the characterization of these phases by introducing order parameters through

$$\begin{aligned}\rho(\mathbf{x})&=\rho(\mathbf{r},\Omega)\\ &=\rho_0\left(1+\sum P_{Lmn}(q)\exp(\mathrm{i}\mathbf{k}_q\cdot\mathbf{r})\,D^L_{mn}(\Omega)\right),\end{aligned} \tag{A 10.2.16}$$

where $\rho_0$ is the average number density, $\mathbf{k}_q$ are reciprocal lattice vectors of the crystalline phase, while $P_{Lmn}(q)$ are the order parameters. Finally, $D^L_{mn}$ represent the Wigner rotation matrices.

As examples one notes that for a monatomic crystalline phase, $L=m=n=0$ while for uniform nematics eqn (A 10.2.16) reduces to

$$\rho(\mathbf{r},\Omega)=\rho_0\left(1+\sum P_{Lmn}(0)\,D^L_{mn}(\Omega)\right). \tag{A 10.2.17}$$

For smectic phases with positional order in one dimension only (e.g. a smectic A or C phase) one can restrict the series (A 10.2.16) to

$$\rho(\mathbf{r},\Omega)=\rho_0\left(1+\sum P_{Lmn}(q)\exp(\mathrm{i}k_q z)\,D^L_{mn}(\Omega)\right) \tag{A 10.2.18}$$

where $k_q=k_{qz}$ and $z$ is parallel to the layer normal.

Apart from the example in the main text, the reader is referred to refs 83 and 84 for further applications of the treatment set out in this Appendix.*

*In a uniaxial nematic phase with cylindrically symmetric molecules, eqn (A 10.2.17) reduces to $\rho(\mathbf{r},\Omega)=\rho_0(1+\Sigma_{L\geqslant 2\text{ even}}(2L+1)\bar{P}_L P_L(\cos\theta))$, where $2\rho_0 P_L=\int\rho(\mathbf{r},\Omega)P_L(\cos\theta)\sin\theta\,\mathrm{d}\theta$ defines the orientated order parameter of the nematic phase used in eqn (10.235).

## References

1. Brown, R.C. and March, N.H. (1973). *J. Phys.*, **C6**, L363.
2. Rice, S.A., Guidotti, D., Lemberg, H.L., Murphy, W.C. and Bloch, A.N. (1974). *Adv. Chem. Phys.*, **27**, 543.
3. Singwi, K.S. and Tosi, M.P. (1980). *Solid St. Commun.*, **34**, 209; (1981). *Phys. Rev.*, **B23**, 1640.
4. March, N.H. and Parrinello, M. (1982). *Collective Effects in Solids and Liquids*, Adam Hilger, Bristol.
5. Rickayzen, G. (1985). In *Amorphous Solids and the Liquid State*, eds March, N.H., Street, R.A. and Tosi, M.P., p. 157, Plenum, New York.
6. Gouy, G. (1910). *J. Phys.*, **9**, 457; (1917). *Ann. Phys.*, **7**, 129; Chapman, D.L. (1913). *Phil. Mag.*, **25**, 475.
7. See, for example, Mermin, N.D. (1965). *Phys. Rev.*, **A137**, 1441.
8. Abraham, F.F. (1979). *Phys. Rep.*, **53**, 93.
9. Evans, R. (1979). *Adv. Phys.*, **28**, 143; see, however, Davis, H.T. and Scriven, L.E. (1982). *Adv. Chem. Phys.*, **49**, 357.
10. Brown, R.C. and March, N.H. (1976). *Phys. Rep.*, **24**, 77.
11. Tosi, M.P. see ref. 5, p. 125.
12. Faber, T.E. (1973). *An Introduction to the Theory of Liquid Metals*, Cambridge University Press.
13. See, for example, Buff, F.P. and Lovett, R.A. (1968). In *Simple Dense Fluids*, Academic Press, New York.
14. Janz, G.J. (1967), *Molten Salts Handbook*, Academic Press, New York.
15. Cahn, J.W. and Hilliard, J.E. (1958). *J. Chem. Phys.*, **28**, 258.
16. Egelstaff, P.A. and Widom, B. (1970). *J. Chem. Phys.*, **53**, 2667.
17. Alonso, J.A. and March, N.H. (1985). *Surf. Sci.*, **160**, 509.
18. Bhatia, A.B. and March, N.H. (1978). *J. Chem. Phys.*, **68**, 1999.
19. Yang, A.J.M., Fleming, P.D. and Gibbs, J.H. (1977). *J. Chem, Phys.*, **67**, 74.
20. Stott, M.J. and Young, W.H. (1981). *Phys. Chem. Liq.*, **11**, 95.
21. Morita, T. and Hiroike, K. (1961). *Prog. Theor. Phys.*, **25**, 537.
22. De Dominicis, C. (1962). *J. Math. Phys.*, **3**, 983.
23. Stillinger, F.H. and Buff, F.P. (1962). *J. Chem. Phys.*, **37**, 1.
24. Lebowitz, J.L. and Percus, J.K. (1963). *J. Math. Phys.*, **4**, 116.
25. Triezenberg, D.G. and Zwanzig, R. (1972). *Phys. Rev. Lett.*, **28**, 1183.
26. Lovett, R.A., Dehaven, P.W., Vieceli, J.J. and Buff, F.P. (1973). *J. Chem. Phys.*, **58**, 1880.
27. Wertheim, M.S. (1976). *J. Chem. Phys.*, **65**, 2377.
28. Yang, A.J.M., Fleming, P.D. and Gibbs, J.H. (1976). *J. Chem. Phys.*, **64**, 3732.
29. Bhatia, A.B., March, N.H. and Tosi, M.P. (1980). *Phys. Chem. Liq.*, **9**, 229.
30. Ebner, C., Saam, W.F. and Stroud, D. (1976). *Phys. Rev.*, **A14**, 2264.
31. Kirkwood, J.G. and Buff, F.P. (1949). *J. Chem. Phys.*, **17**, 338.
32. Fowler, R.H. (1937). *Proc. Roy. Soc.*, **A159**, 229.
33. Freeman, K.S.C. and McDonald, I.R. (1973). *Molec. Phys.*, **26**, 529.
34. Berry, M.V., Durrans, R.F. and Evans, R. (1972). *J. Phys.*, **A5**, 166.
35. Schofield, P. (1979). *Chem. Phys. Lett.*, **62**, 413.
36. Bhatia, A.B. and March, N.H. (1979). *Phys. Chem. Liq.*, **9**, 1.
37. Longuet-Higgins, H.C. (1951). *Proc. Roy. Soc.*, **A205**, 247.
38. Bhatia, A.B., Hargrove, W.H. and March, N.H. (1973). *J. Phys.*, **C6**, 621.
39. Bhatia, A.B. and March, N.H. (1978). *J. Chem. Phys.*, **68**, 4651.

40. Pugachevich, P.P. and Timofeevicheva, O.A. (1951). *Dokl. Akad. Nauk SSSR*, **79**, 831; (1954). *Ibid.*, **94**, 285.
41. See, for example, March, N.H. (1986). *Chemical Bonds Outside Metal Surfaces*, Plenum, New York.
42. Friedel, J. (1954). *Adv. Phys.*, **3**, 446.
43. Wilson, J.R. (1965). *Metall. Rev.*, **10**, 381.
44. March, N.H. (1984). *Phys. Chem. Liq.*, **14**, 79.
45. Gschneidner, K.A. (1964). *Solid State Physics*, Vol. 16, eds Seitz, F. and Turnbull, D. p. 275, Academic Press, New York.
46. Lovett, R. and Buff, F.P. (1980). *J. Chem. Phys.*, **72**, 2425.
47. Kirkwood, J.G. and Monroe, E. (1941). *J. Chem. Phys.*, **9**, 514.
48. Ramakrishnan, T.V. and Yussouff, M. (1977). *Solid St. Commun.*, **21**, 389, (1979). *Phys. Rev.*, **B19**, 2775.
49. March, N.H. and Tosi, M.P. (1981). *Phys. Chem. Liq.*, **11**, 79, 89.
50. Haymet, A.D.J. and Oxtoby, D.W. (1981). *J. Chem. Phys.*, **74**, 2559.
51. Verlet, L. (1968). *Phys. Rev.*, **165**, 201.
52. Ferraz, A. and March, N.H. (1980). *Solid St. Commun.*, **36**, 977.
53. Bhatia, A.B. and March, N.H. (1984). *Phys. Chem. Liq.*, **13**, 313.
54. Yarnell, J.L., Katz, M.J., Wenzel, R.G. and Koenig, S.H. (1973). *Phys. Rev.*, **A7**, 2130.
55. Greenfield, A.J., Wellendorf, J. and Wiser, N. (1971). *Phys. Rev.*, **A4**, 1607.
56. March, N.H. and Tosi, M.P. (1981). *Phys. Chem. Liq.*, **11**, 79.
57. See also March, N.H. (1985). *Proc. Int. School Phys. Enrico Fermi, Course* 89, eds Bassani, F., Fumi, F. and Tosi, M.P., p. 684, North-Holland, Amsterdam.
58. Rovere, M., Tosi, M.P. and March, N.H. (1982). *Phys. Chem. Liq.*, **12**, 177.
59. March, N.H. and Tosi, M.P. (1981). *Phys. Chem. Liq.*, **11**, 89.
60. Edwards, F.G., Howe, R.A., Enderby, J.E. and Page, D.I. (1978). *J. Phys.*, **C11**, 1053.
61. D'Aguanno, B., Rovere, M., Tosi, M.P. and March, N.H. (1983). *Phys. Chem. Liq.*, **13**, 113.
62. Abraham, F.F. (1975). *J. Chem. Phys.*, **63**, 157, 1316; (1976). *Ibid.*, **64**, 2660.
63. Cahn, J.W. (1961). *Acta Metall.*, **9**, 795; (1962). *Ibid.*, **10**, 179; (1965). *J. Chem. Phys.*, **42**, 93; (1968). *Trans. Metall. Soc. A.I.M.E.*, **242**, 166.
64. Hilliard, J.E. (1970). In *Phase Transformations*, ed. Aaronson, H.I., Am. Soc. Metals, Metals Park, Ohio.
65. Abraham, F.F. (1977). *Comments Solid St. Phys.*, **7**, 159.
66. Stern, O. (1924) *Z. Elektrochem.*, **30**, 508.
67. Bockris, J.O'M., Conway, B.E. and Yeager, E. (eds) (1980). *Comprehensive Treatise of Electrochemistyr*, Vol. 1, p. 83, Plenum, New York.
68. Liu, S.H. (1980). *Surf. Sci.*, **101**, 49.
69. Kornyshev, A.A. and Vorotyntsev, M.A. (1980). *Surf. Sci.*, **101**, 23.
70. Henderson, D. and Schmickler, W. (1985). *J. Chem. Phys.*, **82**, 2825.
71. Blum, L. and Stell, G. (1976). *J. Statist. Phys.*, **15**, 439; Blum, L. (1977). *J. Phys. Chem.*, **81**, 136; Henderson, D. and Blum, L. (1981). *Can. J. Chem.*, **59**, 1903.
72. Grimson, M.J. and Rickayzen, G. (1981). *Molec. Phys.*, **42**, 767; (1981). *Ibid.*, **44**, 817; (1982). *Ibid.*, **45**, 221.
73. Painter, K.R., Ballone, P., Tosi, M.P., Grout, P.J. and March, N.H. (1983). *Surf. Sci.*, **133**, 89.
74. Roman, E., Tosi, M.P., Grout, P.J. and March, N.H. (1985). *Phys. Chem. Liq.*, **15**, 123.
75. Ballone, P., Pastore, G. and Tosi, M.P. (1986). *J. Chem. Phys.*, **85**, 2943.

76. Torrie, G.M. and Valleau, J.P. (1979). *Chem. Phys. Lett.*, **65**, 343; (1980). *J. Chem. Phys.*, **73**, 5807.
77. van Megen, W. and Snook, I. (1980). *J. Chem. Phys.*, **73**, 4656.
78. Luckhurst, G.R. and Gray, G.W. (eds) (1979). *The Molecular Physics of Liquid Crystals*, Academic Press, New York.
79. de Gennes, P.G. (1975). *The Physics of Liquid Crystals*, p. 77, Clarendon, Oxford.
80. See, for example, Senbetu, L. and Woo, C.-W. (1982). *Molec. Cryst. Liq. Cryst.*, **84**, 101.
81. Maier, W. and Saupe, A. (1958). *Z. Naturforsch.*, **13**, 564; (1959). *Ibid.* **14**, 882.
82. McMillan, W.L. (1971). *Phys. Rev.*, **A4**, 1238; (1972). *Ibid.*, **A6**, 936.
83. Singh, U.P. and Singh, Y. (1986). *Phys. Rev.*, **A33**, 2725.
84. Singh, Y. (1984). *Phys. Rev.*, **A30**, 583.
85. Stephen, M.J. and Straley, J.P. (1974). *Rev. Mod. Phys.*, **46**, 617.
86. Gelbart, W.M. and Ben-Shaul, A. (1982). *J. Chem. Phys.*, **77**, 916.
87. Priest, R.G. (1973). *Phys. Rev.*, **A7**, 720.
88. Straley, J.P. (1973). *Phys. Rev.*, **A8**, 2181.
89. Frenkel, D., Mulder, B.M. and McTague, J.P. (1984). *Phys. Rev. Lett.*, **52**, 287.
90. Krotscheck, E., Qian, G.-X. and Kohn, W. (1985). *Phys. Rev.*, **B31**, 4245.
91. Pieper, S.C., Wiringa, R.B. and Pandharipande, V.R. (1985). *Phys. Rev.*, **B32**, 3341.
92. Edwards, D.O. and Saam, W.F. (1978). *Prog. Low. Temp. Phys.*, **7A**, 283.
93. Krotscheck, E., Kohn, W. and Qian, G.-X. (1985). *Phys. Rev.*, **B32**, 5693; see also Krotscheck, E. and Kohn, W. (1986). *Phys. Rev. Lett.*, **57**, 862.
94. Ceperley, D.M. and Alder, B.J. (1980). *Phys. Rev. Lett.*, **45**, 566.
95. Mackie, F.D. and Woo, C.-W. (1978). *Phys. Rev.*, **B17**, 2877; Sun, X., Li, T., Farjam, M. and Woo, C.-W. (1983). *Phys. Rev.*, **B27**, 3913; see also (1983). *Ibid.*, **28**, 5599.
96. Lang, N.D. and Kohn, W. (1970). *Phys. Rev.*, **B1**, 4555.
97. See, for example, March, N.H. (1968). *Liquid Metals*, Pergamon, Oxford.
98. Rosenfeld, Y. and Ashcroft, N.W. (1979). *Phys. Rev.*, **A20**, 1208.
99. March, N.H. (1986). *Phys. Chem. Liq.*, **16**, 117.
100. Blum, L. and Stell, G. (1976). *J. Statist. Phys.*, **15**, 439.
101. Sullivan, D.E. and Stell, G. (1977). *J. Chem. Phys.*, **67**, 2567.
102. Bacquet, R. and Rossky, P.J. (1982). *J. Chem. Phys.*, **79**, 1419.
103. Iyetomi, H. and Ichimaru, S. (1983). *Phys. Rev.*, **A27**, 1734.
104. See also March, N.H. and Tosi, M.P. (1984). *Coulomb Liquids*, p. 70, Academic Press, New York.
105. March, N.H. (1987). *Can J. Phys.*, **65**, 219.
106. Toxvaerd, S. (1971). *J. Chem. Phys.*, **55**, 3116.
107. Berre, B.J. and Pechukas, P. (1972). *J. Chem. Phys.*, **56**, 4213.

# Index